Mechanics of Steadily Propagating Cracks

Patrick N. B. Anongba

Mechanics of Steadily Propagating Cracks

Springer

Patrick N. B. Anongba
Université F.H.B. de Cocody, UFR-SSMT
Abidjan, Côte d'Ivoire

ISBN 978-981-96-5702-5 ISBN 978-981-96-5703-2 (eBook)
https://doi.org/10.1007/978-981-96-5703-2

This Springer imprint is published by the registered company Springer Nature Singapore Pte Ltd.
The registered company address is: 152 Beach Road, #21-01/04 Gateway East, Singapore 189721, Singapore

In memory of Patrick N. B. Anongba.
His children are proud of him.

Preface

This monograph deals with crack motion in solids under load with no acceleration term in the equations of motion. This is qualified as steadily propagating cracks. A useful quantity is G, the crack extension force per unit length of the crack front. Physical quantities associated with the crack, measured under the conditions$\langle G \rangle$ (G, averaged over the length of the crack front) positive maxima, conform to predictions. This monograph tells how the evolution of fracture in loaded solids is described. Chapter 1 deals with general procedures for investigating crack propagation. The framework is linear elasticity with dislocations as crystal defects and basic elements in the representation of loaded cracks. Chapter 2 treats special steadily propagating cracks: planar, non-planar, circular, elliptical, rough conoidal cracks in isotropic and cubic materials (Sects. 2.1–2.5); applications of the treatment to twinning and cross-slip in copper and fracture in $CoSi_2$ are given. Sections 2.6 and 2.7 are devoted to electromagnetic fields generated by propagating cracks and crack front vibration in thermal equilibrium, respectively. Chapter 3 describes the propagation of cracks under compression and bending. The only way for a crack to propagate under compression is to align itself parallel to the applied compression direction. In that orientation, Poisson's stresses are tensile and entirely responsible for the opening of the cracked faces. Under bending, the conditions for crack propagation in the compression region control the complete failure of the specimen. Chapter 4 is concerned with crack propagation in three-dimensional elastic half-space in presence of gravitational forces. A planar crack model is introduced where the crack moves under forces applied along a line at the surface and gravitational forces. Analytical expressions for G are obtained, making it possible to account for the commonly observed fact that cracks can break into two parts very large blocks of stone, in absence of forces applied at the surface. Chapter 5 concerns interfacial fracture at bi-material interface. Depending on continuity conditions when crossing the interface, oscillatory singularities (Sect. 5.1) and non-oscillatory singularities of the types obtained in homogeneous materials (Sects. 5.2–5.4) are obtained in the elastic fields (displacement and stress) at the tip of the interface crack. In all cases, the crack extension force G is finite and regular. However, G displays different expressions for the two types of singularities when account is taken of induced Poisson's stresses.

Further confrontations with experiments are welcome. It is remarkable and acknowledged that this monograph has benefited from experimental works performed earlier during the Switzerland period (1984–1992) under Profs. Jean-Luc Martin (EPFL) and Samuel G. Steinemann (UNIL). It is intended to researchers concerned with fracture mechanics and mechanical behaviors of materials. It may be used from undergraduate to graduate levels in universities and engineering schools.

Abidjan, Côte d'Ivoire Patrick N. B. Anongba

Contents

Abbreviations

a	Crack half-length; atomic spacing along crack front
a_1, a_2	Semiaxes of the elliptical crack; crack half-lengths
a_r	Ratio (a_1/a_2) of the semiaxes of the elliptical crack
a_s	Distance between nodes A (medium 1) and B (medium 2) along the interface
$\boldsymbol{b}$, $\boldsymbol{b}_j$	Burgers vector, of crack dislocation j
$\boldsymbol{B}$, $\boldsymbol{B}_C$, $\boldsymbol{B}_{CD}$	Magnetic field, round a crack, and round an elastic-plastic crack
c; c_0	Crack half length; velocity of travelling sine waves along crack fronts
c_l, c_t, c_R	Velocities of longitudinal, transverse, and Rayleigh sound waves $c_*=c_t/c_l$
C, C_0, C_1, C_2	$\mu b/2\pi(1-\nu)$ for edge dislocations; $\mu b/2\pi$ for screw
C_{ij}	Voigt constants
D_J, D_0^J	Dislocation J distribution function; straights
$\bar{e}_{ij}$	Total strain in the medium
E, E_i, E_F	Young's modulus, medium i; Fermi energy (electrons)
$\boldsymbol{E}$, $\boldsymbol{E}_D$, $\boldsymbol{E}_C$, $\boldsymbol{E}_{CD}$	Electric field, round a dislocation, crack, and elastic-plastic crack
f	Crack-front shape in x_2x_3-planes
F	Elliptic integral of the first kind; Gauss's hypergeometric function
g	Acceleration due to gravity
G, G_0^J, $\tilde{G}$, $< \tilde{G} >$	Crack extension force per unit length of the crack front, straight-fronted (mode J), normalised, spatially averaged
h; $\hbar$	Average elevation (height) of the crack front; Planck's constant
H	Heaviside step function
J_n	Bessel function of the first kind
k_B	Boltzmann's constant
K_n	nth-order modified Bessel function
K_J, K_J^0, $K_J^{(0)}$	Stress intensity factors (mode J)

M_f, M_c	Bending moments at failure, with and without shear
M_{12}, M_{13}, M	Applied shears σ_{12}^a and σ_{32}^a reduced by tension σ_{22}^a; $M_{13} \equiv M$ in places
p, p_0, p_A, p_B	$\tan\theta$, $\tan\theta_0$, $\tan\phi_A$, $\tan\phi_B$
P	Load per unit length of the cylinder
P_C, P_S	Spatial positions on cracks, and interfaces
Q	Heat
$\boldsymbol{S}, \boldsymbol{S}_C, \boldsymbol{S}_{CD}$	Poynting's vector S, due to elastic, and elastic-plastic cracks
T	Temperature
$\boldsymbol{u}^{(n)}$	Displacement due to a crack dislocation of family n
$\boldsymbol{u}^{(n)(0)}, \boldsymbol{u}^{(n)\xi}$	Non, and oscillating part of the displacement field $\boldsymbol{u}^{(n)}$
U	Internal energy
v	Velocity of the crack front
W	Work of forces
x_j	Cartesian coordinates
Z	Jacobi's zeta function
α_0	Constant of the order of one
β_0	Line tension of the crack front
$\beta_J^{*(n)}, \bar{\beta}_J^{*(n)}$	Plastic distortion, associated amplitude
γ	Surface energy
δ, δ_{ij}	Dirac, Kronecker delta
$\Delta, \Delta_C, \Delta_D, \Delta_{CD}$	Separation of partial dislocations; lattice dilatation of the lattice due to crack, dislocation, and elastic-plastic crack
$\varepsilon_0, \varepsilon_1$	Electric constant, relative permittivity
$\in_{njh}$	Permutation tensor
θ	$\tan^{-1}(\partial h/\partial x_1)$ measured in x_1x_2-plane at crack tip
θ_0	Inclination angle of planar crack π_0 with respect to the horizontal Ox_1x_3
λ	Lamé's constant
$\mu; \mu_i$	Shear modulus; in medium i
$\mu_0; \mu_1$	Magnetic constant; relative permeability
μ_s	Shear modulus about the interface
$\nu; \nu_i$	Poisson's ratio; in medium i
ν_s	$= \phi/d$: ϕ distance along the interface between node A (medium 1) and B (medium 2) in the welded state underload; d the corresponding distance in the non-welded state
$\nu_A(j)$	Modulus in induced force magnitude (along x_j-direction) due to Poisson's effect
ξ	Crack front shape at the height h in x_2x_3-planes
Π	Elliptic integral of third kind
ρ	Density of the medium
$\sigma_y, \sigma_c, \sigma_t$	Yield stress; failure stress in compression, tension
σ_{ij}^a	Stress applied in the x_i-direction, on a plane perpendicular to the x_j-direction

$\sigma_{ij}^{(n)}, \sigma_{ij}^{(n)(0)}, \sigma_{ij}^{(n)\xi}$	Stress due to a crack dislocation of family n, non-oscillatory, and oscillatory parts
$\bar{\sigma}_{ij}$	Total stress
τ_{ij}^{a}	Induced shear stress due to Poisson's effect
ϕ_j	Relative displacement of the faces of the crack in the x_j-direction
ϕ_A, ϕ_B, ϕ_C	Crack-front inclination angles with respect to the horizontal Ox_1x_3-plane
ω	Frequency

Chapter 1
General Procedures for Crack Propagation in Loaded Materials

1.1 Evolution of Crack Systems

In *Fracture Mechanics*, we are concerned with a continuum solid containing a crack. The medium is subjected to loadings applied externally at the surface. The evolution of the system with time is seek. In thermodynamics equilibrium, the system is characterized by its energy U that is function of various parameters such as crack-front spatial and angular coordinates, crack-length, crack-front speed, applied loadings, elastic constants and temperature T. U depends implicitly on time t trough the dependence on time of its various parameters. U is written as

$$U = W + Q. \tag{1.1}$$

W is expressed as the work of forces and Q is associated to heat. U viewed as a free energy decreases with time during the evolution of the crack system,

$$dU \leq 0. \tag{1.2}$$

It tends to a minimum $U_{\min}$ at which $dU = 0$. We may write, following Bilby and Eshelby (1968),

$$dU = \delta E_{tot} + \delta E_s + \delta Q; \tag{1.3}$$

$$\delta \mathrm{E}_{\mathrm{tot}} = \delta \mathrm{E}_{\mathrm{el}} + \delta \mathrm{E}_{\mathrm{ext}}.$$

δE_{el} is the increase in the elastic energy E_{el} of the fractured body; δE_{ext} is the decrease in the potential energy of the loading mechanism. δE_s is the increase in the energy E_s of the crack surface. Expressing the various physical quantities involved in dU at stationary configuration $dU = 0$ are the objectives in *Fracture Mechanics*; these originate from Griffith (1920, 1924) original idea. Under steady motion (no

P. N.B. Anongba, *Mechanics of Steadily Propagating Cracks*,
https://doi.org/10.1007/978-981-96-5703-2_1

acceleration term in the equation of motion), Bilby and Eshelby (1968) have provided us with an expression of the quantity

$$G(P_0) = \lim_{\Delta s \to 0} - \delta E_{tot}/\Delta s \tag{1.4}$$

Δs is an increment in the crack surface; G is measured at crack-front point P_0. This is called the crack extension force per unit length of the crack front. Hence, we have at hand the relation

$$\langle G \rangle \approx 2\gamma \tag{1.5}$$

in controlled conditions under which $\delta Q \approx 0$; γ is the surface energy. $\langle G \rangle$ is an average value; this is G averaged over the spatial positions along the crack front. The followings may be stressed: (*i*) G (4) is useful under stationary conditions (G positive maximum), only. (*ii*) G contains the crack-front shape at P_0. (*iii*) We do not have at hand an expression for G depending on the crack-front shape, when the equations of motion contain acceleration terms. (*iv*) A crack extending in the medium with no acceleration in the equations of motion is named "a steadily propagating crack". Observed motions of cracks over large distances in materials do correspond to positive G maxima. Hence, instead of using the terminology *Dynamic Fracture* that recall acceleration, it is comfortable to write *Steadily Propagating Fracture.*

1.2 Crack Representations in Loaded Materials

Cracks evolving with time inside fractured solids are expected to take non-planar configurations with crack-fronts lying simultaneously on different planes. This complex behaviour is due to crack-front velocity, applied stresses, crack-front dislocation generation ... In the fracture surface followed by a crack, we distinguish different regions *I*, *II* and *III* illustrated in Fig. 1.1 for a special model.

- Region *I*: the cohesive forces are zero, $|x_1| < c$
- Region *II*: the cohesive forces are important, $c < |x_1| < e$,
- Region *III*: zone of plastic deformation, $e \leq |x_1| \leq a$.

The framework is linear elasticity. The crack system is represented in Region *I* and *III* by continuous distributions of dislocations and Region *II* is free from any dislocations. The distribution functions of dislocations in equilibrium are found under the requirement of zero surface traction. The stresses behave as $1/\sqrt{|x_1| - c}$ in Region *II* but G is finite. Because the evolution of the crack system is performed on the energy level, there is no trouble with the stress singularities. Progresses in representing crack systems by dislocations are due to Friedel (1964), Bilby, Cottrell and Swinden (1963), Bilby and Eshelby (1968), Chang and Ohr (1981, 1982; for a review see Ohr, 1985), Comninou (1977), Anongba and Vitek (2003) and Anongba

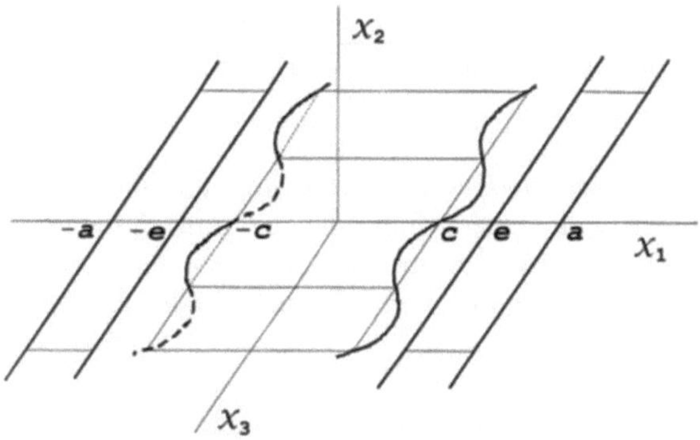

Fig. 1.1 An elastic–plastic crack system located on average in the Ox_1x_3-plane. The crack extends from $x_1 = -c$ to c with a sinusoidal front running indefinitely in the x_3-direction. On both sides of the crack, a dislocation-free zone $c < |x_1| < e$ and plastic domain $e \leq |x_1| \leq a$ are present. The system is subjected to uniform applied tension σ_a in the x_2-direction

(1996, 2003, 2008–2024). The merit of representing a crack by dislocations is that the stresses at the tip of the crack-front are derived from dislocation stress fields. We stress that this representation of crack contours (Fig. 1.1) corresponds to Barenblatt (1962) and Bilby and Eshelby (1968).

1.3 Dislocation Elastic Fields in Single Phase Materials

A way for determining the elastic fields (displacement and stress) due to dislocations is presented below. First the plastic distortions associated with the dislocation are expressed in their Fourier series/integrals forms. These are linear superpositions of plastic distortions β^*_{mn} (at spatial position $\overrightarrow{x} = (x_1, x_2, x_3)$) of form

$$\beta^*_{mn} = \overline{\beta}^*_{mn} e^{i\left(\omega t + \vec{k}.\vec{x}\right)}. \tag{1.6}$$

Here t is time, $\vec{k} = (k_1, k_2, k_3)$ and ω are arbitrary, and $\overline{\beta}^*_{mn}$ are arbitrary functions of constants k_i and ω. Hence, the elastic fields of the dislocation will be similar superpositions of the elastic fields (displacement u_m and stress σ_{pq}) due to β^*_{mn}. These are obtained by Mura (1968) as

$$u_m(\vec{x}, t) = -ik_l C_{klij} L_{mk} \overline{\beta}^*_{ji} e^{i(\omega t + \vec{k}.\vec{x})},$$

$$\sigma_{pq}(\vec{x}, t) = C_{pqmn}\left(k_n k_l C_{klij} L_{mk} \overline{\beta}^*_{ji} - \overline{\beta}^*_{nm}\right) e^{i(\omega t + \vec{k}.\vec{x})}; \tag{1.7}$$

$$L_{mk} = \frac{\in_{kst}\in_{mnr}\left(C_{sjnl}k_lk_j - \rho\omega^2\delta_{sn}\right)\left(C_{tjrl}k_lk_j - \rho\omega^2\delta_{tr}\right)}{2\in_{pqr}\left(C_{pj1l}k_lk_j - \rho\omega^2\delta_{p1}\right)\left(C_{qj2l}k_lk_j - \rho\omega^2\delta_{q2}\right)\left(C_{rj3l}k_lk_j - \rho\omega^2\delta_{r3}\right)}. \tag{1.8}$$

Here forms like $\in_{njh}$ correspond to the permutation unit tensor (all zero except the permutations of $\in_{123}$ with respect to the indices, using 1 for even and -1 for odd permutations). ρ is the density of the medium. We stress that the so-called summation convention is used throughout this work: if a subscript is repeated, summation from 1 to 3 is implied.

$$C_{ijkl} = \lambda\delta_{ij}\delta_{kl} + \mu\delta_{ik}\delta_{jl} + \mu\delta_{il}\delta_{jk} + \mu\prime\delta_{ijkl} \tag{1.9}$$

where all $\delta_{ijkl} = 0$ except $\delta_{1111} = \delta_{2222} = \delta_{3333} = 1$ and

$$\lambda = C_{12},\ \mu = C_{44},\ \mu' = C_{11} - C_{12} - 2C_{44}. \tag{1.10}$$

δ_{ij} is the Kronecker delta; C_{ij} are the Voigt constants. Constants C_{ijkl} are symmetrical as follows,

$$C_{ijkl} = C_{ijlk} = C_{klij} = C_{jikl}. \tag{1.11}$$

For isotropic material,

$$L_{mk}(\vec{k}, \omega) = \frac{\delta_{km}\left((\lambda + 2\mu)k^2 - \rho\omega^2\right) - k_k k_m(\lambda + \mu)}{\left(\mu k^2 - \rho\omega^2\right)\left((\lambda + 2\mu)k^2 - \rho\omega^2\right)} \tag{1.12}$$

where $k^2 = k_1^2 + k_2^2 + k_3^2$ and

$$C_{klji} = \lambda\delta_{kl}\delta_{ji} + \mu\delta_{kj}\delta_{li} + \mu\delta_{ki}\delta_{lj}, \tag{1.13}$$

λ and μ are Lamé's constants.

1.4 Crack Analysis

We consider a crack in elastic material under arbitrarily applied loadings. This corresponds, in Cartesian coordinate system Ox_j ($j = 1, 2, 3$) attached to the laboratory, to stresses (say, uniform) applied at infinity, with tension σ_{22}^a and shears σ_{21}^a and σ_{23}^a in the x_2, x_1 and x_3 directions, respectively. Induced normal Poisson's stresses $-\nu_A(j)\sigma_{22}^a$ along x_1 ($j = 1$) and x_3 ($j = 3$) are considered ($\nu_A(j)$ is Poisson's modulus in the x_j-direction). Under such conditions, the relative displacement of the faces of the crack is a three-dimensional vector with components ϕ_j in the x_j-direction. The crack is represented by a continuous distribution of three dislocation families with

Burgers vectors $\boldsymbol{b}_j = b\vec{x}_j$, directed along positive $\vec{x}_j$, associated with the ϕ_j. To discuss the crack fields and its interactions with other defects (inclusions and heterogeneities, for example), it seems most satisfying treat a crack as an inclusion at whose boundary there is zero surface traction (Bilby and Eshelby 1968). The dislocation distribution function $D_j(x_1)$ gives the number of dislocations of family j in a small interval dx_1 about x_1 as $D_j(x_1)dx_1$. To find the dislocation distributions, we ask the crack faces to be traction-free; this gives

$$\begin{cases} \overline{\sigma}_{12} - \partial x_2/\partial x_1 \overline{\sigma}_{11} - \partial x_2/\partial x_3 \overline{\sigma}_{13} = 0 \\ \overline{\sigma}_{22} - \partial x_2/\partial x_1 \overline{\sigma}_{12} - \partial x_2/\partial x_3 \overline{\sigma}_{23} = 0 \\ \overline{\sigma}_{23} - \partial x_2/\partial x_1 \overline{\sigma}_{13} - \partial x_2/\partial x_3 \overline{\sigma}_{33} = 0. \end{cases} \tag{1.14}$$

$\overline{\sigma}_{ij}$ stands for the total stress at any point (x_1, x_2, x_3) in the medium and is linked to D_j. In (Eq. 1.14), we are concerned with the points of the crack faces only. We write $\overline{\sigma}_{ij}$ as

$$\overline{\sigma}_{ij} = \sigma_{ij}^{A} + \sigma_{ij}^{C(1)} + \sigma_{ij}^{C(2)} + \sigma_{ij}^{C(3)}; \tag{1.15}$$

$(\sigma)^A$ is the externally applied stress tensor including normal induced stresses from Poisson effect,

$$(\sigma)^A = \begin{pmatrix} -\nu_A(1)\sigma_{22}^a & \sigma_{12}^a & 0 \\ \sigma_{12}^a & \sigma_{22}^a & \sigma_{23}^a \\ 0 & \sigma_{23}^a & -\nu_A(3)\sigma_{22}^a \end{pmatrix},$$

$$\sigma_{ij}^{C(n)}(x_1, x_2, x_3) = \int_{-a}^{a} \sigma_{ij}^{(n)}(x_1 - xx_1', x_2, x_3) D_n(x_1')\mathrm{d}x_1' \quad (n, 2 \text{ and } 3). \tag{1.16}$$

The crack is considered of finite extension along $x_1(|x_1| \le a)$. $\sigma_{ij}^{(n)}(n = 1, 2 \text{ and } 3)$ is the stress field produced by a dislocation n located at $x_1 = x_1\prime$ in the distribution. (Eq. 1.14) gives three integral equations the resolution of which yields the D_j. When the dislocation distributions $D_j(x_1)(j = 1 \text{ to } 3)$ are known, the relative displacement of the faces of the crack, the crack-tip stresses and the crack extension force are derived by integrations. The relative displacements ϕ_j of the faces of the crack in the x_j-direction at the position P in the distribution are obtained by integration from the relation $\mathrm{d}\phi_j = -bD_j(x_1')\mathrm{d}x_1'$:

$$\phi_j = \int_{x_1}^{a} bD_j(x_1')\mathrm{d}x_1'. \quad |x_1| \le a \tag{1.17}$$

To obtain expressions for the stress about the crack front at P_0, we proceed as follows. In the neighbourhood of the crack front located at $x_1 = a$, any point P with coordinates (x_1, x_2, x_3) is characterized by x_2 close to $x_2(P_0)$. We can thus consider the Taylor series expansion of $\sigma_{ij}^{(n)}(x_1 - x_1\prime, x_2, x_3)$ about $x_2(P_0)$ to first order with respect to $(x_2 - x_2(P_0))$; this gives

$$\sigma_{ij}^{(n)}(x_1 - x_1', x_2, x_3) = \sigma_{ij}^{(n)}(x_1 - x_1', x_2(P_0), x_3) + \frac{\partial \sigma_{ij}^{(n)}}{\partial x_2}(x_2 - x_2(P_0)) + o(x_2 - x_2(P_0)) \tag{1.18}$$

where $o(x_2 - x_2(P_0))$ is the complementary part of the series form. Writing $x_1 = a + s$, $0 < s \ll a, \overline{\sigma}_{ij}(15)$ is given by the following formula:

$$\overline{\sigma}_{ij}(s, x_2, x_3) = \sum_{n=1}^{3} \int_{a-\delta a}^{a} \sigma_{ij}^{(n)}(a + s - x_1', x_2, x_3) D_n(x_1') \mathrm{d}x_1' \tag{1.19}$$

with $\delta a \ll a$. This stress expression means that only those dislocations located about the crack front in x_1- interval $[a - \delta a, a]$ will contribute significantly to the stress at $x_1 = a + s$ ahead of the crack tip as s tends to zero; any other contribution will be negligible for a sufficiently small value of s. We observe that this formula is precise with no place for any other kind of additional stress term. It is sufficient to identify $\sigma_{ij}^{(n)}$ to unbounded terms with $1/(a + s - x_1\prime)$ in the Taylor expansion. Other terms are bounded and contribute nothing when used to calculate the crack extension force G. This is done in various works (Irwin 1948, 1957; Bilby and Eshelby 1968; Anongba 2010, 2021; among others). Allow the crack front at P_0 to advance from $x_1 = a$ to $a + \delta a$, but apply forces to the freshly formed surfaces to prevent relative displacement; the energy of the system is unaltered. Now allow these forces to relax to zero so that the crack extends effectively from a to $a + \delta a$. The work done by these forces corresponds to a decrease of the energy of the system which we shall estimate (the energy of the system consists of the elastic energy of the medium and the energy of the loading mechanism). The element ds of the newly created fracture surface ahead of the shorter crack, at a point $P = (x_1, x_2 = f(x_1, x_3), x_3)$ in advance of P_0, may be defined by $\mathrm{d}\vec{s} = \vec{\gamma}\mathrm{d}s$ where $\vec{\gamma}$ is the unit vector perpendicular to ds pointing to the positive x_2-direction. We obtain

$$\mathrm{d}\vec{s} = 1/\sqrt{1 + (\partial f/\partial x_1)^2 + (\partial f/\partial x_3)^2}(-\partial f/\partial x_1, 1, -\partial f/\partial x_3)\mathrm{d}s,$$

$$\mathrm{d}s = \sqrt{1 + (\partial f/\partial x_1)^2 + (\partial f/\partial x_3)^2}\mathrm{d}x_1\mathrm{d}x_3.$$

The component of the force acting on ds in the x_i-direction is $\overline{\sigma}_{ij}\mathrm{d}s_j$ (the summation convention on repeated subscripts applies) where $\overline{\sigma}_{ij}$ are stresses ahead of the shorter crack; thus the energy change associated with $\mathrm{d}s$ is $\overline{\sigma}_{ij}\mathrm{d}s_j\Delta u^{(i)}/2$(here a summation is also considered over $i = 1, 2$ and 3) where $\Delta u^{(i)}$ is the difference in displacement across the lengthened crack, just behind its tip, in the x_i-direction. When the crack advances from $x_1 = a$ to $a + \delta a$, the energy decrease associated with a surface element

$$\Delta s = \int_a^{a+\delta a} \mathrm{d}s \cong \sqrt{1 + (\partial f/\partial x_1)^2 + (\partial f/\partial x_3)^2}\delta a \mathrm{d}x_3$$

(δa being small and, when used below, will be let to go to zero) is given as

$$-\delta E = \frac{1}{2} \int_a^{a+\delta a} \sum_i \sum_j \overline{\sigma}_{ij} \mathrm{d}s_j \Delta u^{(i)},$$

the integration being performed with respect to x_1; we stress that Δs is the sum of the surface elements ds taken at the various points $P = (x_1, x_2 = f, x_3)$ as x_1 only changes from a to $a + \delta a$. Let G be a derivative of the energy of the system with respect to crack area. G corresponds to the limiting value taken by $-\delta E/\Delta s$ as δa (as also Δs) decreases to zero: $G(P_0) = \lim_{\delta a \to 0} -\delta E/\Delta s$. Physical quantities ($D_n$, crack-tip stress, G ...) whose mathematical expressions do not invoke the crack-front velocity v correspond to quasistatic conditions ($v = 0$). When calculating dislocation elastic fields, it may be simpler to use a reference frame $(O'; \vec{x}'_j)$ attached to the crack with $\vec{x}'_2$ perpendicular to the average crack plane. The Burgers vectors $\boldsymbol{b}_j$ of the dislocations are defined as $b\vec{x}'_j$. Using $(O; \boldsymbol{x}_j)$ or $(O'; \vec{x}'_j)$ depends on the crack model.

Chapter 2
Special Steadily Propagating Cracks

2.1 Planar Straight-Fronted Cracks in Uniform Motion

The model is illustrated in Fig. 2.1 and refer to Anongba (2020*a*). The medium is elastic and infinitely extended, to which is attached a Cartesian system x_i. It consists of a crack in Ox_1x_3 of finite extension along x_1 with a straight front parallel to x_3. The crack is under load; initially, it is static and extends from $x_1 = -a$ to a. At a given time taken as $t = 0$ and load σ_{ij}^a, it starts moving at constant velocity v. Its extension after time interval t is given by $|x_1| \le c = a + vt$. Uniform tension σ_{22}^a (mode I) and shear σ_{21}^a (mode II) applied at infinity are considered separately.

The crack is represented by a continuous distribution of infinitesimal dislocations parallel to x_3 with Burgers vectors $\vec{b}_I = (0, b, 0)$ for the mode I loading and $\vec{b}_{II} = (b, 0, 0)$ for mode II. Dislocation distributions D_J ($J = I$ and II) are defined such that $D_J(x_1')dx_1'$ represents the number of dislocations J in the infinitesimal x_1-interval dx_1' located about the x_1-spatial position x_1'.

The determination of the displacement and stress fields ($\vec{u}^{(J)}$, $(\sigma)^{(J)}$) of straight edges parallel to x_3 in uniform motion along x_1 in the Ox_1x_3-plane (Fig. 2.2) is concerned. The two types ($J = I$ and II) of dislocation considered have Burgers vectors $\vec{b}_J$ parallel to x_2 and x_1, respectively. The dislocations are located at the origin when $t = 0$ and at spatial x_1-position $x_1' = vt$ after incremental time t. The plastic distortions $\beta_{ij}^{*(J)}(\vec{x}, t)$ associated to the dislocations J ($J = I$ and II), Fig. 2.2, are

$$\beta_{12}^{*(I)}\delta_{JI} + \beta_{21}^{*(II)}\delta_{JII} = b[\delta_{JI}\delta(x_1 - vt)H(x_2) + \delta_{JII}\delta(x_2)H(-(x_1 - vt))]$$
$$= \int_{-\infty}^{\infty}\int_{-\infty}^{\infty} \overline{\beta}^{*(J)} e^{i(k_1y_1 + k_2x_2)} \mathrm{d}k_1 \mathrm{d}k_2 \qquad (2.1.1)$$

P. N.B. Anongba, *Mechanics of Steadily Propagating Cracks*,
https://doi.org/10.1007/978-981-96-5703-2_2

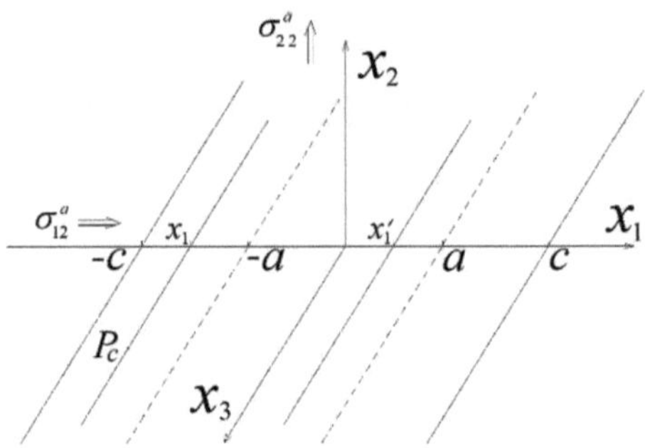

Fig. 2.1 Planar straight-fronted crack in uniform motion at velocity v from initial spreading $|x_1| \leq a$ ($t = 0$) to $|x_1| \leq c = a + vt$ after time interval t. Uniform applied stresses at infinity σ^a_{22} (mode I loading) and σ^a_{21} (mode II) are considered separately; see the text

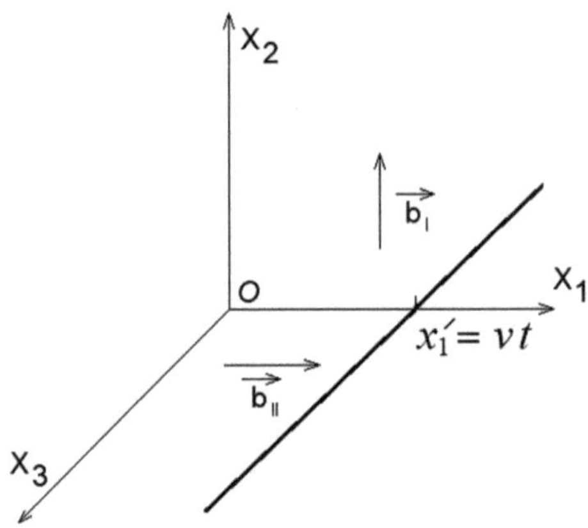

Fig. 2.2 Infinitely long straight dislocation J parallel to x_3 travelling uniformly at constant speed v in the x_1-direction, with Burgers vector $\vec{b}_J$ directed along x_1 ($J = II$) or x_2 ($J = I$)

with $\overline{\beta}^{*(J)} = -(ib/4\pi^2)(\delta_{JI}/k_2 - \delta_{JII}/k_1)$, the other components of the plastic distortions are zero; δ and H are the Dirac delta and Heaviside step functions, respectively; $y_1 = x_1 - vt$.

2.1.1 *Isotropic Materials*

Dislocation Stresses

Dislocation (Fig. 2.2) stresses involved in the crack analyses are, for the subsonic velocity regime ($v < c_t$) where c_t is the velocity of transverse sound wave,

$$\delta_{JI}\sigma_{22}^{(I)} + \delta_{JII}\sigma_{21}^{(II)} = \frac{2\mu b}{\pi \tilde{v}_t^2} y_1 \left(\frac{P_{2t}^2}{y_1^2 + P_t^2 x_2^2} \left[\frac{\delta_{JI}}{P_t} - \frac{P_{2t}^2 \delta_{JII}}{P_t} \right] + \frac{1}{y_1^2 + P_l^2 x_2^2} \left[\delta_{JII} P_l - \delta_{JI} \frac{P_{2t}^2}{P_l} \right] \right),$$

$$\sigma_{21}^{(I)} = \frac{2\mu b}{\pi \tilde{v}_t^2} x_2 \left(\frac{P_l}{y_1^2 + P_l^2 x_2^2} - \frac{P_{2t}^4}{P_t (y_1^2 + P_t^2 x_2^2)} \right); \tag{2.1.2}$$

$\tilde{v}_t = v/c_t$, $P_t^2 = 1 - \tilde{v}_t^2$, $P_{2t}^2 = 1 - \tilde{v}_t^2/2$;
$\tilde{v}_l = v/c_l$, $P_l^2 = 1 - \tilde{v}_l^2$;
the subscript m takes the values 1 and 2 and c_l is the velocity of longitudinal sound wave. For intermediate (transonic) velocity regime ($c_t < v < c_l$),

$$\sigma_{22}^{(I)} = \frac{\mu b(\tilde{v}_t^2 - 2)}{\pi \tilde{v}_t^2} \left(\frac{y_1}{P_l (y_1^2 + P_l^2 x_2^2)} + \frac{\pi}{2\sqrt{\tilde{v}_t^2 - 1}} \times \operatorname{sgn}(y_1) \left[\delta\left(|x_2| \sqrt{\tilde{v}_t^2 - 1} + |y_1| \right) + \delta\left(|x_2| \sqrt{\tilde{v}_t^2 - 1} - |y_1| \right) \right] \right),$$

$$\sigma_{21}^{(II)} = \frac{\mu b}{\pi \tilde{v}_t^2} \left(\frac{2 P_l y_1}{y_1^2 + P_l^2 x_2^2} + \frac{\pi (2 - \tilde{v}_t^2)}{4\sqrt{\tilde{v}_t^2 - 1}} \times \operatorname{sgn}(y_1) \left[\tilde{v}_t^2 \delta\left(|x_2| \sqrt{\tilde{v}_t^2 - 1} + |y_1| \right) + (2 - \tilde{v}_t^2) \delta\left(|x_2| \sqrt{\tilde{v}_t^2 - 1} - |y_1| \right) \right] \right). \tag{2.1.3}$$

Dislocation stresses at special velocities ($v = c_t$ and $v = c_l$) are given by Anongba (2020*a*).

Moving Cracks in Mode I Loading

I.1. *Subsonic velocity regime* ($v < c_t$)

Dislocation distribution D_I:

$$D_I(x_1) = \frac{\sigma_{22}^a}{\pi C_1^{(I)}} \frac{x_1}{\sqrt{c^2 - x_1^2}} \quad (|x_1| < c = a + vt); \tag{2.1.4}$$

$$C_1^{(I)} = \frac{\mu b (2 - \tilde{v}_t^2)}{\pi \tilde{v}_t^2} \left(\frac{1}{P_t} - \frac{1}{P_l} \right). \tag{2.1.5}$$

Relative displacement of the faces of the crack:

$$\phi_I(x_1) = \frac{b \sigma_{22}^a}{\pi C_1^{(I)}} \sqrt{c^2 - x_1^2}. \tag{2.1.6}$$

The total stress $\overline{\sigma}_{22}^{(I)}$ ahead of the crack tip at $x_1 = c$: substituting $x_1 = c + s$, $0 < s \ll c$, we obtain

$$\overline{\sigma}_{22}^{(I)}(s) = \frac{K_I}{\sqrt{2\pi}} \frac{1}{\sqrt{s}}; \tag{2.1.7}$$

$K_I = \sigma_{22}^a \sqrt{\pi c}$. The crack extension force $G^{(I)}$ per unit edge length is

$$G^{(I)} = \frac{bK_I^2}{4\pi C_1^{(I)}}. \tag{2.1.8}$$

At $t = 0$ and $v = 0$, $G^{(I)}$ reduces to

$$G_0^{(I)} = \frac{bK_I^2(0)}{4\pi C_0} \tag{2.1.9}$$

where $C_0 = \mu b/2\pi(1-\nu)$ and $K_I(0) = \sigma_{22}^a\sqrt{\pi a}$. We can write $(c-a)/a = \tilde{v}_t c_t t/a \equiv \tilde{v}_t \tilde{d}_t$; then, the normalized quantity $\tilde{G}^{(I)} = G^{(I)}/G_0^{(I)}$ is reported in Fig. 2.3 as a function of $\tilde{v}_t$ for $\tilde{d}_t = 1$, in other words, when a transverse sound wave emitted at the tip of the crack of length $2a$ travelled a distance equal to half the initial crack length. $\tilde{G}^{(I)}$ increases with $\tilde{v}_t$ from the value 1 ($\tilde{v}_t = 0$) up to a maximum $\tilde{G}_{\max}^{(I)} \cong 1.32$ located at about $\tilde{v}_t^{(e)} = 0.52$; then it decreases to zero as $\tilde{v}_t$ tends to 1. $\tilde{G}^{(I)}$ is continuous. The condition for the maximum for $\tilde{G}^{(I)}$ with respect to $\tilde{v}_t$ is given by $\partial\tilde{G}^{(I)}/\partial\tilde{v}_t = 0$ and this is satisfied when ($c_* = c_t/c_l$)

$$\begin{aligned}&(1+\tilde{d}_t\tilde{v}_t)\tilde{v}_t^2(2-\tilde{v}_t^2)P_tP_l(P_tc_*^2 - P_l) + (P_l - P_t)\\ &\left\{P_t^2P_l^2\tilde{v}_t(2\tilde{d}_t + \tilde{v}_t) + (1+\tilde{d}_t\tilde{v}_t)\times\left[-P_t^2P_l^2\tilde{v}_t^2 + 4P_t^2P_l^2\right.\right.\\ &\left.\left.-4P_l^2\tilde{v}_t^2 + 2(1-c_*^2)\tilde{v}_t^2 + 2P_l^2\tilde{v}_t^4 - (1-c_*^2)\tilde{v}_t^4\right]\right\} = 0.\end{aligned} \tag{2.1.10}$$

I.2. *Intermediate velocity regime* ($c_t < v < c_l$)

Zero surface tractions on the crack faces give:

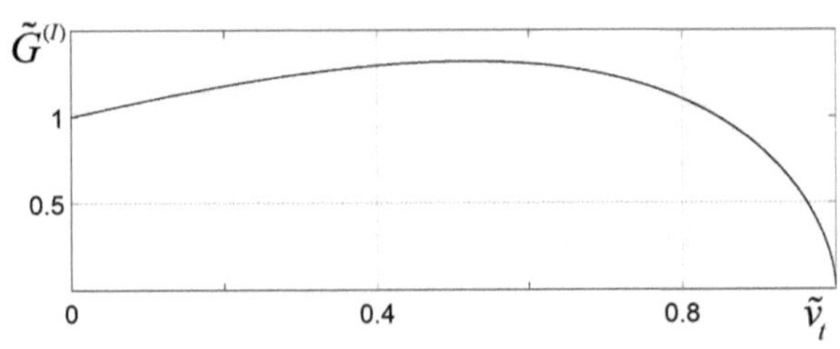

Fig. 2.3 Normalized crack extension force $\tilde{G}^{(I)} = G^{(I)}/G_0^{(I)}$(2.1.8, and 2.1.9) versus $\tilde{v}_t = v/c_t$ in subsonic velocity regime $0 \le \tilde{v}_t \equiv v/c_t < 1$. The curve corresponds to $\tilde{d}_t \equiv c_t t/a = 1$; $\nu = 1/3$

$$\sigma_{22}^{a} + \int_{-c}^{c} \sigma_{22}^{(I)}(x_1 - x_1', x_2 = 0, x_3) D_I(x_1') \mathrm{d}x_1' = 0; \tag{2.1.11}$$

$$\sigma_{22}^{(I)}(x_1 - x_1', x_2 = 0, x_3) = \frac{C_2^{(I)}}{x_1 - x_1'} + \frac{\mu b(\tilde{v}_t^2 - 2)}{\tilde{v}_t^2 \sqrt{\tilde{v}_t^2 - 1}} \mathrm{sgn}(x_1 - x_1') \delta(x_1 - x_1'),$$

$$C_2^{(I)} = \frac{\mu b(\tilde{v}_l^2 - 2c_*^2)}{\pi \tilde{v}_l^2 \sqrt{1 - \tilde{v}_l^2}}.$$

For $\tilde{v}_t^2 = 2$, $\sigma_{22}^{(I)}$ (2.1.3) is identically zero, not only on the fracture surface, but everywhere in the fractured medium. Under such conditions, the condition (2.1.11) gives no information on D_I and cannot be satisfied when $\sigma_{22}^{a} \neq 0$, suggesting that no uniform motion is possible under an applied tension stress. Now, assuming $\tilde{v}_t^2 \neq 2$, (2.1.11) becomes

$$\sigma_{22}^{a} + C_2^{(I)} \int_{-c}^{c} \frac{D_I(x_1')\mathrm{d}x_1'}{x_1 - x_1'} + \frac{\mu b(\tilde{v}_t^2 - 2)}{\tilde{v}_t^2 \sqrt{\tilde{v}_t^2 - 1}} \int_{-c}^{c} \mathrm{sgn}(x_1 - x_1') \delta(x_1 - x_1') D_I(x_1') \mathrm{d}x_1' = 0 \tag{2.1.12}$$

Equation (2.1.12) differs from the familiar form corresponding to a static crack by the third integral term with the Dirac delta function δ. We suggest that this term is zero as follows. For $x_1 \neq x_1'$, it is zero; about x_1, the integral may be written ($0 <$ ε small real number)

$$\int_{x_1-\varepsilon}^{x_1+\varepsilon} \mathrm{sgn}(x_1 - x_1') \delta(x_1 - x_1') D_I(x_1') \mathrm{d}x_1' = D_I(x_1) \int_{-\varepsilon}^{\varepsilon} \mathrm{sgn}(y) \delta(y) \mathrm{d}y = 0, \tag{2.1.13}$$

because $sgn(y)\delta(y)$ is odd function. The usual form is recovered,

$$\sigma_{22}^{a} + C_2^{(I)} \int_{-c}^{c} \frac{D_I(xx_1')}{x_1 - xx_1'} \mathrm{d}x_1' = 0. \quad (\tilde{v}_t^2 \neq 2) \tag{2.1.14}$$

This leads to familiar relations (2.1.4), (2.1.6) and (2.1.7) in which $C_1^{(I)}$ (2.1.5) is replaced by $C_2^{(I)}$ (2.1.11). It is tempting to include the crack extension form (2.1.8) as an indicative estimate. The results are collected:

$$D_I(x_1) = \frac{\sigma_{22}^{a}}{\pi C_2^{(I)}} \frac{x_1}{\sqrt{c^2 - x_1^2}}; \quad \phi_I(x_1) = \frac{b\sigma_{22}^{a}}{\pi C_2^{(I)}} \sqrt{c^2 - x_1^2}; \quad G^{(I)} = \frac{bK_I^2}{4\pi C_2^{(I)}}. \tag{2.1.15}$$

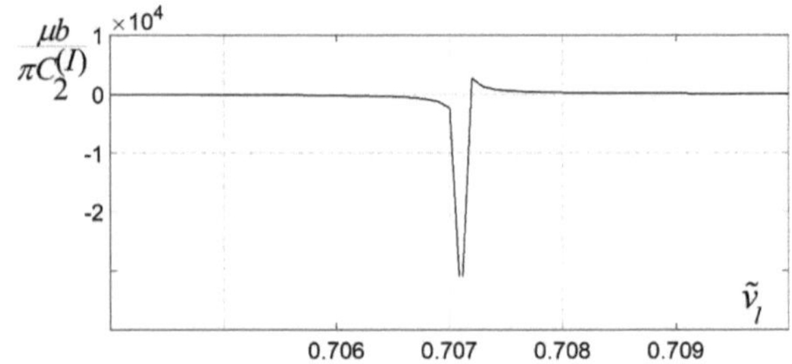

Fig. 2.4 Quantity $\mu b/\pi C_2^{(I)}$ (2.1.11) as a function of $\tilde{v}_l = v/c_l$ in intermediate velocity regime $c_* < \tilde{v}_l < 1$; $c_* = c_t/c_l$, $\nu = 1/3$

Figure 2.4 reports $\mu b/\pi C_2^{(I)}$(2.1.11) as a function of $\tilde{v}_l = v/c_l$. This quantity is negative and decreases with $\tilde{v}_l$ for $\tilde{v}_l < c_*\sqrt{2}$; it is positive above $\tilde{v}_l = c_*\sqrt{2}$ and decreasing with $\tilde{v}_l$. The observed discontinuity is at $\tilde{v}_l = c_*\sqrt{2}$. We stress that a similar analysis can be performed when $v = c_t$ starting from (2.1.14) with $C_2^{(I)}(v = c_t)$ given by (2.1.11).

Moving Cracks in Mode II Loading

II-1. *Subsonic velocity regime* ($v < c_t$)

$$\sigma_{21}^a + \int_{-c}^{c} \sigma_{21}^{(I)}(x_1 - x_1', 0, x_3) D_{II}(x_1')\mathrm{d}x_1' = 0; \tag{2.1.16}$$

$$\sigma_{21}^{(II)} = \frac{C_1^{(II)}}{x_1 - x_1'},$$

$$C_1^{(II)} = \frac{2\mu b(P_t P_l - P_{2t}^4)}{\pi \tilde{v}_t^2 P_t}.$$

Defining $R_w \equiv P_t P_l - P_{2t}^4$, $C_1^{(II)}$ cancels under the condition $R_w = 0$. This condition can be managed to read

$$\tilde{v}_t^6 - 8\tilde{v}_t^4 + 8\tilde{v}_t^2(3 - 2c_*^2) - 16(1 - c_*^2) = 0. \tag{2.1.17}$$

This equation determines Rayleigh wave speed (Landau and Lifshitz 1986). Hence, a crack dislocation *II* travelling at the Rayleigh wave speed c_R produces zero stress $\sigma_{21}^{(II)}$ on the $x_2 = 0$ fracture plane. The equilibrium condition (2.1.16) can't be fulfilled for $\sigma_{21}^a \neq 0$ suggesting no steady motion under such conditions. Except for the Rayleigh wave speed, we obtain similarly as for the mode *I* loading

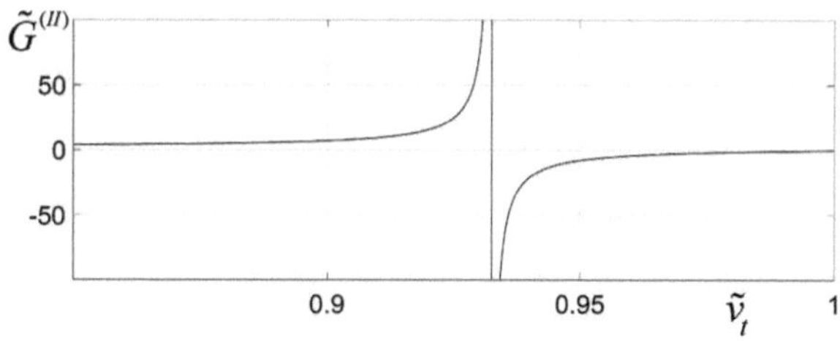

Fig. 2.5 Reduced crack extension force $\tilde{G}^{(II)} = G^{(II)}/G_0^{(II)}$ (2.1.18) versus $\tilde{v}_t$ in subsonic velocity regime. The curve corresponds to $\tilde{d}_t = 1$; $\nu = 1/3$. A vertical asymptote is observed at the Rayleigh wave speed c_R

$$\begin{aligned} D_{II}(x_1) &= \frac{\sigma_{21}^a}{\pi C_1^{(II)}} \frac{x_1}{\sqrt{c^2 - x_1^2}}, \ \phi_{II}(x_1) = \frac{b\sigma_{21}^a}{\pi C_1^{(II)}} \sqrt{c^2 - x_1^2}; \\ \overline{\sigma}_{21}^{(II)}(s) &= \frac{K_{II}}{\sqrt{2\pi}} \frac{1}{\sqrt{s}}, \ K_{II} = \sigma_{21}^a \sqrt{\pi c}; \\ G^{(II)} &= \frac{bK_{II}^2}{4\pi C_1^{(II)}}, \ G_0^{(II)} \equiv G^{(II)}(v = 0) \\ &= \frac{bK_{II}^2(0)}{4\pi C_0}, \ K_{II}(0) = \sigma_{21}^a \sqrt{\pi a}. \end{aligned} \tag{2.1.18}$$

The reduced quantity $\tilde{G}^{(II)} = G^{(II)}/G_0^{(II)}$ is reported in Fig. 2.5 as a function of $\tilde{v}_t$ for $\tilde{d}_t = 1$.

$\tilde{G}^{(II)}$ increases continuously with $\tilde{v}_t$ (below the Rayleigh wave speed) from the value 1 ($\tilde{v}_t = 0$); the increase is very pronounced when one approaches c_R. Above c_R, $\tilde{G}^{(II)}$ is negative and increases to zero with $\tilde{v}_t$.

II-2. *Intermediate velocity regime* ($c_t < v < c_l$)

Similarly, as for the corresponding mode *I* case, it may be written

$$\begin{aligned} \sigma_{21}^a \ &+ \int_{-c}^{c} \sigma_{21}^{(II)}(x_1 - x_1', 0, x_3) D_{II}(x_1') \mathrm{d}x_1' = 0; \\ \sigma_{21}^{(II)} &= \frac{C_2^{(II)}}{x_1 - x_1'} + C_3^{(II)} \mathrm{sgn}(x_1 - x_1') \delta(x_1 - x_1'); \\ C_2^{(II)} &= \frac{2\mu b c_*^2 \sqrt{1 - \tilde{v}_l^2}}{\pi \tilde{v}_l^2}, \ C_3^{(II)} = \frac{\mu b (2 - \tilde{v}_t^2)}{\pi \tilde{v}_t^2 \sqrt{\tilde{v}_t^2 - 1}}. \end{aligned} \tag{2.1.19}$$

Then, follows the integral equation for D_{II}

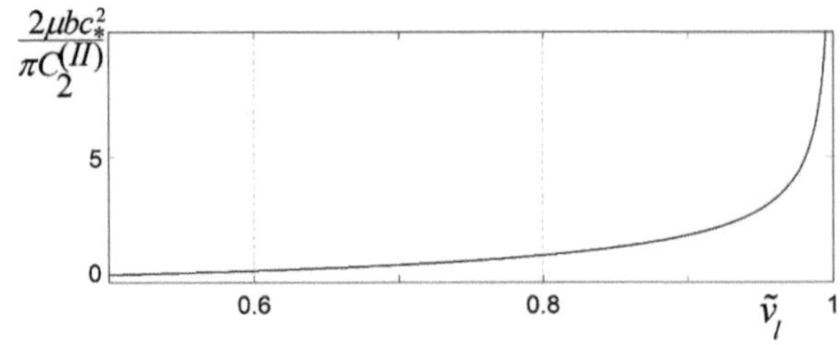

Fig. 2.6 Quantity $2\mu bc_*^2/\pi C_2^{(II)}$(2.1.19) as a function of $\tilde{v}_l$ in intermediate velocity regime $c_* < \tilde{v}_l < 1$; mode II loading, $\nu = 1/3$

$$\sigma_{21}^a + C_2^{(II)} \int_{-c}^{c} \frac{D_{II}(x_1')\mathrm{d}x_1'}{x_1 - x_1'} + C_3^{(II)} \int_{-c}^{c} \mathrm{sgn}(x_1 - x_1')\delta(x_1 - x_1')D_{II}(x_1')\mathrm{d}x_1' = 0. \tag{2.1.20}$$

The third term with $C_3^{(II)}$ cancels for $\tilde{v}_t^2 = 2$ but as shown for the similar mode I Eq. (2.1.12), this term cancels whatever the velocity v. Again, the familiar form is reached

$$\sigma_{21}^a + C_2^{(II)} \int_{-c}^{c} \frac{D_{II}(x_1')}{x_1 - x_1'}\mathrm{d}x_1' = 0; \tag{2.1.21}$$

with associated quantities as in (2.1.18). For $\tilde{v}_t^2 = 2$, the crack extension force is given by the form $G^{(II)}$ (2.1.18) in which $C_1^{(II)}$ is replaced by $C_2^{(II)}$ (2.1.19). Figure 2.6 reports $2\mu bc_*^2/\pi C_2^{(II)}$ as a function of $\tilde{v}_l$. It increases continuously with $\tilde{v}_l$. A vertical asymptote is present at $\tilde{v}_l = 1$.

Summary and Confrontation with Experiments

MODE I LOADING: In the subsonic regime ($v < c_t$), the crack-tip characteristic functions are similar in form with those of a static crack. They depend on v through $C_1^{(J)}$ ($J = I$ and II) and $c = a + vt$; when $v = 0$ and $vt = 0$, the static case is recovered. A special behaviour observed by experimentalists is as follows: an initial static crack ($|x_1| \leq a$, when using Fig. 2.1 as illustration) is under load in tension (mode I); at a given time taken as $t = 0$, it starts to move. It is found experimentally that the crack velocity increases gradually to a constant terminal velocity. This property of tensile cracks is captured on micrographs by Kerkhof and Richter (1969), Fig. 2.7.

The existence of a terminal velocity for tensile cracks in brittle materials is widely acknowledged (Lawn 1993; Freund 1990). This suggests that the quasi-static configuration, corresponding to small v in spatial region located about c_0 in Fig. 2.7, is not an equilibrium one. Subsequent question is: can this terminal velocity be predicted? Interestingly, the answer is yes, from Fig. 2.3. The terminal velocity is $\tilde{v}_t^{(e)}$ corresponding to the maximum value $\tilde{G}_{\max}^{(I)}$ of the reduced crack extension force. The

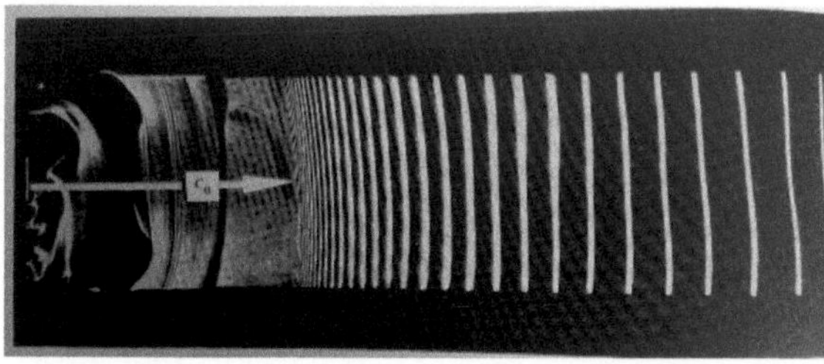

Fig. 2.7 Micrograph by Kerkhof and Richter (1969) showing a crack with initial edge length c_0 moving (from left to right) with a velocity increasing gradually to reach the terminal velocity. The vertical bright lines are the crack front positions captured at identical time interval. The terminal velocity corresponds to equal distance between the lines at the end right. The specimen length is 180 mm

condition to be satisfied is $\partial \tilde{G}^{(I)}/\partial \tilde{v}_t = 0$ and this is given by (2.1.10). The fact that $\tilde{G}^{(I)}$ increases with crack velocity v from zero up to a maximum has an important implication for a macroscopic crack (by "macroscopic", we mean that fracture over large distance is under way) under load in an elastic solid. The Griffith condition $G^{(I)} = 2\gamma$ (γ, surface energy) should be applied at the terminal velocity and not at $v = 0$ as commonly proceeded in laboratory experiments. Under the conditions of Fig. 2.3, $\tilde{v}_t^{(e)} \cong 0.52$ and $\tilde{G}_{\max}^{(I)} \cong 1.32$, in isotropic materials with Poisson ratio $\nu = 1/3$. This corresponds to experiments (see Lawn (1993), for example); above that velocity, crack branching is observed, and this is explained by the rapid decrease to zero of the crack extension force $\tilde{G}^{(I)}$. Because the driving force $\tilde{G}^{(I)}$ is becoming small, the moving crack looks for an alternative to propagate rapidly. We now refer to the overview of dynamic fracture mechanics by Freund (1990) in support of the findings. In mode *II* loading, the crack extension force expression $\tilde{G}^{(II)}$(2.1.18) is the same displayed there (see relation (5.3.10) in Freund (1990)). The corresponding expression for mode *I* exhibits the function R_w (defined in present text about (2.1.16)) in Freund (1990), in contrast to relation (2.1.8). However, both expressions are identical for small v. An attempt to derive the terminal velocity expression is presented by Lawn (1993); this makes use of Mott (1948) extension of the Griffith concept to dynamic fracture. The estimated value falls well below the Rayleigh wave velocity and closer to present findings. From its maximum value, $\tilde{G}^{(I)}$ decreases continuously to zero at $v = c_t$ (Fig. 2.3). This contrasts with many who argue that the theoretical limiting speed of a tensile crack must be the Rayleigh wave speed (see (Freund 1990; Broberg 1999), for example). In the transonic velocity regime ($c_t < v < c_l$), the crack functions (2.1.15) are identical in form with those of the subsonic regime; however, contrary to the latter where $C_1^{(I)}$ is positive, $C_2^{(I)}$ (or equivalently $1/C_2^{(I)}$) changes sign in the former (Fig. 2.4). For $\tilde{v}_l < c_* \sqrt{2}$, $C_2^{(I)}$ is negative. Mathematically, one can transfer the minus sign to the tensile applied stress σ_{22}^a and write, for example,

$$\phi_I(x_1) = \frac{b(-\sigma_{22}^a)}{\pi \left| C_2^{(I)} \right|} \sqrt{c^2 - x_1^2}. \tag{2.1.22}$$

The corresponding physical interpretation is that over $|x_1| < c$, an ellipse of material has been removed so that under the action of a compressive applied stress $(-\sigma_{22}^a)$, the two faces of the elliptical hole close, but remain stress free. In fact, at $t = 0$, just before crack motion, the crack shape is elliptical. The action of the velocity in this region is to close the crack under movement; this suggests that the motion of the crack is inhibited. For $\tilde{v}_l > c_* \sqrt{2}$, $C_2^{(I)}$ becomes positive, suggesting possible motion of the crack.

MODE *II* LOADING: Physical quantities associated to the crack are given in (2.18) in the subsonic speed regime ($v < c_t$), except at the Rayleigh velocity c_R. As can be seen from Fig. 2.5, $C_1^{(II)}$ is negative above c_R. A relation like (2.1.22) can be written

$$\phi_{II}(x_1) = \frac{b(-\sigma_{21}^a)}{\pi \left| C_1^{(II)} \right|} \sqrt{c^2 - x_1^2}. \tag{2.1.23}$$

A similar suggestion in discussing (2.1.22) is applied here: the motion of the crack is impeded in the velocity interval $c_R < v < c_t$. Rayleigh wave velocity appears to be the limiting speed of a moving crack under mode *II* loading in the subsonic regime. In the intermediate regime ($c_t < v < c_l$), the crack functions are similar in form to (2.1.18) with $C_2^{(II)}$ replacing $C_1^{(II)}$. Because $C_2^{(II)}$ is positive (Fig. 2.6), the crack motion is possible.

SUMMARY: An analysis is made of planar cracks located in the Ox_1x_3-plane with a straight front parallel to x_3 in isotropic materials. The crack fronts travel along x_1 with a constant velocity v under mode *I* and *II* loadings (Anongba 2020a). Cracks are represented by a continuous distribution of uniformly moving straight edge dislocations parallel to x_3 with Burgers vectors $\vec{b}_J$ directed along x_1 ($J = II$) or x_2 ($J = I$), according to the loading mode J. Explicit expressions of the elastic fields (displacement and stress) of the crack dislocations are first given for velocities, from zero to values larger than c_l, the velocity of longitudinal sound wave. Then, crack physical quantities are given, namely the dislocation distribution function D_J, the relative displacement ϕ_J of the faces of the crack, the crack-tip stresses and the crack extension force $G^{(J)}$ per unit length of the crack front. These results cover the velocity range $[0, c_l]$. In mode *I* loading and in the subsonic velocity regime ($v < c_t$, the velocity of transverse sound wave), $G^{(I)}$ increases continuously with v from the value in the static case $G_0^{(I)}(v = 0)$ to a maximum $G_{\max}^{(I)} \cong 1.32 G_0^{(I)}$ at $v = v^{(e)} \cong 0.52 c_t$ (Fig. 2.3); then, $G^{(I)}$ decreases rapidly to zero when v tends to c_t. The value $v^{(e)}$ corresponding to the maximum of the crack extension force is identified to the terminal tensile crack velocity, experimentally observed in the fracture of brittle materials. No reference is made to the Rayleigh wave velocity c_R. In the transonic speed regime ($c_t < v < c_l$), the crack characteristic functions (2.1.15) are identical in form with those of the subsonic regime. However, for $v < c_t\sqrt{2}$, it is shown that the faces of the crack, separated under load before the extension of the crack, close under motion; this indicates that the crack movement is hindered. for $v > c_t\sqrt{2}$, the motion of the crack is possible. In mode *II* loading and in the subsonic regime ($v < c_t$), $G^{(II)}$ increases continuously with v (when $v < c_R$) from the calculated value in the static case $G_0^{(II)}(v = 0)$; when v approaches c_R, $G^{(II)}$ increases very rapidly (Fig. 2.5). Above c_R ($c_R < v < c_t$), the relative displacement of the faces of the crack, formed under load before crack motion, closes in movement; this indicates that crack motion is impeded. The velocity of uniformly moving cracks is limited by the Rayleigh wave velocity. In the intermediate speed regime ($c_t < v < c_l$), the crack characteristic functions are similar in form to (2.1.18) with $C_2^{(II)}$ (2.1.19) replacing $C_1^{(II)}$ (2.1.16). Because $C_2^{(II)}$ is positive (Fig. 2.6), the movement of the crack is possible.

2.1.2 Cubic Materials

The uniform motion of planar straight-fronted cracks in cubic materials under Mode *I* loading only is concerned in what follows; the model is that of Fig. 2.1. We restrict ourselves to the dislocation stresses involved in the crack analysis.

$$\sigma_{22}^{(I)}(x_1, x_2 = 0, x_3; t) = \frac{b}{y_1 2\pi^2} \int_{-\pi/2}^{\pi/2} \mathrm{d}\psi \frac{Num\sigma_{22}^{(I)}}{\Delta} \equiv C^{(I)}(v)/y_1; \tag{2.1.24}$$

$$Num\sigma_{22}^{(I)} = \rho C_{44}(C_{12} + C_{44})\left(C_{12}\cos^2\psi + C_{11}\sin^2\psi\right)v^2\cos^2\psi + C_{44}\left\{\begin{array}{l} \cos^4\psi\left[C_{44}C_{11}^2 + C_{12}C_{44}^2 - \rho v^2\left(C_{11}^2 + C_{44}C_{11} + 2C_{44}C_{12}\right)\right. \\ \left. +\left(\rho v^2\right)^2(C_{11} + C_{12})\right] \\ +C_{11}C_{44}^2\sin^4\psi - 2\rho C_{11}C_{44}v^2\cos^2\psi\sin^2\psi \end{array}\right\},$$

$$\Delta = A_1\cos^6\psi + B_2\sin^6\psi + A_{12}\cos^4\psi\sin^2\psi + B_{21}\cos^2\psi\sin^4\psi;$$

$$A_1 = C_{11}C_{44}^2 - \rho C_{44}(C_{44} + 2C_{11})v^2 + \rho^2(C_{11} + 2C_{44})v^4 - \rho^3 v^6,$$

$$B_2 = C_{11}C_{44}^2, \quad A_{12} = \rho v^2\left[2C_{44}C_{12} + \rho(2C_{44} + C_{11})v^2\right],$$

$$B_{21} = -\rho C_{44}(C_{44} + 2C_{11})v^2.$$

Dislocation distribution D_I:

$$D_I(x_1) = \frac{\sigma_{22}^a}{\pi C^{(I)}(v)} \frac{x_1}{\sqrt{c^2 - x_1^2}} \quad (|x_1| < c = a + vt); \tag{2.1.25}$$

Relative displacement of the faces of the crack:

$$\phi_I(x_1) = \frac{b\sigma_{22}^a}{\pi C^{(I)}(v)}\sqrt{c^2 - x_1^2}. \tag{2.1.26}$$

The total stress $\overline{\sigma}_{22}^{(I)}$ ahead of the crack tip at $x_1 = c$ ($x_1 = c + s$, $0 < s \ll c$)

$$\overline{\sigma}_{22}^{(I)}(s) = \frac{K_I}{\sqrt{2\pi}}\frac{1}{\sqrt{s}}; \tag{2.1.27}$$

$K_I = \sigma_{22}^a\sqrt{\pi c}$. The crack extension force $G^{(I)}$ per unit edge length is

$$G^{(I)} = \frac{bK_I^2}{4\pi C^{(I)}(v)}. \tag{2.1.28}$$

The crack-front velocity regimes (subsonic, transonic ...) appear in the dependence with v of $C^{(I)}(v)$. Forbidden velocity regimes correspond to negative $C^{(I)}(v)$ values. Above results lead to the treatment of the subject "Conoidal crack with elliptical bases, within cubic crystals, growing uniformly under arbitrary applied loadings". To do so, average $\langle G\rangle$ obtained in (Anongba Part *I*, 2024) is considered; its normalized value $\langle\tilde{G}\rangle$ is defined as

$$\begin{aligned}\langle\tilde{G}\rangle &= \langle G\rangle/G_C^I = \langle\tilde{G}\rangle(\theta, \theta_0, \phi_0, M_{12}, M_{13}, a_r, C_{nm}),\\ G_C^I &= \frac{2^5\alpha_0^2 a_r^4}{3\pi^2}G^{(I)};\end{aligned} \tag{2.1.29}$$

where $K_I = \sigma_{22}^a\sqrt{\pi c}$ and $c \equiv a_1 = a_1(0) + vt$ is the semiaxis along x_1' of the elliptical base of the conoidal crack after incremental time t from zero. From $t = 0$, the crack starts growing uniformly.

2.2 Non-planar Cracks in Uniform Motion (Isotropic Materials)

2.2.1 Isolated Brittle Cracks

The crack model refers to Anongba (2020*b*). The medium is elastic and infinitely extended to which is attached a Cartesian system x_i. It consists of a non-planar crack of finite extensions along x_1 and x_2 and infinite in the x_3-direction. Initially the crack is static and extends along x_1 from $x_1 = -c_0$ to c_0. It spreads at any x_1-spatial position in the x_2x_3-plane in the form of a Fourier series

$$f = \sum_n(\xi_n \sin\kappa_n x_3 + \delta_n\cos\kappa_n x_3) + h(x_1) \equiv \xi(x_1, x_3) + h(x_1). \tag{2.2.1}$$

Here n is a positive integer; h, ξ_n, δ_n and κ_n are real that are x_1-dependent. Figure 2.8a gives an illustration of the crack front. At a given time taken as $t = 0$ and under general loading, it starts moving at constant velocity v in the x_1-direction. Its extension after incremental time t is given by $|x_1| \leq c = c_0 + vt$. In addition to moving uniformly along x_1, the crack faces are free from any traction. The loadings correspond to tension σ_{22}^a and shears σ_{21}^a and σ_{23}^a applied at infinity in the x_2, x_1 and x_3 directions, respectively. The treatment also includes normal induced stresses $\sigma_{11}^a = \sigma_{33}^a = -\nu\sigma_{22}^a$ originating from the Poisson's effect (ν is Poisson's modulus); these are illustrated in Fig. 2.8b. The crack is represented by a continuous distribution of three types of infinitesimal moving dislocations J ($J = I$, II and III) having the shape f (2.2.1) with Burgers vectors $\vec{b}_I = (0, b, 0)$, $\vec{b}_{II} = (b, 0, 0)$ and $\vec{b}_{III} = (0, 0, b)$,respectively. Types I and II are of edge character on average and type III screw. Dislocation distributions D_J ($J = I$, II and III) are defined such that $D_J(x_1')dx_1'$ represents the number of dislocations J in the infinitesimal x_1-interval dx_1' located about the x_1-spatial position x_1'. The elastic fields (displacement $\vec{u}^{(J)}$ and stress $(\sigma)^{(J)}$) of the dislocation J located at $x_1 = x_1' \equiv vt$ in the medium may be deduced from those (moving sinusoidal dislocations) located at $x_1 = x_1'$ with simple form $A_n = \xi_n \sin \kappa_n x_3$ in x_2x_3-planes (Fig. 2.9). For the later, the elastic fields at $\vec{x} = (x_1, x_2, x_3)$ are (to linear term in amplitude ξ_n)

$$\begin{aligned} \vec{u}^{(J)(n)}(\vec{x}) &= \vec{u}^{(J)(0)}(y_1, x_2) + \vec{u}^{(J)A_n}(y_1, x_2, x_3), \\ (\sigma)^{(J)(n)}(\vec{x}) &= (\sigma)^{(J)(0)}(y_1, x_2) + (\sigma)^{(J)A_n}(y_1, x_2, x_3), \end{aligned} \tag{2.2.2}$$

$y_1 = x_1 - x_1'$; $\vec{u}^{(J)(0)}$ and $(\sigma)^{(J)(0)}$ are of zero order, they correspond to the fields of a moving straight dislocation J (parallel to x_3) at $x_1 = x_1'$ on Ox_1x_3 with Burgers vector $\vec{b}_J$; $\vec{u}^{(J)A_n}$ and $(\sigma)^{(J)A_n}$ are oscillating parts involving A_n or its spatial x_3-derivative $\partial A_n/\partial x_3$.

When the dislocation exhibits shape f (2.2.1), the elastic fields take the form

$$\begin{aligned} \vec{u}^{(J)}(\vec{x}) &= \vec{u}^{(J)(0)}(y_1, y_2) + \sum_n \vec{u}^{(J)A_n}(y_1, y_2, x_3) \equiv \vec{u}^{(J)(0)} + \vec{u}^{(J)\xi}, \\ (\sigma)^{(J)}(\vec{x}) &= (\sigma)^{(J)(0)}(y_1, y_2) + \sum_n (\sigma)^{(J)A_n}(y_1, y_2, x_3) \equiv (\sigma)^{(J)(0)} \\ &\quad + (\sigma)^{(J)\xi}; \end{aligned} \tag{2.2.3}$$

$y_2 = x_2 - h$; here A_n stands for $A_n = \xi_n \sin \kappa_n x_3 + \delta_n \cos \kappa_n x_3$. The plastic distortions $\beta_{ij}^{*(J)}(\vec{x}, t)$ associated to the dislocations J ($J = I$, II and III), are expressed successively to first order in ξ, assuming ξ in (2.2.1) small.

$$\beta_{12}^{*(I)}(\vec{x}, t) = b\delta(y_1)H(y_2) - b\xi\delta(y_1)\delta(y_2), \tag{2.2.4}$$

the other components of the plastic distortion are zero; δ and H are the Dirac delta and Heaviside step functions, respectively; $y_1 = x_1 - vt$ and $y_2 = x_2 - h$. Here, the

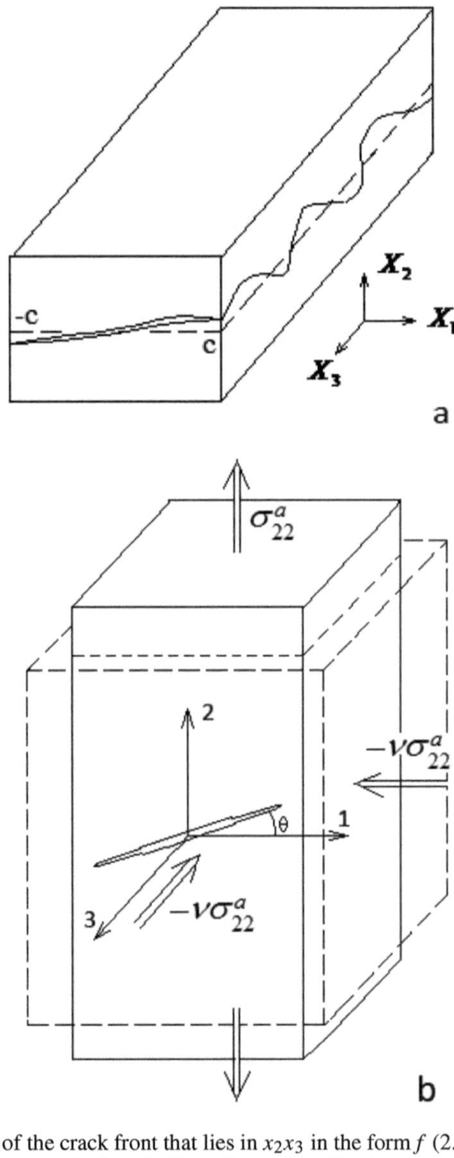

Fig. 2.8 **a** Illustration of the crack front that lies in x_2x_3 in the form f (2.2.1). **b** Illustration of the induced normal stresses due to Poisson' effect. General loading is assumed with loadings σ^a_{22}, σ^a_{21}, σ^a_{23} along x_2, x_1 and x_3 directions (respectively)

first term is due to a straight edge dislocation displaced by $x_2 = h$ from the Ox_1x_3-plane. The corresponding displacement can be derived by replacing x_2 by $(x_2 - h)$ in the displacement of a straight edge I dislocation at $x_1 = x_1' \equiv vt$ (travelling in the Ox_1x_3-plane at velocity v in the x_1-direction; see Sect. 2.1). The Fourier form of the second term denoted $\beta_{12}^{*(I)\xi}$ is taken from Anongba (2010):

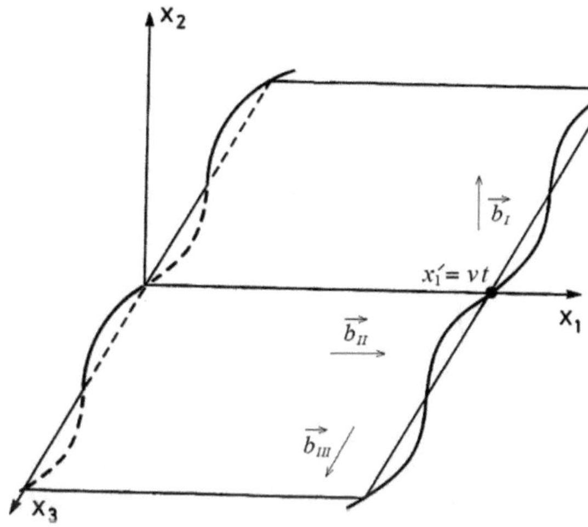

Fig. 2.9 Infinitely long sinusoidal dislocation travelling uniformly at constant speed v in the x_1-direction with Burgers vector $\vec{b}_J$ directed along x_1 $(J = II)$, x_2 $(J = I)$ and x_3 $(J = \text{III})$; the dislocation at the origin $(t = 0)$ and after incremental time t are illustrated

$$\beta_{12}^{*(I)\xi} = -\frac{b}{8\pi^2}\sum_n \int_{-\infty}^{\infty}\int_{-\infty}^{\infty}\left(z_n e^{i(\omega t+\vec{k}'.\vec{x}_h)} + \overline{z}_n e^{i(\omega t+\vec{k}.\vec{x}_h)}\right)\mathrm{d}k_1\mathrm{d}k_2; \qquad (2.2.5)$$

$$\vec{k}' = (k_1' = k_1, k_2' = k_2, k_3' = -\kappa_n), \vec{k} = (k_1, k_2, k_3 = \kappa_n)\vec{x}_h = (x_1, y_2, x_3);$$

$$z_n = \delta_n + i\xi_n, \overline{z}_n = \delta_n - i\xi_n, \omega = -k_1 v.$$

The elastic fields of dislocation $J = II$ with form (2.2.1) at $x_1 = x_1' \equiv vt$ can be derived from those of a sinusoidal dislocation located at $x_1 = x_1'$ with the same Burgers vector, lying in the x_2x_3-plane and defined by $x_2 = A_n \equiv \xi_n \sin \kappa_n x_3$. There are two non-zero components of the plastic distortion (Anongba 2010):

$$\begin{aligned} \beta_{21}^{*(II)(n)} &= \frac{b}{1+(\partial A_n/\partial x_3)^2}\delta(x_2 - A_n)H(-y_1), \\ \beta_{31}^{*(II)(n)} &= -\partial A_n/\partial x_3 \beta_{21}^{*(II)(n)}. \end{aligned} \qquad (2.2.6)$$

We assume that both A_n and $\partial A_n/\partial x_3$ are small in magnitude: corresponding Fourier forms used in the sequel are

$$\beta_{21}^{*(II)(n)} = \beta_{21}^{*(II)(0)} + \beta_{21}^{*(II)A_n}$$

$$\beta_{21}^{*(II)(0)} = \frac{ib}{(2\pi)^2} \int_{-\infty}^{\infty} \int_{-\infty}^{\infty} \frac{1}{k_1} e^{i(\omega t + k_1 x_1 + k_2 x_2)} dk_1 dk_2,$$

$$\beta_{21}^{*(II)A_n} = \frac{ib\xi_n}{8\pi^2} \int_{-\infty}^{\infty} \int_{-\infty}^{\infty} \frac{k_2}{k_1} \left(e^{i(\omega t + \vec{k}'.\vec{x})} - e^{i(\omega t + \vec{k}.\vec{x})} \right) dk_1 dk_2; \tag{2.2.7}$$

$$\beta_{31}^{*(II)(n)} = -\frac{ib\kappa_n \xi_n}{8\pi^2} \int_{-\infty}^{\infty} \int_{-\infty}^{\infty} \frac{1}{k_1} \left(e^{i(\omega t + \vec{k}'.\vec{x})} + e^{i(\omega t + \vec{k}.\vec{x})} \right) dk_1 dk_2. \tag{2.2.8}$$

$\beta_{21}^{*(II)(0)}$ corresponds to a straight edge dislocation located at $x_1 = x_1'$ in Ox_1x_3; the associated elastic fields are known from Sect. 2.1. In these fields, we just replace x_2 by $(x_2 - h)$ to obtain the elastic fields of the straight dislocation at $x_1 = x_1'$ displaced by $x_2 = h$ from the Ox_1x_3-plane. This gives the non-oscillating part of the elastic fields of crack dislocations *II* with form f. For the average screw dislocation *III* with form f at $x_1 = x_1'$, the non-zero component of the plastic distortion used is written to first order in ξ:

$$\beta_{13}^{*(III)}(\vec{x}, t) = b\delta(y_1)H(y_2) - b\xi\delta(y_1)\delta(y_2). \tag{2.2.9}$$

Here, the first term $\beta_{13}^{*(III)(0)}$ is due to a straight screw at $x_1 = x_1'$ with elevation $x_2 = h$ from Ox_1x_3. Its Fourier form reads

$$\beta_{13}^{*(III)(0)} = \int_{-\infty}^{\infty} \int_{-\infty}^{\infty} \overline{\beta}_{13}^{*(III)(0)} e^{i(\omega t + k_1 x_1 + k_2 y_2)} dk_1 dk_2,$$
$$\overline{\beta}_{13}^{*(III)(0)} = -\frac{ib}{4\pi^2 k_2}. \tag{2.2.10}$$

Because $\beta_{13}^{*(III)}$(2.2.9) and $\beta_{12}^{*(I)}$(2.2.4) are identical in form, we can make use of the Fourier form (2.2.5) for the second term $\beta_{13}^{*(III)\xi}$ in (2.2.9). At this stage, we can write down the various displacements $\vec{u}^{(J)}(J = I, II$ and $III)$ associated with the three types J of crack dislocation with form f (2.2.1). The stress fields $(\sigma)^{(J)}$ can be obtained by differentiating the displacements. Figure 2.10 is a schematic representation of simple special cracks captured by the modelling. The cracks extend in the x_1-direction from $x_1 = -c$ to c and must be considered to run indefinitely in the x_3-direction. The crack shape in planes perpendicular to x_1 is described by ξ (Fig. 2.10c for example). The shape f of the crack in planes perpendicular to x_3 is given by both ξ, through the x_1-dependence of positive quantities ξ_n, δ_n and κ_n (2.2.1), and function $h = h(x_1)$. Since ξ is assumed to be small oscillating function, the average fracture plane is described correctly by the equation $x_2 = h(x_1)$. When $\xi = 0$, the crack dislocations are straight

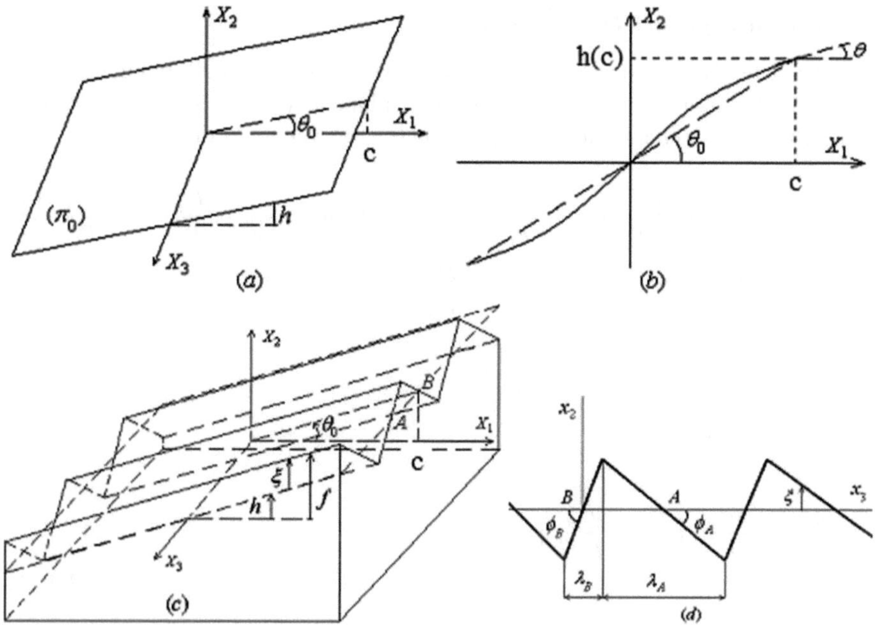

Fig. 2.10 Simple special cracks. **a** Inclined planar crack π_0 (see text). **b** A non-planar crack (parallel to x_3) as h odd function of x_1 ($x_2 = h(x_1)$). **c** Non-planar crack fluctuating about an average inclined plane π_0. The crack consists of planar facets; its fronts at $x_1 = \pm c$ lie in x_2x_3 planes. At $x_1 = c$, the crack front is characterized by inclination angles ϕ_A and ϕ_B (see (**d**)) at points A and B located on the average fracture plane. **d** Sketch of the crack front in (**c**) with B taken as origin. In this geometry (from (**a**) to (**c**)) the general loading of the crack systems corresponds to uniform applied σ_{22}^a, σ_{12}^a and σ_{23}^a at infinity in the x_2, x_1 and x_3 directions, respectively

parallel to x_3 and distributed over the surface $x_2 = h(x_1)$. Specific examples are (Fig. 2.10):

- $h(x_1) = p_0x_1$ ($p_0 \geq 0$) and $\xi = 0$. This corresponds to a planar crack π_0(with a straight front parallel to x_3) rotated around Ox_3 by angle $\theta_0 = \tan^{-1} p_0$ from Ox_1x_3, Fig. 2.10a.
- $h(x_1)$ is an arbitrary function of x_1, $\xi = 0$. The sketch in Fig. 2.10b corresponds to h odd although this is not mandatory. Actually h odd conforms well to homogeneity of the medium, geometry of the applied loadings and D_J approximations adopted in the study.
- $h(x_1) = p_0x_1$ ($p_0 \geq 0$) and $\xi = \xi(x_3)$ independent of x_1. The crack fluctuates about plane π_0 with a front spreading in planes parallel to x_2x_3 in the form ξ. In the example displayed in Fig. 2.10c the crack consists of planar facets with inclination angles ϕ_A and ϕ_B (Fig. 2.10d) at points A and B of the crack front located on the average fracture plane.

Elastic Fields of Crack Dislocation I

The elastic fields of dislocation I, with Burgers vector $\vec{b}_I = (0, b, 0)$ lying indefinitely in the x_3-direction and spreading in the x_2x_3-plane at $x_1 = x_1' \equiv vt$ in the Fourier series form f (2.2.1), take at spatial position (x_1, x_2, x_3) the form:

$$
\begin{aligned}
u_m^{(I)(0)} &= \frac{b}{\pi\tilde{v}_t^2}\left(\delta_{m1}\left[\frac{1}{P_l}\ln r_l - \frac{P_{2t}^2}{P_t}\ln r_t\right] + \delta_{m2}\left[\tan^{-1}\frac{y_1}{P_l y_2} - \frac{P_{2t}^2}{P_t^2}\tan^{-1}\frac{y_1}{P_t y_2}\right]\right);\\
\sigma_{11}^{(I)(0)} &= \frac{\mu b}{\pi\tilde{v}_t^2}y_1\left(\frac{2+(1-2c_*^2)\tilde{v}_t^2}{P_l r_l^2} - \frac{2P_{2t}^2}{P_t r_t^2}\right),\\
\sigma_{22}^{(I)(0)} &= \frac{2\mu b P_{2t}^2}{\pi\tilde{v}_t^2}y_1\left(\frac{1}{P_t r_t^2} - \frac{1}{P_l r_l^2}\right),\\
\sigma_{33}^{(I)(0)} &= \frac{\mu b(1-2c_*^2)}{\pi P_l}\frac{y_1}{r_l^2},\\
\sigma_{21}^{(I)(0)} &= \frac{2\mu b}{\pi\tilde{v}_t^2}y_2\left(\frac{P_l}{r_l^2} - \frac{P_{2t}^4}{P_t r_t^2}\right);
\end{aligned}
\tag{2.2.11a}
$$

$\tilde{v}_t = v/c_t$, $P_t^2 = 1 - \tilde{v}_t^2$, $P_{2t}^2 = 1 - \tilde{v}_t^2/2$;
$\tilde{v}_l = v/c_l$, $P_l^2 = 1 - \tilde{v}_l^2$; $r_s^2 = y_1^2 + P_s^2 y_2^2$, $s = t$ and l; $c_* = c_t/c_l$;

the subscript m takes the values 1 and 2, and c_t and c_l are the velocities of transverse and longitudinal sound waves, respectively. Other elastic field components are zero. The oscillating parts of the elastic fields take the forms:

$$
\begin{aligned}
u_1^{(I)\xi} &= \frac{b}{\pi\tilde{v}_t^2}\sum_n A_n\frac{\partial}{\partial x_2}\left(-\frac{P_{2t}^2}{P_t}K_0[z_n^{(t)}] + \frac{2(1-\nu)c_*^2}{(1-2\nu)P_l}K_0[z_n^{(l)}]\right),\\
u_2^{(I)\xi} &= \frac{b}{2\pi\tilde{v}_t^2}\sum_n A_n\left(\frac{\tilde{v}_t^2}{P_t}\frac{\partial}{\partial y_1}K_0[z_n^{(t)}] + 2\frac{\partial^2}{\partial x_2^2}\left[\frac{1}{P_t}\overline{I}^{(t)} - \frac{2(1-\nu)c_*^2}{(1-2\nu)P_l}\overline{I}^{(l)}\right]\right),\\
u_3^{(I)\xi} &= \frac{b(1-c_*^2)}{\pi\tilde{v}_t^2}\sum_n\frac{\partial A_n}{\partial x_3}\frac{\partial}{\partial x_2}\left(\frac{(1-2\nu)c_*^{-2}}{P_t}\overline{I}^{(t)} - \frac{2(1-\nu)}{P_l}\overline{I}^{(l)}\right);\\
\sigma_{ii}^{(I)\xi} &= \frac{2\mu}{1-2\nu}\left([\delta_{i1}(1-\nu)+\nu(\delta_{i2}+\delta_{i3})]\frac{\partial u_1^{(I)\xi}}{\partial x_1}\right.\\
&\quad + [\delta_{i2}(1-\nu)+\nu(\delta_{i1}+\delta_{i3})]\frac{\partial u_2^{(I)\xi}}{\partial x_2}\\
&\quad \left. + [\delta_{i3}(1-\nu)+\nu(\delta_{i1}+\delta_{i2})]\frac{\partial u_3^{(I)\xi}}{\partial x_3}\right),\ i = 1, 2 \text{ and } 3,\\
\sigma_{ij}^{(I)\xi} &= \mu\left(\frac{\partial u_i^{(I)\xi}}{\partial x_j} + \frac{\partial u_j^{(I)\xi}}{\partial x_i}\right),\ i \neq j,
\end{aligned}
\tag{2.2.11b}
$$

$z_n^{(s)} = (\kappa_n/P_s)\sqrt{y_1^2 + P_s^2 y_2^2}$, $\overline{I}^{(s)} = \int_{y_1}^{\infty} K_0[z_n^{\prime(s)}]\mathrm{d}y$, $s = t$ and l; $K_n[z]$ is the nth-order modified Bessel function usually so denoted and $z_n^{\prime(s)}$ stands for $z_n^{(s)}$ with y instead of y_1.

Elastic Fields of Crack Dislocation II

The dislocation II has Burgers vector $\vec{b}_{II} = (b, 0, 0)$ and spreads in the x_2x_3-plane at $x_1 = x_1' \equiv vt$ in the Fourier series form.

$$
\begin{aligned}
u_m^{(II)(0)} &= \frac{b}{\pi\tilde{v}_t^2}\left(\delta_{m2}\left[P_l \ln r_l - \frac{P_{2t}^2}{P_t}\ln r_t\right] + \delta_{m1}\left[P_{2t}^2 \tan^{-1}\frac{y_1}{P_t y_2} - \tan^{-1}\frac{y_1}{P_l y_2}\right]\right);\\
\sigma_{11}^{(II)(0)} &= \frac{\mu b}{\pi\tilde{v}_t^2}y_2\left(\frac{2P_t P_{2t}^2}{r_t^2} + \frac{P_l[(c_*^{-2}-2)P_l^2 - c_*^{-2}]}{r_l^2}\right),\\
\sigma_{22}^{(II)(0)} &= \frac{\mu b}{\pi\tilde{v}_t^2}y_2\left(\frac{P_l[P_l^2 c_*^{-2} - c_*^{-2} + 2]}{r_l^2} - \frac{2P_t P_{2t}^2}{r_t^2}\right),\\
\sigma_{33}^{(II)(0)} &= -\frac{\mu b(1-2c_*^2)P_l}{\pi}\frac{y_2}{r_l^2},\\
\sigma_{21}^{(II)(0)} &= \frac{2\mu b}{\pi\tilde{v}_t^2}y_1\left(\frac{P_l}{r_l^2} - \frac{P_{2t}^4}{P_t r_t^2}\right);
\end{aligned}
\tag{2.2.12a}
$$

Other elastic fields of zero order with respect to A_n are zero. The oscillating parts of the displacement are:

$$
\begin{aligned}
u_1^{(II)\xi} &= \frac{b}{2\pi\tilde{v}_t^2}\sum_n A_n\left(\frac{\partial}{\partial y_1}\left[\frac{4(1-\nu)c_*^2 P_l}{1-2\nu}K_0[z_n^{(l)}] - (2-\tilde{v}_t^2)P_t K_0[z_n^{(t)}]\right]\right.\\
&\quad - \frac{\kappa_n^2\tilde{v}_t^2[(1-2\nu)c_*^{-2} - 2(1-\nu)]\overline{I}^{(t)}}{P_t}\\
&\quad + \pi\delta(y_2)\mathrm{sgn}y_1\left[\tilde{v}_t^2 - 2 + \frac{4(1-\nu)c_*^2}{1-2\nu}\right.\\
&\quad \left.\left. - y_1^2\frac{\kappa_n^2[(1-2\nu)c_*^{-2} - 2(1-\nu)]}{2}\right]\right) u_2^{(II)\xi}\\
&= \frac{b}{2\pi\tilde{v}_t^2}\sum_n A_n\frac{\partial}{\partial y_2}\left(\left[\frac{2(1-\nu)c_*^2}{1-2\nu} - 1\right]2\pi\delta(y_2)|y_1|\right.\\
&\quad \left(-\frac{(2-\tilde{v}_t^2)K_0[z_n^{(t)}]}{P_t} + \frac{4(1-\nu)c_*^2 P_l}{1-2\nu}K_0[z_n^{(l)}]\right),\\
u_3^{(II)\xi} &= \frac{b}{2\pi\tilde{v}_t^2}\sum_n \frac{\partial A_n}{\partial x_3}(4(1-\nu)(1-c_*^2)P_l K_0[z_n^{(l)}]
\end{aligned}
$$

$$+\frac{(1-2\nu)(1-c_*^{-2})(2-\tilde{v}_t^2)K_0[z_n^{(t)}]}{P_t}$$
$$+\left[(1-2\nu)(2-c_*^{-2})2(1-\nu)(2c_*^2-1)\right]\pi\delta(y_2)|y_1|\Big) \tag{2.2.12b}$$

The associated stress fields are obtained by differentiating the displacements with similar relations as in (2.2.11*b*).

Elastic Fields of Crack Dislocation III

The dislocation *III* has Burgers vector $\vec{b}_{III} = (0, 0, b)$ and spreads in the x_2x_3-plane at $x_1 = x_1' \equiv vt$ in the Fourier series form f. For the non-oscillating part of the displacement, the only non-zero component is

$$u_3^{(III)(0)} = \frac{b}{2\pi\beta_{13(t)}^2}\tan^{-1}\frac{P_t y_2}{y_1};$$

$\beta_{13(t)}^2$ stands for P_t^2. Surprisingly $\beta_{13(t)}^2 = 1$ when the displacement $u_3^{(III)(0)}$ is derived from the plastic distortion $\beta_{23}^*(\vec{x}, t)$ for which it is assumed that a slip $\boldsymbol{b} = (0, 0, b)$ is produced in the x_1x_3-plane for $x_1 < vt$ (Mura 1968). Hence, it is questioned to see the effect of this difference in both values of $u_3^{(III)(0)}$. We observe no change in the reduced crack extension force (see below) for a special non-planar crack with a segmented front.

$$\sigma_{13}^{(III)(0)} = -\frac{\mu b P_t}{2\pi\beta_{13(t)}^2}\frac{y_2}{r_t^2},\quad \sigma_{23}^{(III)(0)} = \frac{\mu b P_t}{2\pi\beta_{13(t)}^2}\frac{y_1}{r_t^2}. \tag{2.2.13a}$$

The compressive $\sigma_{ii}^{(III)(0)}(i = 1, 2 \text{ and } 3)$ stresses are zero, as also $\sigma_{21}^{(III)(0)}$. The oscillating parts of the displacement read:

$$u_1^{(III)\xi} = \frac{b}{2\pi\tilde{v}_t^2}\sum_n \frac{\partial A_n}{\partial x_3}\left(\frac{\tilde{v}_t^2 - 2(c_*^{-2}-1)(1-2\nu)}{P_t}K_0[z_n^{(t)}]\right.$$
$$\left.+\frac{4(1-\nu)(1-c_*^2)}{P_l}K_0[z_n^{(l)}]\right),$$
$$u_2^{(III)\xi} = \frac{b(c_*^{-2}-1)}{\pi\tilde{v}_t^2}\sum_n \frac{\partial A_n}{\partial x_3}\frac{\partial}{\partial x_2}\left(\frac{1-2\nu}{P_t}\overline{I}^{(t)} - \frac{2(1-\nu)c_*^2}{P_l}\overline{I}^{(l)}\right), \tag{2.2.13b}$$
$$u_3^{(III)\xi} = \frac{b}{2\pi\tilde{v}_t^2}\sum_n A_n\left(\frac{\tilde{v}_t^2}{P_t}\frac{\partial K_0[z_n^{(t)}]}{\partial y_1}\right.$$
$$\left.-2\kappa_n^2(c_*^{-2}-1)\left[\frac{(1-2\nu)\overline{I}^{(t)}}{P_t} - \frac{2(1-\nu)c_*^2}{P_l}\overline{I}^{(l)}\right]\right)$$

The associated stress fields $(\sigma)^{(J)\xi}$ are obtained by differentiating the displacements.

Crack Dislocation Distributions

Assume first that the dislocations are straight parallel to the x_3-direction ($\xi = 0$) and $h(x_1) = p_0 x_1$ depends linearly on x_1 with p_0 positive constant (Fig. 2.10a). We thus have a planar crack of finite extension, with straight fronts running indefinitely along x_3, rotated (from Ox_1x_3) about the positive x_3-direction by $\theta_0 = \tan^{-1} p_0$. The crack extends from $x_1 = -c$ to c and is subjected to general loading applied at infinity. Under such conditions, we have $\partial f/\partial x_1 = \partial h/\partial x_1 = p_0$, $\partial f/\partial x_3 = \partial \xi/\partial x_3 = 0$; making use of traction-free crack face condition and stresses from moving straight dislocations J, the dislocation distributions D_J are obtained in the form

$$D_J(x_1) = d_J D_0^{(J)}(x_1), \quad J = I,\ II \text{ and } III; \tag{2.2.14}$$

$D_0^{(J)}$ corresponds to the equilibrium distribution of straight dislocations J when the crack is planar in the Ox_1x_3-plane ($p_0 = 0$), extending from $x_1 = -c$ to c, under pure mode J loading (see Sect. 2.1 for $J = I$ and II). The calculation results are listed ($M_{12} = \sigma_{21}^a/\sigma_{22}^a$):

$$D_0^{(I)}(x_1) = \frac{\sigma_{22}^a}{\pi C_0^{(I)}} \frac{x_1}{\sqrt{c^2 - x_1^2}}, \quad C_0^{(I)} = \frac{\mu b(2 - \tilde{v}_t^2)}{\pi \tilde{v}_t^2}\left(\frac{1}{P_t} - \frac{1}{P_l}\right),$$

$$d_I = \frac{(M_{12} + \nu_A p_0)(C^{(II)(1)}/C^{(II)(2)}) - 1 + p_0 M_{12}}{C^{(I)}/C_0^{(I)}},$$

$$C^{(II)(1)} = \frac{\mu b p_0}{\pi}\left(\frac{P_{2t}^2}{P_t(1 + p_0^2 P_t^2)} - \frac{P_l}{1 + p_0^2 P_l^2}\right),$$

$$C^{(II)(2)} = \frac{\mu b}{\pi \tilde{v}_t^2}\left(\frac{P_l[2 + p_0^2\{2 + (1 - 2c_*^2)\tilde{v}_t^2\}]}{1 + p_0^2 P_l^2} - \frac{2P_{2t}^2(P_{2t}^2 + p_0^2 P_t^2)}{P_t(1 + p_0^2 P_t^2)}\right),$$

$$C^{(I)} = \frac{C^{(I)(2)}C^{(II)(1)} - C^{(II)(2)}C^{(I)(1)}}{C^{(II)(2)}},$$

$$C^{(I)(1)} = \frac{2\mu b}{\pi \tilde{v}_t^2}\left(\frac{P_{2t}^2(1 + p_0^2 P_{2t}^2)}{P_t(1 + p_0^2 P_t^2)} - \frac{P_{2t}^2 + p_0^2 P_l^2}{P_l(1 + p_0^2 P_l^2)}\right),$$

$$C^{(I)(2)} = \frac{\mu b p_0}{\pi}\left(\frac{P_{2t}^2}{P_t(1 + p_0^2 P_t^2)} - \frac{1}{P_l(1 + p_0^2 P_l^2)}\right);$$

$$D_0^{(II)}(x_1) = \frac{\sigma_{21}^a}{\pi C_0^{(II)}} \frac{x_1}{\sqrt{c^2 - x_1^2}}, \quad C_0^{(II)} = \frac{2\mu b(P_t P_l - P_{2t}^4)}{\pi \tilde{v}_t^2 P_t},$$

$$d_{II} = \frac{(1 - p_0 M_{12})(C^{(I)(2)}/C^{(I)(1)}) - M_{12} - \nu_A p_0}{M_{12}(C^{(II)}/C_0^{(II)})}, \quad C^{(II)} = C^{(I)}\frac{C^{(II)(2)}}{C^{(I)(1)}};$$

$$D_0^{(III)}(x_1) = \frac{\sigma_{23}^a}{\pi C_0^{(III)}} \frac{x_1}{\sqrt{c^2 - x_1^2}},\ C_0^{(III)} = \frac{\mu b P_t}{2\pi \beta_{13(t)}^2},\ d_{III} = \frac{1 + p_0^2 P_t^2}{1 + p_0^2}. \tag{2.2.15}$$

The corresponding relative displacements ϕ_J of the faces of the crack, in the x_2 ($J = I$), x_1 ($J = II$) and x_3 ($J = III$) directions, are

$$\phi_J(x_1) = d_J \phi_0^{(J)}(x_1); \tag{2.2.16}$$

$\phi_0^{(J)}$ corresponds to the relative displacement of the crack faces when the crack is in Ox_1x_3-plane under pure mode J loading. They are given in Sect. 2.1 for $J = I$ and II. We have

$$\phi_0^{(J)}(x_1) = \frac{b}{\pi}\left(\frac{\sigma_{22}^a}{C_0^{(I)}}\delta_{JI} + \frac{\sigma_{21}^a}{C_0^{(II)}}\delta_{JII} + \frac{\sigma_{23}^a}{C_0^{(III)}}\delta_{JIII}\right)\sqrt{c^2 - x_1^2},\ J = I, II \text{ and } III. \tag{2.2.17}$$

D_J is unbounded at $x_1 = \pm c$ and the corresponding ϕ_J vertical at these end points. We shall progress further by providing approximate expressions for the stress about the crack front and crack extension force with f given by (2.2.1) using D_J (2.2.14) when the average fracture surface $h = h(x_1)$ can be approximated by plane π_0 of Fig. 2.10a.

Stresses About the Crack-Tip

We consider $P(x_1, x_2, x_3)$ about the crack front located at $x_1 = c$; hence x_2 is close to h since the fracture surface is given by $f = h + \xi$ with ξ small. Writing $x_1 = c + s$ ($0 < s \ll c$), the stress at P is

$$\begin{aligned}\sigma_{ij}^{(C)} &\equiv \sigma_{ij}^{(C)(I)} + \sigma_{ij}^{(C)(II)} + \sigma_{ij}^{(C)(III)},\ \sigma_{ij}^{(C)(J)} = \Bigg(\tilde{\sigma}_{ij}^{(J)(0)} + \tilde{\sigma}_{ij}^{(J)\xi} \\ &+ (x_2 - h(c))\left\{\frac{\partial \tilde{\sigma}_{ij}^{(J)(0)}}{\partial x_2} + \frac{\partial \tilde{\sigma}_{ij}^{(J)\xi}}{\partial x_2}\right\}\Bigg)\frac{d_J K_J^{(0)}}{C_0^{(J)}\sqrt{2\pi s}};\end{aligned} \tag{2.2.18}$$

$$K_J^{(0)} = \sqrt{\pi c}\left(\sigma_{22}^a \delta_{JI} + \sigma_{21}^a \delta_{JII} + \sigma_{23}^a \delta_{JIII}\right).$$

In (2.2.18), x_2 is close to $h(c)$, this means that $(x_2 - h)$ remains small; the various quantities associated with stresses are given in the Appendix to Sect. 2.2.

Crack Extension Force

The crack extension force G per unit length of the crack front is calculated in the way outlined in Chap. 1. We defined a reduced crack extension force $\tilde{G}$ as

$$\tilde{G} = G/\left(G_0^{(I)} + G_0^{(II)} + G_0^{(III)}\right);\ G_0^{(J)} = \frac{b K_J^{(0)^2}}{4\pi C_0^{(J)}}. \tag{2.2.19}$$

$G_0^{(J)}$ is the crack extension force per unit edge length for planar straight-fronted cracks in Ox_1x_3, extending steadily from $x_1 = -c$ to c, under pure mode J loading. We obtain the reduced crack extension force at P_0 ($x_1 = c$, $x_2 = f$, x_3) as (with $M_{12} \equiv \sigma_{12}^a/\sigma_{22}^a$, $M_{13} \equiv \sigma_{23}^a/\sigma_{22}^a$, $M_{23} \equiv \sigma_{23}^a/\sigma_{21}^a$)

$$\tilde{G}(P_0) = \sum_{i,j=1}^{3} \tilde{G}_j^{(i)}(P_0); \tag{2.2.20}$$

$$\begin{aligned}
\tilde{G}_1^{(1)} &= -\frac{\partial f(c, x_3)/\partial x_1}{\sqrt{1 + (\partial f/\partial x_1)^2 + (\partial f/\partial x_3)^2}} \sum_{J=I}^{III} \left(\tilde{\sigma}_{11}^{(J)(0)} + \tilde{\sigma}_{11}^{(J)\xi} \right. \\
&\quad \left. + \xi \left\{ \frac{\partial \tilde{\sigma}_{11}^{(J)(0)}}{\partial x_2} + \frac{\partial \tilde{\sigma}_{11}^{(J)\xi}}{\partial x_2} \right\} \right) \frac{d_J K_J^{(0)} d_{II} C_0^{(I)} M_{12}^2}{C_0^{(J)} K_{II}^{(0)} C_0^{(II)} \Delta_{123}}, \\
\tilde{G}_2^{(1)} &= \frac{1}{\sqrt{1 + (\partial f/\partial x_1)^2 + (\partial f/\partial x_3)^2}} \sum_{J=I}^{III} \left(\tilde{\sigma}_{12}^{(J)(0)} + \tilde{\sigma}_{12}^{(J)\xi} \right. \\
&\quad \left. + \xi \left\{ \frac{\partial \tilde{\sigma}_{12}^{(J)(0)}}{\partial x_2} + \frac{\partial \tilde{\sigma}_{12}^{(J)\xi}}{\partial x_2} \right\} \right) \frac{d_J K_J^{(0)} d_{II} C_0^{(I)} M_{12}^2}{C_0^{(J)} K_{II}^{(0)} C_0^{(II)} \Delta_{123}}, \\
\tilde{G}_3^{(1)} &= -\frac{\partial f/\partial x_3}{\sqrt{1 + (\partial f/\partial x_1)^2 + (\partial f/\partial x_3)^2}} \sum_{J=I}^{III} \left(\tilde{\sigma}_{13}^{(J)(0)} + \tilde{\sigma}_{13}^{(J)\xi} \right. \\
&\quad \left. + \xi \left\{ \frac{\partial \tilde{\sigma}_{13}^{(J)(0)}}{\partial x_2} + \frac{\partial \tilde{\sigma}_{13}^{(J)\xi}}{\partial x_2} \right\} \right) \frac{d_J K_J^{(0)} d_{II} C_0^{(I)} M_{12}^2}{C_0^{(J)} K_{II}^{(0)} C_0^{(II)} \Delta_{123}}, \\
\tilde{G}_1^{(2)} &\equiv -\frac{\partial f}{\partial x_1} \frac{d_I C_0^{(II)}}{d_{II} C_0^{(I)} M_{12}} \tilde{G}_2^{(1)}, \\
\tilde{G}_2^{(2)} &= \frac{1}{\sqrt{1 + (\partial f/\partial x_1)^2 + (\partial f/\partial x_3)^2}} \sum_{J=I}^{III} \left(\tilde{\sigma}_{22}^{(J)(0)} + \tilde{\sigma}_{22}^{(J)\xi} \right. \\
&\quad \left. + \xi \left\{ \frac{\partial \tilde{\sigma}_{22}^{(J)(0)}}{\partial x_2} + \frac{\partial \tilde{\sigma}_{22}^{(J)\xi}}{\partial x_2} \right\} \right) \frac{d_J K_J^{(0)} d_I}{C_0^{(J)} K_I^{(0)} \Delta_{123}}, \\
\tilde{G}_3^{(2)} &= -\frac{\partial f/\partial x_3}{\sqrt{1 + (\partial f/\partial x_1)^2 + (\partial f/\partial x_3)^2}} \sum_{J=I}^{III} \left(\tilde{\sigma}_{23}^{(J)(0)} + \tilde{\sigma}_{23}^{(J)\xi} \right. \\
&\quad \left. + \xi \left\{ \frac{\partial \tilde{\sigma}_{23}^{(J)(0)}}{\partial x_2} + \frac{\partial \tilde{\sigma}_{23}^{(J)\xi}}{\partial x_2} \right\} \right) \frac{d_J K_J^{(0)} d_I}{C_0^{(J)} K_I^{(0)} \Delta_{123}}, \\
\tilde{G}_1^{(3)} &\equiv \frac{\partial f/\partial x_1}{\partial f/\partial x_3} \frac{d_{III} C_0^{(II)} M_{13}}{d_{II} C_0^{(III)} M_{12}} \tilde{G}_3^{(1)}, \ \tilde{G}_2^{(3)} \\
&\equiv -\frac{1}{\partial f/\partial x_3} \frac{d_{III} C_0^{(I)} M_{13}}{d_I C_0^{(III)}} \tilde{G}_3^{(2)},
\end{aligned}$$

$$\tilde{G}_3^{(3)} = -\frac{\partial f/\partial x_3}{\sqrt{1+(\partial f/\partial x_1)^2+(\partial f/\partial x_3)^2}}\sum_{J=I}^{III}\left(\tilde{\sigma}_{33}^{(J)(0)}+\tilde{\sigma}_{33}^{(J)\xi}\right.$$

$$\left.+\,\xi\left\{\frac{\partial\tilde{\sigma}_{33}^{(J)(0)}}{\partial x_2}+\frac{\partial\tilde{\sigma}_{33}^{(J)\xi}}{\partial x_2}\right\}\right)\frac{d_J K_J^{(0)} d_{III} C_0^{(I)} M_{13}^2}{C_0^{(J)} K_{III}^{(0)} C_0^{(III)}\Delta_{123}}; \qquad (2.2.21)$$

$$\Delta_{123} \equiv 1+(C_0^{(I)}/C_0^{(II)})M_{12}^2+(C_0^{(I)}/C_0^{(III)})M_{13}^2.$$

Here, quantities associated with stress are given in the Appendix to Sect. 2.2. Below is given a more detailed description of G for a special plane-fronted non-planar crack with a segmented front (Fig. 2.10c).

Non-planar Crack with a Segmented Front

The described example is given in Fig. 2.10c. This is a non-planar crack with a segmented front whose average fracture surface is plane π_0. The crack front at $x_1 = c$ runs indefinitely in the x_3-direction and is in a x_2x_3-plane. B is chosen as origin. ξ is then odd and $(2\lambda = \lambda_A + \lambda_B)$-periodical with respect to x_3 where λ_A and λ_B(Fig. 2.10d) are the projected length along x_3 of planar facet A and B respectively. ξ is given by:

$$\begin{aligned}\xi &= \tan\phi_B x_3, \quad |x_3| \le \lambda_B/2\\ &= \tan\phi_A(-x_3+\lambda), \quad x_3 \in [\lambda_B/2, \lambda_B/2+\lambda_A].\end{aligned} \qquad (2.2.22)$$

We assume general loading, write $p_0 = p = \tan\theta$ for simplicity and express the spatial average $\langle\tilde{G}\rangle$ of $\tilde{G}$ defined as $\langle\tilde{G}\rangle = (1/2\lambda)\int_0^{2\lambda}\tilde{G}dx_3$.

$$\langle\tilde{G}\rangle = \langle\tilde{G}^{(1)}\rangle + \langle\tilde{G}^{(2)}\rangle + \langle\tilde{G}^{(3)}\rangle;\ \langle\tilde{G}^{(1)}\rangle = \frac{C_0^{(I)} d_{II} M_{12}}{C_0^{(II)}\Delta_{123}}$$

$$\left(\frac{d_{III}M_{13}}{C_0^{(III)}}\left\{-\tilde{\sigma}_{13}^{(III)(0)} - \frac{2\mu p_0}{1-2\nu}\left[(1-\nu)\hat{u}_{1,1}^{(III)} + \nu\hat{u}_{2,2}^{(III)} + \nu\hat{u}_{3,3}^{(III)}\right]\right.\right.$$

$$\left.+\,\mu\left[\hat{u}_{1,2}^{(III)} + \hat{u}_{2,1}^{(III)}\right]\right\}\nu_1 - \mu\left\{\frac{d_I}{C_0^{(I)}}\left(\hat{u}_{1,3}^{(I)} + \hat{u}_{3,1}^{(I)}\right) + \frac{d_{II}M_{12}}{C_0^{(II)}}\left(\hat{u}_{1,3}^{(II)} + \hat{u}_{3,1}^{(II)}\right)\right\}\nu_2$$

$$\left.+\left\{\frac{d_I}{C_0^{(I)}}\left(-p_0\tilde{\sigma}_{11}^{(I)(0)} + \tilde{\sigma}_{12}^{(I)(0)}\right) + \frac{d_{II}M_{12}}{C_0^{(II)}}\left(-p_0\tilde{\sigma}_{11}^{(II)(0)} + \tilde{\sigma}_{12}^{(II)(0)}\right)\right\}\nu_0\right),$$

$$\langle\tilde{G}^{(2)}\rangle = \frac{d_I}{\Delta_{123}}\left(\frac{d_{III}M_{13}}{C_0^{(III)}}\left\{-\tilde{\sigma}_{23}^{(III)(0)} + \frac{2\mu}{1-2\nu}\left[\nu\hat{u}_{1,1}^{(III)} + (1-\nu)\hat{u}_{2,2}^{(III)} + \nu\hat{u}_{3,3}^{(III)}\right]\right.\right.$$

$$\left.-\,p_0\mu\left[\hat{u}_{1,2}^{(III)} + \hat{u}_{2,1}^{(III)}\right]\right\}\nu_1 - \mu\left\{\frac{d_I}{C_0^{(I)}}\left(\hat{u}_{2,3}^{(I)} + \hat{u}_{3,2}^{(I)}\right) + \frac{d_{II}M_{12}}{C_0^{(II)}}\left(\hat{u}_{2,3}^{(II)} + \hat{u}_{3,2}^{(II)}\right)\right\}\nu_2$$

$$+\left\{\frac{d_I}{C_0^{(I)}}\left(-p_0\tilde{\sigma}_{12}^{(I)(0)}+\tilde{\sigma}_{22}^{(I)(0)}\right)+\frac{d_{II}M_{12}}{C_0^{(II)}}\left(-p_0\tilde{\sigma}_{21}^{(II)(0)}+\tilde{\sigma}_{22}^{(II)(0)}\right)\right\}v_0\Bigg),$$

$$\langle\tilde{G}^{(3)}\rangle=\frac{d_{III}C_0^{(I)}M_{13}}{C_0^{(III)}\Delta_{123}}\Bigg(-\frac{d_{III}M_{13}2\mu}{C_0^{(III)}(1-2\nu)}\left\{\nu\hat{u}_{1,1}^{(III)}+\nu\hat{u}_{2,2}^{(III)}+(1-\nu)\hat{u}_{3,3}^{(III)}\right\}v_2$$

$$+\left\{\frac{d_I}{C_0^{(I)}}\left(-\mu p_0\left[\hat{u}_{1,3}^{(I)}+\hat{u}_{3,1}^{(I)}\right]+\mu\left[\hat{u}_{2,3}^{(I)}+\hat{u}_{3,2}^{(I)}\right]-\tilde{\sigma}_{33}^{(I)(0)}\right)\right.$$

$$\left.+\frac{d_{II}M_{12}}{C_0^{(II)}}\left(-\mu p_0\left[\hat{u}_{1,3}^{(II)}+\hat{u}_{3,1}^{(II)}\right]+\mu\left[\hat{u}_{2,3}^{(II)}+\hat{u}_{3,2}^{(II)}\right]-\tilde{\sigma}_{33}^{(II)(0)}\right)\right\}v_1$$

$$+\frac{d_{III}M_{13}}{C_0^{(III)}}\left\{-p_0\tilde{\sigma}_{13}^{(III)(0)}+\tilde{\sigma}_{23}^{(III)(0)}\right\}v_0\Bigg), \quad (2.2.23)$$

$$v_0=(1/(p_A+p_B))\left(p_A/\sqrt{1+p^2+p_B^2}+p_B/\sqrt{1+p^2+p_A^2}\right),$$

$$v_1=(p_Ap_B/(p_A+p_B))\left(-1/\sqrt{1+p^2+p_A^2}+1/\sqrt{1+p^2+p_B^2}\right), \quad (2.2.24)$$

$$v_2=(p_Ap_B/(p_A+p_B))\left(p_A/\sqrt{1+p^2+p_A^2}+p_B/\sqrt{1+p^2+p_B^2}\right)$$

and $p_A=\tan\phi_A, p_B=\tan\phi_B$. Quantities associated to stresses and displacements are given in the Appendix. Hence $\langle\tilde{G}\rangle$ is a function of parameters $(\tilde{v}_t;\, p, p_A, p_B;\, M_{12}, M_{13})$ including Poisson's ratio ν; expressions with Poisson's ratio ν_A originate from normal induced stresses due to Poisson's effect. For $\tilde{v}_t=0$, the static case (Anongba 2009, 2010, 2017, for example) is recovered; as also are the particular cases of Ox_1x_3-plane moving cracks in pure modes of loading *I, II* and *III* (Sect. 2.1). When ϕ_B (or ϕ_A) equals zero, the crack front is essentially straight parallel to the x_3-direction. The corresponding crack is like planar crack π_0 (Fig. 2.10a).

First, we look for non-planar cracks for which $\langle\tilde{G}\rangle$ is larger than 1. Such crack configurations have been found in the quasi-static analyses. They are expected to be fracture paths in broken real materials. For $0<\langle\tilde{G}\rangle<1$, planar cracks are favoured. Negative values suggest that crack motion is impeded: in the case of planar cracks, we have indicated that the relative displacement of the faces of the crack, formed under load, cancels when motion starts (Sect. 2.1). Figure 2.11 shows $\langle\tilde{G}\rangle$ (2.2.23) for non-planar cracks travelling at $\tilde{v}_t=0.3$ ($\phi_B=70°, M_{12}=M_{13}=10^{-4}$).

Positive values of $\langle\tilde{G}\rangle$ are observed in a restricted zone of θ ($40°<\theta<70°$, approximately), with a peak in the vicinity of $\theta=50°$; this peak increases continuously with ϕ_A to reach values $\langle\tilde{G}\rangle\approx150$ at $\phi_A\approx70°$.

Figure 2.12((a) $\theta=55°$, (b) $\theta=54°$ and (c) $\theta=53.5°$) shows a non-planar crack whose front is strongly segmented ($\phi_A=\phi_B=70°$) under dominant mode *I* loading ($M_{12}=M_{13}=10^{-3}$). When the crack is inclined with respect to Ox_1x_3 by $\theta=53.5°$ in (*c*), $\langle\tilde{G}\rangle$ remains negative for practically all velocity. For a small increase of θ (1.5°), $\langle\tilde{G}\rangle$ becomes largely positive over a wide range of speed $0.2\le\tilde{v}_t\le1$ in

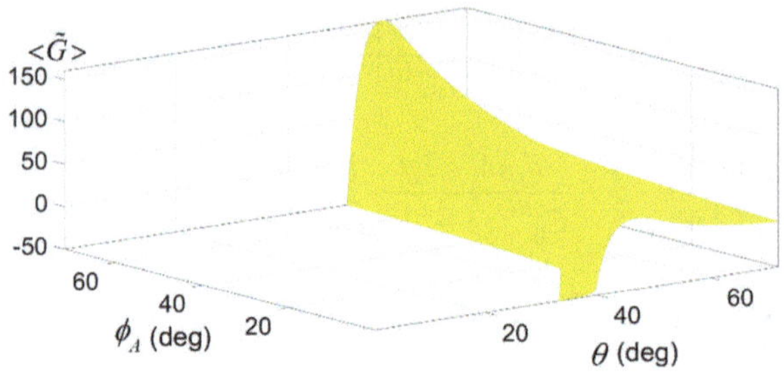

Fig. 2.11 Surface $\langle\tilde{G}\rangle$ (ϕ_A, θ) (2.2.23) for non-planar cracks moving at $\tilde{v}_t = 0.3$; $\phi_B = 70°$, $M_{12} = M_{13} = 10^{-4}$, $\nu = 0.22 = \nu_A$, $\beta_{13(t)} = P_t$

(*a*). In (*b*), a maximum of $\langle\tilde{G}\rangle$ is present at about $\tilde{v}_t = 0.55 \equiv \tilde{v}_1^{(e)}$; at that velocity, the motion of the crack is uniform along x_1(steady motion). This result is a theoretical prediction of a steady motion for this non-planar crack in solids.

Figure 2.13 corresponds to a rather flat crack ($\phi_A = 0.1°$; λ_A large) but with pronounced isolated kinks ($\phi_B = 88°$; λ_B small) (for λ_A and λ_B, see Fig. 2.10d). The average inclination angle of the crack with respect to Ox_1x_3 *is* $\theta = 35°$. Mode *I* loading is dominant ($M_{12} = M_{13} = 10^{-4}$). $\langle\tilde{G}\rangle$ is largely above 1 in $0.3 \le \tilde{v}_t \le 0.55$, approximately. Interestingly, we have checked that $\beta_{13(t)} = 1$ leaves unchanged this curve.

Summary and Confrontation with Experiments

DISCUSSION: The dislocation elastic fields measured by an observer in an inertial reference frame moving with the dislocation, are like those of a static dislocation in the laboratory, particularly in the subsonic regime. This allows to describe static and uniformly moving cracks in a similar way. Hence, there is one–one similarity in form (in both cases) of crack characteristic quantities: crack dislocation distributions D_J (2.2.14) and corresponding relative displacements of the faces of the crack ϕ_J (2.2.16), stresses at the tip of the crack $\sigma_{ij}^{(C)(J)}$ (2.2.18) and crack extension force G (2.2.19–2.2.21); for the static case, we may refer to Anongba (2010) and references therein. Most theoretical analyses in fracture mechanics concern planar cracks, with straight fronts, in static position or in uniform movement (see Fig. 2.10a for illustration). Because the length of the crack front is large in general, the modelling applies in practice to cracking over large distances. Crack propagation is the mechanism controlling fracture with no place for nucleation. For non-planar cracks, the modelling extends to flat-fronted cracks. The crack front f (2.2.1) is arbitrary (see Fig. 2.8); a simple special case is given in Fig. 2.10c. For these non-planar cracks, most of the work refers to the static case. The present study is a first extension to the uniform motion of the x_2x_3-plane front of the crack; the static case is restored when $v = 0$. The numerical application here concerns a front made up of 2 types of inclined segments, with angles ϕ_A and ϕ_B, with respect to the x_3-direction (Fig. 2.10c, d). On average, the crack fluctuates around the plane π_0 (Fig. 2.3a) inclined by $\theta = \theta_0$ with respect to Ox_1x_3 with Ox_3 as the axis of rotation. When the crack front is

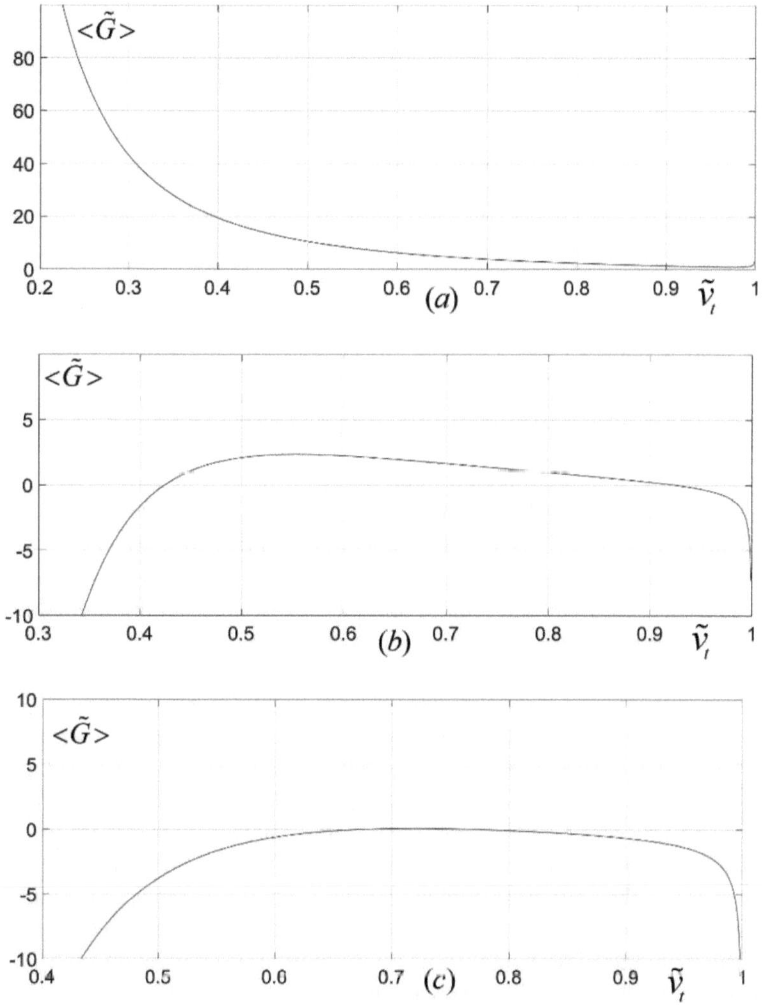

Fig. 2.12 $\langle \tilde{G} \rangle$ (2.2.23) as a function of $\tilde{v}_t$ for **a** $\theta = 55°$, **b** $\theta = 54°$ and **c** $\theta = 53.5°$ with identical values of the other parameters ($\phi_A = \phi_B = 70°$, $M_{12} = M_{13} = 10^{-3}$, $\nu = 1/3 = \nu_A$). A dramatic change in the form of the curves is observed for a small variation 1° of θ

strongly segmented ($\phi_A = \phi_B$ large) as in Fig. 2.12, the highest positive values of $\langle \tilde{G} \rangle$ (2.2.23) are for $\theta > 53.5°$ in dominant mode I loading. A maximum of $\langle \tilde{G} \rangle$, $\langle \tilde{G} \rangle_{\max 1} \cong 2.5$, is observed for $\tilde{v}_t = 0.55 \equiv \tilde{v}_1^{(e)}$ with $\theta = 54°$ (Fig. 2.12b). For the moving planar crack in Ox_1x_3, the steady motion velocity found is at $\tilde{v}_t = 0.52$ for $\nu = 1/3$ (Sect. 2.1). We may safely say that, for the crack corresponding to Fig. 2.12, the steady motion is predicted at the velocity $\tilde{v}_1^{(e)}$. A second example (in dominant mode I loading and $\theta = 35°$) where the crack front is flat over large distance λ_A ($\phi_A = 0.1°$) with isolated kinks strongly inclined ($\phi_B = 88°$) over short distance

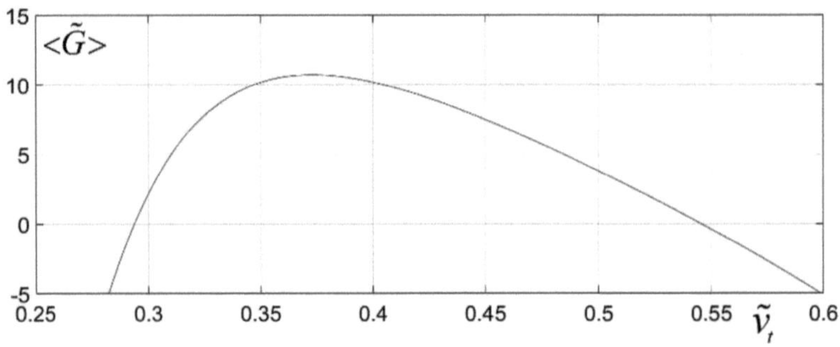

Fig. 2.13 Normalized crack extension force $\langle\tilde{G}\rangle$ $(\tilde{v}_t)$ (2.2.23) averaged over the length of the segmented crack-front under dominant mode I loading ($M_{12} = M_{13} = 10^{-4}$). The average inclination angle with respect to Ox_1x_3 is $\theta_0 = 35°$. The crack face is almost flat ($\phi_A = 0.1°$; $\phi_B = 88°$); $\nu = 0.22 = \nu_A$; $\beta_{13(t)} = P_t$. $\beta_{13(t)} = 1$ leaves this curve unchanged

λ_B is also presented (Fig. 2.13). The velocity interval where $\langle\tilde{G}\rangle$ is positive and larger than 1 is $\tilde{v}_t \in [0.3, 0.54]$, approximately. Out of that interval, the motion of the crack is not favoured. A maximum of $\langle\tilde{G}\rangle$, $\langle\tilde{G}\rangle_{\max 2} \cong 11.2$, is observed at $\tilde{v}_t = 0.37 \equiv \tilde{v}_2^{(e)}$. We can also safely say this is the steady motion condition. We shall use Fig. 2.13 to provide a qualitative explanation of crack branching observed in glass. Figure 2.14 shows the broken surface of a glass rod in tension (Johnson and Holloway 1966). Fracture propagates from bottom to top from a surface flaw. Initially, the crack is flat and smooth over a relatively large area; then branching occurs. The latter covers a region that is flat on average ($\phi_A \approx 0$) but comprising pronounced short, spaced segmentations ($\phi_B \approx 90°$). The crack progresses upwards with an increasing speed v from zero; for a certain value v_1, it turns into a non-plane crack. The transformation takes place without any attenuation of the speed regime. Using Fig. 2.13, the following explanation may be given: for $\tilde{v}_t < 0.3$, the planar starter crack is favoured; its speed increases towards the terminal velocity. An estimate of the terminal velocity depending on Poisson's ratio is given in Sect. 2.1 where it is shown to be about $v^{(e)} \cong 0.52c_t$ for $\nu = 1/3$; it is also shown there that above the terminal velocity, the crack extension force for the tensile planar crack decreases rapidly towards zero. Figure 2.13 tells us that in the velocity range $\tilde{v}_t \in [0.3, 0.54]$ non-planar crack configurations (like that observed in Fig. 2.14) exist with a much larger value of the crack extension force. Hence, the starter planar crack transforms itself into a non-planar configuration to maintain its motion.

SUMMARY: Non-planar cracks of finite extension in the x_1 and x_2 directions and infinite along x_3, inside an infinitely extended elastic medium, subjected to arbitrarily applied loadings, have been investigated. The loadings σ_{22}^a, σ_{21}^a and σ_{23}^a are applied along the x_2, x_1 and x_3 directions, respectively. The front of the crack is planar in x_2x_3 and travels at a constant velocity v along the x_1-direction whilst its faces remain stress free. The crack front has an average elevation $h = h(x_1)$ from Ox_1x_3 and fluctuates weakly about that position in the form $\xi = \xi(x_1, x_3)$ (2.2.1). The crack is represented by a continuous distribution of 3 types J ($J = I, II$ and III) of infinitesimal dislocation

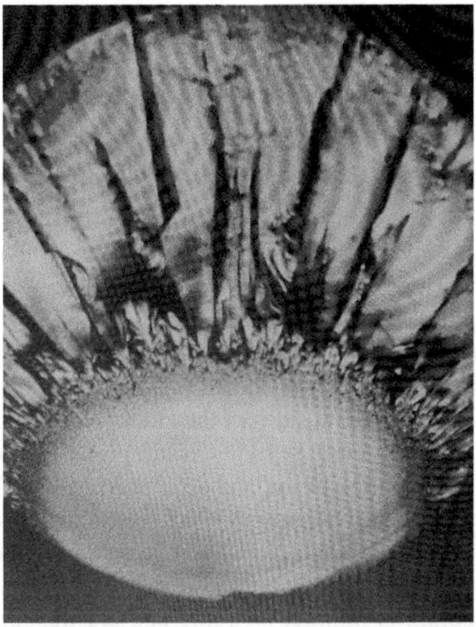

Fig. 2.14 Fracture surface of broken rod of glass under tension viewed in optical microscopy. Fracture propagated from bottom to top. Rod diameter 4.5 mm (Johnson and Holloway 1966)

having the shape of the crack front. The associated Burgers vectors $\vec{b}_I = (0, b, 0)$, $\vec{b}_{II} = (b, 0, 0)$ and $\vec{b}_{III} = (0, 0, b)$ are directed along the applied loadings, tension and shears directions, respectively. Adopting the method of Fourier series (Mura 1968; see Sect. 2.1), explicit expression of the elastic fields (displacement and stress) of the dislocations J are given. Then, distribution functions D_J of straight dislocation arrays corresponding to a planar crack π_0, inclined by angle θ_0 with respect to Ox_1x_3 are calculated. Adopting these D_J, explicit expressions are provided for the crack-tip stresses and crack extension force per unit length of the crack front with general form f (2.2.1). The analysis is then applied to a simple special non-planar crack with a segmented front. This type of crack is given on Fig. 2.10c, d; the average surface is plane π_0 (Fig. 2.10a); its front is segmented, at the average elevation $h(x_1)$ from Ox_1x_3 in the x_2x_3 located at x_1, in the form (2.2.22). Two types of segmentation characterized by angles ϕ_A and ϕ_B of the segments with respect to x_3 direction are distinguished. It is shown first that for this type of special cracks, configurations exist for which values $\langle G \rangle$, averaged over the length of the crack front, are larger than those corresponding to the planar crack travelling in Ox_1x_3 (Fig. 2.11) in similar velocity intervals. Then, two types of segmentation are investigated under dominant mode I loading: (1) Strong segmentation of the crack front ($\phi_A = \phi_B = 70°$), Fig. 2.12. Steady motion sets in at the velocity $v_1^{(e)} = 0.55c_t$ for an inclination $\theta = 54°$. (2) Weak segmentation, flat average fracture surface with isolated strong kinks ($\phi_A = 0.1°$, $\phi_B = 88°$). A steady motion is also evidenced at the velocity $v_2^{(e)} = 0.37c_t$ when

$\theta = 35°$. The latter configuration (Fig. 2.13) explains crack branching observed in the fracture of glass.

2.2.2 *Brittle Cracks with Crack-Front Plastic Yielding*

The non-planar crack model and associated treatment are like those of Anongba (2010, 2017, 2021a). Along x_1, the crack extends from $x_1 = -c$ to c with a x_2x_3-planar front. It is represented by a continuous distribution of three families J ($J =$ *I, II* and *III*) of infinitesimal dislocations (types *I* and *II* are edges on average and type *III* screws). The shape of the dislocation spreads in x_2x_3 and depends on x_1 and x_3 in the form (2.2.1). On both sides of the crack, a dislocation free zone (DFZ) $c \le |x_1| \le e$ and plastic region $e \le |x_1| \le a$ are present. In the plastic zone, the dislocations J are straight parallel to x_3 and cover the average fracture surface $x_2 = h(x_1)$. Friction stresses oppose their motion in the plastic region while in the crack the dislocations move freely. In both the crack and plastic region, the dislocation J ($J =$ *I, II* and *III*) Burgers vectors are $\vec{b}_I = (0, b, 0)$, $\vec{b}_{II} = (b, 0, 0)$ and $\vec{b}_{III} = (0, 0, b)$, respectively. The medium is infinite, isotropic and elastic, subjected at infinity to uniform applied tension σ_{22}^a and shears σ_{12}^a and σ_{23}^a. In addition, the treatment includes normal induced stresses $\sigma_{11}^a = -\nu_A\sigma_{22}^a$ and $\sigma_{33}^a = -\nu_A\sigma_{22}^a$ ($\nu_A = \nu$). The notation ν_A permits to identify in the various listed mathematical expressions below, those associated with these induced normal stresses. The dislocation distribution functions D_J ($J = I$, *II* and *III*) are defined such that $D_J(x_1')dx_1'$ represents the number of dislocations J in the infinitesimal x_1-interval dx_1' located about the x_1-spatial position x_1'. The equilibrium condition is equivalent to asking that at any spatial position on the crack and plastic zone, the tractions are zero. Additional requirement is that the stresses at the boundary of the plastic zone are bounded. This corresponds to $D_J(\pm e) = D_J(\pm a) = 0$. Figures 2.15 and 2.16 are schematical representations of simple special elastic–plastic cracks captured by the modelling. The crack and plastic zone cover the x_1-intervals $|x_1| \le c$ and $e \le |x_1| \le a$ with DFZ in between; these must be considered running indefinitely in the x_3-direction. The crack shape in planes perpendicular to x_1 is described by ξ (Fig. 2.15a for example). The shape f of the crack in planes perpendicular to x_3 is given by both ξ, through the x_1-dependence of positive quantities ξ_n, δ_n and κ_n(2.2.1), and function $h = h(x_1)$. Since ξ is assumed to be small oscillating function, the average fracture surface is described correctly by the equation $x_2 = h(x_1)$. When $\xi = 0$, the crack dislocations are straight parallel to x_3 and distributed over the surface $x_2 = h(x_1)$. The plastic region dislocations are straight parallel to x_3 and cover the average fracture surface $x_2 = h(x_1)$. Specific examples are given where the plastic region for positive x_1 only is shown:

- $h(x_1) = p_0x_1$ ($p_0 \ge 0$) and $\xi = 0$. This corresponds to a planar elastic–plastic crack π_0(with a straight front parallel to x_3) rotated around Ox_3 by angle $\theta_0 = \tan^{-1} p_0$ from Ox_1x_3, Fig. 2.15a.

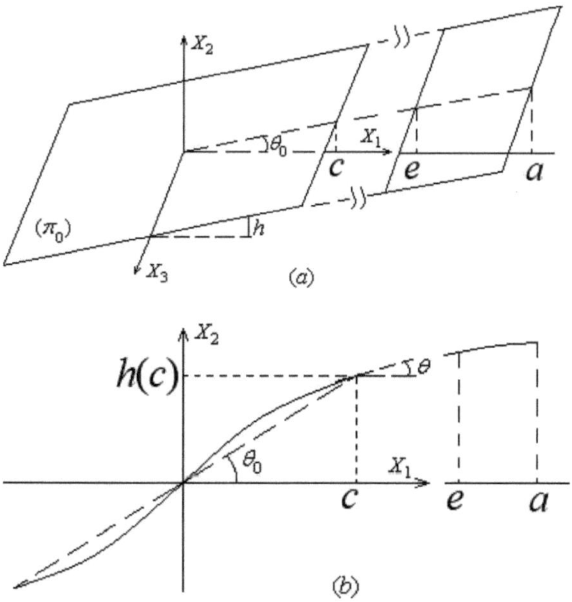

Fig. 2.15 Simple special elastic–plastic cracks. **a** Inclined planar elastic–plastic crack π_0(see text). **b** A non-planar elastic–plastic crack (parallel to x_3) as h odd function of x_1 ($x_2 = h(x_1)$)

- $h(x_1)$ is an arbitrary function of x_1, $\xi = 0$. The sketch in Fig. 2.15b corresponds to h odd although this is not mandatory.

- $h(x_1) = p_0 x_1$ ($p_0 \geq 0$) and $\xi = \xi(x_3)$ independent of x_1. The crack fluctuates about plane π_0 with a front spreading in planes parallel to $x_2 x_3$ in the form ξ. In the example displayed in Fig. 2.16a the crack consists of planar facets with inclination angles ϕ_A and ϕ_B (Fig. 2.16b) at points A and B of the crack front located on the average fracture plane. Points A and B are indicated in Fig. 2.16a.

To derive physical quantities of the elastic–plastic crack system, it is sufficient to consider the treatment made earlier (Sect. 2.2.1) for isolated brittle cracks and give to $K_J^{(0)}$, in various expressions such as $\sigma_{ij}^{(C)(J)}$ (2.2.18) and $G_0^{(J)}$ (2.2.19) there, the following value (Anongba 2003, 2021*a*; See also Sect. 2.2.4)

$$K_J^{(0)} = \sqrt{\pi c} \frac{F(\pi/2, k)\sqrt{a^2 - c^2}}{\sqrt{e^2 - c^2}\,\Pi(\pi/2, e^2k^2/c^2, k)} \left(\sigma_{22}^a \delta_{JI} + \sigma_{21}^a \delta_{JII} + \sigma_{23}^a \delta_{JIII}\right),$$

$$k^2 = \frac{c^2(a^2 - e^2)}{e^2(a^2 - c^2)};$$

F and Π are elliptic integrals of first and third kind, respectively. $c \equiv c_0 + vt$ (c_0 and v are constants) where c_0 is the half-length of the crack in Ox_1x_3 at time $t =$

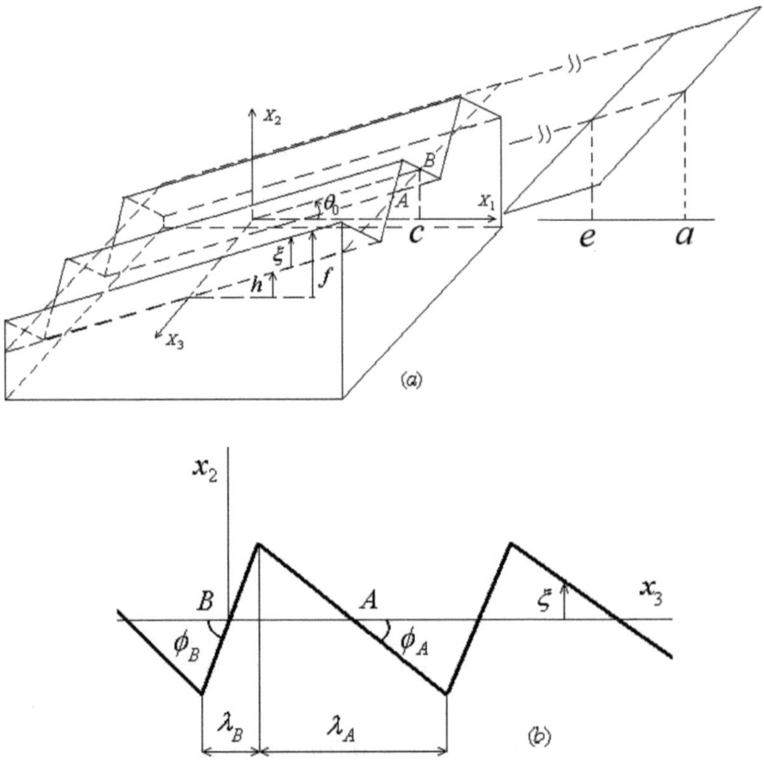

Fig. 2.16 Special elastic–plastic crack. **a** Non-planar elastic crack fluctuating about an average inclined plane π_0. The crack consists of planar facets; its fronts at $x_1 = \pm c$ lie in x_2x_3-planes. At $x_1 = c$, the crack front is characterized by inclination angles ϕ_A and ϕ_B (see (**b**)) at points A and B located on the average fracture plane. The plastic zone lies on π_0. **b** Sketch of the crack front in (**a**) with B taken as origin

0 measured in the x_1-direction. Uniform motion of the elastic–plastic crack system implies that a and e also depend linearly on time t so that it can be assumed a/e, a/c and e/c constant with time. Under such conditions $G_0^{(J)}$, G and $\langle G \rangle$ depend linearly on time through c.

2.2.3 Appendix to Sect. 2.2.1

The various quantities involved in the stresses at the tip of the crack (2.2.18) and crack extension force G (2.2.21), Sect. 2.2.1, are given in the following. For terms with the superscript ξ in (2.2.18), similar relations as in (2.2.11) between stress and displacement are used here ($J = I$, II and III):

$$\begin{aligned}
\tilde{\sigma}_{ii}^{(J)\xi} &= \frac{2\mu}{1-2\nu}\Big([\delta_{i1}(1-\nu)+\nu(\delta_{i2}+\delta_{i3})]\tilde{u}_{1,1}^{(J)\xi} \\
&\quad + [\delta_{i2}(1-\nu)+\nu(\delta_{i1}+\delta_{i3})]\tilde{u}_{2,2}^{(J)\xi} \\
&\quad + [\delta_{i3}(1-\nu)+\nu(\delta_{i1}+\delta_{i2})]\tilde{u}_{3,3}^{(J)\xi}\Big),
\end{aligned}$$

$$\begin{aligned}
\frac{\partial\tilde{\sigma}_{ii}^{(J)\xi}}{\partial x_2} &= \frac{2\mu}{1-2\nu}\Big([\delta_{i1}(1-\nu)+\nu(\delta_{i2}+\delta_{i3})]\tilde{u}_{1,21}^{(J)\xi} \\
&\quad + [\delta_{i2}(1-\nu)+\nu(\delta_{i1}+\delta_{i3})]\tilde{u}_{2,22}^{(J)\xi} \\
&\quad + [\delta_{i3}(1-\nu)+\nu(\delta_{i1}+\delta_{i2})]\tilde{u}_{3,23}^{(J)\xi}\Big),\ i=1,2 \text{ and } 3;
\end{aligned}$$

$$\tilde{\sigma}_{ij}^{(J)\xi} = \mu\Big(\tilde{u}_{i,j}^{(J)\xi}+\tilde{u}_{j,i}^{(J)\xi}\Big),$$

$$\partial\tilde{\sigma}_{ij}^{(J)\xi}/\partial x_2 = \mu\Big(\tilde{u}_{i,2j}^{(J)\xi}+\tilde{u}_{j,2i}^{(J)\xi}\Big),\quad i\neq j. \tag{A1}$$

For $J = I$, II and III, we listed below the various quantities associated to the displacements in (A.1).

For $J = I$:

$$\begin{aligned}
\tilde{\sigma}_{11}^{(I)(0)} &= \frac{\mu b}{\pi\tilde{v}_t^2}\left(\frac{2+(1-2c_*^2)\tilde{v}_t^2}{P_l[1+p^2P_l^2]} - \frac{2P_{2t}^2}{P_t[1+p^2P_t^2]}\right), \\
\tilde{\sigma}_{22}^{(I)(0)} &= \frac{2\mu bP_{2t}^2}{\pi\tilde{v}_t^2}\left(\frac{1}{P_t[1+p^2P_t^2]} - \frac{1}{P_l[1+p^2P_l^2]}\right),\ \tilde{\sigma}_{33}^{(I)(0)} = \frac{\mu b(1-2c_*^2)}{\pi P_l[1+p^2P_l^2]}; \\
&\partial\tilde{\sigma}_{ii}^{(I)(0)}/\partial x_2 = 0,\ i=1,\ 2 \text{ and } 3; \\
\tilde{\sigma}_{21}^{(I)(0)} &= \frac{2\mu bp}{\pi\tilde{v}_t^2}\left(\frac{P_l}{1+p^2P_l^2} - \frac{P_{2t}^4}{P_t[1+p^2P_t^2]}\right),\ \partial\tilde{\sigma}_{21}^{(I)(0)}/\partial x_2 = 0; \\
\tilde{\sigma}_{j3}^{(I)(0)} &= 0,\ \partial\tilde{\sigma}_{j3}^{(I)(0)}/\partial x_2 = 0,\ j=1 \text{ and } 2.
\end{aligned} \tag{A2}$$

$$\tilde{u}_{i,i}^{(I)\xi} = 0,\ i=1,\ 2 \text{ and } 3;$$

$$\begin{aligned}
\tilde{u}_{1,21}^{(I)\xi} &= -\frac{b}{\pi\tilde{v}_t^2}\left(\frac{p_{2t}^2[1-p^2P_t^2]}{2P_t[1+p^2P_t^2]^2}\right. \\
&\quad \left. - \frac{(1-\nu)c_*^2[1-p^2P_l^2]}{(1-2\nu)P_l[1+p^2P_l^2]^2}\right)\frac{\partial^2\xi(c,x_3)}{\partial x_3^2} \equiv \hat{u}_{1,21}^{(I)}\frac{\partial^2\xi}{\partial x_3^2}, \\
\tilde{u}_{2,22}^{(I)\xi} &= -\frac{b}{2\pi\tilde{v}_t^2}\left(\frac{\tilde{v}_t^2[5+3p^2P_t^2]-2[3+p^2P_t^2]}{2P_t[1+p^2P_t^2]^2}\right. \\
&\quad \left. + \frac{2(1-\nu)c_*^2P_l[3+p^2P_l^2]}{(1-2\nu)[1+p^2P_l^2]^2}\right)\frac{\partial^2\xi}{\partial x_3^2} \equiv \hat{u}_{2,22}^{(I)}\frac{\partial^2\xi}{\partial x_3^2},
\end{aligned}$$

$$\tilde{u}_{3,23}^{(I)\xi} = \frac{b(c_*^{-2}-1)}{\pi\tilde{v}_t^2}\left(\frac{2(1-\nu)c_*^2P_l}{1+p^2P_l^2} - \frac{(1-2\nu)P_t}{1+p^2P_t^2}\right)\frac{\partial^2\xi}{\partial x_3^2} \equiv \hat{u}_{3,23}^{(I)}\frac{\partial^2\xi}{\partial x_3^2};$$
$$\tilde{u}_{3,2}^{(I)\xi} = \frac{b(c_*^{-2}-1)}{\pi\tilde{v}_t^2}\left(\frac{2(1-\nu)c_*^2P_l}{1+p^2P_l^2} - \frac{(1-2\nu)P_t}{1+p^2P_t^2}\right)\frac{\partial\xi}{\partial x_3}$$
$$\equiv \hat{u}_{3,2}^{(I)}\frac{\partial\xi}{\partial x_3}, \tilde{u}_{3,22}^{(I)\xi} = 0;$$
$$\tilde{u}_{2,3}^{(I)\xi} = \frac{b}{\pi\tilde{v}_t^2}\left(\frac{2(1-\nu)c_*^2P_l}{(1-2\nu)[1+p^2P_l^2]} - \frac{P_{2t}^2}{P_t[1+p^2P_t^2]}\right)\frac{\partial\xi}{\partial x_3}$$
$$\equiv \hat{u}_{2,3}^{(I)}\frac{\partial\xi}{\partial x_3}, \tilde{u}_{2,23}^{(I)\xi} = 0;$$
$$\tilde{u}_{1,2}^{(I)\xi} = 0, \tilde{u}_{1,22}^{(I)\xi} = \frac{bp}{2\pi\tilde{v}_t^2}\left(\frac{2(1-\nu)c_*^2P_l[3+p^2P_l^2]}{(1-2\nu)[1+p^2P_l^2]^2}\right.$$
$$\left. - \frac{P_{2t}^2P_t[3+p^2P_t^2]}{[1+p^2P_t^2]^2}\right)\frac{\partial^2\xi}{\partial x_3^2} \equiv \hat{u}_{1,22}^{(I)}\frac{\partial^2\xi}{\partial x_3^2};$$
$$\tilde{u}_{2,1}^{(I)\xi} = 0, \tilde{u}_{2,21}^{(I)\xi} = \frac{bp}{4\pi\tilde{v}_t^2}\left(\frac{4(1-\nu)c_*^2P_l[3+p^2(1-3\tilde{v}_l^2)]}{(1-2\nu)[1+p^2P_l^2]^2}\right.$$
$$\left. - \frac{6-5\tilde{v}_t^2+p^2P_t^2[2-7\tilde{v}_t^2]}{P_t[1+p^2P_t^2]^2}\right)\frac{\partial^2\xi}{\partial x_3^2} \equiv \hat{u}_{2,21}^{(I)}\frac{\partial^2\xi}{\partial x_3^2};$$
$$\tilde{u}_{1,3}^{(I)\xi} = \frac{bp}{\pi\tilde{v}_t^2}\left(\frac{P_{2t}^2P_t}{1+p^2P_t^2} - \frac{2(1-\nu)c_*^2P_l}{(1-2\nu)[1+p^2P_l^2]}\right)\frac{\partial\xi}{\partial x_3}$$
$$\equiv \hat{u}_{1,3}^{(I)}\frac{\partial\xi}{\partial x_3}, \tilde{u}_{1,23}^{(I)\xi} = 0;$$
$$\tilde{u}_{3,1}^{(I)\xi} = \frac{b(c_*^{-2}-1)p}{\pi\tilde{v}_t^2}\left(\frac{(1-2\nu)P_t}{1+p^2P_t^2} - \frac{2(1-\nu)c_*^2P_l}{1+p^2P_l^2}\right)\frac{\partial\xi}{\partial x_3}$$
$$\equiv \hat{u}_{3,1}^{(I)}\frac{\partial\xi}{\partial x_3}, \tilde{u}_{3,21}^{(I)\xi} = 0. \quad \text{(A3)}$$

For $J = II$:

$$\tilde{\sigma}_{11}^{(II)(0)} = \frac{\mu bp}{\pi\tilde{v}_t^2}\left(\frac{2P_tP_{2t}^2}{1+p^2P_t^2} + \frac{P_l[(c_*^{-2}-2)P_l^2 - c_*^{-2}]}{1+p^2P_l^2}\right),$$
$$\tilde{\sigma}_{22}^{(II)(0)} = \frac{\mu bp}{\pi\tilde{v}_t^2}\left(\frac{P_l[c_*^{-2}P_l^2 - c_*^{-2}+2]}{1+p^2P_l^2} - \frac{2P_tP_{2t}^2}{1+p^2P_t^2}\right),$$
$$\tilde{\sigma}_{33}^{(II)(0)} = -\frac{\mu b(1-2c_*^2)pP_l}{\pi[1+p^2P_l^2]};$$
$$\partial\tilde{\sigma}_{ii}^{(II)(0)}/\partial x_2 = 0, i = 1, 2 \text{ and } 3;$$
$$\tilde{\sigma}_{21}^{(II)(0)} = \frac{2\mu b}{\pi\tilde{v}_t^2}\left(\frac{P_l}{1+p^2P_l^2} - \frac{P_{2t}^4}{P_t[1+p^2P_t^2]}\right), \partial\tilde{\sigma}_{21}^{(II)(0)}/\partial x_2 = 0; \quad \text{(A4)}$$

The other stresses and their derivatives of order zero with respect to ξ corresponding to the straight dislocation are zero. We have
$\tilde{u}_{i,i}^{(II)\xi} = 0$, $i = 1$, 2 and 3;

$$\tilde{u}_{1,21}^{(II)\xi} = \frac{bp}{2\pi\tilde{v}_t^2}\left(\frac{P_{2t}^2 P_t[1-p^2P_t^2]}{[1+p^2P_t^2]^2} - \frac{2(1-\nu)c_*^2 P_l[1-p^2P_l^2]}{(1-2\nu)[1+p^2P_l^2]^2}\right)\frac{\partial^2\xi(c,x_3)}{\partial x_3^2} \equiv \hat{u}_{1,21}^{(II)}\frac{\partial^2\xi}{\partial x_3^2},$$

$$\tilde{u}_{2,22}^{(II)\xi} = \frac{bp}{2\pi\tilde{v}_t^2}\left(\frac{2(1-\nu)c_*^2 P_l^3[3+p^2P_l^2]}{(1-2\nu)[1+p^2P_l^2]^2} - \frac{P_{2t}^2 P_t[3+p^2P_t^2]}{[1+p^2P_t^2]^2}\right)\frac{\partial^2\xi}{\partial x_3^2} \equiv \hat{u}_{2,22}^{(II)}\frac{\partial^2\xi}{\partial x_3^2},$$

$$\tilde{u}_{3,23}^{(II)\xi} = -\frac{bp}{\pi\tilde{v}_t^2}\left(\frac{(1-2\nu)(1-c_*^{-2})P_{2t}^2 P_t}{1+p^2P_t^2} + \frac{2(1-\nu)(1-c_*^2)P_l^3}{1+p^2P_l^2}\right)\frac{\partial^2\xi}{\partial x_3^2} \equiv \hat{u}_{3,23}^{(II)}\frac{\partial^2\xi}{\partial x_3^2};$$

$$\tilde{u}_{1,2}^{(II)\xi} = 0,\ \tilde{u}_{1,22}^{(II)\xi} = \frac{b}{2\pi\tilde{v}_t^2}\left(\frac{2(1-\nu)c_*^2 P_l[1-p^2P_l^2]}{(1-2\nu)[1+p^2P_l^2]^2}\right.$$
$$\left. - \frac{\tilde{v}_t^2 P_t[(1-2\nu)c_*^{-2} - 2(1-\nu)]}{1+p^2P_t^2} - \frac{P_{2t}^2 P_t[1-p^2P_t^2]}{[1+p^2P_t^2]^2}\right)\frac{\partial^2\xi}{\partial x_3^2} \equiv \hat{u}_{1,22}^{(II)}\frac{\partial^2\xi}{\partial x_3^2};$$

$$\tilde{u}_{1,3}^{(II)\xi} = \frac{b}{\pi\tilde{v}_t^2}\left(\frac{P_{2t}^2 P_t}{1+p^2P_t^2} - \frac{2(1-\nu)c_*^2 P_l}{(1-2\nu)][1+p^2P_l^2]}\right)\frac{\partial\xi}{\partial x_3} \equiv \hat{u}_{1,3}^{(II)}\frac{\partial\xi}{\partial x_3},\ \tilde{u}_{1,23}^{(II)\xi} = 0;$$

$$\tilde{u}_{2,3}^{(II)\xi} = \frac{bp}{\pi\tilde{v}_t^2}\left(\frac{P_{2t}^2 P_t}{1+p^2P_t^2} - \frac{2(1-\nu)c_*^2 P_l^3}{(1-2\nu)[1+p^2P_l^2]}\right)\frac{\partial\xi}{\partial x_3} \equiv \hat{u}_{2,3}^{(II)}\frac{\partial\xi}{\partial x_3},\ \tilde{u}_{2,23}^{(II)\xi} = 0;$$

$$\tilde{u}_{3,1}^{(II)\xi} = \frac{b}{\pi\tilde{v}_t^2}\left(\frac{(1-2\nu)(c_*^{-2}-1)P_{2t}^2}{P_t[1+p^2P_t^2]} - \frac{2(1-\nu)(1-c_*^2)P_l}{1+p^2P_l^2}\right)\frac{\partial\xi}{\partial x_3}$$
$$\equiv \hat{u}_{3,1}^{(II)}\frac{\partial\xi}{\partial x_3},\ \tilde{u}_{3,21}^{(II)\xi} = 0;\ \tilde{u}_{3,2}^{(II)\xi} = \frac{bp}{\pi\tilde{v}_t^2}\left(\frac{(1-2\nu)(c_*^{-2}-1)P_{2t}^2 P_t}{1+p^2P_t^2}\right.$$
$$\left. - \frac{2(1-\nu)(1-c_*^2)P_l^3}{1+p^2P_l^2}\right)\frac{\partial\xi}{\partial x_3} \equiv \hat{u}_{3,2}^{(II)}\frac{\partial\xi}{\partial x_3},\ \tilde{u}_{3,22}^{(II)\xi} = 0. \tag{A5}$$

For $J = III$:

$$\tilde{\sigma}_{13}^{(III)(0)} = -\frac{\mu b p P_t}{2\pi\beta_{13(t)}^2[1+p^2P_t^2]},\ \partial\tilde{\sigma}_{13}^{(III)(0)}/\partial x_2 = 0;$$
$$\tilde{\sigma}_{23}^{(III)(0)} = \frac{\mu b P_t}{2\pi\beta_{13(t)}^2[1+p^2P_t^2]},\ \partial\tilde{\sigma}_{23}^{(III)(0)}/\partial x_2 = 0. \tag{A6}$$

All the other similar quantities associated with the stresses, of zero order with respect to ξ, are zero. We have

$$\tilde{u}_{1,1}^{(III)\xi} = \frac{b}{2\pi\tilde{v}_t^2}\left(\frac{2(1-2\nu)(c_*^{-2}-1) - \tilde{v}_t^2}{P_t[1+p^2P_t^2]} - \frac{4(1-\nu)(1-c_*^2)}{P_l[1+p^2P_l^2]}\right)\frac{\partial\xi}{\partial x_3} \equiv \hat{u}_{1,1}^{(III)}$$

$$\frac{\partial\xi}{\partial x_3}, \tilde{u}_{2,2}^{(III)\xi} = \frac{b(c_*^{-2}-1)}{\pi\tilde{v}_t^2}\left(\frac{2(1-\nu)c_*^2P_l}{1+p^2P_l^2} - \frac{(1-2\nu)P_t}{1+p^2P_t^2}\right)\frac{\partial\xi}{\partial x_3} \equiv \hat{u}_{2,2}^{(III)}$$

$$\frac{\partial\xi}{\partial x_3}, \tilde{u}_{3,3}^{(III)\xi} = -\frac{b}{2\pi}\frac{1}{P_t[1+p^2P_t^2]}\frac{\partial\xi}{\partial x_3} \equiv \hat{u}_{3,3}^{(III)}\frac{\partial\xi}{\partial x_3}, \tilde{u}_{i,2i}^{(III)\xi} = 0, i = 1,\ 2 \text{ and } 3;$$

$$\tilde{u}_{1,2}^{(III)\xi} = \frac{bp}{2\pi\tilde{v}_t^2}\left(\frac{[2(1-2\nu)(c_*^{-2}-1)-\tilde{v}_t^2]P_t}{1+p^2P_t^2}\right.$$

$$\left.-\frac{4(1-\nu)(1-c_*^2)P_l}{1+p^2P_l^2}\right)\frac{\partial\xi}{\partial x_3} \equiv \hat{u}_{1,2}^{(III)}\frac{\partial\xi}{\partial x_3}, \tilde{u}_{1,22}^{(III)\xi} = 0;$$

$$\tilde{u}_{2,1}^{(III)\xi} = \frac{b(c_*^{-2}-1)p}{\pi\tilde{v}_t^2}\left(\frac{(1-2\nu)P_t}{1+p^2P_t^2} - \frac{2(1-\nu)c_*^2P_l}{1+p^2P_l^2}\right)\frac{\partial\xi}{\partial x_3} \equiv \hat{u}_{2,1}^{(III)}\frac{\partial\xi}{\partial x_3}, \tilde{u}_{2,21}^{(III)\xi} = 0;$$

$$\tilde{u}_{2,23}^{(III)\xi} = \frac{b(c_*^{-2}-1)}{\pi\tilde{v}_t^2}\left(\frac{2(1-\nu)c_*^2P_l}{1+p^2P_l^2} - \frac{(1-2\nu)P_t}{1+p^2P_t^2}\right)\frac{\partial^2\xi}{\partial x_3^2} \equiv \hat{u}_{2,23}^{(III)}\frac{\partial^2\xi}{\partial x_3^2}, \tilde{u}_{2,3}^{(III)\xi} = 0.$$

$$\tilde{u}_{3,1}^{(III)\xi} = 0, \tilde{u}_{3,21}^{(III)\xi} = \frac{bp}{2\pi}\left(\frac{2(c_*^{-2}-1)}{\tilde{v}_t^2}\left[\frac{(1-2\nu)P_t}{1+p^2P_t^2} - \frac{2(1-\nu)c_*^2P_l}{1+p^2P_l^2}\right]\right.$$

$$\left.-\frac{1-p^2P_t^2}{2P_t[1+p^2P_t^2]^2}\right)\frac{\partial^2\xi}{\partial x_3^2} \equiv \hat{u}_{3,21}^{(III)}\frac{\partial^2\xi}{\partial x_3^2};$$

$$\tilde{u}_{3,2}^{(III)\xi} = 0, \tilde{u}_{3,22}^{(III)\xi} = \frac{b}{2\pi}\left(\frac{2(c_*^{-2}-1)}{\tilde{v}_t^2}\left[\frac{2(1-\nu)c_*^2P_l}{1+p^2P_l^2} - \frac{(1-2\nu)P_t}{1+p^2P_t^2}\right]\right.$$

$$\left.+\frac{1-p^2P_t^2}{2P_t[1+p^2P_t^2]^2}\right)\frac{\partial^2\xi}{\partial x_3^2} \equiv \hat{u}_{3,22}^{(III)}\frac{\partial^2\xi}{\partial x_3^2}; \tilde{u}_{1,23}^{(III)\xi} = \frac{bp}{2\pi\tilde{v}_t^2}$$

$$\left(\frac{[2(1-2\nu)(c_*^{-2}-1)-\tilde{v}_t^2]P_t}{1+p^2P_t^2} - \frac{4(1-\nu)(1-c_*^2)P_l}{1+p^2P_l^2}\right)\frac{\partial^2\xi}{\partial x_3^2} \equiv \hat{u}_{1,23}^{(III)}\frac{\partial^2\xi}{\partial x_3^2},$$

$$\tilde{u}_{1,3}^{(III)\xi} = 0. \quad \text{(A7)}$$

2.2.4 *Quasistatic Brittle Cracks*

Cracks with Crack-Front Plastic Yielding

The general crack model is described in Sect. 2.2.2 with $\nu = 0$. First, considering in Ox_1x_3, a planar crack of finite extension $|x_1| \leq c$ along x_1 surrounded by plastic zone of extension $e \leq |x_1| \leq a$, $(e > c)$, both running indefinitely along x_3 under mode of applied loading J ($J = I, II$ and III), we have (Chang and Ohr 1981, 1982; Ohr 1985; Anongba and Vitek 2003; Anongba 2021*a*)

$$D_0^J = \frac{K_J^0}{\pi[C_1(\delta_{JI} + \delta_{JII}) + \delta_{JIII}C_2]\sqrt{2\pi s_1}},$$

$$\phi_0^J = \frac{2bK_J^0\sqrt{s_1}}{\pi[C_1(\delta_{JI} + \delta_{JII}) + \delta_{JIII}C_2]\sqrt{2\pi}},$$

$$\sigma_0^J = \frac{K_J^0}{\sqrt{2\pi s}}, G_0^J = \frac{1-\nu^2}{E}K_J^{02}(\delta_{JI} + \delta_{JII}) + \frac{1+\nu}{E}K_{III}^{02}\delta_{JIII}; \qquad (2.2.25)$$

E, μ and ν are Young and shear modules and Poisson's ratio respectively; $C_1 = \mu b/2\pi(1-\nu)$, $C_2 = \mu b/2\pi$, $s_1 = c - x_1$ $(0 < s_1 \ll c)$, $s = x_1 - c$ $(0 < s \ll c)$; δ_{ij} is the Kronecker delta; K_J^0 is the stress intensity factor, it reads

$$K_J^0 = \left\{\sigma_J^a(0)\sqrt{\pi c}\right\}\frac{F(\pi/2, k)\sqrt{a^2 - c^2}}{\sqrt{e^2 - c^2}\,\Pi\left(\pi/2, e^2k^2/c^2, k\right)}, \quad k^2 = \frac{c^2\left(a^2 - e^2\right)}{e^2\left(a^2 - c^2\right)}; \quad (2.2.26)$$

F and Π are elliptic integrals of first and third kind, respectively; $\sigma_J^a(0)$ takes the values $\sigma_{22}^a(J = I)$, $\sigma_{12}^a(J = II)$, and $\sigma_{23}^a(J = III)$ corresponding to the remote applied tension along x_2 and shears parallel to x_1 and x_3, respectively. In K_J^0 (2.2.26), the factor in curly brackets $\{-\}$ is the stress intensity factor for an isolated crack of finite length 2*c*. From above, it is convincing that brittle isolated crack and crack under crack-tip yielding can be analysed in the same way, equating G_0^J to twice the surface energy under pure mode *J* of applied loading. So far, only planar cracks are considered; however as stressed by Anongba and Vitek (2003), dislocation generation from the tip of the crack modifies the shape of the crack from planar to non-planar. Equally well, under mixed mode $I + II + III$ loading, the applied shearing stresses promote non-planar crack motion (we may refer to Anongba (2010, 2017) and references there in). In these cases, results like (2.2.25) and (2.2.26) now depend on the shape of the non-planar crack-front and orientation of the average fracture plane with respect of the applied loadings. Second, are presented below results corresponding to non-planar cracks of Sect. 2.2.2 with $\nu = 0$ (Anongba 2021*a*).

DISLOCATION DISTRIBUTIONS: Assume first that the dislocations are straight parallel to the x_3-direction ($\xi = 0$) and $h(x_1) = p_0 x_1$ depends linearly on x_1 with p_0 positive constant. This corresponds to a planar elastic–plastic crack (π_0 in Fig. 2.15a) of finite extension (with straight fronts running indefinitely along x_3) rotated about the positive x_3-direction by $\theta_0 = \tan^{-1} p_0$ from Ox_1x_3. The elastic–plastic crack ($|x_1| \leq c$, $e \leq |x_1| \leq a$) is subjected to mixed mode $I + II + III$ with loadings applied at infinity. Under such conditions $\partial f/\partial x_1 = \partial h/\partial x_1 = p_0$ and $\partial f/\partial x_3 = \partial \xi/\partial x_3 = 0$; the condition for the elastic–plastic crack faces to be free from tractions becomes:

$$\sigma_J + [C_1(\delta_{JI} + \delta_{JII}) + C_2\delta_{JIII}]\left(\int_{-a}^{-e} + \int_{-c}^{c} + \int_{e}^{a}\right)\frac{D_J(x_1')}{x_1 - x_1'}\mathrm{d}x_1' = 0(J = I, II\,\text{and}\,III). \tag{2.2.27}$$

For $|x_1| \leq c$

$$\sigma_I = \sigma_{22}^a - p_0\sigma_{12}^a \equiv \sigma_I^a,\, \sigma_{II} = \sigma_{12}^a + p_0\nu_A\sigma_{22}^a \equiv \sigma_{II}^a\,\text{and}\,\sigma_{III} = \sigma_{23}^a \equiv \sigma_{III}^a$$

For $e \leq |x_1| \leq a$

$$\sigma_I = \sigma_I^a - (\sigma_{22}^{fr} - p_0\sigma_{12}^{fr}) \equiv \sigma_I^a - \sigma_I^{fr},\, \sigma_{II} = \sigma_{II}^a - (\sigma_{12}^{fr} - p_0\sigma_{11}^{fr})$$
$$\equiv \sigma_{II}^a - \sigma_{II}^{fr}\,\text{and}\,\sigma_{III} = \sigma_{III}^a - \sigma_{23}^{fr} \equiv \sigma_{III}^a - \sigma_{III}^{fr}$$

The solution to (2.2.27) is known in terms of the Jacobi's zeta function $Z\,(\varphi, k)$ (Anongba and Vitek 2003). For $|x_1| < c$

$$D_J(x_1) \equiv D_J^C(x_1) = \frac{2\sigma_J^{fr}F(\pi/2, k)}{\pi^2[C_1(\delta_{JI} + \delta_{JII}) + C_2\delta_{JIII}]}\left(\frac{(e^2 - c^2)x_1}{e\sqrt{a^2 - c^2}}\right.$$
$$\left.\times\sqrt{\frac{a^2 - x_1^2}{(c^2 - x_1^2)(e^2 - x_1^2)}} + \mathrm{sgn}(x_1)Z(\beta_1(x_1), k)\right) \tag{2.2.28}$$

For $e \leq |x_1| \leq a$

$$D_J(x_1) \equiv D_J^{ea}(x_1) = \frac{2\sigma_J^{fr}F(\pi/2, k)}{\pi^2[C_1(\delta_{JI} + \delta_{JII}) + C_2\delta_{JIII}]}\mathrm{sgn}(x_1)Z(\beta_2(x_1), k) \tag{2.2.29}$$

$$\beta_1(x_1) = \sin^{-1}\left(1/\sqrt{n(x_1)}\right),\, \beta_2(x_1) = \sin^{-1}\left(\sqrt{n(x_1)}/k\right),\, k^2 = \frac{c^2(a^2 - e^2)}{e^2(a^2 - c^2)},$$

$$n(x_1) = \frac{c^2(e^2 - x_1^2)}{e^2(c^2 - x_1^2)}.$$

It must be attached to this solution of D_J the following relation between the applied stresses, friction stresses, and size of the elastic–plastic crack

$$\frac{\pi\sigma_J^a}{2\sigma_J^{fr}} = \frac{e^2 - c^2}{e\sqrt{a^2 - c^2}}\Pi\left(\pi/2, e^2k^2/c^2, k\right). \tag{2.2.30}$$

Because $\sigma_J^a = \sigma_J^a(p_0)$ and $\sigma_J^{fr} = \sigma_J^{fr}(p_0)$ (see about (2.2.27)) depend on p_0, relation (2.2.30) is valid for all p_0; this leads to $\sigma_J^a(p_0)/\sigma_J^{fr}(p_0) = \sigma_J^a(0)/\sigma_J^{fr}(0)$ constant with p_0. It can thus be written

$$D_J(x_1) = \frac{\sigma_J^a(p_0)}{\sigma_J^a(0)} D_0^J(x_1) \tag{2.2.31}$$

where D_0^J is the value of D_J (2.2.28, and 2.2.29) at $p_0 = 0$; $D_0^J(J = I, II$ and $III)$ corresponds to the equilibrium distribution of straight dislocations J when the elastic–plastic crack is planar in Ox_1x_3.

$$\frac{\sigma_I^a(p_0)}{\sigma_I^a(0)} = 1 - p_0\frac{\sigma_{12}^a}{\sigma_{22}^a}, \frac{\sigma_{II}^a(p_0)}{\sigma_{II}^a(0)} = 1 + p_0\frac{\nu_A\sigma_{22}^a}{\sigma_{12}^a}, \frac{\sigma_{III}^a(p_0)}{\sigma_{III}^a(0)} = 1. \tag{2.2.32}$$

The relative displacement ϕ_J^C of the faces of the crack, in the x_2 $(J = I)$, x_1 $(J = II)$ and x_3 $(J = III)$ directions reads

$$\phi_J^C(x_1) = \int_{x_1}^{c} bD_J^C(x_1')\mathrm{d}x_1', \ |x_1| \leq c,$$

which gives after integration

$$\phi_J^C(x_1) = \frac{\sigma_J^a(p_0)}{\sigma_J^a(0)} \phi_0^{CJ}(x_1) \tag{2.2.33}$$

ϕ_0^{CJ} is the value of ϕ_J^C at $p_0 = 0$ (see relation (16) in (Anongba and Vitek 2003) for $J = I$). We have

$$\begin{aligned}\phi_0^{CJ} &= \frac{2b\sigma_J^{fr}(0)}{\pi^2[C_1(\delta_{JI} + \delta_{JII}) + C_2\delta_{JIII}]}\left(\frac{F(\pi/2, k)(e^2 - c^2)}{e\sqrt{a^2 - c^2}}\sqrt{\frac{(a^2 - x_1^2)(c^2 - x_1^2)}{e^2 - x_1^2}}\right. \\ &- F\left(\frac{\pi}{2}, k\right)|x_1|Z(\beta_1(x_1), k) - ek^2\left[F\left(\frac{\pi}{2}, k\right)\hat{D}\left(\beta_3(x_1), \frac{ek}{c}\right)\right. \\ &- \left.\left. F\left(\beta_3(x_1), \frac{ek}{c}\right)\hat{D}\left(\frac{\pi}{2}, k\right)\right]\right), \ |x_1| \leq c\end{aligned} \tag{2.2.34}$$

$\hat{D}$ is a function of variables (φ, k) defined as $\hat{D}(\varphi, k) = (F(\varphi, k) - E(\varphi, k))/k^2$ in which E is the elliptic integral of the second kind, $\beta_3(x_1) = \sin^{-1}\left(c/e\sqrt{n(x_1)}\right)$. The number of dislocations J in the plastic zone between e and x_1 is

$$N_J^{ea}(x_1) = \frac{\sigma_J^a(p_0)}{\sigma_J^a(0)} N_0^{ea(J)}(x_1) \tag{2.2.35}$$

where $N_0^{ea(J)}$ is the value of N_J^{ea} for $p_0 = 0$.

$$N_0^{ea(J)} = \frac{2\sigma_J^{fr}(0)}{\pi^2[C_1(\delta_{JI}+\delta_{JII})+C_2\delta_{JIII}]}\left(F\left(\frac{\pi}{2},k\right)x_1 Z(\beta_2(x_1),k)\right.$$
$$+\,ek^2\left(F\left(\frac{\pi}{2},k\right)\hat{D}\left(\beta_2(x_1),\frac{ek}{c}\right)\right)$$
$$\left.-\,F\left(\beta_2(x_1),\frac{ek}{c}\right)\hat{D}\left(\frac{\pi}{2},k\right)\right)\Bigg),\ e\le x_1\le a. \tag{2.2.36}$$

The total number $\overline{N}_J^{ea}$ of dislocations J between e and a is thus

$$\bar{N}_J^{ea} = \frac{\sigma_J^a(p_0)}{\sigma_J^a(0)}\frac{2e\sigma_J^{fr}(0)k^2}{\pi^2[C_1(\delta_{JI}+\delta_{JII})+C_2\delta_{JIII}]}$$
$$\left(F(\frac{\pi}{2},k)\hat{D}(\frac{\pi}{2},\frac{ek}{c}) - F(\frac{\pi}{2},\frac{ek}{c})\hat{D}(\frac{\pi}{2},k)\right)$$
$$= \frac{\sigma_J^a(p_0)(a^2-e^2)c^2}{\pi[C_1(\delta_{JI}+\delta_{JII})+C_2\delta_{JIII}]}$$
$$\left(F(\frac{\pi}{2},k)\hat{D}(\frac{\pi}{2},\frac{ek}{c}) - F(\frac{\pi}{2},\frac{ek}{c})\hat{D}(\frac{\pi}{2},k)\right)$$
$$\times\frac{1}{(e^2-c^2)\sqrt{a^2-c^2}\,\Pi(\pi/2,e^2k^2/c^2,k)} \tag{2.2.37}$$

As performed below, we can reasonably give approximate expressions for the stress about the crack front in the DFZ and crack extension force with f given by (2.2.1) when the average fracture surface h can be approximated by plane π_0 of Fig. 2.15a.

CRACK-TIP STRESSES and CRACK EXTENSION FORCE: We would like to express the total stress $\overline{\sigma}_{ij}$, ahead of the front of the crack with shape f (2.2.1), at $x_1 = c$ in the DFZ. Writing $x_1 = c + s$, $0 < s \ll c$, $\overline{\sigma}_{ij}$ is given by the following formula

$$\overline{\sigma}_{ij}(s,x_2,x_3) = \sum_{J=I}^{III}\int_{c-\delta c}^{c}\sigma_{ij}^{(J)}(c+s-x_1',x_2,x_3)D_J(x_1')\mathrm{d}x_1' \tag{2.2.38}$$

with $\delta c \ll c$ and x_2 close to $h(c)$. Then, proceeding exactly as in previous works (Anongba 2010, 2017) taking for D_J the straight edge and screw dislocation distributions (2.2.31) corresponding to an elastic–plastic planar crack π_0 with a straight front parallel to x_3 (Fig. 2.15a), it follows ($\overline{\sigma}_{ij} \equiv \overline{\sigma}_{ij}^{(I)} + \overline{\sigma}_{ij}^{(II)} + \overline{\sigma}_{ij}^{(III)}$):

$$\bar{\sigma}_{ii}^{(I)}(s,x_2,x_3) = \frac{1}{(1+p^2)^3}([\delta_{i1}+\delta_{i2}+2\nu\delta_{i3}$$
$$+(-\delta_{i1}+3\delta_{i2}+2\nu\delta_{i3})p^2](1+p^2)+\frac{1}{2}(x_2-h(c))$$

$$\big[\delta_{i1} - \delta_{i2} + 2(1+\nu)\delta_{i3} - 6(\delta_{i1} - \delta_{i2})p^2$$

$$- \ (-\delta_{i1} + \delta_{i2} + 2(1+\nu)\delta_{i3})p^4\big]\frac{\partial^2 \xi}{\partial x_3^2}(c, x_3)\Big)$$

$$\left(1 - p_0\frac{\sigma_{12}^a}{\sigma_{22}^a}\right)\frac{K_I^0}{\sqrt{2\pi}}\frac{1}{\sqrt{s}}\bar{\sigma}_{ii}^{(II)}(s, x_2, x_3) = \frac{(-\delta_{i1} + \delta_{i2} - \delta_{i3})p}{(1+p^2)^3}$$

$$\big([3\delta_{i1} + \delta_{i2} + 2\nu\delta_{i3} + (\delta_{i1} - \delta_{i2} + 2\nu\delta_{i3})p^2](1+p^2)$$

$$+ \frac{1}{2}(x_2 - h(c))[3\delta_{i1} + (3+4\nu)\delta_{i2} + 2(1+3\nu)\delta_{i3}$$

$$+ (-6\delta_{i1} - 2(3-4\nu)\delta_{i2} + 8\nu\delta_{i3})p^2$$

$$+ \ (-\delta_{i1} - (1-4\nu)\delta_{i2} - 2(1-\nu)\delta_{i3})p^4\big]\frac{\partial^2 \xi}{\partial x_3^2}(c, x_3)\Big)$$

$$\left(1 + p_0\frac{\nu_A\sigma_{22}^a}{\sigma_{12}^a}\right)\frac{K_{II}^0}{\sqrt{2\pi}}\frac{1}{\sqrt{s}}, \bar{\sigma}_{ii}^{(III)}(s, x_2, x_3)$$

$$= \frac{\big[(p^2-1)\delta_{i1} + \big(1 - 2\nu - (1+2\nu)p^2\big)\delta_{i2} - 2(1+p^2)\delta_{i3}\big]}{(1-\nu)(1+p^2)^2}$$

$$\frac{\partial \xi}{\partial x_3}\frac{K_{III}^0}{\sqrt{2\pi}}\frac{1}{\sqrt{s}}, \bar{\sigma}_{j3}^{(I)}(s, x_2, x_3)$$

$$= \frac{p[-(2-\nu) + \nu p^2]\delta_{j1} + [\nu - (2-\nu)p^2]\delta_{j2}}{(1+p^2)^2}$$

$$\frac{\partial \xi}{\partial x_3}\left(1 - p_0\frac{\sigma_{12}^a}{\sigma_{22}^a}\right)\frac{K_I^0}{\sqrt{2\pi}}\frac{1}{\sqrt{s}}, \bar{\sigma}_{j3}^{(II)}(s, x_2, x_3)$$

$$= -\frac{[1 - (1-\nu)p^2 - (2-\nu)p^4]\delta_{j1} + p[1 + 2\nu - (1-2\nu)p^2]\delta_{j2}}{(1+p^2)^2}$$

$$\times \frac{\partial \xi}{\partial x_3}\left(1 + p_0\frac{\nu_A\sigma_{22}^a}{\sigma_{12}^a}\right)\frac{K_{II}^0}{\sqrt{2\pi}}\frac{1}{\sqrt{s}}\bar{\sigma}_{j3}^{(III)}(s, x_2, x_3)$$

$$= \frac{1}{(1-\nu)(1+p^2)^2}\Big((1-\nu)[-p\delta_{j1} + \delta_{j2}](1+p^2) - \frac{1}{2}(x_2 - h(c))$$

$$\times \big[p\big(5 - 3\nu - (1+\nu)p^2\big)\delta_{j1} - \big(3 - \nu - (3-\nu)p^2\big)\delta_{j2}\big]\frac{\partial^2 \xi}{\partial x_3^2}\Big)$$

$$\frac{K_{III}^0}{\sqrt{2\pi}}\frac{1}{\sqrt{s}}, \bar{\sigma}_{12}^{(I)}(s, x_2, x_3) = \frac{p}{(1+p^2)^3}\Big(1 - p^4 + \frac{1}{2}(x_2 - h(c))$$

$$\times \big[5 - 2\nu - 2(1+2\nu)p^2 + (1-2\nu)p^4\big]\frac{\partial^2 \xi}{\partial x_3^2}\Big)$$

$$\left(1 - p_0\frac{\sigma_{12}^a}{\sigma_{22}^a}\right)\frac{K_I^0}{\sqrt{2\pi}}\frac{1}{\sqrt{s}}, \bar{\sigma}_{12}^{(II)} = \frac{1}{(1+p^2)^3}$$

$$\left(1-p^4+\frac{1}{2}(x_2-h(c))\left[1+2(13+2\nu)p^2+p^4\right]\frac{\partial^2\xi}{\partial x_3^2}\right)$$
$$\times\left(1+p_0\frac{\nu_A\sigma_{22}^a}{\sigma_{12}^a}\right)\frac{K_{II}^0}{\sqrt{2\pi}}\frac{1}{\sqrt{s}},$$
$$\bar{\sigma}_{12}^{(III)}(s,x_2,x_3)=-\frac{p(2-\nu-\nu p^2)}{(1-\nu)(1+p^2)^2}\frac{\partial\xi}{\partial x_3}\frac{K_{III}^0}{\sqrt{2\pi}}\frac{1}{\sqrt{s}} \tag{2.2.39}$$

where subscripts i and j take the values (1, 2 and 3) and (1 and 2) respectively; $p=\partial h(c)/\partial x_1$, $K_J^0(J=I, II$ and III respectively) is given by (2.2.26). It is stressed again that s, x_2 and x_3 are arbitrary, $s=x_1-c\ll c$ $(s>0)$ and $(x_2-h(c))$ is small. The parameter p_0 in (2.2.39) originates from a planar crack π_0 (Fig. 2.15a) hypothetically assumed to approximate the average fracture surface $x_2=h(x_1)$. One observes that (2.2.39) and the relation (10) in Anongba (2010) are identical. The crack extension force G (per unit length of the crack front) is calculated in the same way as for the isolated non-planar crack (Anongba 2010, 2017). Reduced crack extension force $\tilde{G}$ is defined as $\tilde{G}=G/(G_0^I+G_0^{II}+G_0^{III})$ with G_0^J given in (2.2.25) and obtain at $P_0(c,x_2=f,x_3)$(with $M_{12}\equiv\sigma_{12}^a/\sigma_{22}^a$, $M_{13}\equiv\sigma_{23}^a/\sigma_{22}^a$, $M_{23}\equiv\sigma_{23}^a/\sigma_{12}^a$)

$$\tilde{G}(P_0)=\sum_{i,j=1}^{3}\tilde{G}_j^{(i)}(P_0) \tag{2.2.40}$$

where

$$\tilde{G}_1^{(1)}=-\frac{1}{2(1+p^2)^3}\frac{\partial f(c,x_3)/\partial x_1}{\sqrt{1+(\partial f/\partial x_1)^2+(\partial f/\partial x_3)^2}}(2(1-p^4)(1-p_0M_{12})$$
$$2p(1+p^2)\times(3+p^2)(p_0\nu_A+M_{12})+\frac{2}{1-\nu}(p^4-1)M_{13}\frac{\partial\xi}{\partial x_3}$$
$$+\left[(1-6p^2+p^4)(1-p_0M_{12})\ -p(3-6p^2-p^4)\right)(p_0\nu_A+M_{12})\Big]\xi\frac{\partial^2\xi}{\partial x_3^2}\bigg)$$
$$\frac{(1-\nu)(p_0\nu_A+M_{12})}{\left[1-\nu+(1-\nu)M_{12}^2+M_{13}^2\right]},\ \tilde{G}_2^{(1)}=\frac{1}{2(1+p^2)^3}$$
$$\frac{1}{\sqrt{1+(\partial f/\partial x_1)^2+(\partial f/\partial x_3)^2}}(2(1-p^4)\left[p(1-p_0M_{12})+M_{12}+p_0\nu_A\right]$$
$$-\frac{2}{1-\nu}p(1+p^2)(2-\nu-\nu p^2)M_{13}\frac{\partial\xi}{\partial x_3}$$
$$+\left[p(5-2\nu-2(1+2\nu)p^2+(1-2\nu)p^4\right)$$
$$\times(1-p_0M_{12})+\left(1+2(13+2\nu)p^2+p^4\right)(M_{12}+p_0\nu_A)\Big]$$
$$\times\ \xi\frac{\partial^2\xi}{\partial x_3^2}\bigg)\frac{(1-\nu)(p_0\nu_A+M_{12})}{\left[1-\nu+(1-\nu)M_{12}^2+M_{13}^2\right]}$$

$$\tilde{G}_3^{(1)} = -\frac{1}{(1+p^2)^2}\frac{\partial f/\partial x_3}{\sqrt{1+(\partial f/\partial x_1)^2+(\partial f/\partial x_3)^2}}$$
$$\Big(-p(1+p^2)M_{13} + \big[p\big(-2+\nu+\nu p^2\big)$$
$$\times\ (1-p_0M_{12}) - \big(1-(1-\nu)p^2-(2-\nu)p^4\big)(p_0\nu_A+M_{12})\big]\frac{\partial\xi}{\partial x_3}$$
$$-\ p\big(5-3\nu-(1+\nu)p^2\big)\frac{M_{13}}{2(1-\nu)}\xi\frac{\partial^2\xi}{\partial x_3^2}\Big)$$
$$\frac{(1-\nu)(p_0\nu_A+M_{12})}{\big[1-\nu+(1-\nu)M_{12}^2+M_{13}^2\big]},\ \tilde{G}_1^{(2)} = -\tilde{G}_2^{(1)}\frac{\partial f}{\partial x_1}\frac{1-p_0M_{12}}{p_0\nu_A+M_{12}},$$
$$\tilde{G}_2^{(2)} = \frac{1}{2(1+p^2)^3}\frac{1}{\sqrt{1+(\partial f/\partial x_1)^2+(\partial f/\partial x_3)^2}}$$
$$\Big(2(1+p^2)\big[(1+3p^2)(1-p_0M_{12}) + p(1-p^2)(M_{12}+p_0\nu_A)\big]$$
$$+\ \frac{2}{1-\nu}(1+p^2)(1-2\nu-(1+2\nu)p^2)M_{13}\frac{\partial\xi}{\partial x_3}$$
$$+\big[(-1+6p^2-p^4)(1-p_0M_{12}) + p\big(3+4\nu-2(3-4\nu)p^2$$
$$-\ (1-4\nu)p^4\big)(M_{12}+p_0\nu_A)\big]$$
$$\times\ \xi\frac{\partial^2\xi}{\partial x_3^2}\Big)\frac{(1-\nu)(1-p_0M_{12})}{\big[1-\nu+(1-\nu)M_{12}^2+M_{13}^2\big]},$$
$$\tilde{G}_3^{(2)} = -\frac{1}{(1+p^2)^2}\frac{\partial f/\partial x_3}{\sqrt{1+(\partial f/\partial x_1)^2+(\partial f/\partial x_3)^2}}\Big((1+p^2)M_{13} + \Big[\Big(\nu\ (2-\nu)p^2\Big)$$
$$\times\ (1-p_0M_{12}) - p\Big(1+2\nu-(1-2\nu)p^2\Big)(M_{12}+p_0\nu_A)\Big]\frac{\partial\xi}{\partial x_3} + \frac{(3-\nu)(1-p^2)M_{13}}{2(1-\nu)}$$
$$\times\ \xi\frac{\partial^2\xi}{\partial x_3^2}\Big)\frac{(1-\nu)(1-p_0M_{12})}{\big[1-\nu+(1-\nu)M_{12}^2+M_{13}^2\big]},$$
$$\tilde{G}_1^{(3)} = \tilde{G}_3^{(1)}\frac{\partial f/\partial x_1}{\partial f/\partial x_3}\frac{M_{13}}{(1-\nu)(p_0\nu_A+M_{12})},\ \tilde{G}_2^{(3)} = -\tilde{G}_3^{(2)}\frac{1}{\partial f/\partial x_3}\frac{M_{13}}{(1-\nu)(1-p_0M_{12})},$$
$$\tilde{G}_3^{(3)} = -\frac{1}{2(1+p^2)^3}\frac{\partial f/\partial x_3}{\sqrt{1+(\partial f/\partial x_1)^2+(\partial f/\partial x_3)^2}}\Big(4\nu(1+p^2)^2$$
$$\big[1-(p+p_0)M_{12}-pp_0\nu_A\big] - \frac{4}{1-\nu}(1+p^2)^2M_{13}\frac{\partial\xi}{\partial x_3} + \big[2(1+\nu)(1-p^4)(1-p_0M_{12})$$
$$-\ p\Big(2(1+3\nu)+8\nu p^2-2(1-\nu)p^4\Big)(M_{12}+p_0\nu_A)\Big]\xi\frac{\partial^2\xi}{\partial x_3^2}\Big)\frac{M_{13}}{\big[1-\nu+(1-\nu)M_{12}^2+M_{13}^2\big]} \tag{2.2.41}$$

Equation (2.2.41) is identical to the corresponding one for the isolated crack (see relation (16) in Anongba 2017). Expression (2.2.41) gives $\tilde{G}$ for any arbitrary shape f (2.2.1) of the crack front. Since fracture over large distance proceeds through the motion of a macroscopic length of the crack front, the relevant quantity is $\langle\tilde{G}\rangle$, the value of $\tilde{G}$ averaged over the length of the crack front. $\langle\tilde{G}\rangle$ for several special isolated crack fronts have been provided: straight, sinusoidal, segmented (Anongba 2008, 2010, 2017). All these results apply here i.e. when the crack front is associated with localized plastic yielding. The expected crack configuration after fracture propagation over large distance is the one that maximizes $\langle\tilde{G}\rangle$ under the Griffith condition $\langle G\rangle_{\max} = 2\gamma$ where γ is the surface energy.

DISCUSSION: The reduced crack extension force $\tilde{G}$ (2.2.41) for the elastic–plastic crack is identical to that of the isolated crack given in Anongba (2017). Consequently, the various graphical representations of $\langle\tilde{G}\rangle$ there, as a function of parameters (p, p_0, p_A, p_B; M_{12}, M_{13}; ν_A), for simple special isolated cracks are unchanged for the elastic–plastic cracks (see Figs. 2.15 and 2.16, for example). From these works, non-planar crack configurations exist for which the crack extension force $\langle G\rangle$ is larger than that of the planar crack in Ox_1x_3, thus corroborating the occurrence of non-planar fracture abundantly observed in real materials. One more observation on the value of $\langle\tilde{G}\rangle$ on inverting the sign of the applied shears σ_{12}^a and σ_{23}^a is stressed. First, consider the planar crack π_0 (Fig. 2.15a) under mode $I + II$ loading ($\sigma_{23}^a = 0$):

$$\tilde{G}(P_0) = \frac{1 + p_0^2\nu_A^2 + (1 + p_0^2)M_{12}^2 - 2(1 - \nu_A)p_0 M_{12}}{\sqrt{1 + p_0^2(1 + M_{12}^2)}}(\xi = 0;\, h = p_0 x_1). \quad (2.2.42)$$

Inverting the sign of σ_{12}^a from positive to negative (i.e. $M_{12} < 0$) increases $\tilde{G}$. $M_{12} < 0$ corresponds to applying the shear in the negative x_1-direction when the specimen is suffering tension ($\sigma_{22}^a > 0$) along x_2. If the specimen is under fatigue by inverting the sign of σ_{12}^a only, this asymmetry implies that this is the condition $G = 2\gamma$ ($M_{12} > 0$) that controls the complete failure of the fracture material. A similar behaviour is found with σ_{23}^a for the non-planar crack with a segmented crack front (Fig. 2.16a). We have in tension under mixed mode $I + III$ ($\sigma_{12}^a = 0$) (see also (20) in Anongba (2017)):

$$\langle\tilde{G}\rangle(\sigma_{23}^a < 0) - \langle\tilde{G}\rangle(\sigma_{23}^a > 0) = \frac{2\nu\nu_1|M_{13}|\left(2 + 4p_0^2 + (2 - \nu)p_0^4 + (2 - \nu)p_0^6\right)}{(1 + p_0^2)^2\left(1 - \nu + M_{13}^2\right)},$$
$$\nu_1 = (p_A p_B/(p_A + p_B))\left(1/\sqrt{1 + p_0^2 + p_B^2} - 1/\sqrt{1 + p_0^2 + p_A^2}\right); \quad (2.2.43)$$

$p_A = \tan\phi_A, p_B = \tan\phi_B$ (see Fig. 2.16b for ϕ_A and ϕ_B). If $p_A = p_B$, $\nu_1 = 0$ and there is no asymmetry on inverting the sign of the shear σ_{23}^a. However if $p_A \neq p_B$, $\nu_1 \neq 0$ and an asymmetry does exist. The controlling $\langle\tilde{G}\rangle$ will be the smallest one depending on the values of p_A compared with p_B. Similarly, one may invert the sign of σ_{22}^a (< 0) and maintains the shearing stresses positive. Under such conditions (2.2.42) and (2.2.43) are unchanged; this leads to same asymmetries (note that σ_{22}^a replaces σ_{23}^a in the left-hand side in (2.2.43)). One notes that Poisson's stress $\sigma_{11}^a = -\nu_A\sigma_{22}^a$ is tensile positive and opens the crack faces when p_0 is different from zero.

SUMMARY: Non-planar elastic–plastic cracks have been studied. These are of finite extension along x_1 and x_2 and infinite in the x_3-direction, inside an infinitely extended elastic medium, subjected to general loading. The loadings, tension σ_{22}^a and shears σ_{12}^a and σ_{23}^a, are applied along the x_2, x_1 and x_3 directions at infinity, respectively. The front of the crack is planar in x_2x_3, has an average elevation $h = h(x_1)$ from Ox_1x_3 and fluctuates weakly there in the form ξ (2.2.1). The plastic region is described by $x_2 = h(x_1)$ with a straight front parallel to x_3. The crack and plastic zone are represented by a continuous distribution of three types J of infinitesimal dislocation with Burgers vectors directed along the applied loadings. Distribution functions D_J of straight dislocation arrays corresponding to an elastic–plastic crack π_0 (Fig. 2.15a), inclined by angle θ_0 with respect to Ox_1x_3 are calculated. Adopting these D_J, explicit expressions of the crack-tip stresses and crack extension force per unit length of the crack front, for the general crack front f (2.2.1), are evaluated. Except for a difference in the value of the stress intensity factor, that now depends on the size of the elastic–plastic crack, these quantities agree with those of the isolated crack. Hence both types of cracks can be treated in the equal manner.

Isolated Brittle Cracks

The general crack model is described in Sect. 2.2.1 with $v = 0$. Expressions of the physical quantities obtained above for the elastic–plastic crack apply entirely here for the isolated brittle crack. It is sufficient to take $K_J^0 = \sigma_J^a(0)\sqrt{\pi c}$ in (2.2.26) for the stress intensity factor. $\tilde{G}$ (2.2.40) is unchanged.

PLANAR STRAIGHT-FRONTED CRACKS. First, the crack in Fig. 2.10b is considered; it extends from $x_1 = -c$ to c and runs indefinitely in the x_3- direction. The crack front is straight parallel to x_3 ($\xi = 0$) and $f = h(x_1)$ independent of x_3. We assume the crack to fluctuate about plane π_0 (Fig. 2.10a). $\tilde{G}$ under general loading takes the form

$$\tilde{G}(P_0) = \frac{(p_0\nu_A + M_{12})^2 + (1 - p_0M_{12})^2 + M_{13}^2/(1-\nu)}{\sqrt{1+p^2}[1 + M_{12}^2 + M_{13}^2/(1-\nu)]} \quad (\xi = 0; h = h(x_1)) \tag{2.2.44}$$

at point $P_0(c, x_2 = h(c), x_3)$ of the crack front located at $x_1 = c; p = \partial h(c)/\partial x_1 = \tan\theta$ (for θ, see Fig. 2.10b). A similar relation (2.2.45) below will be given a more detailed description. The second crack is given in Fig. 2.10a. This is a tilted planar crack corresponding to the rotation of plane Ox_1x_3 about Ox_3 by angle $\theta_0 = \tan^{-1}(p_0)$. The crack front is straight ($\xi = 0$) and $h = p_0x_1$. The normalized crack extension force $\tilde{G}$ then reads

$$\tilde{G}(P_0) = \frac{(p_0\nu_A + M_{12})^2 + (1 - p_0M_{12})^2 + M_{13}^2/(1-\nu)}{\sqrt{1+p_0^2}[1 + M_{12}^2 + M_{13}^2/(1-\nu)]} \quad (\xi = 0; h = p_0x_1) \tag{2.2.45}$$

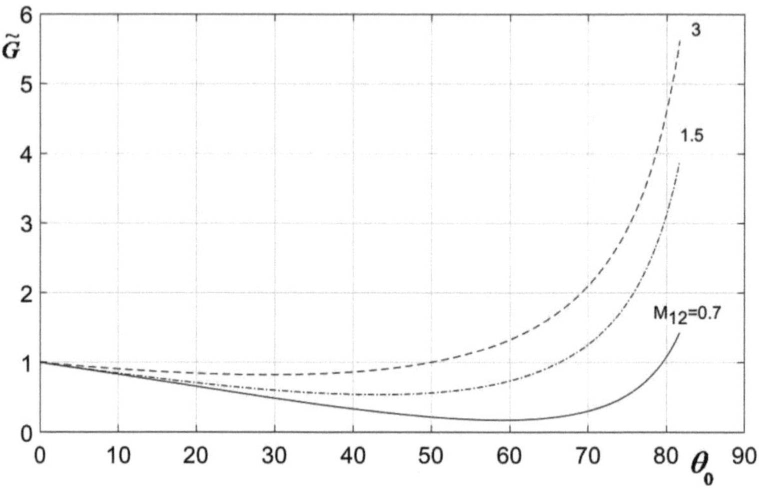

Fig. 2.17 Normalized crack extension force $\tilde{G}$ versus θ_0 ($\nu_A = 0$ absence of Poisson effect) for a planar crack π_0 with a straight front parallel to Ox_3 (Fig. 2.10a), tilted about Ox_3 by an angle $\theta_0 = \tan^{-1} p_0$ from Ox_1x_3 ($M_{13} = 0$). The curves correspond from bottom to top to $M_{12} = 0.7$, 1.5 and 3

at point $P_0(c, x_2 = p_0c, x_3)$ of the crack front located at $x_1 = c$. Figures 2.17 and 2.18 are plots of $\tilde{G}$ as a function of θ_0 for different M_{12} values when $M_{13} = 0$ under $\nu_A = 0$(absence of Poisson effect) and $\nu_A \neq 0$(presence of Poisson effect), respectively. We expect that for a tilted plane crack π_0 (Fig. 2.10a) to be observable experimentally, it is necessary that $\tilde{G}$ be larger than 1. As we can see from Fig. 2.17, this occurs for sufficiently large M_{12} and θ_0 values; the larger the M_{12}(dominant mode II loading) the smaller the θ_0 values for which $\tilde{G} > 1$. As we can see from Fig. 2.18, Poisson effect increases significantly $\tilde{G}$ values.

SINUSOIDAL CRACK FRONT (Anongba 2008): Here the crack model is assumed to have a sinusoidal front $f = \xi$ (2.2.1) and $h = 0$ (Fig. 2.19)

$$\xi/\xi_c = \sin \kappa_c x_3 = \sin 2\pi \overline{x}_3 \tag{2.2.46}$$

where $\overline{x}_3 = x_3/\lambda_c$ and $\lambda_c = 2\pi/\kappa_c$ is the wavelength. With $M_{12} = 0$ and $M = \sigma_{23}^a/\sigma_{22}^a$,

$$\begin{aligned}\tilde{G}(P_0) = {} & \frac{1}{\sqrt{1 + (\partial\xi/\partial x_3)^2}}\left(1 - \frac{2\nu M}{1 - \nu + M^2}\frac{\partial \xi}{\partial x_3} + \frac{(3-\nu)M^2 - (1-\nu)^2}{2(1-\nu)(1-\nu+M^2)}\xi\frac{\partial^2\xi}{\partial x_3^2}\right. \\ & \left. + \frac{2M^2 - \nu(1-\nu)^2}{(1-\nu)(1-\nu+M^2)}\left(\frac{\partial\xi}{\partial x_3}\right)^2 - \frac{(5+\nu)M}{2(1-\nu+M^2)}\xi\frac{\partial\xi}{\partial x_3}\frac{\partial^2\xi}{\partial x_3^2}\right).\end{aligned} \tag{2.2.47}$$

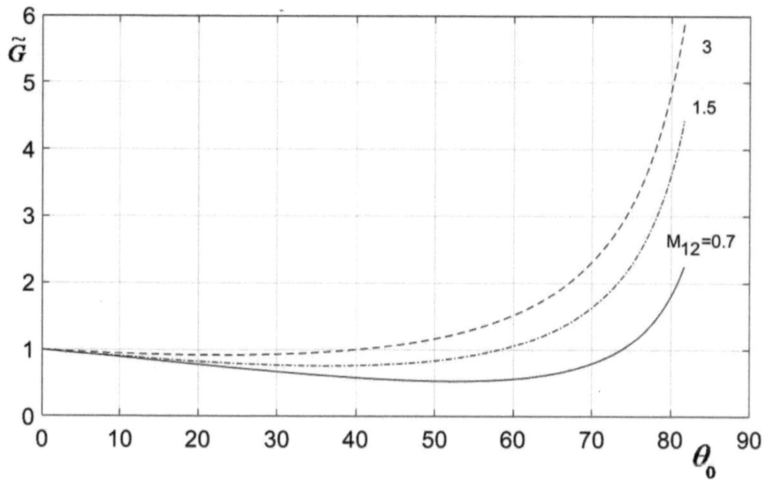

Fig. 2.18 Normalized crack extension force $\tilde{G}$ versus θ_0 ($\nu_A \neq 0$ presence of Poisson effect) for a planar crack π_0(other conditions as in Fig. 2.17)

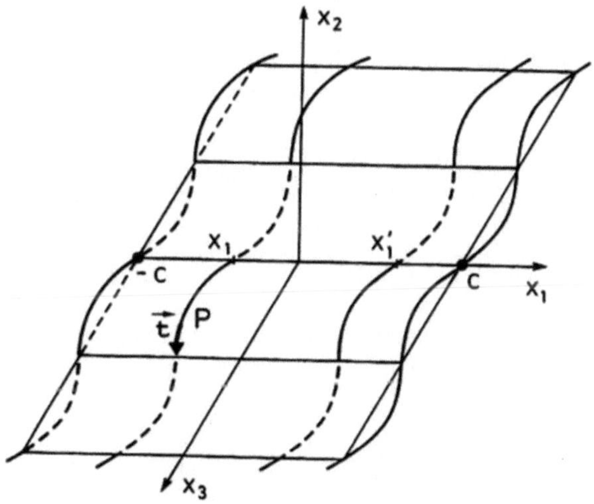

Fig. 2.19 A wavy crack with a sinusoidal front extending from $x_1 = -c$ to c and subjected to uniform σ_{22}^a and σ_{23}^a at infinity. The crack fluctuates about a mean plane, which is illustrated, with small amplitude (say roughly of the order of 1 µm or smaller)

In the following are expressed the reduced crack extension force (2.2.47), denoted $\tilde{G}_c$ for the sinusoidal crack, the spatial average $\langle \tilde{G}_c \rangle$ of $\tilde{G}_c$ defined as $\langle \tilde{G}_c \rangle = (1/\lambda_c) \int_0^{\lambda_c} \tilde{G}_c \mathrm{d}x_3$ and finally conditions under which $\langle \tilde{G}_c \rangle$ has extremum. First, are provided associated formulae and then, with the help of plots, corresponding behaviours and properties of the model.

The reduced value $\tilde{G}$ of the crack extension force depends on the product $\xi_c \kappa_c = \tan\phi_c$; ϕ_c, the crack-front inclination angle, is the acute angle measured in the plane perpendicular to the crack propagation direction between the crack front and the average fracture plane:

$$\begin{aligned}\tilde{G}_c(P_0) = \frac{1}{\sqrt{1+\tan^2\varphi_c\cos^2 2\pi\overline{x}_3}}\Bigg(1 + \frac{2M^2 - \nu(1-\nu)^2}{(1-\nu)\big(1-\nu+M^2\big)}\tan^2\varphi_c \\ - \frac{2\nu M}{1-\nu+M^2}\tan\varphi_c\cos 2\pi\overline{x}_3 \\ + \frac{(1+2\nu)(1-\nu)^2 - (7-\nu)M^2}{2(1-\nu)\big(1-\nu+M^2\big)}\tan^2\varphi_c\sin^2 2\pi\overline{x}_3 \\ + \frac{(5+\nu)M}{2\big(1-\nu+M^2\big)}\tan^3\varphi_c\cos 2\pi\overline{x}_3\sin^2 2\pi\overline{x}_3\Bigg)\end{aligned} \tag{2.2.48}$$

For given M value and crack profile ϕ_c, (2.2.48) gives the reduced crack extension force at an arbitrary point $P_0(c, \xi_c \sin\kappa_c x_3, x_3)$ on the crack front.

In the calculation of $\langle\tilde{G}_c\rangle$, the average values of terms with $\tan\phi_c$ and $\tan^3\phi_c$ (odd powers) in the brackets () in (2.2.48) contribute nothing; the two other involved integrals are given by the complete elliptic integral of the first kind and formulae 3.671 intable of integrals by Gradshteyn and Ryzhik (1980).

$$\begin{aligned}\langle\tilde{G}_c\rangle = \cos\phi_c\Bigg(\left[1 + \frac{2M^2 - \nu(1-\nu)^2}{(1-\nu)(1-\nu+M^2)}\tan^2\phi_c\right]F(\frac{1}{2},\frac{1}{2};1;\sin^2\phi_c) \\ + \frac{(1+2\nu)(1-\nu)^2 - (7-\nu)M^2}{4(1-\nu)(1-\nu+M^2)}\tan^2\phi_c F(\frac{3}{2},\frac{1}{2};2;\sin^2\phi_c)\Bigg)\end{aligned} \tag{2.2.49}$$

where F is Gauss's hypergeometric function. The condition for an extremum of $\langle\tilde{G}_c\rangle(\phi_c)$ is $\partial\langle\tilde{G}_c\rangle/\partial\phi_c = 0$. We have

$$\begin{aligned}\frac{\partial\langle\tilde{G}_c\rangle}{\partial\phi_c} = \cos\phi_c\Bigg(u_0 + u_1\frac{2M^2 - \nu(1-\nu)^2}{(1-\nu)(1-\nu+M^2)} \\ + u_2\frac{(1+2\nu)(1-\nu)^2 - (7-\nu)M^2}{4(1-\nu)(1-\nu+M^2)}\Bigg)\end{aligned} \tag{2.2.50}$$

where

$$\begin{aligned}u_0 &= -\tan\phi_c F(\frac{1}{2},\frac{1}{2};1;\sin^2\phi_c) + \frac{1}{4}\sin 2\phi_c F(\frac{3}{2},\frac{3}{2};2;\sin^2\phi_c),\\ u_1 &= \tan\phi_c\Bigg((2+\tan^2\phi_c)F(\frac{1}{2},\frac{1}{2};1;\sin^2\phi_c) \\ &\quad + \frac{1}{4}\sin 2\phi_c\tan\phi_c F(\frac{3}{2},\frac{3}{2};2;\sin^2\phi_c)\Bigg),\end{aligned}$$

$$u_2 = \tan\phi_c\left((2+\tan^2\phi_c)F(\frac{3}{2},\frac{1}{2};2;\sin^2\phi_c) + \frac{3}{8}\sin 2\phi_c \tan\phi_c F(\frac{5}{2},\frac{3}{2};3;\sin^2\phi_c)\right). \quad (2.2.51)$$

$\partial\langle\tilde{G}_c\rangle/\partial\phi_c = 0$ corresponds to finding the roots of a polynomial of order 2 in M;

$$M^2 = \frac{(1-\nu)^2(-4u_0 + 4\nu u_1 - (1+2\nu)u_2)}{4(1-\nu)u_0 + 8u_1 - (7-\nu)u_2}. \quad (2.2.52)$$

For a given M value, $\langle\tilde{G}_c\rangle$ (2.2.49) is a function of ϕ_c that may contain extremums (maximum and minimum), the associated ϕ_c values of which are given by (2.2.52). Figures 2.20, 2.21 and 2.22 are plots of $\tilde{G}_c$ (2.2.48), $\langle\tilde{G}_c\rangle$ (2.2.49) and points (ϕ_c, M)(2.2.52) respectively. Gauss's hypergeometric functions are restricted to the nine first terms of their associated series. One can distinguish two types of behaviour depending on M. When $M > 1-\nu \cong 0.67$ for $\nu = 1/3$ (note that $1-\nu$ is obtained from (2.2.52) as ϕ_c tends to zero) $\langle\tilde{G}_c\rangle$ presents a maximum larger than 1, Fig. 2.21. For a given M, $\langle\tilde{G}_c\rangle$ increases continuously with ϕ_c from 1 (value corresponding to $\phi_c = 0$) up to a maximum (about 1.07 at $\phi_c = 41°$ for $M = 3$) and then decreases indefinitely for larger ϕ_c. The value of ϕ_c at which $\langle\tilde{G}_c\rangle$ is maximum is given by (2.2.52) and reported as a function of M in Fig. 2.22. ϕ_c values are located between zero and a value denoted $\phi_c^L \cong 41.30°$ ($\nu = 1/3$). ϕ_c^L cancels the denominator in (2.2.52) and the associated M tends to infinity. From $\phi_c = 0$ and $M = 1-\nu$, Fig. 2.22, M increases first slowly with ϕ_c and then, as ϕ_c approaches ϕ_c^L, very rapidly to infinity. ϕ_c^L corresponds to a vertical asymptote. The behaviour of $\tilde{G}_c$ along the crack front is illustrated in Fig. 2.20a. $\tilde{G}_c$ is maximum on the average fracture plane with a slight asymmetry: points at $x_3/\lambda_c = p/2$ (p odd natural number) correspond to a $\tilde{G}_c$ slightly larger than those for which p is even; $\tilde{G}_c$ increases with ϕ_c on the average fracture plane. $\tilde{G}_c$ is minimum about the wave peaks corresponding to $x_3/\lambda_c = (j+1/2)/2$ (j natural number), Fig. 2.20a, and decreases there as ϕ_c increases. When $M < 1-\nu$, $\langle\tilde{G}_c\rangle$ presents a minimum smaller than 1, Figs. 2.21 and 2.22. For a given M, $\langle\tilde{G}_c\rangle$ decreases with ϕ_c from 1 ($\phi_c = 0$) to a minimum (close to 0.94 at $\phi_c \cong 47.30°$ for $M = 0.2$) and then increases continuously for larger ϕ_c. The ϕ_c values at which $\langle\tilde{G}_c\rangle$ is minimum are given by (2.2.52) and reported as a function of M on Fig. 2.22. These are larger than a minimum $\phi_c^H \cong 47°$. ϕ_c^H cancels the numerator in (2.2.52) (i.e., $M = 0$). From ϕ_c^H, Fig. 2.22, M increases slowly with ϕ_c. The behaviour of $\tilde{G}_c$ along the crack front is given in Fig. 2.20b. $\tilde{G}_c$ takes larger values about the wave peaks and increases there with ϕ_c. $\tilde{G}_c$ takes smaller values about the mean fracture plane and decreases there with ϕ_c. If non-planar crack propagation occurs when $\langle\tilde{G}_c\rangle$ is larger than 1, the condition for crack motion corresponds to $M > 1-\nu$; the sites for the initiation of crack motion are located on the average fracture plane. When $M \leq 1-\nu$, non-planar crack propagation is unexpected.

SEGMENTED CRACK FRONT ($h = 0$, $M_{12} = 0$, $M_{13} = M = \sigma_{23}^a/\sigma_{22}^a$). Here the crack front is segmented as illustrated in Fig. 2.10d and B is taken as origin. ξ is

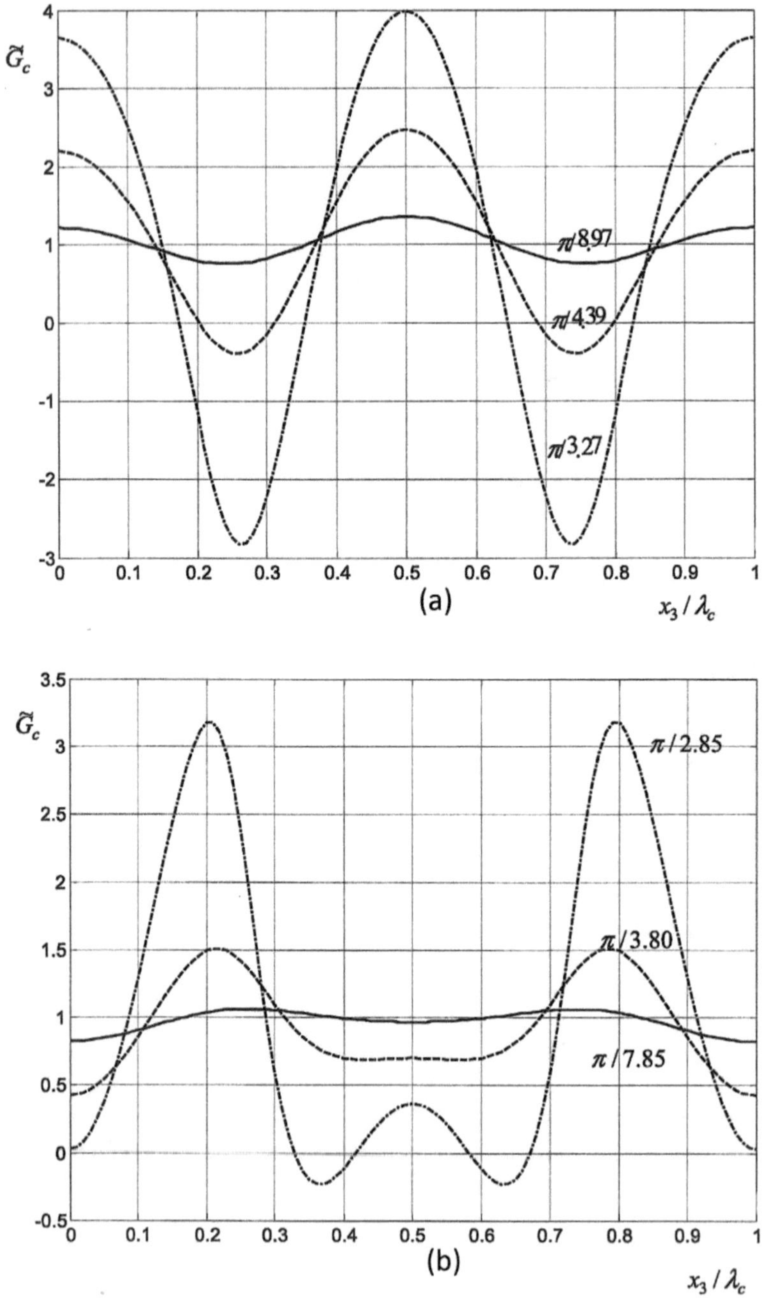

Fig. 2.20 Normalized crack extension force $\tilde{G}_c$(2.2.48) for the sinusoidal crack front as a function of x_3 (reduced by the wavelength λ_c). **a** $M = 3$: the curves are associated with crack profiles $\phi_c = \pi/8.97$, $\pi/4.39$ and $\pi/3.27$. **b** $M = 0.2$: $\phi_c = \pi/2.85$, $\pi/3.80$ and $\pi/7.85$. $\nu = 1/3$

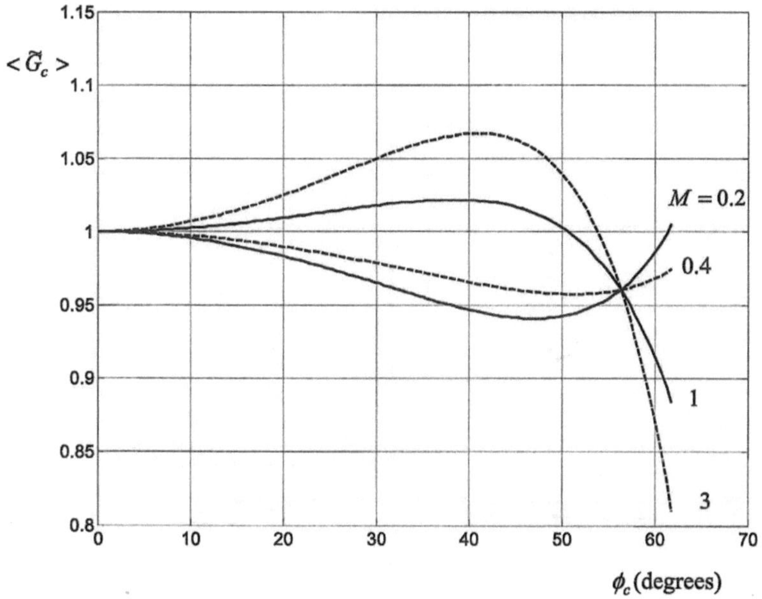

Fig. 2.21 Reduced crack extension force averaged over x_3, $\langle \tilde{G}_c \rangle$ (2.2.49), as a function of crack profile ϕ_c(sinusoidal crack front). Curves with a maximum are associated with $M = 1$ and 3 ($M > 1 - \nu$), those with a minimum with $M = 0.2$ and 0.4 ($M < 1 - \nu$); $\nu = 1/3$. The values of ϕ_c at which $\langle \tilde{G}_c \rangle$ is maximum or minimum are given by (2.2.52) and plotted in Fig. 2.22

then odd and ($2\lambda = \lambda_A + \lambda_B$)-periodical with respect to x_3 where λ_A and λ_B are the projected length along x_3 of facet A and B respectively. ξ is given over a wavelength as

$$\begin{aligned} \xi &= \tan\phi_B x_3, \quad |x_3| \le \lambda_B/2 \\ &= \tan\phi_A(-x_3 + \lambda), \quad x_3 \subset [\lambda_B/2, \lambda_B/2 + \lambda_A]. \end{aligned} \tag{2.2.53}$$

As in the case of a sinusoidal crack front above, we next consider successively the normalized crack extension force, now denoted $\tilde{G}_v$, the average $\langle \tilde{G}_v \rangle = (1/2\lambda)\int_0^{2\lambda} \tilde{G}_v \mathrm{d}x_3$ and ultimately the condition for an extremum of $\langle \tilde{G}_v \rangle$.

Using (2.2.47), $\tilde{G}_v$ on a period reads

$$\begin{aligned} \tilde{G}_v(P_0) &= \cos\varphi_B\left(1 - \frac{2\nu M}{1-\nu+M^2}\tan\varphi_B + \frac{2M^2 - \nu(1-\nu)^2}{(1-\nu)\left(1-\nu+M^2\right)}\tan^2\varphi_B\right), \\ |x_3|\lambda_B/2 &= \cos\varphi_A\left(1 + \frac{2\nu M}{1-\nu+M^2}\tan\varphi_A + \frac{2M^2 - \nu(1-\nu)^2}{(1-\nu)\left(1-\nu+M^2\right)}\tan^2\varphi_A\right), \\ x_3 &\in\,]\lambda_B/2, \lambda_B/2 + \lambda_A[. \end{aligned} \tag{2.2.54}$$

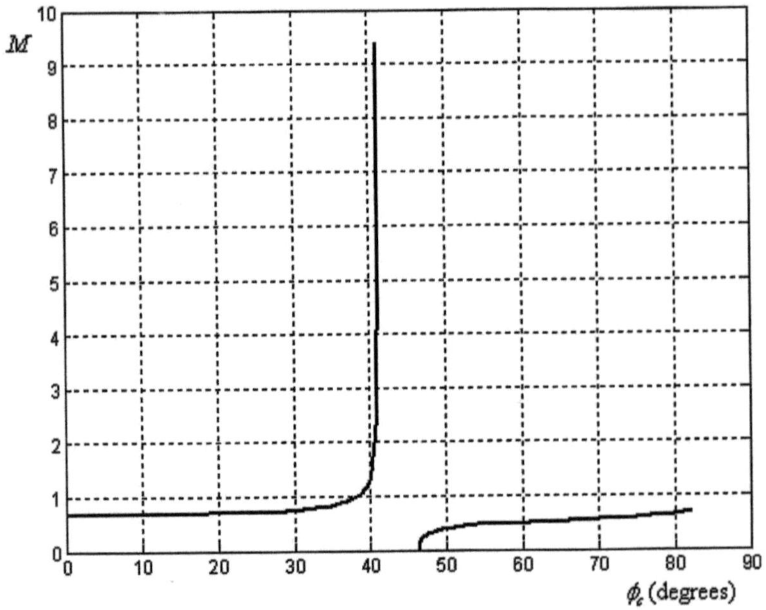

Fig. 2.22 Points (ϕ_c, M)(2.2.52) for which $<\tilde{G}_c>$(2.2.49) is maximum and larger than 1 ($M > 1-\nu \cong 0.67$) and minimum and smaller than 1 ($M < 1-\nu$) when plotted against ϕ_c as in Fig. 2.21; $\nu = 1/3$; sinusoidal crack front

For given M and crack profile (ϕ_A, ϕ_B), (2.2.54) provides the reduced crack extension force at an arbitrary point $P_0(c, \xi, x_3)$ on the segmented crack front. $\tilde{G}_v$ takes constant values on facet A and B.

$$< \tilde{G}_v >= v_0 - v_1 \frac{2\nu M}{1-\nu+M^2} + v_2 \frac{2M^2 - \nu(1-\nu)^2}{(1-\nu)(1-\nu+M^2)} \tag{2.2.55}$$

where

$$\begin{aligned} v_0 &= (1/1+\varepsilon)(\cos\phi_A + \varepsilon\cos\phi_B), \\ v_1 &= (1/1+\varepsilon)(-\sin\phi_A + \varepsilon\sin\phi_B), \\ v_2 &= (1/1+\varepsilon)\tan\phi_A(\sin\phi_A + \sin\phi_B) \end{aligned} \tag{2.2.56}$$

and $\varepsilon = \lambda_B/\lambda_A = \tan\phi_A/\tan\phi_B$.

We restrict ourselves to the condition for an extremum for $\langle\tilde{G}_v\rangle$ with respect to ϕ_A by cancelling $\partial\langle\tilde{G}_v\rangle/\partial\phi_A$. From (2.2.55)

$$\partial\langle\tilde{G}_v\rangle/\partial\phi_A = v_0' - v_1' \frac{2\nu M}{1-\nu+M^2} + v_2' \frac{2M^2 - \nu(1-\nu)^2}{(1-\nu)(1-\nu+M^2)} \tag{2.2.57}$$

where

$$\begin{aligned} v_0' &= \partial v_0/\partial\varphi_A = \partial f_\varepsilon/\partial\varphi_A(\cos\varphi_A - \cos\varphi_B) - f_\varepsilon \sin\varphi_A, \\ v_1' &= \partial v_1/\partial\varphi_A = -\partial f_\varepsilon/\partial\varphi_A(\sin\varphi_A + \sin\varphi_B) - f_\varepsilon \cos\varphi_A, \\ v_2' &= \partial v_2/\partial\varphi_A = \partial f_\varepsilon/\partial\varphi_A \tan\varphi_A(\sin\varphi_A + \sin\varphi_B) \\ &\quad + f_\varepsilon\big(2\sin\varphi_A + \sin\varphi_B - \sin^3\varphi_A\big)/\cos^2\varphi_A \end{aligned} \tag{2.2.58}$$

and

$$f_\varepsilon = 1/1 + \varepsilon, \quad \partial f_\varepsilon/\partial\phi_A = -f_\varepsilon^2/\tan\phi_B\cos^2\phi_A. \tag{2.2.59}$$

$\partial\langle\tilde{G}_v\rangle/\partial\phi_A = 0$ leads to finding the roots of a polynomial of order 2 in M; this gives

$$M = (1-\nu)\frac{v_1'\nu + \sqrt{\nu^2 v_1'^2 + (v_2'\nu - v_0')\big((1-\nu)v_0' + 2v_2'\big)}}{(1-\nu)v_0' + 2v_2'}. \tag{2.2.60}$$

Equation (2.2.60) is the required solution. It leads to positive M values and agrees with extremums observed on $\langle\tilde{G}_v\rangle$ as a function of ϕ_A. Figures 2.23, 2.24 and 2.25 are plots of $\tilde{G}_v$ (2.2.54), $\langle\tilde{G}_v\rangle$ (2.2.55) and points (ϕ_A, M) (2.2.60) for ϕ_B = constant. For a given ϕ_B, one can distinguish different behaviours as a function of M. This description applies to any ϕ_B. Consider $\phi_B = \pi/2.57 = 70°$ to illustrate. Figure 2.25 gives M in the interval [0.3022, 0.3992]. First M increases with ϕ_A from 0.3022 ($\phi_A = 0$) to 0.3992 ($\phi_A \cong 40°$); these (ϕ_A, M) correspond to a maximum larger than 1 for $\langle\tilde{G}_v\rangle$ when plotted against ϕ_A; this is the case for the pair ($\phi_A = 21.15°, M = 0.38$), Fig. 2.24. Then, from maximum $M = 0.3992(\phi_A \cong 40°)$, Fig. 2.25, M decreases with ϕ_A; these (ϕ_A, M) correspond to a minimum for $\langle\tilde{G}_v\rangle$; this is the case for the point ($\phi_A = 59.10°, M = 0.38$), Fig. 2.24. Outside [0.3022, 0.3992], $\langle\tilde{G}_v\rangle$ decreases continuously with ϕ_A for $M < 0.3022$ ($M = 0.3$, Fig. 2.24, for instance) and increases indefinitely with ϕ_A for $M > 0.3992$ ($M = 0.5$, Fig. 2.24). Figure 2.23 shows $\tilde{G}_v$ along the crack front for 3 different M (0.3, 0.38 and 0.5) corresponding to values located below, inside and above [0.3022, 0.3992] with $\phi_A = \pi/7.85 \cong 23°$ and $\phi_B = \pi/2.57 \cong 70°$. $\tilde{G}_v$ takes constant different values, given by Eq. (2.2.54), in x_3-interval$]-\lambda_B/2, \lambda_B/2[$ and$]\lambda_B/2, \lambda_B/2+\lambda_A[$ and increases with M, Fig. 2.23.

In summary, for a given ϕ_B there exists a value M, say M_t (given by (2.2.60) when $\phi_A = 0$), such that:

- When $M > M_t$ there are ϕ_A values limited by zero (as discussed above) corresponding to $\langle\tilde{G}_v\rangle > 1$; under such conditions non-planar crack motion is possible.
- When $M \leq M_t$, $\langle\tilde{G}_v\rangle \leq 1$ for any ϕ_A; non-planar crack propagation is unexpected.

COMPARISON WITH EXPERIMENTS. Focus first on the situation where the average value of the reduced crack extension force is larger than 1 and the crack-front inclination angle (ϕ_A or ϕ_c) takes moderate values (i.e. ranging from zero to 50° approximately). From above, it is seen that this occurs when the ratio (shear/

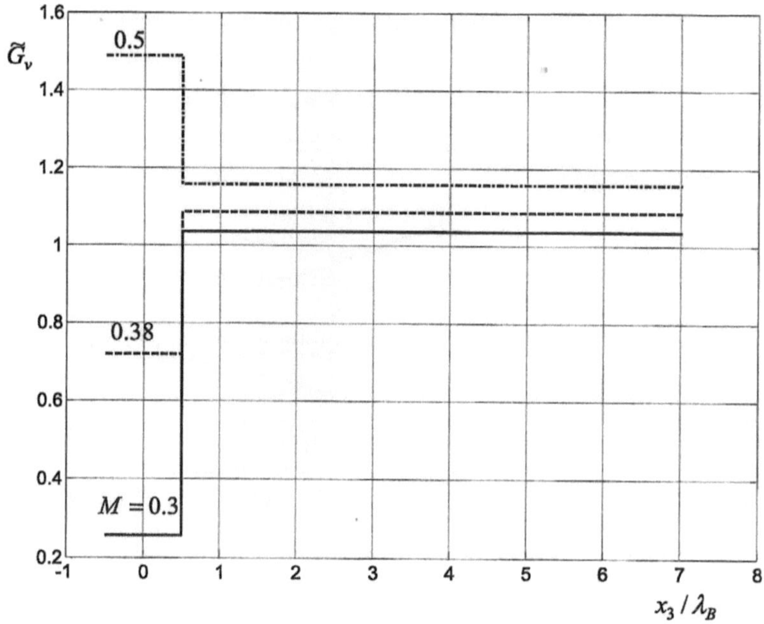

Fig. 2.23 Normalized crack extension force $\tilde{G}_v$ (2.2.5) for the segmented crack front as a function of x_3(reduced by λ_B defined in Fig. 2.10d). The curves correspond to $M = 0.3$, 0.38 and 0.5; they take different constant values in intervals$]-1/2, 1/2[$ and $]1/2, 1/2 + 1/\varepsilon[$ where $\varepsilon = \lambda_B/\lambda_A = \tan\phi_A/\tan\phi_B$. $\phi_A = \pi/7.85$, $\phi_B = \pi/2.57$ and $\nu = 1/3$

tension) M of the applied stresses is larger than $1-\nu \cong 0.67$ ($\nu = 1/3$) in the case of a crack with a sinusoidal front (Figs. 2.21 and 2.22); however when the crack has a segmented front as in Fig. 2.10d, the corresponding M is notably smaller ($M \geq 0.183$ for $\phi_B = \pi/2$ and $M \geq 0.302$ for $\phi_B = \pi/2.57$, Figs. 2.24 and 2.25, for example). If crack motion may occur when the average value of the normalized crack extension force is larger than 1, analysis suggests that non-planar cracks with segmented front with $\phi_B \approx \pi/2$ are favoured at small M. This is in full agreement with experiments (Sommer 1969; Cooke and Pollard 1996). We have reported on Fig. 2.26 (curve 2) the points (ϕ_c, M) (2.2.52) for which $\langle\tilde{G}_c\rangle$ (2.2.49) is maximum and larger than 1 ($M > 1-\nu \cong 0.62$, $\nu = 0.38$). We have also plotted on Fig. 2.26 the points (ϕ_A, M) ($\phi_B =$ constant, Eq. (2.2.60)) for which $\langle\tilde{G}_v\rangle$(2.2.55) is maximum and larger than 1: curve (3) corresponds to $\phi_B = \pi/2$ and curve (4) to $\phi_B = \pi/4$. These additional curves, provided by the present study, fall inside experimental points for relatively small M. It is hazardous to go further into comparison because, for the experimental points given in Fig. 2.26, we have no information about corresponding ϕ_B. A detailed comparison with experiments, with the aim of improving existing models, should take full account of the whole observed shape of the crack front. Several theoretical criteria were proposed in an attempt to explain the dependence of ϕ_A on the ratio M of the applied stresses (Cooke and Pollard 1996): these are based on the crack-tip

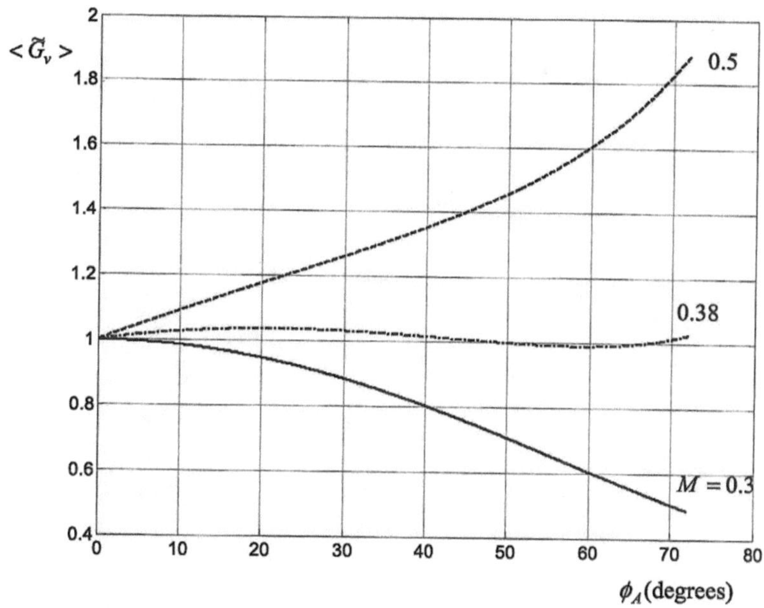

Fig. 2.24 Reduced crack extension force averaged over x_3, $\langle \tilde{G}_v \rangle$ (2.2.55), as a function of the inclination angle ϕ_A of crack-front segment A (segmented crack front); $\phi_B = \pi/2.57 \cong 70°$; $\nu = 1/3$. $M = 0.3$: the corresponding curve decreases continuously as ϕ_A increases. $M = 0.38$: the associated curve displays a local maximum (above 1) and a local minimum (below 1) at ϕ_A values given by (2.2.60) (see also Fig. 2.25, curve $\pi/2.57$). $M = 0.5$: the corresponding curve increases continuously with ϕ_A

stress field of a planar crack subjected to mixed mode I + III loading. They all lead to the same Eq. (2.2.61) for moderate M; a simple explanation is that incipient cracks grow from the parent crack front at twist angle ϕ_A, from the original crack plane, for which they suffer no shear stresses;

$$\phi_A = \frac{1}{2}\tan^{-1}\left(\frac{M}{1/2 - \nu}\right). \tag{2.2.61}$$

Curve (1) corresponds to this equation. Unfortunately, as can be seen from Fig. 2.26, measured angles fall markedly below prediction (2.2.61).

SEGMENTED CRACK FRONT UNDER GENERAL LOADING ($h \neq 0$). The example is given in Fig. 2.10c. This is a non-planar crack with a segmented front (Fig. 2.10d) whose average fracture surface is plane π_0 (illustrated in Fig. 2.10a). The crack front at $x_1 = c$ runs indefinitely in the x_3-direction and is in a x_2x_3-plane. We describe ξ below taking locally B as origin as in Fig. 2.10d. ξ is then odd and ($2\lambda = \lambda_A + \lambda_B$)-periodical with respect to x_3 where λ_A and λ_B(Fig. 2.10d) are the projected length along x_3 of planar facet A and B respectively. ξ is given by:

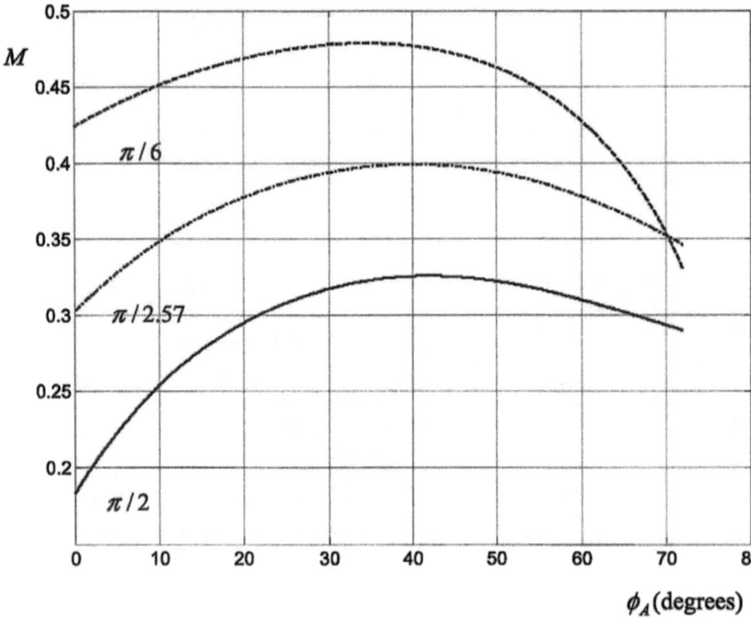

Fig. 2.25 Points (ϕ_A, M) (ϕ_B = constant, Eq. (2.2.60)) for which $\langle \tilde{G}_v \rangle$ (2.2.55) is extremum when plotted against ϕ_A as in Fig. 2.24 with $M = 0.38$; $\nu = 1/3$; segmented crack front. The curves correspond to $\phi_B = \pi/2, \pi/2.57$ and $\pi/6$: in any of these curves, points before (from the left) the maximum correspond to $\langle \tilde{G}_v \rangle$ maximum and larger than 1 and points after the maximum to $\langle \tilde{G}_v \rangle$ minimum and smaller than 1

$$\xi = \tan \phi_B x_3, \quad |x_3| \leq \lambda_B/2$$
$$= \tan \phi_A(-x_3 + \lambda), \quad x_3 \in [\lambda_B/2, \lambda_B/2 + \lambda_A].$$

We assume general loading, write $p_0 = p = \tan\theta$ for simplicity in (2.2.41) for the reduced crack extension force, now denoted $\tilde{G}_v$, and express the spatial average $\langle \tilde{G}_v \rangle$ of $\tilde{G}_v$ defined as $\langle \tilde{G}_v \rangle = (1/2\lambda) \int_0^{2\lambda} \tilde{G}_v \mathrm{d}x_3$. We obtain

$$\langle \tilde{G}_v \rangle = \frac{1}{(1+p^2)^2} \Big\{ (1+p^2)^2 \Big[M_{13}^2 + (1-\nu)\Big((p\nu_A + M_{12})^2 + (1-pM_{12})^2 \Big) \Big] \nu_0$$
$$+ \Big[p^3 \Big(\nu - (2-\nu)p^2 \Big)(p\nu_A + M_{12}) - 2\nu(1+p^2)^2(1-pM_{12}) \Big] M_{13}\nu_1$$
$$+ \Big[\Big\{ p\Big(3 + \nu - p^2(1-\nu)\Big)(1-pM_{12}) + \Big(1 - (1-\nu)p^2 - (2-\nu)p^4\Big)(p\nu_A + M_{12}) \Big\}$$
$$\times (p\nu_A + M_{12}) - \Big(\nu - (2-\nu)p^2\Big)(1-pM_{12})^2 + 2(1+p^2)M_{13}^2/(1-\nu)^2 \Big](1-\nu)\nu_2 \Big\}$$
$$\times \frac{1}{(1-\nu) + (1-\nu)M_{12}^2 + M_{13}^2} \tag{2.2.62}$$

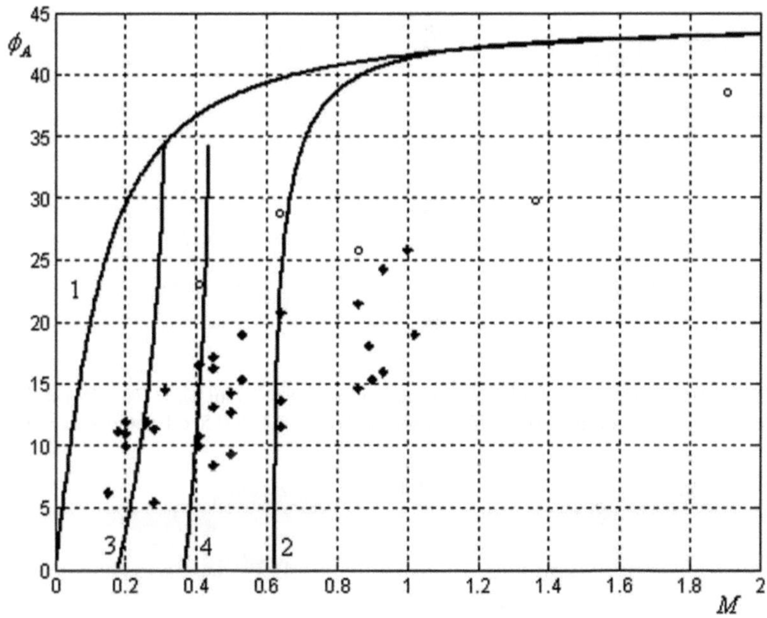

Fig. 2.26 Crack-front inclination angle ϕ_A (degrees) as a function of the ratio M (shear/tension) of the applied stresses in polymethyl methacrylate (PMMA or Plexiglas: symbol $*$) and steel (o) as reproduced from Cooke and Pollard (1996). Curve (1) corresponds to Eq. (2.2.61) in the text, curve (2) to Eq. (2.2.52) with $M > 1 - \nu$; curve (3) and (4) correspond to Eq. (2.2.60) with $\phi_B = \pi/2$ and $\pi/4$ respectively, plots being restricted to (M, ϕ_A) for which $< \tilde{G}_v >$ is maximum and larger than 1. $\nu = 0.38$ (PMMA)

where

$$\begin{aligned}
v_0 &= (1/(p_A + p_B))\left(p_A/\sqrt{1+p^2+p_B^2} + p_B/\sqrt{1+p^2+p_A^2}\right),\\
v_1 &= (p_A p_B/(p_A + p_B))\left(-1/\sqrt{1+p^2+p_A^2} + 1/\sqrt{1+p^2+p_B^2}\right),\\
v_2 &= (p_A p_B/(p_A + p_B))\left(p_A/\sqrt{1+p^2+p_A^2} + p_B/\sqrt{1+p^2+p_B^2}\right)
\end{aligned} \tag{2.2.63}$$

and $p_A = \tan\phi_A$, $p_B = \tan\phi_B$. Hence $\langle\tilde{G}_v\rangle$ is a function of parameters $(p, p_A, p_B; M_{12}, M_{13})$. When ϕ_B (or ϕ_A) equals zero, the crack front is essentially straight parallel to the x_3-direction. The corresponding crack is like planar crack π_0(Fig. 2.10a); under such conditions $\langle\tilde{G}_v\rangle$ (2.2.62) is identical to (2.2.45). $\langle\tilde{G}_v\rangle$ has been reported on Figs. 2.27, 2.28, 2.29 and 2.30 as a function of (θ, ϕ_A) for constant $\phi_B = 70°$; Figs. 2.27 and 2.28 correspond to $M_{13} = 0$ and Figs. 2.29 and 2.30 correspond to general loading, in absence or presence of Poisson effect.

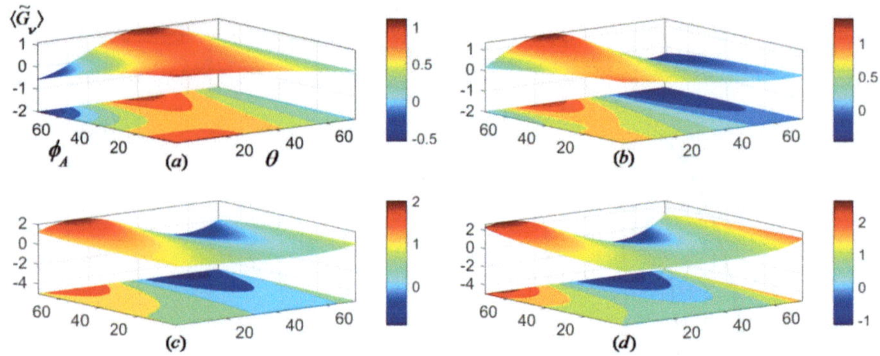

Fig. 2.27 Surfaces $\langle \tilde{G}_v \rangle(\theta, \phi_A)$ with associated contours at constant $\phi_B = 70°$, $M_{13} = 0$ ($\nu_A = 0$ absence of Poisson effect) and four different M_{12}: **a** $M_{12} = 0.1$, **b** 0.5, **c** 1 and **d** 2; $\nu = 1/3$

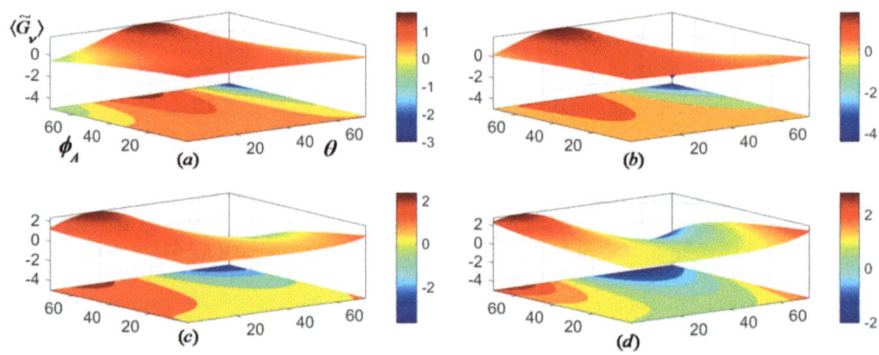

Fig. 2.28 Surfaces $\langle \tilde{G}_v \rangle(\theta, \phi_A)$ with associated contours ($\nu_A \neq 0$ presence of Poisson effect; other conditions as in Fig. 2.27)

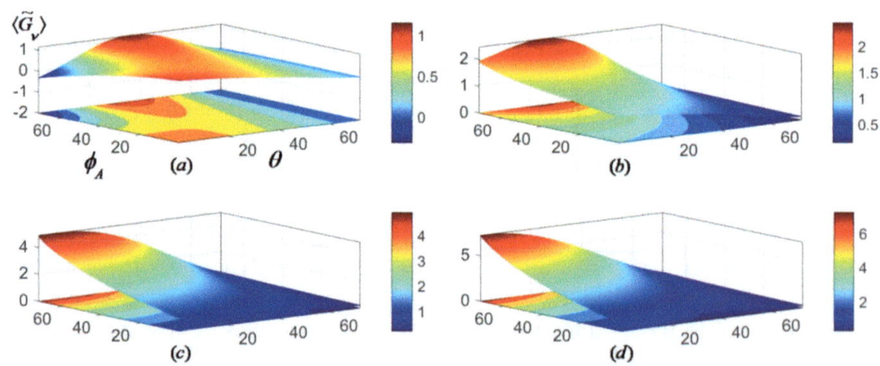

Fig. 2.29 Surfaces $\langle \tilde{G}_v \rangle(\theta, \phi_A)$ with associated contours at constant $\phi_B = 70°$, $M_{12} = 0.2$ ($\nu_A = 0$ absence of Poisson effect) and four different M_{13}: **a** $M_{13} = 0.1$, **b** 0.5, **c** 1 and **d** 2; $\nu = 1/3$

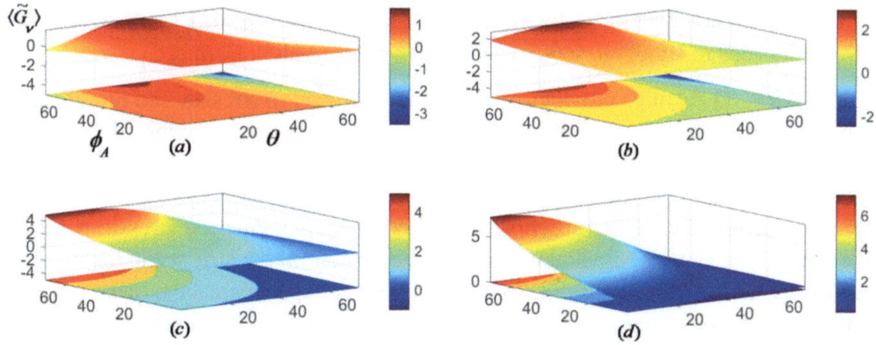

Fig. 2.30 Surfaces $\langle\tilde{G}_v\rangle(\theta, \phi_A)$ with associated contours ($\nu_A \neq 0$ presence of Poisson effect; other conditions as in Fig. 2.29)

DISCUSSION: The normal induced stress $\sigma_{11}^a = -\nu\sigma_{22}^a$ (originating from Poisson effect) increases clearly the crack extension force for the planar crack tilted around x_3 by angle θ_0 with respect to Ox_1x_3. This is observed on Figs. 2.17 and 2.18. For $M_{12} = 3$, from values less than 1 and lower angle θ_0 values, $\tilde{G}$ passes above 1 at $\theta_0 = 40°$ in presence of normal stresses (Fig. 2.18), while in their absence (Fig. 2.17), this is only done from $\theta_0 = 50°$. The value $\tilde{G} = 1$ corresponds to that of the planar crack in Ox_1x_3. Higher values of $\tilde{G}$ (larger than the value 1 as observed under certain conditions, in Fig. 2.18 for example) means that inclined cracks π_0(Fig. 2.10a) are in favourable conditions to move, hence to exist in the broken specimen. The possibility that $\tilde{G}$ may achieved values larger than 1 agrees with experiments showing that when a mode II loading (in addition to mode I) is applied to a planar crack located in Ox_1x_3, the subsequent fracture propagation path departs from x_1x_3 (Cooke and Pollard 1996; Erdogan and Sih 1963; Radon et al. 1977). In general, the normal stresses induced by the Poisson effect increase the force of extension of the cracks. This is very clear in Fig. 2.20 compared with Fig. 2.27 for all values of M_{12}. Figures 2.29 and 2.30 are also in agreement with this conclusion. Next, agreement of (2.2.45) with previous work (Cotterell and Rice 1980) is emphasized. We consider a planar crack in Ox_1x_3 whose shape about one tip is in the form of an infinitesimal kink inclined by angle θ_0 with respect to Ox_1x_3; Poisson's effect is neglected, $M_{13} = 0$ and a straight crack-front parallel to x_3. Cotterell and Rice (1980) have established an expression for the stress intensity factors at the tip of the infinitesimal branch (i.e. their relation (31)) valid for small θ_0. Since their formula does not depend on the shape of the crack, it applies also to an inclined crack π_0 (Fig. 2.10a) as demonstrated below. The stress intensity factors given by these authors can be expanded to second order in θ_0; these combined with Irwin's expression (i.e. the usual plane strain relation) lead to the following formula for the crack extension force:

$$G = \left(1 - \frac{\theta_0^2}{2} - 2\theta_0 M_{12} + \left(1 + \frac{\theta_0^2}{2}\right) M_{12}^2\right) G_0^I$$

It is easy to show that in absence of mode III loading ($M_{13} = 0$), (2.2.45) developed to second order in small $p_0 \equiv \theta_0$ gives an identical G value.

SUMMARY: This study provides expressions of crack-tip stresses and crack extension force for non-planar cracks of arbitrary shapes and macroscopic sizes, corresponding to cracking over large distances under arbitrary external loadings. These incorporate induced normal stresses that result from the contraction, due to Poisson effect, perpendicular to the direction of applied tension. General expressions for the extension force G of the crack are sought that consider all the stresses in the material under load. This makes it possible to best compare to experiments, the idea that crack configurations that maximize G are those observed on broken materials.

2.3 Circular Cracks in Uniform Expansion

2.3.1 *Elliptical Cracks in Isotropic Medium (Quasistatic Condition)*

Model and Crack Analysis

The objective is to investigate the expansion of an elliptical crack in its own plane under arbitrary applied loadings by representing the crack by a continuous distribution of elliptical dislocations. Figure 2.31 serves to illustrate the modelling (Anongba 2021*b*). Is considered infinitely extended isotropic elastic medium containing in its interior an elliptical crack with centre O, with semiaxes a_1 and a_2 along x'_1 and x'_3. The crack is in the $(\pi) = Ox'_1x'_3$ plane. It is made up of three families j ($j = 1, 2$ and 3) of elliptical dislocations with Burgers vectors $\vec{b}_j$ along x'_1. The medium is stressed uniformly at infinity with a tension σ^a_{22} in the vertical x_2-direction and shears σ^a_{21} and σ^a_{23} (parallel to the horizontal x_1x_3-plane) in the x_1 and x_3 directions. The plane (π) $= Ox'_1x'_3$ is inclined around the x_3-direction by the angle θ from Ox_1x_3. The equation of the ellipse is given by:

$$\left(\frac{x'_1}{a_1}\right)^2 + \left(\frac{x'_3}{a_2}\right)^2 = 1. \tag{2.3.1}$$

Distribution functions D_j of the dislocations j are defined such that $D_j(\overline{x}_1)\mathrm{d}\overline{x}_1$ represents the number of dislocations j in a small interval $\mathrm{d}\overline{x}_1$ located at the position $x'_1 = \overline{x}_1$ on the Ox'_1-axis. To that position of the dislocations correspond the position $x'_3 = \overline{x}_3$ on the Ox'_3-axis. The following proportionality relation is assumed:

$$a_1\overline{x}_3 = a_2\overline{x}_1. \tag{2.3.2}$$

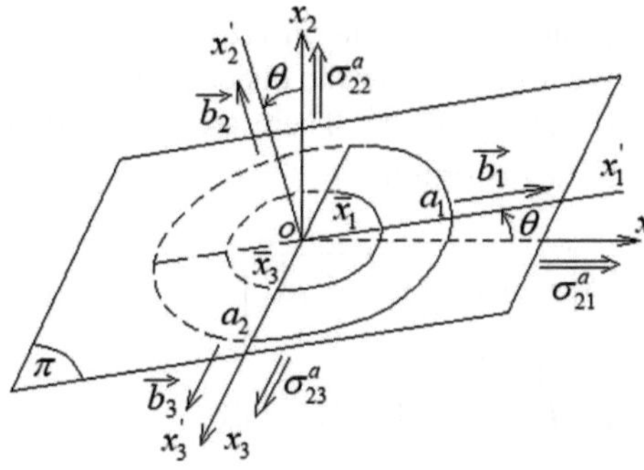

Fig. 2.31 Elliptical crack of centre O and semiaxes a_1 and a_2 along x_1' and x_3'. The applied tension σ_{22}^a acts in the vertical x_2-direction and the shears σ_{21}^a and σ_{23}^a (parallel to the horizontal x_1x_3-plane) act in the x_1 and x_3 directions. The crack is in the plane $(\pi) = Ox_1'x_3'$ tilted around $Ox_3 = Ox_3'$ by an angle θ from Ox_1x_3. The portions of the ellipse above and below the horizontal plane are solid and dashed as also are an elliptical crack dislocation with semiaxes $x_1' = \bar{x}_1$ and $x_3' = \bar{x}_3$. Burgers vectors $\vec{b}_j$ of the crack dislocations (j) are directed along x_j'

Equation (2.3.2) is used in the calculation of the crack extension force G (per unit length of the crack front) as explained below. Hence, it is admitted that $D_j(\bar{x}_1) = D_j(\bar{x}_3)$; this conforms with the result that the relative displacement of the faces of the crack ϕ_j is an ellipse (Green and Sneddon 1950; Eshelby 1957). The three types j (j = 1, 2 and 3) of crack dislocation considered have Burgers vectors $\vec{b}_1 = (b, 0, 0)$, $\vec{b}_2 = (0, b, 0)$ and $\vec{b}_3 = (0, 0, b)$ along x_j'; they spread in the $Ox_1'x_3'$-plane in the form (2.3.1). The plastic distortions $\beta_{ij}^{*(l)}$ associated to the dislocations l ($l = 1$, 2 and 3), are expressed successively.

$$\beta_{21}^{*(1)} = b\delta(x_2')\left(H\left[x_1' + a_1\sqrt{1 - x_3'^2/a_2^2}\right] - H\left[x_1' - a_1\sqrt{1 - x_3'^2/a_2^2}\right]\right), \; |x_3'| \le a_2$$
$$= \frac{ba_1a_2}{(2\pi)^2} \int_{-\infty}^{\infty}\int_{-\infty}^{\infty}\int_{-\infty}^{\infty} \frac{J_1[\eta_2]}{\eta_2} e^{i\vec{k}'.\vec{x}'} \mathrm{d}\vec{k}'; \qquad (2.3.3)$$

$J_1[\eta_2]$ is the Bessel function of the first kind; $\beta_{21}^{*(1)} = 0$ for $|x_3'| \ge a_2$, $\eta_2^2 = a_1^2k_1'^2 + a_2^2k_3'^2$, $\mathrm{d}\vec{k}' = \mathrm{d}k_1'\mathrm{d}k_2'\mathrm{d}k_3'$ and, δ and H are the Dirac delta and Heaviside step functions, respectively. The other components of the plastic distortion are zero. Associated displacements $u_m^{(1)}$ ($m = 1$, 2 and 3) are

$$u_m^{(1)} = -i\frac{ba_1a_2}{(2\pi)^2} \int_{-\infty}^{\infty}\int_{-\infty}^{\infty}\int_{-\infty}^{\infty} \frac{1}{k'^2}\left(k_2'\delta_{m1} + k_1'\delta_{m2} - \frac{k_1'k_2'}{(1-\nu)k'^2}\right.$$
$$\times \left.\left[k_1'\delta_{m1} + k_2'\delta_{m2} + k_3'\delta_{m3}\right]\right)\frac{J_1[\eta_2]}{\eta_2} e^{i\vec{k}'.\vec{x}'} \mathrm{d}\vec{k}'. \qquad (2.3.4)$$

Performing the necessary integrations in (2.3.4) provide the displacements in Cartesian coordinates. We now consider the dislocation $j = 2$ with form (2.3.1) and Burgers vector $\vec{b}_2 = (0, b, 0)$ along x_2'. There are only two non-zero components $\beta_{12}^{*(2)}$ and $\beta_{32}^{*(2)}$ of the plastic distortion.

$$\begin{aligned}
\beta_{12}^{*(2)} &= bH(x_2')\left(\delta\left[x_1' + a_1\sqrt{1 - x_3'^2/a_2^2}\right] - \delta\left[x_1' - a_1\sqrt{1 - x_3'^2/a_2^2}\right]\right)\\
&= \frac{ba_1a_2}{(2\pi)^2}\int_{-\infty}^{\infty}\int_{-\infty}^{\infty}\int_{-\infty}^{\infty}\frac{k_1'}{k_2'}\frac{J_1[\eta_2]}{\eta_2}e^{i\vec{k}'.\vec{x}'}\mathrm{d}\vec{k}';\\
\beta_{32}^{*(2)} &= bH(x_2')\left(\delta\left[x_3' + a_2\sqrt{1 - x_1'^2/a_1^2}\right] - \delta\left[x_3' - a_2\sqrt{1 - x_1'^2/a_1^2}\right]\right)\\
&= \frac{ba_1a_2}{(2\pi)^2}\int_{-\infty}^{\infty}\int_{-\infty}^{\infty}\int_{-\infty}^{\infty}\frac{k_3'}{k_2'}\frac{J_1[\eta_2]}{\eta_2}e^{i\vec{k}'.\vec{x}'}\mathrm{d}\vec{k}'.
\end{aligned} \tag{2.3.5}$$

These give

$$\begin{aligned}
u_m^{(2)} &= -i\frac{ba_1a_2}{(2\pi)^2}\int_{-\infty}^{\infty}\int_{-\infty}^{\infty}\int_{-\infty}^{\infty}\frac{k_1'\delta_{m1} + k_3'\delta_{m3}}{1-\nu}\\
&\quad\times\left(-\frac{\nu}{k'^2} + \frac{k_2'^2}{k'^4}\right)\frac{J_1[\eta_2]}{\eta_2}e^{i\vec{k}'.\vec{x}'}\mathrm{d}\vec{k}',\ m = 1 \text{ and } 3\\
u_2^{(2)} &= -i\frac{ba_1a_2}{(2\pi)^2}\int_{-\infty}^{\infty}\int_{-\infty}^{\infty}\int_{-\infty}^{\infty}\left(\frac{1}{k_2'} - \frac{(2-\nu)k_2'}{(1-\nu)k'^2}\right.\\
&\quad\left.+\frac{k_2'^3}{(1-\nu)k'^4}\right)\frac{J_1[\eta_2]}{\eta_2}e^{i\vec{k}'.\vec{x}'}\mathrm{d}\vec{k}'.
\end{aligned} \tag{2.3.6}$$

We consider the dislocation $j = 3$ with form (2.3.1) and Burgers vector $\vec{b}_3 = (0, 0, b)$ along x_3'. For the plastic distortion, we have

$$\begin{aligned}
\beta_{23}^{*(3)} &= b\delta(x_2')H\left(\left[x_1' + a_1\sqrt{1 - x_3'^2/a_2^2}\right]\left[a_1\sqrt{1 - x_3'^2/a_2^2} - x_1'\right]\right),\ |x_3'| \le a_2\\
&= \frac{ba_1a_2}{(2\pi)^2}\int_{-\infty}^{\infty}\int_{-\infty}^{\infty}\int_{-\infty}^{\infty}\frac{J_1[\eta_2]}{\eta_2}e^{i\vec{k}'.\vec{x}'}\mathrm{d}\vec{k}';
\end{aligned} \tag{2.3.7}$$

$\beta_{23}^{*(3)} = 0$ for $|x_3'| \ge a_2$ and the other components of the plastic distortion are zero. Equations (2.3.7) and (2.3.3) are identical although their Cartesian form is written differently. Associated displacement fields are

$$u_m^{(3)} = -i\frac{ba_1a_2}{(2\pi)^2}\int_{-\infty}^{\infty}\int_{-\infty}^{\infty}\int_{-\infty}^{\infty}\frac{1}{k'^2}\left(k'_3\delta_{m2} + k'_2\delta_{m3} - \frac{k'_2k'_3}{(1-\nu)k'^2}\right.$$
$$\left.\times\left[k'_1\delta_{m1} + k'_2\delta_{m2} + k'_3\delta_{m3}\right]\right)\frac{J_1[\eta_2]}{\eta_2}e^{i\vec{k}'.\vec{x}'}\mathrm{d}\vec{k}'. \tag{2.3.8}$$

The dislocation stress fields $(\sigma)^{(j)}$ can be obtained by differentiating the displacements. The condition that the crack faces remain free from any traction is adopted; this gives

$$\begin{aligned}\overline{\sigma}_{12} &= 0\\ \overline{\sigma}_{22} &= 0\\ \overline{\sigma}_{23} &= 0.\end{aligned} \tag{2.3.9}$$

$\overline{\sigma}_{ij}(\vec{x}')$ stands for the total stress at any point $\vec{x}' = (x'_1, x'_2, x'_3)$ in the medium and is linked to D_j. In (2.3.9), we are concerned with the positions on the crack faces only. We write $\overline{\sigma}_{ij}$ as

$$\overline{\sigma}_{ij} = \sigma_{ij}^A + \sigma_{ij}^{(C)(1)} + \sigma_{ij}^{(C)(2)} + \sigma_{ij}^{(C)(3)}. \tag{2.3.10}$$

$(\sigma)^A$ is the externally applied stress including normal induced stresses from Poisson effect; relatively to x'_i, its components are

$$\begin{aligned}\sigma_{11}^A &= \left[-\nu_A + (1+\nu_A)\sin^2\theta\right]\sigma_{22}^a + \sin 2\theta\sigma_{21}^a\\ \sigma_{21}^A &= (1+\nu_A)\sin 2\theta\sigma_{22}^a/2 + \cos 2\theta\sigma_{21}^a\\ \sigma_{13}^A &= \sin\theta\sigma_{23}^a\\ \sigma_{22}^A &= \left[1 - (1+\nu_A)\sin^2\theta\right]\sigma_{22}^a - \sin 2\theta\sigma_{21}^a\\ \sigma_{23}^A &= \cos\theta\sigma_{23}^a\\ \sigma_{33}^A &= -\nu_A\sigma_{22}^a.\end{aligned} \tag{2.3.11}$$

ν_A is Poisson's ratio ν so denoted to track the contributions of the Poisson's stress.

$$\sigma_{ij}^{(C)(m)}(\vec{x}') = \int_0^{a_1}\sigma_{ij}^{(m)}(\vec{x}'; \overline{x}_1)D_m(\overline{x}_1)\mathrm{d}\overline{x}_1 (m = 1, 2 \text{ and } 3) \tag{2.3.12}$$

$\sigma_{ij}^{(m)}$ is the stress field at $\vec{x}' = (x'_1, x'_2, x'_3)$ due to the elliptical crack dislocation m with position $x'_1 = \overline{x}_1$ along x'_1 in the distribution (Fig. 2.31). Equation (2.3.9) gives three integral equations the resolution of which yields the D_m. The relative displacements ϕ_m of the faces of the crack in the x'_m-direction ($m = 1, 2$ and 3) are obtained by integration from the relation $d\phi_m = -bD_m(x'_1)dx'_1$:

$$\phi_m = \int_{x_1'}^{a_1} bD_m(x_1)\mathrm{d}x_1,\ |x_1'| \le a_1. \tag{2.3.13}$$

From (2.3.10) to (2.3.12), one can obtain the crack-tip stresses. We shall refer to Fig. 2.32 as illustration to write down an expression for G (the crack extension force per unit length of the crack front) under the requirement that the crack loop expands in its own plane. Allow the elliptical crack with semiaxes a_1 and a_2 on the x_1' and x_3' axes to advance steady from a_1 to $a_1 + \delta a_1$ and a_2 to $a_2 + \delta a_2$ but apply forces to the freshly formed surfaces to prevent relative displacements; the energy of the system is unaltered. Now allow these forces to relax to zero so that the crack extends effectively from a_1 to $a_1 + \delta a_1$ and a_2 to $a_2 + \delta a_2$ on the x_1' and x_3' axes. The work done by these forces corresponds to a decrease of the energy of the system that we shall estimate (the energy of the system consists of the elastic energy plus the potential energy of the loading mechanism). We take a position P on the freshly formed surface (Fig. 2.32) to which we attach a surface element $\mathrm{d}s = \left| d\rho \vec{e}_\rho \wedge \overrightarrow{PP'} \right|$ with $\vec{e}_\rho = \overrightarrow{OP}/OP$. The energy change associated with ds is $(\mathrm{d}s \sum_i \overline{\sigma}_{i2} \Delta u^{(i)}/2)$ where $\Delta u^{(i)}$ is the difference in displacement across the lengthened crack just behind its tip, in the x_1'-direction. When the crack advances from $x_1' = a_1$ to $a_1 + \delta a_1$ and $x_3' = a_2$ to $a_2 + \delta a_2$ on the x_1' and x_3' axes, the energy decrease $(-\delta E)$ associated with a surface element

$$\Delta s = \int_{\rho_0}^{\rho_f} \mathrm{d}s \cong \rho_0^2 \left(\sin^2\phi + a_r^2 \cos^2\phi\right) \mathrm{d}\phi \delta a_1 / a_1 \tag{2.3.14}$$

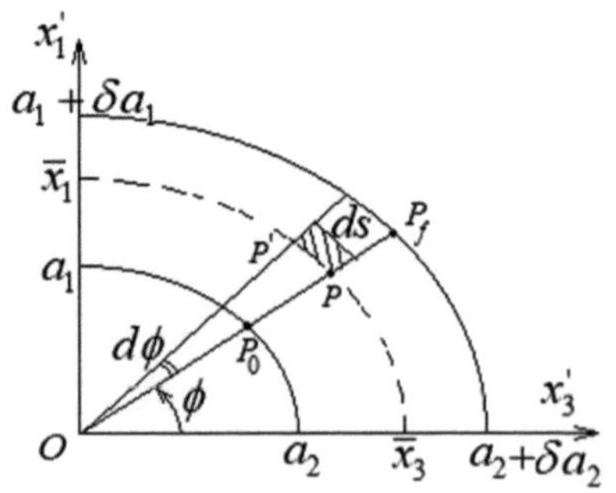

Fig. 2.32 To illustrate the calculation of the crack extension force G. Elliptical crack front allowed to advance steadily from a_1 to $a_1 + \delta a_1$ and a_2 to $a_2 + \delta a_2$ along x_1' and x_3' axes. An arbitrary point P ($x_1' = \rho \sin\phi$, $x_3' = \rho\cos\phi$) on the newly created surface is indicated to which is associated a surface element ds (hatched). At fixed ϕ, as the crack expands, P moves from P_0 ($x_3' = \rho_0 \sin\phi$, $x_3' = \rho_0 \cos\phi$) on the shorter crack to P_f ($x_1' = \rho_f \sin\phi$, $x_3' = \rho_f \cos\phi$) on the lengthened crack front. P' is the position of P on an elliptical crack dislocation after a change dϕ of the polar angle ϕ; the crack dislocation meets the x_1' and x_3' axes at $\overline{x}_1$ and $\overline{x}_3$ such that $a_2\overline{x}_1 = a_1\overline{x}_3$

($a_r = a_1/a_2$ and δa_1 being small and, when used below, will be let to go to zero.) is given as

$$-\delta E = \frac{1}{2}\int_{\rho_0}^{\rho_f}\sum_{i=1}^{3}\overline{\sigma}_{i2}\Delta u^{(i)}\rho\left(\sin^2\phi + a_r^2\cos^2\phi\right)\mathrm{d}\rho\mathrm{d}\phi, \tag{2.3.15}$$

the integration being performed with respect to ρ only. The crack extension force G (per unit length of the crack front) at P_0 is defined as

$$G = \lim_{\delta a_1\to 0} -\delta E/\Delta s. \tag{2.3.16}$$

Next, are given successively dislocation stresses involved in the crack analysis, equilibrium distribution functions D_j of the crack dislocations and corresponding relative displacement ϕ_j of the faces of the crack, crack-tip stresses and crack extension force G per unit length of the crack front, assuming the crack to expand in its own plane.

Stress Fields of Crack Dislocations

Dislocation stresses used in the crack analysis are ($x_2' = 0$):

$$\begin{aligned}
\sigma_{12}^{(1)} &= -C_1a_1a_2\int_{-\pi/2}^{\pi/2}\left[1-\nu\cos^2\psi\right]M^*\mathrm{d}\psi\\
\sigma_{23}^{(1)} &= -\frac{1}{2}\nu C_1a_1a_2\int_{-\pi/2}^{\pi/2}\sin 2\psi M^* d\psi = \sigma_{12}^{(3)}\\
\sigma_{22}^{(1)} &= 0 = \sigma_{22}^{(3)}\\
\sigma_{23}^{(3)} &= -C_1a_1a_2\int_{-\pi/2}^{\pi/2}\left[1-\nu+\nu\cos^2\psi\right]M^*\mathrm{d}\psi\\
\sigma_{22}^{(2)} &= C_1a_1a_2\int_{-\pi/2}^{\pi/2}M^*\mathrm{d}\psi\\
\sigma_{12}^{(2)} &= 0 = \sigma_{23}^{(2)}
\end{aligned} \tag{2.3.17}$$

where $C_1 = \mu b/2\pi(1-\nu)$ and

$$M^* = \begin{cases}(a^{*2}-\Omega^2)^{-3/2} & for\ \Omega^2/a^{*2} < 1\\ \infty & \Omega^2/a^{*2} = 1\\ 0 & \Omega^2/a^{*2} > 1\end{cases}. \tag{2.3.18}$$

$$a^{*2} = a_2^2\left(1 - \left[1 - a_r^2\right]\sin^2\psi\right),\ a_r = a_1/a_2,\ \Omega = \sin\psi x_1' + \cos\psi x_3'.$$

Crack Dislocation Distributions

We use the condition (2.3.9) for traction-free at the crack faces and associated stress quantities (2.3.10–2.3.12) and (2.3.17) at positions $P_C(x_1', x_2' = 0, x_3' = 0)$, $|x_1'| < a_1$, to obtain the following integral equations for the D_j:

$$\sigma_{12}^A + 2C_1 a_r^2 \int_{x_1'}^{a_1} d\overline{x}_1 D_1(\overline{x}_1)\frac{1}{\overline{x}_1}\left(\frac{\nu F(\pi/2, \overline{y}_1)}{\overline{y}_1^2} + \left[\nu - 1 - \nu/\overline{y}_1^2\right]\Pi(\pi/2, \overline{y}_1^2, \overline{y}_1)\right) = 0$$

$$\sigma_{22}^A + 2C_1 a_r^2 \int_{x_1'}^{a_1} d\overline{x}_1 D_2(\overline{x}_1)\frac{1}{\overline{x}_1}\Pi(\pi/2, \overline{y}_1^2, \overline{y}_1) = 0$$

$$\sigma_{23}^A - 2C_1 a_r^2 \int_{x_1'}^{a_1} d\overline{x}_1 D_3(\overline{x}_1)\frac{1}{\overline{x}_1}\left(\frac{\nu F(\pi/2, \overline{y}_1)}{\overline{y}_1^2} + \left[1 - \nu/\overline{y}_1^2\right]\Pi(\pi/2, \overline{y}_1^2, \overline{y}_1)\right) = 0; \quad (2.3.19)$$

$$F(\pi/2, \bar{y}_1) = \int_0^{\pi/2} \frac{d\psi}{\sqrt{1 - \bar{y}_1^2 \sin^2\psi}},\ \Pi(\pi/2, \bar{y}_1^2, \bar{y}_1) = \int_0^{\pi/2} \frac{d\psi}{(1 - \bar{y}_1^2 \sin^2\psi)\sqrt{1 - \bar{y}_1^2 \sin^2\psi}}$$

are complete elliptic integral of first and third kind, respectively; $\overline{y}_1^2 = a_r^2 x_1'^2/\overline{x}_1^2 + 1 - a_r^2$. As P_C moves closer to the crack-tip, (i.e. $x_1' \to a_1$), (2.3.19) becomes

$$\sigma_{12}^A - C_1 \int_{1-\delta\overline{y}_1}^{1} d\overline{y}_1 D_1(\overline{y}_1)\frac{1}{1 - \overline{y}_1} = 0$$

$$\sigma_{22}^A + C_1 \int_{1-\delta\overline{y}_1}^{1} d\overline{y}_1 D_2(\overline{y}_1)\frac{1}{1 - \overline{y}_1} = 0 \quad (2.3.20)$$

$$\sigma_{23}^A - (1 - \nu)C_1 \int_{1-\delta\overline{y}_1}^{1} d\overline{y}_1 D_3(\overline{y}_1)\frac{1}{1 - \overline{y}_1} = 0;$$

$0 < \delta\overline{y}_1 \ll 1$, $\delta\overline{y}_1$ is a finite sufficiently small real, a fraction of unity. Using Muskhelishvili (1992) (his relations (88.8 to 88.11), p. 251), we arrive at

$$D_j \cong \alpha_0\alpha_j\frac{1}{\sqrt{1-y_1'}}; \quad (j = 1,\ 2 \text{ and } 3) \tag{2.3.21}$$

$$y_1'^2 = a_r^2 x_1'^2/a_1^2 + 1 - a_r^2,$$

$$\alpha_j = \frac{1}{\pi C_1}\left(-\sigma_{12}^A\delta_{j1} + \sigma_{22}^A\delta_{j2} - \delta_{j3}\sigma_{23}^A/(1-\nu)\right),$$

$$\alpha_0 = \sqrt{\delta y_1'}$$

$\delta y_1'$ is of the order (or a fraction) of unity. Equation (2.3.21) is the value of the dislocation j distribution closer to the crack tip. The associated relative displacement of the faces of the crack is

$$\phi_j \cong 2b\alpha_0\alpha_j a_1\sqrt{1-y_1'}. \tag{2.3.22}$$

Crack-Tip Stress and Crack Extension Force

To obtain the crack extension force G, it is required to express the crack-tip $\overline{\sigma}_{ij}$ stresses at P (Fig. 2.32) ahead of the shorter crack or equivalently as performed below behind the lengthened crack. The crack-tip stresses may be identified to the following expression

$$\overline{\sigma}_{ij}(P) = \sum_{m=1}^{3}\int_{a_1}^{a_1+\delta a_1}\sigma_{ij}^{(m)}(P;\overline{x}_1')D_m(\overline{x}_1')\mathrm{d}\overline{x}_1', \quad \delta a_1 \ll a_1$$

$$= \sum_{m=1}^{3}\int_{\overline{x}_1}^{a_1+\delta a_1}\sigma_{ij}^{(m)}(P;\overline{x}_1')D_m(\overline{x}_1')\mathrm{d}\overline{x}_1' \equiv \sum_{m=1}^{3}\overline{\sigma}_{ij}^{(m)}. \tag{2.3.23}$$

The lower limit of integration goes from a_1 to $\overline{x}_1$ because, from the stress expressions (2.3.17), the elliptical crack dislocations passing through positions between P_0 and P (Fig. 2.32) contribute nothing. The relation (2.3.2) applies to any crack dislocation. Equation (2.3.21) is used for $D_m(\overline{x}_1')$ where for the lengthened crack $y_1'^2 = a_r^2\overline{x}_1'^2/(a_1+\delta a_1)^2 + 1 - a_r^2$; this applies also to ϕ_m (2.3.22) that will be used for $\Delta u^{(m)}$ (2.3.15) in the calculation of G (2.3.16). We may write in contracted form

$$\delta_{m1}\overline{\sigma}_{12} + \delta_{m2}\overline{\sigma}_{22} + \delta_{m3}\overline{\sigma}_{23} = -2C_1\alpha_0 a_r\sqrt{\frac{\delta a_1}{a_1}}\left(\int_{-\pi/2}^{0} + \int_{0}^{\pi/2}\right)\frac{\mathrm{d}\psi}{\overline{a}^{*3}\left(1-b^{*2}\right)^{3/2}}$$

$$\times \left(\delta_{m1}\left[\alpha_1(1-\nu\cos^2\psi)+\alpha_3\nu\sin 2\psi/2\right]-\delta_{m2}\alpha_2 + \delta_{m3}\left[\alpha_3(1-\nu+\nu\cos^2\psi)+\alpha_1\nu\sin 2\psi/2\right]\right),\ m=1,2 \text{ and } 3 \quad (2.3.24)$$

where $\overline{a}^{*2} = a_r^{-2}\left(1-\left[1-a_r^2\right]\sin^2\psi\right)$ and $b^{*2} = 1-a_r^2+a_r^2\Omega^2/\left(\overline{a}^*[a_1+\delta a_1]\right)^2$. We can write G (2.3.16) at P_0 (Fig. 2.32) as

$$G(P_0) = 4bC_1\alpha_0^2\frac{a_1}{a_r\rho_0}\lim_{\delta a_1\to 0}\frac{1}{2}\left(\int_{-\pi/2}^{0}+\int_{0}^{\pi/2}\right)\frac{d\psi}{\overline{a}^{*2}} \times \left(\alpha_2^2-\alpha_1\left[\alpha_1(1-\nu\cos^2\psi)+\alpha_3\nu\sin 2\psi/2\right] - \alpha_3\left[\alpha_3(1-\nu+\nu\cos^2\psi)+\alpha_1\nu\sin 2\psi/2\right]\right)\int_{\rho_0}^{\rho_f}\frac{d\rho}{\left(1-b^{*2}\right)^{3/2}}. \quad (2.3.25)$$

Next, we specialize the calculation to two positions $P_1(x_1' = a_1, x_3' = 0)$ and $P_2(x_1' = 0, x_3' = a_2)$ on the front of the elliptical crack lying in $Ox_1'x_3'$ with semiaxes a_1 and a_2 along x_1' and x_3'. We obtain $G_1 = G(P_1)$ and $G_2 = G(P_2)$ as

$$G_m = \frac{8\alpha_0^2}{\pi^2}G_0^I\ln\left(\frac{a_1}{\overline{a}_1}\right)\frac{1}{a_r^2}\left[\delta_{m2}+\delta_{m1}/a_r^2\right]\left(\left[1-(1+\nu_A)\sin^2\theta - \sin 2\theta M_{12}\right]^2 - \left[\delta_{m1}+(1-\nu)\delta_{m2}\right]\left((1+\nu_A)\sin 2\theta/2+\cos 2\theta M_{12}\right)^2 - \cos^2\theta M_{13}^2/\left[(1-\nu)\{\delta_{m1}+(1-\nu)\delta_{m2}\}\right]\right); \quad (2.3.26)$$

$M_{12} = \sigma_{21}^a/\sigma_{22}^a$, $M_{13} = \sigma_{23}^a/\sigma_{22}^a$, $G_0^I = K_I^{0^2}(1-\nu^2)/E$, $K_I^0 = \sigma_{22}^a\sqrt{a_1\pi}$.

The quantity in the logarithm is dimensionless, hence $\overline{a}_1$ is introduced with this respect; E is Young's modulus. We defined $\langle G\rangle = (G_1 + G_2)/2$ as an average value of the crack extension force per unit length of the crack front, and

$$\langle\tilde{G}\rangle = \langle G\rangle \Big/ \left(G_0^I\frac{8\alpha_0^2}{\pi^2 a_r^4}\ln(a_1/\overline{a}_1)\right) \quad (2.3.27)$$

as a normalized quantity.

$$2\langle\tilde{G}\rangle = (1+a_r^2)\left[1-(1+\nu_A)\sin^2\theta-\sin 2\theta M_{12}\right]^2 - \left[a_r^2(1-\nu)+1\right]\left((1+\nu_A)\sin 2\theta/2+\cos 2\theta M_{12}\right)^2 - (a_r^2+1-\nu)\cos^2\theta M_{13}^2/(1-\nu)^2; \quad (2.3.28)$$

$$2\langle\tilde{G}\rangle(\theta=0) = 1+a_r^2-\left[1+a_r^2(1-\nu)\right]M_{12}^2-(a_r^2+1-\nu)M_{13}^2/(1-\nu)^2.$$

The condition for an extremum of $\langle \tilde{G} \rangle$ with respect to θ is given by $\partial \langle \tilde{G} \rangle / \partial \theta = 0$. We obtain

$$2(1-\nu)^2 \partial \langle \tilde{G} \rangle / \partial \theta = AM_{12}^2 + BM_{12} + C \tag{2.3.29}$$

Here $A = 2(1-\nu)^2 \left[2 + (2-\nu)a_r^2\right] \sin 4\theta$,

$$B = -2(1-\nu)^2 \left((1-\nu_A)(1+a_r^2)\cos 2\theta + (1+\nu_A)\left[2+(2-\nu)a_r^2\right]\cos 4\theta\right),$$

$$C = -\sin 2\theta \Big((1-\nu)^2(1-\nu_A^2)(1+a_r^2) + (1-\nu)^2(1+\nu_A)^2\left[2+a_r^2(2-\nu)\right]\cos 2\theta + (a_r^2+1-\nu)M_{13}^2\Big)$$

For fixed M_{13}, $\partial \langle \tilde{G} \rangle / \partial \theta = 0$ corresponds to finding the roots of a polynomial of degree 2 in M_{12}. We obtain ($\sigma_{22}^a > 0$ for tension and $\sigma_{22}^a < 0$ for compression)

$$M_{12} = \operatorname{sgn}(\sigma_{22}^a) \frac{-B + \sqrt{\Delta}}{2A}; \tag{2.3.30}$$

$$\Delta \doteq B^2 - 4AC.$$

In Fig. 2.33a, b are exhibited $\langle \tilde{G} \rangle$ (2.3.27) as a function of $M_{12} = \sigma_{21}^a / \sigma_{22}^a$ and θ the crack inclination angle (see Fig. 2.31) at $M_{13} = \sigma_{23}^a / \sigma_{22}^a = 0$. For moderate fixed $|M_{12}| \leq 6$, positive maximums are observed both in tension and compression at angles 53° and 42° in tension and compression, approximately. Figure 2.34 is a plot of the couples (M_{12}, θ) (2.3.30) at which $\langle \tilde{G} \rangle$ exhibits extremums for fixed M_{13}. M_{12} ($\sigma_{22}^a > 0$, tension) increases strongly as θ decreases to zero. The smallest positive values of M_{12} are at about 53°. The region $\theta \leq 53°$(approximately) are associated with positive $\langle \tilde{G} \rangle$ maximums. As θ increases towards 90°, M_{12} becomes negative, decreasing strongly as θ moves closer to 90°. The extremums in these θ regions are presumably minimums in tension. We should be aware that our analysis also cover compression ($\sigma_{22}^a < 0$; $M_{12} < 0$). Results corresponding to $\sigma_{22}^a < 0$ (compression) are obtained by inverting the sign of M_{12} in Fig. 2.34. In this latter situation, as θ increases towards 90°, Poisson's stress $-\nu_A \sigma_{22}^a$ that is tensile becomes more and more effective; this indicates that positive $\langle \tilde{G} \rangle$ maximums exist above 50°.

Discussion

Approximately, one can write from (2.3.26) that $G(P_1) \simeq G(P_2)/a_r^2$ for the two positions $P_1(x_1' = a_1, x_3' = 0)$ and $P_2(x_1' = 0, x_3' = a_2)$ on the front of the crack lying in $Ox_1'x_3'$ with semiaxes a_1 and a_2 along x_1' and x_3'. $a_r = a_1/a_2 < 1$ leads to $G_1 > G_2$; the crack expansion would begin first from P_1 and a_1 will increase towards a_2. Under such conditions, the crack evolves towards a circle ($a_r = 1$). We arrive at the same conclusion when $a_r > 1$. Hence, $\langle G \rangle = (G_1 + G_2)/2$ is valuable as a measure of the crack extension force averaged over the length of the crack front. Another

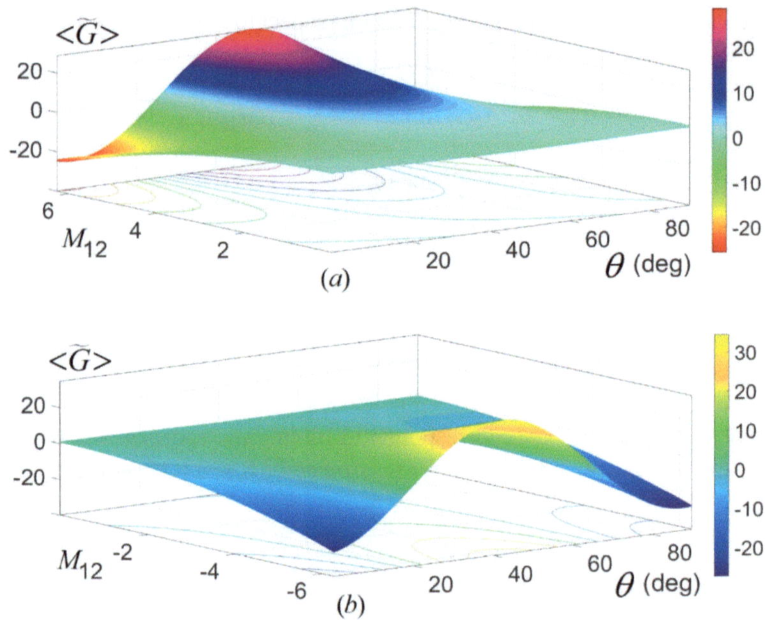

Fig. 2.33 Normalized crack extension force $\langle \tilde{G} \rangle$ (2.3.27) as a function of the reduced applied shearing stress $M_{12} = \sigma^a_{21}/\sigma^a_{22}$ and crack inclination angle θ. **a** $\sigma^a_{22} > 0$ for tension: $\langle \tilde{G} \rangle$ displays positive maximums in the region about $\theta = 53°$ for moderate $M_{12} \le 6$. These maximums increase with M_{12}. **b** $\sigma^a_{22} < 0$ for compression: positive maximums $\langle \tilde{G} \rangle$ are at about $\theta = 42°$ for moderate $|M_{12}| \le 6$. These maximums increase with $|M_{12}|$. $M_{13} = 0$, $\nu = 1/3$ and $a_r = 3/4$

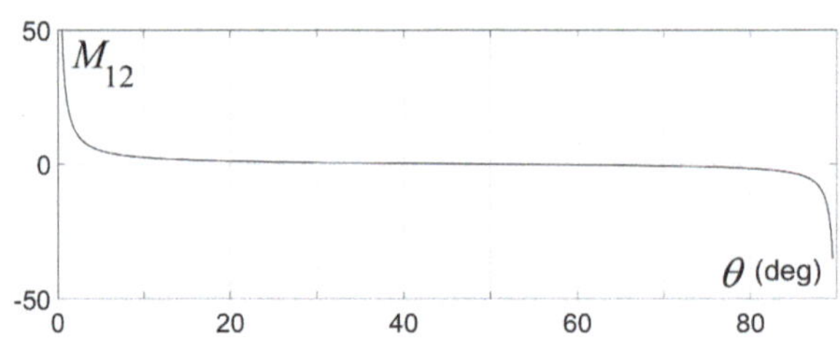

Fig. 2.34 Couples (M_{12}, θ) (2.3.30), $\sigma^a_{22} > 0$ for tension, at which $\langle \tilde{G} \rangle$ (2.3.27) displays extremums when plotted against θ (at fixed M_{13}). To smaller θ correspond higher M_{12}. The positive smallest values of M_{12} are at about 53°. Negative values are further observed as θ moves toward 90°. $M_{13} = 0$, $\nu = 1/3$ and $a_r = 3/4$. Results corresponding to $\sigma^a_{22} < 0$ (compression) are obtained by inverting the sign of M_{12}

property of the crack system comes from the value of $\langle G \rangle$ at $\theta = 0$ (2.3.28). $\langle \tilde{G} \rangle$ ($\theta = 0$) is positive under pure tension ($M_{12} = 0$ and $M_{13} = 0$) and equal to $(1 + a_r^2)$. Positive $\langle \tilde{G} \rangle$ indicates that the expansion of the crack corresponds to a decrease of the energy of the crack system (see Bilby and Eshelby (1968), for example). Negative $\langle \tilde{G} \rangle$ means the contrary. As can be seen from (2.3.28) at $\theta = 0$, the shearing stresses produce negative $\langle \tilde{G} \rangle$ indicating that the crack is unable to expand under pure shears parallel to the plane of the loop. This is also apparent on Fig. 2.34 that displays the couple (M_{12}, θ) at which $\langle \tilde{G} \rangle$ is positive maximum in tension for $\theta < 53°$ approximately. M_{12} increases indefinitely as θ goes to zero. In support of our findings, we shall refer to Eshelby (1957) who investigated the elastic fields due to an ellipsoidal inclusion (a crack is an inclusion at whose boundary there are zero surface tractions). Eshelby defined a parameter γ such that

$$\frac{\text{energy in matrix}}{\text{energy in inclusion}} = \frac{1-\gamma}{\gamma}$$

He showed that γ is about 1 for shearing stresses parallel to the plane of the loop and concluded that there is no accommodation by the matrix of the expansion of the inclusion in its own plane. In the case displayed by Eshelby (1957) (see his relation (5.7), p. 393), the shearing applied stress S makes an angle α with the plane of the loop. An important implication of the observation that an elliptical crack cannot be expanded (in its own plane) under externally applied shearing stresses parallel to its plane may be that the planar elliptical crack is not the correct crack model to provide the right crack configuration that corresponds to the largest decrease of the energy of the system, under arbitrarily applied loadings. A way to get this exact configuration is to begin the crack analysis with a non-planar loop whose front should be locally of arbitrary shape, calculate the average crack extension force $\langle G \rangle$ and look for crack configuration (the expected one) that maximises $\langle G \rangle$. It is noticed for small positive M_{12} that positive $\langle \tilde{G} \rangle$ maximums are observed at $\theta < 53°$ approximately (θ values decrease with increasing M_{12} as in Fig. 2.34). When the crack is infinitely long (under general loading mixed mode $I + II + III$), works exist that predict values of the crack inclination angle θ for positive average G maximums (see Anongba (2010 and 2020 *b*), as examples). Above predictions are for brittle fracture with no slip dislocations (i.e. plasticity). In many materials where there are both plasticity and crack initiation, an overlap is anticipated between brittle fracture prediction and maximum 45° shear direction inclination with respect to the fatigue load direction. However, situations exist where this confusion cannot be made. We shall refer to the work by Zhao et al. (2020) (see their Fig. 5). The average angle of the initial fracture surface was measured as 52° with respect to the [001] far-field loading direction. This angle 52° has nothing to do with a (111)-slip plane because the angle between a [001]-direction and a (111) plane is only 35° in face-centred-cubic metals. This 52° angle observation belongs to the predictions of brittle elastic crack propagation (present analysis or Anongba (2020*b*) as a most general analysis for large cracks). Now assume that the crack inclination angle θ is close to 90° (see Fig. 2.31). Under positive σ_{22}^a no tension stress is applied to the crack loop. However, in compression ($\sigma_{22}^a < 0$), a tension $-\nu_A \sigma_{22}^a$ normal to the crack plane is suffered by the crack. Under such conditions, for sufficiently applied compression, a prediction can be anticipated that the crack will expand. Positive $\langle \tilde{G} \rangle$ maximums should exist for angles θ closer to 90°. Lastly, we assume that $\theta = 0$ and no shearing stress (*i.e.* $M_{12} = 0 = M_{13}$). Assuming expansion of the circular crack in its own plane under pure tension, the critical stress σ_T for crack propagation is given by the Griffith $G = 2\gamma$ condition; this leads with (2.3.27) to

$$\sigma_T = \sqrt{\frac{\pi E \gamma}{4(1-\nu^2)\alpha_0^2 a_1 \ln(a_1/\overline{a}_1)}},\ a_r = 1 \tag{2.3.31}$$

α_0 is of the order or a fraction of 1. Its rigorous derivation requires to solve exactly the integral Eq. (2.3.19) for D_j. Eventually, a value to α_0 may be obtained from the expressions of the relative displacement of the faces of the crack under tension or shear as provided by different methods (see Green and Sneddon (1950), Eshelby (1957), for example). In the case of a finite planar straight-fronted crack under load inside an isotropic medium, $\alpha_0 = 1/\sqrt{2}$. The classical decrease of stress as the crack length increases is present in (2.3.31) except for a coefficient $\ln(a_1/\overline{a}_1)$. a_1 must be larger than $\overline{a}_1$. The physical meaning of $\overline{a}_1$ is wanted. One can speculate that the representation of the elliptical crack by continuous distributions of Volterra dislocations requires a minimum size to the crack loop.

Conclusion

An elliptical crack with center O, inside an infinitely extended isotropic elastic medium, is considered. The medium is stressed uniformly at infinity in tension σ_{22}^a along the vertical x_2-direction and shears σ_{21}^a and σ_{23}^a (parallel to the horizontal x_1x_3-plane) in the x_1 and x_3 directions, respectively. Poisson's normal stresses $-\nu_A\sigma_{22}^a$ acting in the x_1 and x_3 directions are incorporated into the analysis. The crack is in the plane $(\pi) = Ox_1'x_3'$ tilted around $Ox_3 = Ox_3'$ by an angle θ from Ox_1x_3. The objective of the study is to analyze the conditions of expansion of this crack in its own plane. The approach used is to represent the crack by a continuous distribution of three families j ($j = 1, 2$ and 3) of elliptic dislocations of Burgers vectors $\boldsymbol{b}_j$ attached to the crack and oriented in the x_j' directions. The displacement and stress fields of the dislocations are first provided. The method used consists in giving the plastic distortions of the dislocations in their form in Fourier series; then by the superposition of the elastic fields due to plastic distortions of simple sinusoidal shape, one arrives at the elastic fields of the considered dislocations. The distribution functions D_j of the dislocations at equilibrium satisfy individually a singular integral equation whose expression closer to the crack front is of the simple Cauchy type; this makes it possible to give simple mathematical expressions of D_j as well as the associated relative displacement ϕ_j of the faces of the crack, crack-tip stress, and average crack extension force $\langle G\rangle$ per unit length of the crack front (averaged over the crack-front points). The tension stress σ_{22}^a gives a positive contribution to $\langle G\rangle$ suggesting that an expansion of the crack in its own plane is feasible in tension. Both shears σ_{21}^a and σ_{23}^a give a negative contribution to $\langle G\rangle$ when they are parallel to the crack, which indicates that an expansion of the loop is not possible under such conditions. Under general applied loading, $\langle G\rangle$ (θ) exhibits positive maximums in tension $\sigma_{22}^a > 0$ for θ less than 53° (approximately) and in compression $\sigma_{22}^a < 0$ for θ greater than 40° (approximately) the origin of which is associated with the Poisson's stress $-\nu_A\sigma_{22}^a$

which acts in tension on the crack loop. The observation that an elliptic crack cannot be expanded in its own plane under externally applied shearing stresses parallel to the plane of the loop would mean that a small planar elliptical crack is not the right configuration to deal with crack nucleation and initiation in loaded brittle solids under general loading. We would have to begin the crack expansion study with a non-planar loop whose front is locally of arbitrary shape (thinking about an expansion into a Fourier series). Then look for the configuration that corresponds to the largest energy decrease. This is under this configuration that we should apply the Griffith condition $\langle G \rangle_{\max} = 2\gamma$.

2.3.2 *Circular Cracks in Uniform Expansion*

From the results on the elliptical cracks in static condition (Sect. 2.3.1), those for the circular crack of radius R in uniform expansion can be derived as follows. The various physical quantities displayed there (Sect. 2.3.1) depend implicitly on time t through the dependence on time of the circular crack radius $R \equiv a_0 + vt$ (v and a_0 are constants). We take $a_r = a_1/a_2 = 1$ where a_i are the semiaxes of the elliptical crack; $R = a_1$. In the subsonic velocity regime ($v < c_t$), we replace G_0^I in crack extension force expressions (2.3.26 and 2.3.27) by $G^{(I)} = bK_I^2/4\pi C_1^{(I)}$ taken from (2.1.8) (Sect. 2.1); for the transonic velocity regime ($c_t < v < c_l$), $G^{(I)} = bK_I^2/4\pi C_2^{(I)}$ taken from (2.1.15) (Sect. 2.1). From $t = 0$, the crack starts expanding uniformly with increasing time t.

2.4 Rough Conoidal Cracks Growing Uniformly (Isotropic Medium)

2.4.1 *Rough Conoidal Crack (Quasistatic Condition)*

Model and Crack Analysis

Section 2.3.1 (see also Anongba 2021*b*) has shown what follows: assume an elliptical crack of centre O, inside an infinitely extended isotropic elastic medium (Fig. 2.31). The stresses are applied uniformly at infinity with tension σ_{22}^a along x_2 and shears σ_{21}^a and σ_{23}^a along x_1 and x_3. The plane of the crack is tilted around Ox_3 by an angle θ. The tension σ_{22}^a is in general out of the loop plane and provides positive values to $\langle G \rangle$ the average crack extension force G per unit length of the crack front (averaged over all the positions on the crack front). The shears when parallel to the plane of the loop provide negative contributions. These suggest that an elliptical crack is unable to expand in its own plane under a shearing stress that lies in its plane. Shearing stresses provide positive values to $\langle G \rangle$ only when they are inclined with respect to the plane of the crack loop. Under such conditions, the planar elliptic crack is not the right configuration to deal with crack propagation in brittle solids under applied mixed mode $I + II + III$ loading (shearing stresses in the plane of the crack). We would have to start, from the beginning of crack expansion analysis, with a non-planar crack loop whose front should be locally of arbitrary shape. Consequently, it is more appropriate to investigate the expansion of the rough circular crack of mean radius R whose front lies on an edge-on cylinder (radius R and vertical axis). In other words, the front $\xi(P)$ (i.e. the vertical position of P with respect to the average crack front plane) of the rough crack at an arbitrary position P lies on a vertical cylinder and is expanded in the form of a Fourier series. By doing so the following crack problem that we call "the rough conoidal crack under general loading" is the subject of the present study. The rough conoidal crack is represented by a continuous distribution of rough circular crack dislocations. The elastic fields in the fractured medium read as a superposition of the elastic fields of the crack dislocations. Figure 2.35 serves to illustrate the modelling. We consider an infinitely extended isotropic elastic medium containing in its interior a rough conoidal crack (vertex O, axis Ox_2 and height $h(a) = a \tan\theta$) whose basis is in the form of a rough circular line (centre O', average radius $R = a$) displaced vertically from the origin O by $x_2 = h(a)$. This rough crack is made of three families j ($j = 1, 2$ and 3) of rough circular crack dislocations with Burgers vectors $\boldsymbol{b}_j$ along x_j. Figure 2.35 gives an illustration of a crack dislocation of arbitrary shape ξ located at the elevation $h = h(R)$ in the crack dislocation distribution.

The running point P_D along the dislocation is given with respect to (O, x_j) by

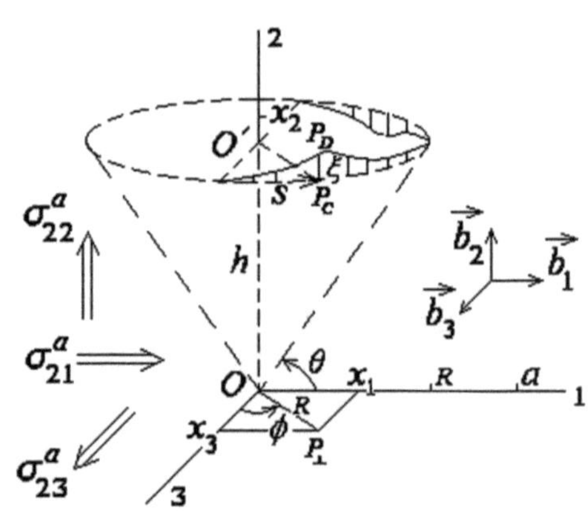

Fig. 2.35 Schematic illustration of a rough conoidal crack dislocation (Burgers vectors $\boldsymbol{b}_j$ along $x_j, j = 1, 2$ and 3) at the elevation $h = h(R)$ in the dislocation distribution of a rough conoidal crack with vertex O, axis Ox_2, average radius and height $R = a$ and $h(a) = a \tan\theta$. In this geometry, applied loadings correspond to tension σ^a_{22} along x_2 and shears σ^a_{21} and σ^a_{23} along x_1 and x_3; see text for details

$$\overrightarrow{OP_D} = \begin{pmatrix} x_1 = R\sin\phi \\ x_2 = f = \sum_n A_n(R,\phi) + h(R) \equiv \xi + h \\ x_3 = R\cos\phi \end{pmatrix} \quad (2.4.1a)$$

assuming that $A_n(R, \phi) = \xi_n(R) \sin \kappa_n |S|$; here n is a positive integer, κ_n and ξ_n (small in magnitude) are wave number and amplitude depending on R. $S = R\phi$ is the curvilinear abscissa of P_C (vertical projection of P_D on x_1x_3); $S \geq 0$ for $0 \leq \phi \leq \pi$ and $S \leq 0$ for $-\pi \leq \phi \leq 0$. We shall impose the condition $\sin \kappa_n |S| = 0$ for $\phi = 0$ and $\pm\pi$ that gives $\kappa_n R \equiv n$ positive integer. In this way, the series form of ξ in $\overrightarrow{OP_D}(2.4.1a)$ reads

$$\xi = \sum_n \xi_n(R) \sin n|\phi| \quad (2.4.1b)$$

$h(R) = R \tan\theta$ and $0 < R \leq a$. Only positions P_D with positive ϕ values are illustrated (Fig. 2.35). We have $OO' = h(R)$; P_C and $P_\uparrow$ are the projections along x_2 of P_D on horizontal planes $x_2 = h(R)$ and $x_2 = 0$; θ is the angle complementary to the half angle at the vertex of average conoidal crack surface $(0 \leq \theta \leq \pi/2)$. The medium is stressed uniformly at infinity with a tension σ_{22}^a in the vertical x_2-direction and shears σ_{21}^a and σ_{23}^a (parallel to the horizontal x_1x_3-plane) in the x_1 and x_3 directions. Distribution functions D_j of the crack dislocations j are defined such that $D_j(R)dR$ represents the number of dislocations j, in a small interval dR location at the position $x_1 = R$ on the Ox_1-axis $(-a < R < a)$; $D_j(R) = D_j(-R)$, by symmetry, this restricts us to positive x_1 values. The elastic fields (displacement $\vec{u}^{(j)}$ and stress $(\sigma)^{(j)}$) of the dislocation j located at $x_1 = R$ and average elevation $x_2 = h(R)$, in the rough conoidal crack dislocation distribution, may be deduced from those due to a rough circular sinusoidal dislocation located on average on Ox_1x_3 $(h = 0)$ and defined by $x_2 = A_n (R, \phi) = \xi_n(R) \sin n|\phi|$(identical Burgers vector $\vec{b}_j$). For the later, the elastic fields at $\vec{x} = (x_1, x_2, x_3)$ are (to linear term in ξ_n)

$$\begin{aligned} \vec{u}^{(j)(n)}(\vec{x}) &= \vec{u}^{(j)(0)} + \vec{u}^{(j)A_n} \\ (\sigma)^{(j)(n)}(\vec{x}) &= (\sigma)^{(j)(0)} + (\sigma)^{(j)A_n} \end{aligned} \quad (2.4.2)$$

$\vec{u}^{(j)(0)}$ and $(\sigma)^{(j)(0)}$ are of zero order corresponding to the fields due to a circular dislocation with centre O at the origin, radius R in the horizontal Ox_1x_3 plane with Burgers vector $\vec{b}_j$; these are available in Sect. 2.3.1; $\vec{u}^{(j)A_n}$ and $(\sigma)^{(j)A_n}$ are oscillating parts involving ξ_n. When the dislocation exhibits shape (2.4.1), the elastic fields take the forms

$$\vec{u}^{(j)}(\vec{x}) = \vec{u}^{(j)(0)}(x_1, y_2, x_3) + \sum_n \vec{u}^{(j)A_n}(x_1, y_2, x_3) \equiv \vec{u}^{(j)(0)} + \vec{u}^{(j)\xi} \quad (2.4.3)$$

$$(\sigma)^{(j)}(\vec{x}) = (\sigma)^{(j)(0)}(x_1, y_2, x_3) + \sum_n (\sigma)^{(j)A_n}(x_1, y_2, x_3) \equiv (\sigma)^{(j)(0)} + (\sigma)^{(j)\xi}$$

$y_2 = x_2 - h$. Elastic fields of rough conoidal crack dislocations (Fig. 2.35) are provided using the method called "Method of Fourier series or integrals" in review works by Mura (1968, 1987): the method consists in writing the plastic distortions associated with the dislocation in Fourier integral series forms. The displacement associated with a single wave (i.e., simple sinusoidal) plastic distortion is available. Those of the crack dislocations are derived by superposition. The three types j ($j = 1$, 2 and 3) of crack dislocation considered have Burgers vectors $\vec{b}_1 = (b, 0, 0)$, $\vec{b}_2 = (0, b, 0)$ and $\vec{b}_3 = (0, 0, b)$ along x_j; they spread on vertical cylinders (axis Ox_2, radii R with $0 < R \leq a$) with running point P_D coordinates given by (2.4.1) (Fig. 2.35). The non-zero plastic distortions $\beta_{rs}^{*(j)}$ associated with a rough circular dislocation j in Ox_1x_3 on average, centred at the origin with radius R and running point coordinates $x_2 = A_n \equiv \xi_n(R)\sin n\phi$(see (2.4.1) for the others) take the forms (at arbitrary spatial position $\vec{x} = (x_1, x_2, x_3)$)

$$\begin{aligned}
\beta_{21}^{*(1)} &= \frac{bR^2\sin^2\phi}{R^2\sin^2\phi + (\partial A_n/\partial\phi)^2}\delta(x_2 - A_n)(H[\rho_1] - H[\rho_2]) = \beta_{23}^{*(3)} \\
\beta_{31}^{*(1)} &= \frac{bR\sin\phi\partial A_n/\partial\phi}{R^2\sin^2\phi + (\partial A_n/\partial\phi)^2}\delta(x_2 - A_n)(H[\rho_1] - H[\rho_2]) = \beta_{33}^{*(3)} \\
\beta_{12}^{*(2)} &= bH(x_2 - A_n)(\delta[\rho_1] - \delta[\rho_2]) \\
\beta_{32}^{*(2)} &= bH(x_2 - A_n)\left(\delta\left[x_3 + \sqrt{R^2 - x_1^2}\right] - \delta\left[x_3 - \sqrt{R^2 - x_1^2}\right]\right)
\end{aligned} \tag{2.4.4}$$

where $\rho_1 = x_1 + \sqrt{R^2 - x_3^2}$, $\rho_2 = x_1 - \sqrt{R^2 - x_3^2}$, $0 \leq \phi \leq \pi$ and $R > 0$. δ and H are the Dirac delta function and Heaviside step function, respectively. To linear terms with respect to A_n and $\partial A_n/\partial\phi$, the plastic distortions of these crack dislocations j read

$$\beta_{rs}^{*(j)} = \beta_{rs}^{*(j)(0)} + \beta_{rs}^{*(j)A_n}. \tag{2.4.5}$$

Here $\beta_{rs}^{*(j)(0)}$ corresponds to the circular dislocation j in Ox_1x_3 with centre O and radius R the elastic fields ($\vec{u}^{(j)(0)}$ and $(\sigma)^{(j)(0)}$) of which are available (Sect. 2.3.1) and used in (2.4.3) to provide the zero order terms with respect to ξ_n in the elastic fields of the rough conoidal crack dislocation j with shape (2.4.1). $\beta_{rs}^{*(j)A_n}$ is the complementary term with ξ_n that is used to provide $\vec{u}^{(j)A_n}$ and $(\sigma)^{(j)A_n}$. Their used Fourier forms are

$$\begin{aligned}
\beta_{21}^{*(1)A_n} &= \frac{bR\xi_n}{4(2\pi)^3}\sum_{m_1,m_3=-\infty}^{\infty}(-1)^{m_1+m_3}i^{m_3+1}b_{21}^{*(1)}(n; m_1, m_3) \\
&\times \int_{-\infty}^{\infty}\int_{-\infty}^{\infty}\int_{-\infty}^{\infty}\frac{k_2}{k_1}J_{m_1}[k_1R]J_{m_3}[k_3R]e^{i\vec{k}.\vec{x}}d\vec{k} = \beta_{23}^{*(3)A_n}
\end{aligned} \tag{2.4.6}$$

$$b_{21}^{*(1)}(n; m_1, m_3) = -8n[(-1)^{m_1+m_3+n} + 1]\left(\frac{m_1 + m_3}{[(m_1 + m_3 - n)^2 - 1][(m_1 + m_3 + n)^2 - 1]} + \frac{m_1 - m_3}{[(m_1 - m_3 - n)^2 - 1][(m_1 - m_3 + n)^2 - 1]}\right)$$
$$\equiv -16n\overline{b}_{21}^{*(1)}(n; m_1, m_3) \tag{2.4.7}$$

m_1 and m_3 are natural numbers taking positive and negative values, $\mathrm{d}\vec{k} = \mathrm{d}k_1\mathrm{d}k_2\mathrm{d}k_3$, J_{m_1} and J_{m_3} are Bessel functions of the first kind with arguments k_1R and k_3R, respectively.

$$\beta_{31}^{*(1)A_n} = \frac{ibR\xi_n\kappa_n}{2(2\pi)^3}\sum_{m_1,m_3=-\infty}^{\infty}(-1)^{m_1+m_3}i^{m_3+1}b_{31}^{*(1)}(n; m_1, m_3) \times \int_{-\infty}^{\infty}\int_{-\infty}^{\infty}\int_{-\infty}^{\infty}\frac{1}{k_1}J_{m_1}[k_1R]J_{m_3}[k_3R]e^{i\vec{k}.\vec{x}}\mathrm{d}\vec{k} = \beta_{33}^{*(3)A_n} \tag{2.4.8}$$

$$b_{31}^{*(1)} = 2[(-1)^{m_1+m_3+n} - 1]\left(\frac{m_1 + m_3}{(m_1 + m_3)^2 - n^2} + \frac{m_1 - m_3}{(m_1 - m_3)^2 - n^2}\right) \equiv -4\overline{b}_{31}^{*(1)}, \tag{2.4.9}$$

$$\beta_{12}^{*(2)A_n} = \frac{bR\xi_n}{4(2\pi)^3}\sum_{m_1,m_3=-\infty}^{\infty}(-1)^{m_1+m_3}i^{m_3+1}b_{21}^{*(1)}(n; m_1, m_3) \times \int_{-\infty}^{\infty}\int_{-\infty}^{\infty}\int_{-\infty}^{\infty}J_{m_1}[k_1R]J_{m_3}[k_3R]e^{i\vec{k}.\vec{x}}\mathrm{d}\vec{k}, \tag{2.4.10}$$

$$\beta_{32}^{*(2)A_n} = \frac{bR\xi_n}{4(2\pi)^3}\sum_{m_1,m_3=-\infty}^{\infty}(-1)^{m_1}i^{m_3}b_{32}^{*(2)}(n; m_1, m_3) \times \int_{-\infty}^{\infty}\int_{-\infty}^{\infty}\int_{-\infty}^{\infty}J_{m_1}[k_1R]J_{m_3}[k_3R]e^{i\vec{k}.\vec{x}}\mathrm{d}\vec{k} \tag{2.4.11}$$

$$b_{32}^{*(2)} = i\big(1 + (-1)^{m_3}\big)[(-1)^{m_1+m_3+n} + 1]i^{m_1+m_3+n}$$
$$\left(\frac{(-1)^{m_3}}{(m_3 - m_1 - n - 1)(m_3 - m_1 - n + 1)} - \frac{(-1)^{m_1}}{(m_3 - m_1 + n - 1)(m_3 - m_1 + n + 1)}\right.$$

$$
\left. + \frac{(-1)^n}{(m_3 + m_1 - n - 1)(m_3 + m_1 - n + 1)} - \frac{1}{(m_3 + m_1 + n - 1)(m_3 + m_1 + n + 1)} \right) \tag{2.4.12}
$$

The associated elastic fields $\vec{u}^{(j)A_n}$ and $(\sigma)^{(j)A_n}$ are readily obtained. It is then possible to write down the elastic fields from (2.4.3) for the rough conoidal crack dislocation at the elevation $x_2 = h(R)$ in the rough crack dislocation distribution as illustrated in Fig. 2.35. The results are displayed in the following. In the crack (Fig. 2.35) modelling, in addition to the externally applied tension σ_{22}^a along x_2 and shears σ_{21}^a and σ_{23}^a along x_1 and x_3, induced Poisson's normal stresses $-\nu_A \sigma_{22}^a$ (ν_A is Poisson's ratio ν so denoted to track the contributions of Poisson's stress) acting in the x_1 and x_3 directions are considered. The traction-free boundary condition at any spatial position of the fracture surface is required:

$$
\begin{cases}
\overline{\sigma}_{12} - \partial f/\partial x_1 \overline{\sigma}_{11} - \partial f/\partial x_3 \overline{\sigma}_{13} = 0 \\
\overline{\sigma}_{22} - \partial f/\partial x_1 \overline{\sigma}_{12} - \partial f/\partial x_3 \overline{\sigma}_{23} = 0 \\
\overline{\sigma}_{23} - \partial f/\partial x_1 \overline{\sigma}_{13} - \partial f/\partial x_3 \overline{\sigma}_{33} = 0
\end{cases} \tag{2.4.13}
$$

$\overline{\sigma}_{ij}$ stands for the total stress at any point P (x_1, x_2, x_3) in the medium and is linked to the rough crack dislocation distribution D_j. Equation (2.4.13) concerns the positions on the crack faces only. $\overline{\sigma}_{ij}$ is written as

$$
\overline{\sigma}_{ij} = \sigma_{ij}^A + \sigma_{ij}^{(C)(1)} + \sigma_{ij}^{(C)(2)} + \sigma_{ij}^{(C)(3)} \tag{2.4.14}
$$

$(\sigma)^A$ is the externally applied stress including normal induced stresses from Poisson effect,

$$
(\sigma)^A = \begin{pmatrix} -\nu_A \sigma_{22}^a & \sigma_{12}^a & 0 \\ \sigma_{12}^a & \sigma_{22}^a & \sigma_{23}^a \\ 0 & \sigma_{23}^a & -\nu_A \sigma_{22}^a \end{pmatrix} \tag{2.4.15}
$$

$$
\sigma_{ij}^{(C)(m)}(\vec{x}) = \int_0^a \sigma_{ij}^{(m)}(\vec{x}; R) D_m(R) \mathrm{d}R \quad (m = 1, 2 \text{ and } 3) \tag{2.4.16}
$$

$\sigma_{ij}^{(m)}$ is the stress field at $\vec{x}$ due to a conoidal rough crack dislocation m with position $x_1 = R$ along x_1 in the distribution. Equation (2.4.13) give three integral equations the resolution of which yields the D_m. The relative displacements ϕ_m of the faces of the crack in the x_m-direction at the position P_D (2.4.1a) in the distribution are obtained by integration from the relation $d\phi_m = -bD_m(R)dR$:

$$\phi_m = \int_R^a bD_m(R')\mathrm{d}R', \quad 0 \leq R \leq a \text{ and } 0 \leq \phi \leq \pi. \tag{2.4.17}$$

Special cracks captured by the modelling are the conoidal smooth crack ($\xi = 0$) and the rough conoidal crack with a sinusoidal front $\xi = \xi_n \sin n|\phi|$ ($n = 1, 2 \ldots$). From (2.4.14) to (2.4.16), one can obtain the crack-tip stresses. The crack extension force G per unit length of the crack front is obtained as follows. The front of the rough crack (Fig. 2.35) with shape (say P_D, (2.4.1)) at (R, ϕ) is assumed to advance quasi statically from ($R = a$, ϕ) (shorter crack) to ($R = a + \delta a$, ϕ) (lengthened crack). At an arbitrary position P_D (x_1, x_2, x_3)(2.4.1) with $a \leq R \leq a + \delta a$, is attached a surface element $\overrightarrow{ds}$ that reads

$$\overrightarrow{\mathrm{d}s} = \begin{pmatrix} \mathrm{d}s_1 \\ \mathrm{d}s_2 \\ \mathrm{d}s_3 \end{pmatrix} = \mathrm{d}R\mathrm{d}\phi \begin{pmatrix} -x_1 \frac{\partial x_2}{\partial R} - \frac{x_3}{R}\frac{\partial x_2}{\partial \phi} \\ R \\ \frac{x_1}{R}\frac{\partial x_2}{\partial \phi} - x_3 \frac{\partial x_2}{\partial R} \end{pmatrix} \tag{2.4.18}$$

The component of the force acting on ds in the x_i-direction is $\overline{\sigma}_{ij}\mathrm{d}s_j$(the summation convention on repeated subscripts applies) where $\overline{\sigma}_{ij}$ are stresses ahead of the shorter or behind the lengthened crack; thus the energy change associated with ds is $\overline{\sigma}_{ij}\mathrm{d}s_j\Delta u^{(i)}/2$(here a summation is also considered over $i = 1$, 2 and 3) where $\Delta u^{(i)}$ is the difference in displacement across the lengthened crack, just behind its tip, in the x_i-direction. When the crack advances from $R = a$ to $a+\delta a$, the energy decrease associated with a surface element

$$\Delta s = \int_{R=a}^{a+\delta a} \mathrm{d}s(P_D) \cong a\delta a\mathrm{d}\phi\sqrt{1 + (\partial x_2(P_0)/\partial R)^2 + (\partial x_2(P_0)/\partial S)^2} \tag{2.4.19}$$

(δa being small and, when used below, will be let to go to zero) is given as

$$-\delta E = \frac{1}{2}\int_a^{a+\delta a} \sum_i \sum_j \overline{\sigma}_{ij}\mathrm{d}s_j\Delta u^{(i)}, \tag{2.4.20}$$

the integration being performed with respect to R; we stress that Δs is the sum of the surface elements ds taken at the various points P_D (2.4.1) as R only changes from a to $a + \delta a$. The crack extension force per unit length of the crack front at $P_0 = P_D$ (with $R = a$, ϕ given) is defined as

$$G(P_0) = \lim_{\delta a \to 0} -\delta E/\Delta s. \tag{2.4.21}$$

Elastic Fields of Rough Conoidal Crack Dislocations

$$\frac{u_1^{(1)A_n}(\vec{x})}{u_0(R)} = \sum_{m_1=1}^{\infty}((-1)^{m_1}-1)\Bigg(\Bigg\| -b_{21}^{*(1)}(n;m_1,m_3=0)\left[2(1-\nu)\frac{\partial^2}{\partial x_2^2}\right.$$

$$\left. -\frac{\partial^2}{\partial x_1^2} - y_2\frac{\partial^3}{\partial x_2 \partial x_1^2}\right] + 4(1-\nu)\kappa_n b_{31}^{*(1)}(n;m_1,m_3=0)\frac{\partial}{\partial x_3}\Bigg\| B_2^{*(1)}(m_1,m_3=0)$$

$$+i2\kappa_n b_{31}^{*(1)}(n;m_1,m_3=0)\frac{\partial^2}{\partial x_3 \partial x_1}\left[1-y_2\frac{\partial}{\partial x_2}\right]B_1^{*(1)}(m_1,m_3=0)$$

$$-\sum_{m_3=1}^{\infty} 2i^{m_3}(-1)^{m_3}\Bigg\{\Bigg\| b_{21}^{*(1)}(n;m_1,m_3)\left[2(1-\nu)\frac{\partial^2}{\partial x_2^2} - \frac{\partial^2}{\partial x_1^2} - y_2\frac{\partial^3}{\partial x_2 \partial x_1^2}\right]$$

$$-4(1-\nu)\kappa_n b_{31}^{*(1)}(n;m_1,m_3)\frac{\partial}{\partial x_3}\Bigg\| B_2^{*(1)}(m_1,m_3) - i2\kappa_n b_{31}^{*(1)}(n;m_1,m_3)$$

$$\times\frac{\partial^2}{\partial x_3 \partial x_1}\left[1-y_2\frac{\partial}{\partial x_2}\right]B_1^{*(1)}(m_1,m_3)\Bigg\}\Bigg),$$

$$(2.4.22)$$

$$\frac{u_2^{(1)A_n}(\vec{x})}{u_0(R)} = \sum_{m_1=1}^{\infty}((-1)^{m_1}-1)\Bigg(i\Bigg\| b_{21}^{*(1)}(n;m_1,m_3=0)\left[2\nu\frac{\partial}{\partial x_2} + y_2\frac{\partial^2}{\partial x_2^2}\right]$$

$$-2\kappa_n b_{31}^{*(1)}(n;m_1,m_3=0)y_2\frac{\partial}{\partial x_3}\Bigg\| B^{*(2)}(m_1,m_3=0)$$

$$+\sum_{m_3=1}^{\infty} 2i^{m_3+1}(-1)^{m_3}\Bigg\| b_{21}^{*(1)}(n;m_1,m_3)\left[2\nu\frac{\partial}{\partial x_2} + y_2\frac{\partial^2}{\partial x_2^2}\right]$$

$$-2\kappa_n b_{31}^{*(1)}(n;m_1,m_3)y_2\frac{\partial}{\partial x_3}\Bigg\| B^{*(2)}(m_1,m_3)\Bigg),$$

$$\frac{u_3^{(1)A_n}(\vec{x})}{u_0(R)} = \sum_{m_1=1}^{\infty}((-1)^{m_1}-1)\Bigg(i\Bigg\{\Bigg\| b_{21}^{*(1)}(n;m_1,m_3=0)\frac{\partial}{\partial x_3}\left[1+y_2\frac{\partial}{\partial x_2}\right]$$

$$+\,4(1-\nu)\kappa_n b_{31}^{*(1)}(n;m_1,m_3=0)\Bigg\| B^{*(2)}(m_1,m_3=0)$$

$$+\,2\kappa_n b_{31}^{*(1)}(n;m_1,m_3=0)\frac{\partial^2}{\partial x_3^2}\left[1-y_2\frac{\partial}{\partial x_2}\right]B_1^{*(1)}(m_1,m_3=0)\Bigg\}$$

$$+\sum_{m_3=1}^{\infty} 2i^{m_3+1}(-1)^{m_3}\Bigg\{\Bigg\| b_{21}^{*(1)}(n;m_1,m_3)\frac{\partial}{\partial x_3}\left[1+y_2\frac{\partial}{\partial x_2}\right]$$

$$+\,4(1-\nu)\kappa_n b_{31}^{*(1)}(n;m_1,m_3)\Bigg\| B^{*(2)}(m_1,m_3) + 2\kappa_n b_{31}^{*(1)}(n;m_1,m_3)$$

$$\times\,\frac{\partial^2}{\partial x_3^2}\left[1-y_2\frac{\partial}{\partial x_2}\right]B_1^{*(1)}(m_1,m_3)\Bigg\}\Bigg);$$

$$B_2^{*(1)}(m_1, m_3) = \int_{-\infty}^{\infty}\int_{-\infty}^{\infty} \frac{e^{-|y_2|\eta_1}}{\eta_1}\frac{J_{m_1}[k_1 R]}{k_1} J_{m_3}[k_3 R] e^{i(k_1 x_1 + k_3 x_3)} \mathrm{d}k_1 \mathrm{d}k_3,$$

$$B_1^{*(1)}(m_1, m_3) = \int_{-\infty}^{\infty}\int_{-\infty}^{\infty} \frac{e^{-|y_2|\eta_1}}{\eta_1^3} J_{m_1}[k_1 R] J_{m_3}[k_3 R] e^{i(k_1 x_1 + k_3 x_3)} \mathrm{d}k_1 \mathrm{d}k_3,$$

$B^{*(2)} = -i\frac{\partial B_2^{*(1)}}{\partial x_1}$, $\eta_1^2 = k_1^2 + k_3^2$, $B_1^{*(1)} = -i\frac{\partial B_3^{*(1)}}{\partial x_1}$(used below) and

$$u_0(R) = \frac{bR\xi_n}{16(2\pi)^2(1-\nu)}. \tag{2.4.23}$$

$$\frac{u_1^{(3)A_n}(\vec{x})}{u_0(R)} = \sum_{m_1=1}^{\infty} \left((-1)^{m_1} - 1\right)\Bigg(i\Bigg\{\Bigg\| b_{21}^{*(1)}(n; m_1, m_3 = 0)\frac{\partial}{\partial x_3}\left[1 + y_2\frac{\partial}{\partial x_2}\right]$$
$$+4\nu\kappa_n b_{31}^{*(1)}(n; m_1, m_3 = 0)\Bigg\| B^{*(2)}(m_1, m_3 = 0) + 2\kappa_n b_{31}^{*(1)}(n; m_1, m_3 = 0)$$
$$\times\frac{\partial^2}{\partial x_3^2}\left[1 - y_2\frac{\partial}{\partial x_2}\right] B_1^{*(1)}(m_1, m_3 = 0)\Bigg\}$$
$$+\sum_{m_3=1}^{\infty} 2i^{m_3+1}(-1)^{m_3}\Bigg\{\Bigg\| b_{21}^{*(1)}(n; m_1, m_3)\frac{\partial}{\partial x_3}\left[1 + y_2\frac{\partial}{\partial x_2}\right] + 4\nu\kappa_n b_{31}^{*(1)}(n; m_1, m_3)\Bigg\|$$
$$B^{*(2)}(m_1, m_3) + 2\kappa_n b_{31}^{*(1)}(n; m_1, m_3) \times\frac{\partial^2}{\partial x_3^2}\left[1 - y_2\frac{\partial}{\partial x_2}\right] B_1^{*(1)}(m_1, m_3)\Bigg\}\Bigg), \tag{2.4.24}$$

$$\frac{u_2^{(3)A_n}(\vec{x})}{u_0(R)} = 2\sum_{m_1=1}^{\infty} \left((-1)^{m_1} - 1\right)\Bigg(\Bigg\| b_{21}^{*(1)}(n; m_1, m_3 = 0)\Bigg[(1+\nu)\frac{\partial}{\partial x_3}$$
$$+ y_2\frac{\partial^2}{\partial x_3 \partial x_2}\Bigg]\frac{\partial}{\partial x_2} + 2\kappa_n b_{31}^{*(1)}(n; m_1, m_3 = 0)\left[\nu\frac{\partial}{\partial x_2} - y_2\frac{\partial^2}{\partial x_3^2}\right]\Bigg\|$$
$$B_2^{*(1)}(m_1, m_3 = 0) + \sum_{m_3=1}^{\infty} 2i^{m_3}(-1)^{m_3}$$
$$\Bigg\{\Bigg\| b_{21}^{*(1)}(n; m_1, m_3)\left[(1+\nu)\frac{\partial}{\partial x_3} + y_2\frac{\partial^2}{\partial x_3 \partial x_2}\right]$$
$$\frac{\partial}{\partial x_2} + 2\kappa_n b_{31}^{*(1)}(n; m_1, m_3)\left[\nu\frac{\partial}{\partial x_2} - y_2\frac{\partial^2}{\partial x_3^2}\right]\Bigg\| B_2^{*(1)}(m_1, m_3)\Bigg\}\Bigg),$$

$$\frac{u_3^{(3)A_n}(\vec{x})}{u_0(R)} = \sum_{m_1=1}^{\infty} \left((-1)^{m_1} - 1\right)\Bigg(\Bigg\| b_{21}^{*(1)}(n; m_1, m_3 = 0)\Bigg[\frac{\partial^2}{\partial x_3^2} - 2(1-\nu)\frac{\partial^2}{\partial x_2^2}$$

$$+ y_2 \frac{\partial^3}{\partial x_2 \partial x_3^2} \Bigg] + 4(2-\nu)\kappa_n b_{31}^{*(1)}(n; m_1, m_3 = 0) \frac{\partial}{\partial x_3} \Bigg\| B_2^{*(1)}(m_1, m_3 = 0)$$

$$+ 4(1-\nu)\kappa_n b_{31}^{*(1)}(n; m_1, m_3 = 0) \Bigg\| \frac{\partial^3}{\partial x_3^3} - y_2 \frac{\partial^4}{\partial x_2 \partial x_3^3} \Bigg\| B_3^{*(1)}(m_1, m_3 = 0)$$

$$+ \sum_{m_3=1}^{\infty} 2i^{m_3}(-1)^{m_3} \Bigg\{ \Bigg\| b_{21}^{*(1)}(n; m_1, m_3) \Bigg[\frac{\partial^2}{\partial x_3^2} - 2(1-\nu) \frac{\partial^2}{\partial x_2^2} + y_2 \frac{\partial^3}{\partial x_2 \partial x_3^2} \Bigg]$$

$$+ 4(2-\nu)\kappa_n b_{31}^{*(1)}(n; m_1, m_3) \frac{\partial}{\partial x_3} \Bigg\| B_2^{*(1)}(m_1, m_3) + 4(1-\nu)\kappa_n b_{31}^{*(1)}(n; m_1, m_3)$$

$$\times \Bigg[\frac{\partial^3}{\partial x_3^3} - y_2 \frac{\partial^4}{\partial x_2 \partial x_3^3} \Bigg] B_3^{*(1)}(m_1, m_3) \Bigg\} \Bigg).$$

$$\frac{u_1^{(3)A_n}(\vec{x})}{u_0(R)} = \sum_{m_1=1}^{\infty} \left((-1)^{m_1} - 1\right) \Bigg(i \Bigg\{ \Bigg\| b_{21}^{*(1)}(n; m_1, m_3 = 0) \frac{\partial}{\partial x_3} \Bigg[1 + y_2 \frac{\partial}{\partial x_2} \Bigg]$$

$$+ 4\nu\kappa_n b_{31}^{*(1)}(n; m_1, m_3 = 0) \Bigg\| B^{*(2)}(m_1, m_3 = 0) + 2\kappa_n b_{31}^{*(1)}(n; m_1, m_3 = 0)$$

$$\times \frac{\partial^2}{\partial x_3^2} \Bigg[1 - y_2 \frac{\partial}{\partial x_2} \Bigg] B_1^{*(1)}(m_1, m_3 = 0) \Bigg\}$$

$$+ \sum_{m_3=1}^{\infty} 2i^{m_3+1}(-1)^{m_3} \Bigg\{ \Bigg\| b_{21}^{*(1)}(n; m_1, m_3) \frac{\partial}{\partial x_3} \Bigg[1 + y_2 \frac{\partial}{\partial x_2} \Bigg]$$

$$+ 4\nu\kappa_n b_{31}^{*(1)}(n; m_1, m_3) \Bigg\| B^{*(2)}(m_1, m_3) + 2\kappa_n b_{31}^{*(1)}(n; m_1, m_3)$$

$$\times \frac{\partial^2}{\partial x_3^2} \Bigg[1 - y_2 \frac{\partial}{\partial x_2} \Bigg] B_1^{*(1)}(m_1, m_3) \Bigg\} \Bigg), \tag{2.4.25}$$

$$u_2^{(2)A_n}(\vec{x}) = u_0(R) \Bigg(- \sum_{m_3=1}^{\infty} i^{m_3} \Big\{ 2b_{32}^{*(2)}(n; m_1 = 0, m_3)$$

$$+ \Delta b_{32}^{*(2)}(n; m_1 = 0, m_3) \Big\} \times \left((1 - 2\nu - y_2 \frac{\partial}{\partial x_2} \right) \frac{\partial}{\partial x_3}$$

$$B^{*(2)}(m_1 = 0, m_3) + \sum_{m_1=1}^{\infty} (-1)^{m_1+1} \| i \big[1 + (-1)^{m_1+1} \big]$$

$$\times b_{21}^{*(1)}(n; m_1, m_3 = 0) \Bigg[1 - 2\nu - y_2 \frac{\partial}{\partial x_2} \Bigg] \frac{\partial}{\partial x_1} B^{*(2)}$$

$$(m_1, m_3 = 0) + \sum_{m_3=1}^{\infty} i^{m_3+1} \Bigg\| \Bigg[1 - 2\nu - y_2 \frac{\partial}{\partial x_2} \Bigg] \{ 2(-1)^{m_3}$$

$$\Bigg\{ \left(1 + (-1)^{m_1+1}\right) b_{21}^{*(1)}(n; m_1, m_3) \frac{\partial}{\partial x_1} - i \Big[2\left(1 + (-1)^{m_1}\right) b_{32}^{*(2)}$$

$$(n; m_1, m_3) + (-1)^{m_1} \Delta b_{32}^{*(2)}(n; -m_1, m_3) + \Delta b_{32}^{*(2)}(n; m_1, m_3) \Big]$$

$$\left.\frac{\partial}{\partial x_3}\right\}\left\|B^{*(2)}(m_1,m_3)\right\|\Bigg),$$

$$\frac{u_3^{(3)A_n}(\vec{x})}{u_0(R)} = \sum_{m_1=1}^{\infty}((-1)^{m_1}-1)\Bigg(\left\|b_{21}^{*(1)}(n;m_1,m_3=0)\left[\frac{\partial^2}{\partial x_3^2}-2(1-\nu)\frac{\partial^2}{\partial x_2^2}\right.\right.$$

$$\left.+\,y_2\frac{\partial^3}{\partial x_2\partial x_3^2}\right]+4(2-\nu)\kappa_n b_{31}^{*(1)}(n;m_1,m_3=0)\frac{\partial}{\partial x_3}\left\|B_2^{*(1)}(m_1,m_3=0)\right.$$

$$+\,4(1-\nu)\kappa_n b_{31}^{*(1)}(n;m_1,m_3=0)\left\|\frac{\partial^3}{\partial x_3^3}-y_2\frac{\partial^4}{\partial x_2\partial x_3^3}\right\|B_3^{*(1)}(m_1,m_3=0)$$

$$+\sum_{m_3=1}^{\infty}2i^{m_3}(-1)^{m_3}\Bigg\{\left\|b_{21}^{*(1)}(n;m_1,m_3)\left[\frac{\partial^2}{\partial x_3^2}-2(1-\nu)\frac{\partial^2}{\partial x_2^2}+y_2\frac{\partial^3}{\partial x_2\partial x_3^2}\right]\right.$$

$$+\,4(2-\nu)\kappa_n b_{31}^{*(1)}(n;m_1,m_3)\frac{\partial}{\partial x_3}\left\|B_2^{*(1)}(m_1,m_3)\right.$$

$$+\,4(1-\nu)\kappa_n b_{31}^{*(1)}(n;m_1,m_3)\times\left[\frac{\partial^3}{\partial x_3^3}-y_2\frac{\partial^4}{\partial x_2\partial x_3^3}\right]B_3^{*(1)}(m_1,m_3)\Bigg\}\Bigg)$$

$$b_{32}^{*(2)}(n;m_1,m_3)-b_{32}^{*(2)}(n;m_1,-m_3)\equiv-\Delta b_{32}^{*(2)}(n;m_1,m_3). \tag{2.4.26}$$

$$\sigma_{ii}^{(m)A_n} = \frac{2\mu}{1-2\nu}\Bigg([\delta_{i1}(1-\nu)+\nu(\delta_{i2}+\delta_{i3})]\frac{\partial u_1^{(m)A_n}}{\partial x_1}$$

$$+[\delta_{i2}(1-\nu)+\nu(\delta_{i1}+\delta_{i3})]\frac{\partial u_2^{(m)A_n}}{\partial x_2}+[\delta_{i3}(1-\nu)+\nu(\delta_{i1}+\delta_{i2})]\frac{\partial u_3^{(m)A_n}}{\partial x_3}\Bigg),$$

$$\sigma_{ij}^{(m)A_n} = \mu\left(\frac{\partial u_i^{(m)A_n}}{\partial x_j}+\frac{\partial u_j^{(m)A_n}}{\partial x_i}\right),\ i\neq j; \tag{2.4.27}$$

Conoidal Crack Dislocation Distributions

Assume first that the crack (Fig. 2.35) is smooth ($\xi = 0$) and write the traction-free boundary condition (2.4.13) for positions P_C ($x_1 = R, x_2 = f = R\tan\theta, x_3 = 0$) with $x_3 = 0$. Hence in (2.4.13), $\partial f/\partial x_1 = \tan\theta$, $\partial f/\partial x_3 = x_3\tan\theta/R = 0$ and $0 < R \leq a$. Denoting the associated dislocation distributions by $D_m^{(0)}$,

$$\sigma_{21}^{a} + \tan\theta\nu_{A}\sigma_{22}^{a} + \int_{0}^{a}\left(\sigma_{21}^{(1)(0)}(P_C;\overline{x}_1) - \tan\theta\sigma_{11}^{(1)(0)}(P_C;\overline{x}_1)\right)D_1^{(0)}(\overline{x}_1)\mathrm{d}\overline{x}_1$$
$$+\int_{0}^{a}\left(\sigma_{21}^{(2)(0)}(P_C;\overline{x}_1) - \tan\theta\sigma_{11}^{(2)(0)}(P_C;\overline{x}_1)\right)D_2^{(0)}(\overline{x}_1)\mathrm{d}\overline{x}_1 = 0,$$
$$\sigma_{22}^{a} + \tan\theta\sigma_{21}^{a} + \int_{0}^{a}\left(\sigma_{22}^{(1)(0)}(P_C;\overline{x}_1) - \tan\theta\sigma_{21}^{(1)(0)}(P_C;\overline{x}_1)\right)D_1^{(0)}(\overline{x}_1)\mathrm{d}\overline{x}_1$$
$$+\int_{0}^{a}\left(\sigma_{22}^{(2)(0)}(P_C;\overline{x}_1) - \tan\theta\sigma_{21}^{(2)(0)}(P_C;\overline{x}_1)\right)D_2^{(0)}(\overline{x}_1)\mathrm{d}\overline{x}_1 = 0,$$
$$\sigma_{23}^{a} + \int_{0}^{a}\left(\sigma_{32}^{(3)(0)}(P_C;\overline{x}_1) - \tan\theta\sigma_{31}^{(3)(0)}(P_C;\overline{x}_1)\right)D_3^{(0)}(\overline{x}_1)\mathrm{d}\overline{x}_1 = 0.$$
(2.4.28)

Stress terms in (2.4.28) are given in *Appendix A*. Relations (2.4.28) consist of three singular integral equations the resolution of which provides the $D_m^{(0)}$ that depend on θ (see Fig. 2.35 for θ). $D_m^{(0)}$ may be used to approximate D_m as given by (2.4.13). Similar approximations have been used in previous works (Anongba 2008, 2009, 2010, 2018, 2020) to provide expressions for the crack extension force G when the crack fronts run indefinitely along the x_3-direction and spread in the plane perpendicular to the x_1 fracture propagation direction. For $\theta = 0$, $D_m^{(0)}(\theta = 0)$ correspond to the circular crack dislocation distributions that are available from (Sect. 2.3.1) and will be used in the following.

Crack-Tip Stresses

The crack-tip stresses are required for the calculation of the crack extension force G. At an arbitrary position P (R, ϕ) (given by (2.4.1) with R between $R = a$ (shorter crack) and $R = a + \delta a$ (lengthened crack), $\delta a \ll a$) the crack-tip stresses separate into non-oscillatory and oscillatory terms as in (2.4.3). The formers are identified to the following formulae

$$\overline{\sigma}_{ij}^{(0)}(P) \equiv \sum_{m=1}^{3}\int_{a}^{a+\delta a}\sigma_{ij}^{(m)(0)}(P;\overline{x}_1)D_m(\overline{x}_1)\mathrm{d}\overline{x}_1 \equiv \sum_{m=1}^{3}\overline{\sigma}_{ij}^{(m)(0)} \quad (m = 1, 2 \text{ and } 3)$$
(2.4.29a)

in a similar way as in Sect. 2.3.1; the oscillatory parts are given by

$$\overline{\sigma}^{\xi}_{ij}(P) \equiv \sum_{n} \sum_{m=1}^{3} \int_{a-\delta a}^{a} \sigma_{ij}^{(m)A_n}(P;\overline{x}_1)D_m(\overline{x}_1)\mathrm{d}\overline{x}_1 \tag{2.4.29b}$$

with following notations $\overline{\sigma}^{\xi}_{ij}(P) = \sum_n \sum_{m=1}^{3} \overline{\sigma}_{ij}^{(m)A_n} = \sum_n \overline{\sigma}_{ij}^{A_n} = \sum_{m=1}^{3} \overline{\sigma}_{ij}^{(m)\xi}$. These stress expressions mean that only those crack dislocations located about the crack front will contribute significantly to the stress at P; any other contribution becomes negligible for a sufficiently small value of δa. $\sigma_{ij}^{(m)(0)}(P;\overline{x}_1)$ is the non-oscillatory part of the stress at P due to a crack dislocation at an average elevation $\overline{h} = \overline{x}_1 \tan\theta$ (refer to Fig. 2.35); this corresponds to the stress of a circular dislocation. $\sigma_{ij}^{(m)A_n}(P;\overline{x}_1)$ is the oscillatory part of the stress at P due to a rough conoidal crack dislocation with form $A_n(\overline{x}_1,\phi)$ (refer to (2.4.1a)) at an average elevation $\overline{h}$. D_m is obtained from (2.4.13); $D_m^{(0)}$(2.4.28) may be used instead, writing

$$\overline{\sigma}_{ij}^{(m)(0)}(P) = \int_{a}^{a+\delta a} \sigma_{ij}^{(m)(0)}(P;\overline{x}_1)D_m^{(0)}(\overline{x}_1)\mathrm{d}\overline{x}_1$$

$$\overline{\sigma}_{ij}^{(m)A_n}(P) = \int_{a-\delta a}^{a} \sigma_{ij}^{(m)A_n}(P;\overline{x}_1)D_m^{(0)}(\overline{x}_1)\mathrm{d}\overline{x}_1 \cong \sigma_{ij}^{(m)A_n}(P;a)\int_{a-\delta a}^{a} D_m^{(0)}(\overline{x}_1)\mathrm{d}\overline{x}_1. \tag{2.4.30}$$

δa ($\ll a$) is small and will be made to go to zero in the calculation of the crack extension force. $\sigma_{ij}^{(m)(0)}(P;\overline{x}_1)$ and $\sigma_{ij}^{(m)A_n}(P;\overline{x}_1)$ are given in *Appendix B*.

Crack Extension Force for the Rough Circular Crack ($\theta = 0$)

The crack extension force G (per unit length of the crack front) at an arbitrary position P_0 on the crack front is given by (2.4.21). For $\Delta u^{(i)}$ (2.4.20), we use (2.4.17) in which $D_m(\theta)$ is approximated by $D_m^{(0)}(\theta)$ solution of the singular integral Eqs. (2.4.28). In the present work, G using $D_m^{(0)}(\theta = 0)$ is provided. Hence, this corresponds to a circular rough crack with average radius $R = a$ in the Ox_1x_3-plane. It is shown that only the non-oscillatory parts of the crack-tip stresses $\overline{\sigma}_{ij}^{(0)}$ (2.4.29a) contribute non-zero values to G. The oscillatory terms (2.4.29b) contribute nothing because they don't contain the appropriate singularity (of the Cauchy type in the crack dislocation stress fields) closer to the crack-tip. The rough circular crack extension force per unit length of the crack front shall be denoted $G_{RC}(P_0)$ at $P_0 \equiv P_D$ (1) with $h(a) = 0$ and $R = a$. We provide an associated normalized quantity $\tilde{G}_{RC}(P_0)$ defined as

$$\tilde{G}_{RC}(P_0) = G_{RC}(P_0)/G_C^{(I)},$$
$$G_C^{(I)} = \frac{8\alpha_0^2}{\pi^2}\ln\left(\frac{a}{\overline{a}}\right)G_0^I; \tag{2.4.31}$$

$G_0^I = K_I^{0^2}(1-\nu^2)/E, K_I^0 = \sigma_{22}^a\sqrt{a\pi}$. The quantity in the logarithm is dimensionless, hence $\overline{a}$ is introduced with this respect; E is Young's modulus. $G_C^{(I)}$ (Sect. 2.3.1) represents a value of the crack extension force for the circular crack (centre O, radius a) in Ox_1x_3 under applied mode I loading. α_0 is a constant (Sect. 2.3.1). We write ($M_{12} = \sigma_{21}^a/\sigma_{22}^a, M_{13} = \sigma_{23}^a/\sigma_{22}^a$)

$$\tilde{G}_{RC}(P_0) = \sum_{j=1}^{3}\left(\tilde{G}_{1j}^{(1)(0)} + \tilde{G}_{2j}^{(2)(0)} + \tilde{G}_{3j}^{(3)(0)}\right); \tag{2.4.32a}$$

$$\begin{aligned}
\tilde{G}_{11}^{(1)(0)} &= \frac{(2\ln 2-1)M_{12}}{8\ln(a/\bar{a})}\Bigg(2(1-2\nu)\sin^2\phi W_1^*(a,\phi) + \frac{\xi(a,\phi)}{a}\{6W_1^*(a,\phi)\\
&\times\left[-\sin\phi(1-4\cos^2\phi)M_{12} + \cos\phi(1-4\sin^2\phi)M_{13}/(1-\nu)\right]\\
&- LN_1^*(a,\phi)\left[\sin\phi(3-\cos^2\phi)M_{12} + \cos\phi(2\nu+\sin^2\phi)M_{13}/(1-\nu)\right]\}\Bigg),\\
\tilde{G}_{12}^{(1)(0)} &= -\frac{(2\ln 2-1)M_{12}}{4\ln(a/\bar{a})}\big(\overline{W}(a,\phi)\big[(1-\nu\cos^2\phi)M_{12}\\
&+ \nu\sin 2\phi M_{13}/(2(1-\nu))\big] + (\xi(a,\phi)/a)(1-\nu)\sin\phi\overline{LN}(a,\phi)\big),\\
\tilde{G}_{13}^{(1)(0)} &= -\frac{(2\ln 2-1)M_{12}}{8\ln(a/\bar{a})}\Bigg((1-2\nu)\sin 2\phi W_2^*(a,\phi) + \frac{\xi(a,\phi)}{a}\{6W_2^*(a,\phi)\\
&\times\left[\cos\phi(\cos^2\phi-3\sin^2\phi)M_{12} + \sin\phi(\sin^2\phi-3\cos^2\phi)M_{13}/(1-\nu)\right]\\
&- LN_2^*(a,\phi)\left[\cos\phi(1-\nu+\sin^2\phi)M_{12} + \sin\phi(1-\nu+\cos^2\phi)M_{13}/(1-\nu)\right]\}\Bigg),\\
\tilde{G}_{21}^{(2)(0)} &= -\frac{(2\ln 2-1)}{4\ln(a/\bar{a})}\big(W_1^*(a,\phi)\left[(1-\nu\cos^2\phi)M_{12} + \nu\sin 2\phi M_{13}/(2(1-\nu))\right]\\
&+ (\xi(a,\phi)/a)(1-\nu)\sin\phi LN_1^*(a,\phi)\big),
\end{aligned} \tag{2.4.32b}$$

$$\begin{aligned}
\tilde{G}_{22}^{(2)(0)} &= -\frac{2\ln 2-1}{8\ln(a/\bar{a})}\left(2(1-2\nu)\bar{W} - \frac{\xi(a,\phi)}{a}\overline{LN}\left[\sin\phi M_{12} + \frac{\cos\phi M_{13}}{1-\nu}\right]\right),\\
\tilde{G}_{23}^{(2)(0)} &= \frac{(2\ln 2-1)}{4\ln(a/\bar{a})}\big(W_2^*(a,\phi)\big[\nu\sin 2\phi M_{12}/2 + (1-\nu+\nu\cos^2\phi)\\
&M_{13}/(1-\nu)\big] + (\xi(a,\phi)/a)(1-\nu)\cos\phi LN_2^*(a,\phi)\big),\\
\tilde{G}_{31}^{(3)(0)} &= \frac{(2\ln 2-1)M_{13}}{(1-\nu)8\ln(a/\bar{a})}\Bigg((1-2\nu)\sin 2\phi W_1^*(a,\phi) + \frac{\xi(a,\phi)}{a}\{6W_1^*(a,\phi)\\
&\times\left[\cos\phi(\cos^2\phi-3\sin^2\phi)M_{12} + \sin\phi(\sin^2\phi-3\cos^2\phi)M_{13}/(1-\nu)\right]\\
&- LN_1^*(a,\phi)\left[\cos\phi(1-\nu+\sin^2\phi)M_{12} + \sin\phi(1-\nu+\cos^2\phi),\ M_{13}/(1-\nu)\right]\}\Bigg)\\
\tilde{G}_{32}^{(3)(0)} &= -\frac{(2\ln 2-1)M_{13}}{(1-\nu)4\ln(a/\bar{a})}\left(\bar{W}\left[\nu\sin 2\phi\frac{M_{12}}{2} + \left(1+\frac{\nu\cos^2\phi}{1-\nu}\right)M_{13}\right]\right.\\
&\left.+ \frac{\xi(a,\phi)}{a}(1-\nu)\cos\phi\overline{LN}\right),
\end{aligned}$$

$$\tilde{G}_{33}^{(3)(0)} = -\frac{(2\ln 2 - 1)M_{13}}{(1-\nu)8\ln(a/\overline{a})}\Bigg(2(1-2\nu)\cos^2\phi W_2^*(a,\phi) + \frac{\xi(a,\phi)}{a}\{6W_2^*(a,\phi)$$
$$\times\big[\sin\phi(\sin^2\phi - 3\cos^2\phi)M_{12} + \cos\phi(3\sin^2\phi - \cos^2\phi)M_{13}/(1-\nu)\big]$$
$$- LN_2^*(a,\phi)\big[\sin\phi(2\nu + \cos^2\phi)M_{12} + \cos\phi(3-\sin^2\phi)M_{13}/(1-\nu)\big]\}\Bigg);$$
$$\overline{W}(a,\phi) = \frac{\cos\phi(1+\text{sgn}(\sin\phi)) - \sin\phi(1-\text{sgn}(\cos\phi))}{\sqrt{1+(\partial\xi(a,\phi)/\partial R)^2 + (\partial\xi(a,\phi)/\partial S)^2}},$$
$$\overline{LN}(a,\phi) = \frac{\ln[\![\{(1+\cos\phi)(1-\sin\phi)\}/\{(1-\cos\phi)(1+\sin\phi)\}]\!]}{\sqrt{1+(\partial\xi(a,\phi)/\partial R)^2 + (\partial\xi(a,\phi)/\partial S)^2}},$$
$$W_1^*(a,\phi) = \left(\sin\phi\frac{\partial\xi}{\partial R}(a,\phi) + \frac{\cos\phi}{a}\frac{\partial\xi}{\partial\phi}(a,\phi)\right)\overline{W}(a,\phi),$$
$$LN_1^*(a,\phi) = \left(\sin\phi\frac{\partial\xi}{\partial R}(a,\phi) + \frac{\cos\phi}{a}\frac{\partial\xi}{\partial\phi}(a,\phi)\right)\overline{LN}(a,\phi),$$
$$W_2^*(a,\phi) = \left(-\cos\phi\frac{\partial\xi}{\partial R}(a,\phi) + \frac{\sin\phi}{a}\frac{\partial\xi}{\partial\phi}(a,\phi)\right)\overline{W}(a,\phi),$$
$$LN_2^*(a,\phi) = \left(-\cos\phi\frac{\partial\xi}{\partial R}(a,\phi) + \frac{\sin\phi}{a}\frac{\partial\xi}{\partial\phi}(a,\phi)\right)\overline{LN}(a,\phi). \tag{2.4.32c}$$

To graphical plots of $\tilde{G}_{RC}(P_0)$ (2.4.32), the special case of a sinusoidal crack front (use (1) for running point $P_0 \equiv P_D$ with $R = a$ and $h(a) = 0$) with

$$\xi(a,\phi) \equiv A_n(a,\phi) = \xi_n(a)\sin n|\phi|, \quad |\phi| \le \pi, \tag{2.4.33}$$

is analysed. ϕ_C, the crack-front inclination angle, is used with the following meaning: ϕ_C is the acute angle, at crack-front positions P_M located on the average crack plane, between the crack-front and the average fracture plane. Hence, for given R

$$\tan\phi_C = \frac{\partial A_n}{\partial S}(R,\phi) = \frac{1}{R}\frac{\partial A_n}{\partial\phi} = \frac{1}{R}\xi_n(R)n\text{sgn}\phi\cos n\phi$$

Since $A_n(R,\phi) = 0$ at P_M and $\tan\phi_C$ positive, these lead to

$$\tan\phi_C = \frac{n}{R}\xi_n(R).$$

At any position with arbitrary R

$$A_n(R,\phi) = \frac{R}{n}\tan\phi_C\sin n|\phi|. \tag{2.4.34}$$

We then write (2.4.33) as

$$\xi(a,\phi) = A_n(a,\phi) = \frac{a}{n}\tan\phi_C \sin n|\phi|. \tag{2.4.35}$$

Under such conditions, the following average values for $\tilde{G}_{RC}$ (2.4.32) now denoted $\langle \tilde{G}_{RC}^{(S)} \rangle$ for the sinusoidal crack-front are evaluated:

$$\langle \tilde{G}_{RC}^{(S)} \rangle = \frac{1}{2\pi}\int_{-\pi}^{\pi} \tilde{G}_{RC}^{(S)}(P_0)\mathrm{d}\phi = \frac{1}{\pi}\int_{-\pi/2}^{\pi/2} \tilde{G}_{RC}^{(S)}(P_0)\mathrm{d}\phi. \tag{2.4.36}$$

$\langle \tilde{G}_{RC}^{(S)} \rangle$ then depends on various parameters (ϕ_C; M_{12}, M_{13}; n, $a/\overline{a}$; ν). Extrema with respect to ϕ_C may also be measured using the relation

$$\frac{\partial}{\partial \phi_C}\langle \tilde{G}_{RC}^{(S)} \rangle = 0. \tag{2.4.37}$$

Figure 2.36a–c show plots of $\langle \tilde{G}_{RC}^{(S)} \rangle$ (2.4.36) as a function of the couple (ϕ_C, M_{13}) for different values of the other parameters. No coloured regions of (ϕ_C, M_{13}) correspond to negative $\langle \tilde{G}_{RC}^{(S)} \rangle$ (< 0); the associated crack configurations are not favoured. The coloured regions correspond to positive $\langle \tilde{G}_{RC}^{(S)} \rangle$, indicating that the associated rough cracks can expand. (***b***) corresponds to (***a***) viewed in M_{13} magnitudes close to zero. Because $M_{12} = 0.001$ is small, it becomes apparent that rough sinusoidal circular cracks can expand under nearly pure applied tension for any ϕ_C. When M_{13} increases from zero, the non-planar crack motion becomes impossible except for large ϕ_C (say larger than 65°, approximately, in Fig. 2.36a, b). Similar behaviours are present for larger $n = 10$ in Fig. 2.36c. Hence applied shearing stresses have the property of impeding the expansion of rough circular cracks. These latter cracks are mostly created under low shearing stresses.

DISCUSSION: A modelling of the rough conoidal crack (Fig. 2.35) is presented that consists in representing the crack by a continuous distribution of dislocations with infinitesimal Burgers vectors b. The running position P_D along the dislocations is described by (2.4.1a and b). Their elastic fields are given to linear terms with respect to the oscillations A_n or their spatial derivative $\partial A_n/\partial S$ along the crack front. The crack dislocation distributions $D_m(\theta)$ provided by the traction-free crack face condition are required to give explicit expressions of the crack-tip stresses and crack extension force. $D_m(\theta)$ is approximated by $D_m^{(0)}(\theta)$ due to the smooth conoidal crack. The later are solutions of (2.4.28). The crack-tip stresses (2.4.30) are written as a function of $D_m^{(0)}(\theta)$ which remain to be determined at present. In the present work, to proceed further, we use $D_m^{(0)}(\theta = 0)$ to calculate the crack extension force (2.4.32) for the rough circular cracks. For plots, the special case of a sinusoidal crack front (2.4.33) is adopted involving the crack-front inclination angle ϕ_C. Similar ϕ_C have been used in earlier works (Sect. 2.2.4) where the non-planar crack front is also sinusoidal; in these works, applied M_{13} have been found under which average crack extension force $\langle G \rangle$ plotted as a function of ϕ_C exhibit positive maxima. The graphical representations

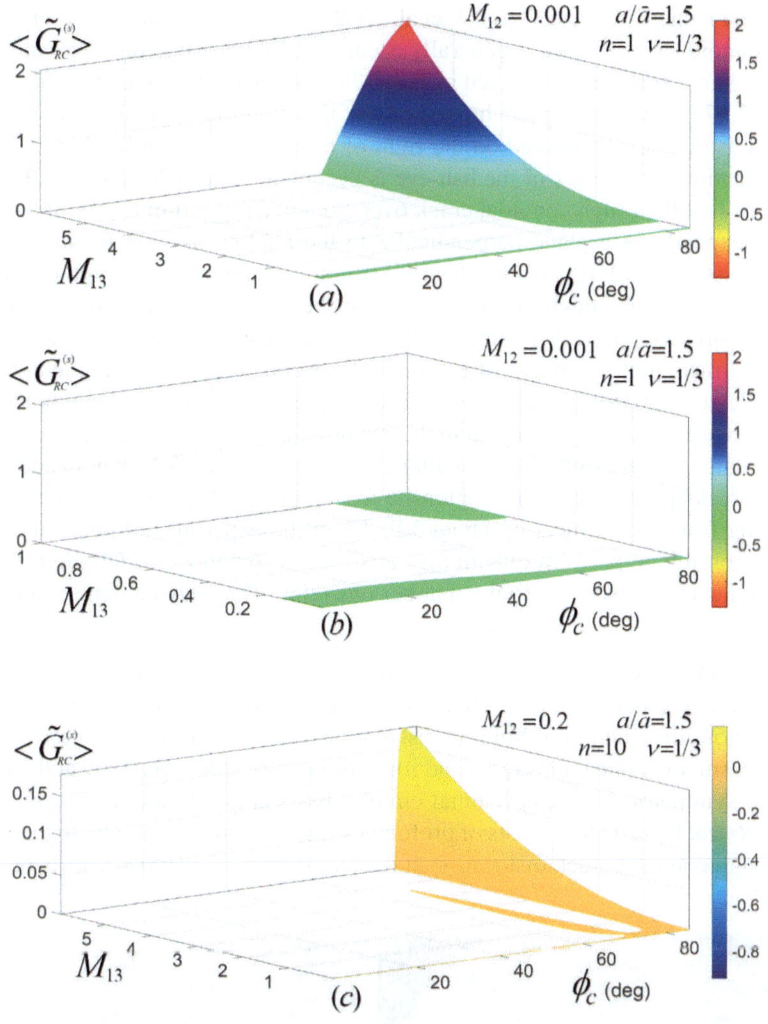

Fig. 2.36 **a–c** Plots of the reduced crack extension force $\langle \tilde{G}_{RC}^{(S)} \rangle$ (2.4.36) for the rough circular sinusoidal crack as a function of the pair of variables (ϕ_C, M_{13}) for different values of the other parameters M_{12}, n, ν and $a/\overline{a}$. (**b**) corresponds to (**a**) viewed in magnitudes of M_{13} closer to zero. (**c**) corresponds to a larger $n = 10$ value

of $\tilde{G}_{RC}^{(S)}$ (2.4.36) reveal that these types of non-planar closed crack loop can expand at small shearing stresses for all ϕ_C. With increasing M_{13} and M_{12}, their motions are unfavoured expect for large ϕ_C ($\geq 65°$, approximately). Characteristics that derive from Fig. 2.36, $D_m^{(0)}(\theta = 0)$, are expected to prevail for the general rough conoidal crack ($\theta \neq 0$), after inspection of the mathematics involved. On the experimental side, fracture surfaces of fatigued high strength steels in long-life regimes generally exhibit conoidal rough cracks, at the beginning of fracture initiation. These cracks have been named "fish-eyes" (Sakai et al. 2002; Abdesselam et al. 2018). The vertex of the conoidal rough crack is generally occupied by heterogeneities. An example of such fractured surface is displayed in Fig. 2.37 taken from Abdesselam et al. (2018). Figure 2.37a shows a complete broken part of the cylindrical specimen in which the fish-eye is distinctly observed on the right of the micrograph and in Fig. 2.37b, its magnification. The vertex of the fish-eye is occupied by an AlN inclusion. Fracture develops first the rough conoidal crack over a diameter of 1 mm approximately and further expands on a plane perpendicular to the loading axis. In the experiments by Abdesselam et al. (2018), no shearing stresses are applied. This agrees with the conclusion that the rough circular crack can be created under vanishing applied shearing stresses (Fig. 2.36). Very similar fracture surface characteristics are present on SEM fractographic images of René 88DT specimens tested in fatigue at 650 °C (Stinville et al. 2018): in their Fig. 2c fracture produces, in order, a conoidal rough crack, a planar crack perpendicular to the tensile axis and lastly another planar surface that deviates notably from the preceding. It is shown (Sect. 2.3.1) that, in addition to planar cracks perpendicular to the tensile axis, there exists planar fracture surfaces deviating from the formers by about 52°. From the experimental point of view, it is then possible to perform quantitative analyses of this mode of fatigue failure by linking the applied stresses to the number of cycles. This may be done empirically, first.

SUMMARY: A way to analyse mathematically the rough conoidal crack under general loading is introduced. This consists in representing the complex crack by a continuous distribution of Volterra dislocations with infinitesimal Burgers vectors. This is a smooth cone on average, and for definiteness, with a vertex O at the origin, vertical symmetrical axis, horizontal circular basis at an elevation $x_2 = h(a)$ with radius $R = a$. Its surface consists of profound striations running radially to the vertex. The rough conoidal crack dislocation at the elevation $x_2 = h(R)$ $(R \leq a)$ is obtained

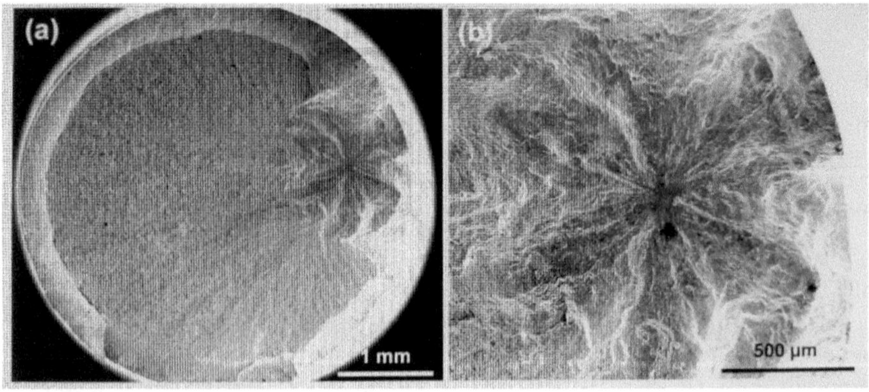

Fig. 2.37 **a** Fracture surface of broken high strength steel specimen tested in fatigue at 20 °C with the formation of a rough conoidal crack with an AlN inclusion located at its vertex. **b** Magnification of the conoidal crack (fish-eye). Micrographs are taken from Abdesselam et al. (2018)

as the intersection of a vertical cylinder of radius R (axis x_2) and the rough crack. The dislocation distribution consists of three families (m) (distribution function D_m; $m = 1$, 2 and 3) with burgers vectors $\boldsymbol{b}_m$ directed along the positive x_m-directions. The stresses are applied uniformly at infinity with tension σ_{22}^a along x_2 and shears σ_{21}^a and σ_{23}^a along x_1 and x_3. Poisson's normal stresses $-\nu\sigma_{22}^a$ (ν is Poisson's ratio) acting in the x_1 and x_3 directions are incorporated into the analysis. Plastic distortions associated with these dislocations are first given and corresponding expressions of the associated elastic fields (displacement and stress) are also provided. Then distribution function $D_m^{(0)}$ of circular horizontal dislocations (with the identical Burgers vectors $\boldsymbol{b}_m$) covering the smooth conoidal crack are considered and three singular integral equations that determine $D_m^{(0)}$ are written down. $D_m^{(0)} = D_m^{(0)}(\theta)$ depends on angle θ, complementary to the half angle at the vertex of the cone. $D_m^{(0)}$ may be used to approximate D_m. The crack-tip stresses have been given as functions of $D_m^{(0)}$. Explicit expressions of the rough circular crack extension force G_{RC} per unit length of the crack front are given that are associated with $D_m^{(0)}(\theta = 0)$. Expressions $\langle \tilde{G}_{RC}^{(S)} \rangle$ of $G_{RC}^{(S)}$ (crack extension force) averaged over the length of the oscillatory crack-front are plotted for the special case of sinusoidal fronts. These plots reveal that the rough circular cracks can expand under small shearing stresses. Agreements are found with long-life fatigue experiments in high strength materials. In the meanwhile, the crack extension force G_{RCo} of the rough conoidal crack has been obtained as a function of the angle θ. The results are given in the *Appendix C*.

2.4.2 Rough Conoidal Crack Growing Uniformly

From the results on the rough conoidal crack in static condition (Sect. 2.4.1), those for the rough conoidal crack growing uniformly can be obtained as follows. The various physical quantities displayed there depend implicitly on time t through the dependence on time of the horizontal circular base with radius $R \equiv a = a_0 + vt$ (v and a_0 are constants) of the conoidal crack. In the subsonic velocity regime ($v < c_t$), we replace G_0^I in crack extension force expressions $\tilde{G}_{RC}(P_0)$ (2.4.31) and $\tilde{G}_{RCo}(P_0)$ (relation C7, *Appendix C*, Sect. 2.4.3) by $G^{(I)} = bK_I^2/4\pi C_1^{(I)}$ taken from (2.1.8) (Sect. 2.1); for the transonic velocity regime ($c_t < v < c_l$), $G^{(I)} = bK_I^2/4\pi C_2^{(I)}$ taken from (2.1.15) (Sect. 2.1). From $t = 0$, the crack starts growing uniformly with increasing time t.

2.4.3 Appendix to Rough Conoidal Crack (Sect. 2.4.1)

APPENDIX A: STRESS TERMS IN (2.4.28)

Under $x_1 \neq \overline{x}_1$, we have

$$\sigma_{21}^{(1)(0)}(P_C;\overline{x}_1) - \tan\theta\sigma_{11}^{(1)(0)}(P_C;\overline{x}_1)$$

$$= C_1\frac{\overline{x}_1}{2}\int_{-\infty}^{\infty}\int_{-\infty}^{\infty} e^{-\tan\theta|x_1-\overline{x}_1|\eta_1}J_1[\overline{x}_1\eta_1]e^{ik_1x_1}\mathrm{d}k_1\mathrm{d}k_3$$

$$\times\left(\nu - 1 + \frac{1}{\eta_1}\{\tan\theta|x_1-\overline{x}_1|k_1^2 - i2\tan\theta\,\mathrm{sgn}(x_1-\overline{x}_1)k_1\right.$$

$$\left.+\frac{1}{\eta_1}\left[i\tan^2\theta(x_1-\overline{x}_1)k_1^3 - \nu k_1^2\right]\}\right),$$

$$\sigma_{21}^{(2)(0)}(P_C;\overline{x}_1) - \tan\theta\sigma_{11}^{(2)(0)}(P_C;\overline{x}_1)$$

$$= C_1\overline{x}_1\int_{-\infty}^{\infty}\int_{-\infty}^{\infty}\frac{J_1[\overline{x}_1\eta_1]}{\eta_1}e^{ik_1x_1}\mathrm{d}k_1\mathrm{d}k_3$$

$$\times((1-\nu)H[x_1-\overline{x}_1]ik_1 + e^{-\tan\theta|x_1-\overline{x}_1|\eta_1}\{(1-\nu)\mathrm{sgn}(x_1-\overline{x}_1)ik_1$$

$$+\tan^2\theta|x_1-\overline{x}_1|(k_1^2+2\nu k_3^2)/[2(1-2\nu)] - (1-2\nu)\tan\theta k_1^2/(2\eta_1)\});$$

$$\sigma_{22}^{(1)(0)}(P_C;\overline{x}_1) - \tan\theta\sigma_{21}^{(1)(0)}(P_C;\overline{x}_1)$$

$$= C_1\frac{\overline{x}_1}{2}\int_{-\infty}^{\infty}\int_{-\infty}^{\infty} e^{-\tan\theta|x_1-\overline{x}_1|\eta_1}J_1[\overline{x}_1\eta_1]e^{ik_1x_1}\mathrm{d}k_1\mathrm{d}k_3$$

$$\times\tan\theta\left(1-\nu+(x_1-\overline{x}_1)ik_1 + \frac{k_1^2}{\eta_1}\left\{\frac{\nu}{\eta_1} - \tan\theta|x_1-\overline{x}_1|\right\}\right),$$

$$\sigma_{22}^{(2)(0)}(P_C;\overline{x}_1) - \tan\theta\sigma_{21}^{(2)(0)}(P_C;\overline{x}_1)$$

$$= C_1\overline{x}_1\int_{-\infty}^{\infty}\int_{-\infty}^{\infty}\frac{J_1[\overline{x}_1\eta_1]}{\eta_1}e^{ik_1x_1}\mathrm{d}k_1\mathrm{d}k_3\times((\nu-1)\tan\theta H[x_1-\overline{x}_1]ik_1$$

$$-e^{-\tan\theta|x_1-\overline{x}_1|\eta_1}\{(1-\nu)\tan\theta\,\mathrm{sgn}(x_1-\overline{x}_1)ik_1$$

$$+\eta_1\left[(1-2\nu)^2+\tan\theta|x_1-\overline{x}_1|\eta_1\right]/[2(1-2\nu)]\});$$

$$\sigma_{32}^{(3)(0)}(P_C;\overline{x}_1) - \tan\theta\sigma_{31}^{(3)(0)}(P_C;\overline{x}_1)$$

$$= C_1\frac{\overline{x}_1}{2}\int_{-\infty}^{\infty}\int_{-\infty}^{\infty} e^{-\tan\theta|x_1-\overline{x}_1|\eta_1}\frac{J_1[\overline{x}_1\eta_1]}{\eta_1}e^{ik_1x_1}\mathrm{d}k_1\mathrm{d}k_3$$

$$\times\left((\nu-1)\tan\theta\,\mathrm{sgn}(x_1-\overline{x}_1)ik_1 + \tan\theta|x_1-\overline{x}_1|k_3^2 - \eta_1\right.$$

$$\left.+\frac{k_1}{\eta_1}\left[\nu k_1 + \tan^2\theta(x_1-\overline{x}_1)ik_3^2\right]\right);$$

$C_1 = \mu b/2\pi(1-\nu)$, μ is the shear modulus and H the Heaviside step function.

APPENDIX B: STRESSES FOR THE CRACK-TIP

$\sigma_{ij}^{(m)(0)}(P;\overline{x}_1)$ (2.4.30) are listed below; these are written to linear terms with respect to $(x_2-\overline{h})$, small quantity different from zero $(x_2-\overline{h}\neq 0)$

$$\sigma_{11}^{(1)(0)}(P;\bar{x}_1)=C_1\bar{x}_12\left(-\mathrm{sgn}(x_2-\bar{h})2\int\limits_0^{\pi/2}\sin\psi L_1^*\mathrm{d}\right.$$

$$\left.+\,(x_2-\bar{h})\int\limits_0^{\pi/2}\sin\psi(2+\sin^2\psi)M_2^*\mathrm{d}\psi\,\psi\right),$$

$$\sigma_{22}^{(1)(0)}(P;\overline{x}_1)=-C_1\overline{x}_1(x_2-\overline{h})2\int\limits_0^{\pi/2}\sin\psi M_2^*\mathrm{d}\psi,$$

$$\sigma_{33}^{(1)(0)}(P;\bar{x}_1)=C_1\bar{x}_12\left(-2\nu\,\mathrm{sgn}(x_2-\bar{h})\int\limits_0^{\pi/2}\sin\psi L_1^*\mathrm{d}\psi\right.$$

$$\left.+\,(x_2-\bar{h})\int\limits_0^{\pi/2}\sin\psi\left[1+2\nu-\sin^2\psi\right]M_2^*\mathrm{d}\psi\right),$$

$$\sigma_{21}^{(1)(0)}(P;\bar{x}_1)=C_1\bar{x}_12\left(-\int\limits_0^{\pi/2}(1-\nu+\nu\sin^2\psi)M_1^*\mathrm{d}\psi\right.$$

$$\left.+\,\left|x_2-\bar{h}\right|\int\limits_0^{\pi/2}\left[1-\nu+(1+\nu)\sin^2\psi\right]L_2^*\mathrm{d}\psi\right),$$

$$\sigma_{13}^{(1)(0)}(P;\bar{x}_1)=C_1\bar{x}_12\left(-(1-\nu)\mathrm{sgn}(x_2-\bar{h})\int\limits_0^{\pi/2}\cos\psi L_1^*\mathrm{d}\psi\right.$$

$$\left.+\,(x_2-\bar{h})\int\limits_0^{\pi/2}\cos\psi\left[1-\nu+\sin^2\psi\right]M_2^*\mathrm{d}\psi\right),$$

$$\sigma_{23}^{(1)(0)}(P;\overline{x}_1)=-C_1\overline{x}_1\left(\nu\int\limits_0^{\pi/2}\sin 2\psi M_1^*\mathrm{d}\psi+(1-\nu)\left|x_2-\overline{h}\right|\int\limits_0^{\pi/2}\sin 2\psi L_2^*\mathrm{d}\psi\right);$$

$$\sigma_{11}^{(2)(0)}(P;\bar{x}_1)=-\frac{2C_1\bar{x}_1}{1-2\nu}\left(-(1-2\nu)^2\int\limits_0^{\pi/2}\sin^2\psi M_1^*\mathrm{d}\psi\right.$$

$$+ 2\left|x_2 - \bar{h}\right| \int\limits_0^{\pi/2} \left[\nu + (1-\nu)(1-2\nu)\sin^2\psi\right] L_2^* \mathrm{d}\psi \Bigg),$$

$$\sigma_{22}^{(2)(0)}(P;\bar{x}_1) = -\frac{2C_1\bar{x}_1}{1-2\nu}\left((1-2\nu)^2 \int\limits_0^{\pi/2} M_1^* \mathrm{d}\psi + 4\nu(1-\nu)\left|x_2 - \bar{h}\right| \int\limits_0^{\pi/2} L_2^* \mathrm{d}\psi\right),$$

$$\sigma_{33}^{(2)(0)}(P;\bar{x}_1) = -\frac{2C_1\bar{x}_1}{1-2\nu}\Bigg(-(1-2\nu)^2 \int\limits_0^{\pi/2} \cos^2\psi M_1^* \mathrm{d}\psi$$

$$+2\left|x_2 - \bar{h}\right| \int\limits_0^{\pi/2} \left[\nu^2 + (1-\nu)\left\{\nu + (1-2\nu)\cos^2\psi\right\}\right] L_2^* \mathrm{d}\psi \Bigg),$$

$$\sigma_{12}^{(2)(0)}(P;\bar{x}_1) = C_1 4(1-\nu)\bar{x}_1\Bigg(-\left\{H(x_2 - \bar{h}) + \mathrm{sgn}(x_2 - \bar{h})\right\} \int\limits_0^{\pi/2} \sin\psi L_1^* \mathrm{d}\psi$$

$$+(x_2 - \bar{h}) \int\limits_0^{\pi/2} \sin\psi M_2^* \mathrm{d}\psi \Bigg),$$

$$\sigma_{13}^{(2)(0)}(P;\bar{x}_1) = C_1\bar{x}_1\Bigg((1-2\nu) \int\limits_0^{\pi/2} \sin 2\psi M_1^* \mathrm{d}\psi$$

$$- 2(1-\nu)\left|x_2 - \bar{h}\right| \int\limits_0^{\pi/2} \sin 2\psi L_2^* \mathrm{d}\psi \Bigg),$$

$$\sigma_{23}^{(2)(0)}(P;\bar{x}_1) = C_1 4(1-\nu)\bar{x}_1\Bigg(-\left\{H(x_2 - \bar{h}) + \mathrm{sgn}(x_2 - \bar{h})\right\}$$

$$\int\limits_0^{\pi/2} \cos\psi L_1^* \mathrm{d}\psi + (x_2 - \bar{h}) \int\limits_0^{\pi/2} \cos\psi M_2^* \mathrm{d}\psi \Bigg);$$

$$\sigma_{11}^{(3)(0)}(P;\bar{x}_1) = C_1 2\bar{x}_1\Bigg(-2\nu\,\mathrm{sgn}(x_2 - \bar{h}) \int\limits_0^{\pi/2} \cos\psi L_1^* \mathrm{d}\psi$$

$$+(x_2 - \bar{h}) \int\limits_0^{\pi/2} \cos\psi\left[2\nu + \sin^2\psi\right] M_2^* \mathrm{d}\psi \Bigg),$$

$$\sigma_{22}^{(3)(0)}(P;\overline{x}_1) = -C_1 2\overline{x}_1(x_2-\overline{h})\int\limits_0^{\pi/2}\cos\psi M_2^*\mathrm{d}\psi,$$

$$\sigma_{33}^{(3)(0)}(P;\bar{x}_1) = C_1 2\bar{x}_1\left(-2\mathrm{sgn}(x_2-\bar{h})\int\limits_0^{\pi/2}\cos\psi L_1^*\mathrm{d}\psi\right.$$
$$\left. + (x_2-\bar{h})\int\limits_0^{\pi/2}\cos\psi(2+\cos^2\psi)M_2^*\mathrm{d}\psi\right),$$
$$\sigma_{12}^{(3)(0)}(P;\overline{x}_1) = \sigma_{23}^{(1)(0)},$$

$$\sigma_{13}^{(3)(0)}(P;\overline{x}_1) = C_1 2\overline{x}_1\left(-(1-\nu)\mathrm{sgn}(x_2-\overline{h})\int\limits_0^{\pi/2}\sin\psi L_1^*\mathrm{d}\psi\right.$$
$$\left.+(x_2-\overline{h})\int\limits_0^{\pi/2}\sin\psi\left[1-\nu+\cos^2\psi\right]M_2^*\mathrm{d}\psi\right),$$

$$\sigma_{23}^{(3)(0)}(P;\overline{x}_1) = C_1 2\overline{x}_1\left(-\int\limits_0^{\pi/2}\left(1-\nu+\nu\cos^2\psi\right)M_1^*\mathrm{d}\psi + \left|x_2-\overline{h}\right|\right.$$
$$\left.\int\limits_0^{\pi/2}\left[1-\nu+(1+\nu)\cos^2\psi\right]L_2^*\mathrm{d}\psi\right); \tag{B1}$$

$$M_1^* = \begin{cases}\overline{x}_1/\left(\overline{x}_1^2-\Omega^2\right)^{3/2} & for\ \Omega^2/\overline{x}_1^2 < 1 \\ 0 & \Omega^2/\overline{x}_1^2 \ge 1\end{cases},$$
$$M_2^* = \begin{cases}-3\overline{x}_1\Omega/\left(\overline{x}_1^2-\Omega^2\right)^{5/2} & for\ \Omega^2/\overline{x}_1^2 < 1 \\ 0 & \Omega^2/\overline{x}_1^2 \ge 1\end{cases},$$
$$L_1^* = \begin{cases}0 & \Omega^2/\overline{x}_1^2 \le 1 \\ -\overline{x}_1/\left(\Omega^2-\overline{x}_1^2\right)^{3/2} & for\ \Omega^2/\overline{x}_1^2 > 1\end{cases}, \tag{B2}$$
$$L_2^* = \begin{cases}0 & \Omega^2/\overline{x}_1^2 \le 1 \\ 3\overline{x}_1\Omega/\left(\Omega^2-\overline{x}_1^2\right)^{5/2} & for\ \Omega^2/\overline{x}_1^2 > 1\end{cases},$$
$$\Omega = R\cos(\phi-\psi).$$

For $x_2-\overline{h}=0$ corresponding to positions $P\equiv P_M$ ($x_1 = R\sin\phi$, $x_2=\overline{h}=\overline{x}_1\tan\theta$, $x_3 = R\cos\phi$) (see (2.4.30)), the stresses $\sigma_{ij}^{(m)(0)}(P_M,\overline{x}_1)$ are due to a circular smooth dislocation parallel to x_1x_3 with radius $\overline{x}_1$, displaced vertically by $x_2=\overline{h}(\overline{x}_1)$ from the initial centre O. They may be calculated from Sect. 2.3.1. The oscillatory parts of the crack-tip stress fields $\overline{\sigma}_{ij}^{(m)An}(P)$ (2.4.30) ($m = 1$, 2 and 3) contribute zero to the crack extension force. This is because the stresses $\sigma_{ij}^{(m)An}(P;a)$ do not contain a singularity of the Cauchy-type; these are bounded. Below are displayed results for crack dislocation family $m = 3$ only. Expressions for the others are similar. To the linear order with respect to ξ_n,

$\sigma_{ii}^{(3)An}(P;\overline{x}_1) = 0 (i = 1, 2 \text{ and } 3)$, $\sigma_{13}^{(3)An}(P;\overline{x}_1) = 0$,

$$
\begin{aligned}
\overline{\sigma}_{23}^{(3)\xi}(P) = \frac{C_1}{\pi} \sum_{s,s_1,s_3=0}^{\infty} & (-1)^{s_3} \| a2(2s+1)\xi_{n=2s+1}(a)\mathrm{sgn}A_{2s+1}(R,\phi) \\
& \times (2-\delta_{0(2s_3)})\overline{b}_{21}^{*(1)}(n=2s+1; m_1, m_3 = 2s_3) I_{23}^{(3)}(1) - 2\{2s\xi_{2s}(a)\mathrm{sgn}A_{2s}(R,\phi) \\
& \times (2-\delta_{0(2s_3)})\overline{b}_{31}^{*(1)}(n=2s; m_1, m_3 = 2s_3) I_{23}^{(3)}(3) + 2(2s+1)\xi_{2s+1}(a)\mathrm{sgn}A_{2s+1}(R,\phi) \\
& \times \overline{b}_{31}^{*(1)}(n=2s+1; m_1, m_3 = 2s_3+1) I_{23}^{(3)}(4)\} \| \int_{a-\delta a}^{a} D_3^{(0)}(\overline{x}_1)\mathrm{d}\overline{x}_1, \\
\overline{\sigma}_{21}^{(3)\xi}(P) = \frac{C_1}{\pi} \sum_{s,s_1,s_3=0}^{\infty} & (-1)^{s_3} \| a(2+\nu)2(2s+1)\xi_{n=2s+1}(a)\mathrm{sgn}A_{2s+1}(R,\phi) \\
& \times (2-\delta_{0(2s_3)})\overline{b}_{21}^{*(1)}(n=2s+1; m_1, m_3 = 2s_3) I_{21}^{(3)}(1) + 2\nu\{2s\xi_{2s}(a)\mathrm{sgn}A_{2s}(R,\phi) \\
& \times (2-\delta_{0(2s_3)})\overline{b}_{31}^{*(1)}(n=2s; m_1, m_3 = 2s_3) I_{21}^{(3)}(3) + 2(2s+1)\xi_{2s+1}(a)\mathrm{sgn}A_{2s+1}(R,\phi) \\
& \times \overline{b}_{31}^{*(1)}(n=2s+1; m_1, m_3 = 2s_3+1) I_{21}^{(3)}(4)\} \| \int_{a-\delta a}^{a} D_3^{(0)}(\overline{x}_1)\mathrm{d}\overline{x}_1;
\end{aligned}
\tag{B3}
$$

$$
m_1 = 2s_1 + 1, \quad \delta_{0(2s_3)} = \begin{cases} 1 \; for \; s_3 = 0 \\ 0 \quad s_3 \neq 0 \end{cases},
$$

$$
I_{23}^{(3)}(1) = 4 \int_0^{\pi/2} \frac{(2+\nu)\cos^2\psi + 1 - \nu}{\sin\psi} I_1^*(m_1 = 2s_1+1, m_3 = 2s_3)\mathrm{d}\psi,
$$

$$
I_{23}^{(3)}(3) = 4 \int_0^{\pi/2} \frac{\cos\psi}{\sin\psi} I_2^*(m_1 = 2s_1+1, m_3 = 2s_3)\mathrm{d}\psi,
$$

$$
I_{23}^{(3)}(4) = 4 \int_0^{\pi/2} \frac{\cos\psi}{\sin\psi} I_2^*(m_1 = 2s_1+1, m_3 = 2s_3+1)\mathrm{d}\psi,
$$

$$I_{21}^{(3)}(1) = 4\int_0^{\pi/2} \cos\psi I_1^*(m_1 = 2s_1 + 1, m_3 = 2s_3)\mathrm{d}\psi,$$

$$I_{21}^{(3)}(3) = 4\int_0^{\pi/2} I_2^*(m_1 = 2s_1 + 1, m_3 = 2s_3)\mathrm{d}\psi,$$

$$I_{21}^{(3)}(4) = 4\int_0^{\pi/2} I_2^*(m_1 = 2s_1 + 1, m_3 = 2s_3 + 1)\mathrm{d}\psi,$$

$$I_1^*(P; a; m_1, m_3) = \frac{(-1)^{s_1+s_3}a^{m_1+m_3}}{2^{m_1+m_3}m_3!}$$
$$\frac{1}{(x_1 + (\tan\psi)^{-1}x_3)^{m_1}(\tan\psi x_1 + x_3)^{m_3}\Omega^3}$$
$$\times\sum_{k=0}^{\infty}(m_1 + m_3 + 2k + 2)!\frac{a^{2k}}{2^{2k}(x_1 + (\tan\psi)^{-1}x_3)^{2k}}$$
$$\frac{F\left(-k, -m_1 - k; m_3 + 1; (\tan\psi)^{-2}\right)}{k!(m_1 + k)!},$$

$$I_2^*(P; a; m_1, m_3) = \frac{(-1)^{s_1+s_3+1}a^{m_1+m_3}}{2^{m_1+m_3}m_3!}$$
$$\frac{1}{(x_1 + (\tan\psi)^{-1}x_3)^{m_1}(\tan\psi x_1 + x_3)^{m_3}\Omega^2}$$
$$\times\sum_{k=0}^{\infty}(m_1 + m_3 + 2k + 1)!\frac{a^{2k}}{2^{2k}(x_1 + (\tan\psi)^{-1}x_3)^{2k}}$$
$$\frac{F\left(-k, -m_1 - k; m_3 + 1; (\tan\psi)^{-2}\right)}{k!(m_1 + k)!};$$

F is Gauss's hypergeometric function.

APPENDIX C: CRACK EXTENSION FORCE AS A FUNCTION OF THE AVERAGE CONE ANGLE

Here $G_{RCo}(\theta)$ for arbitrary θ is provided. To do so, we start with the singular integral Eqs. (2.4.28) that characterize the equilibrium distribution functions $D_m^{(0)} = D_m^{(0)}(\theta)$ ($m = 1, 2$ and 3) of the circular crack dislocations m. As P_C moves closer to the crack tip (i.e., $x_1 \to a$), (2.4.28) becomes

$$\sigma_{21}^{A*} - C_1 \int\limits_{1-\delta a/a}^{1} \mathrm{d}\overline{y}_1 D_1^{(0)}(\overline{y}_1)\frac{1}{1-\overline{y}_1} = 0$$
$$\sigma_{22}^{A*} + C_1 \int\limits_{1-\delta a/a}^{1} \mathrm{d}\overline{y}_1 D_2^{(0)}(\overline{y}_1)\frac{1}{1-\overline{y}_1} = 0 \tag{C1}$$
$$\sigma_{23}^{A*} - (1-\nu)C_1 \int\limits_{1-\delta a/a}^{1} \mathrm{d}\overline{y}_1 D_3^{(0)}(\overline{y}_1)\frac{1}{1-\overline{y}_1} = 0;$$

$$\begin{aligned} \sigma_{21}^{A*} &= \frac{2}{2+9\tan^2\theta}\left(\sigma_{21}^{a} + \nu_A \tan\theta\sigma_{22}^{a} - (2-\nu)\tan\theta\sigma_{22}^{A*}\right) \\ \sigma_{22}^{A*} &= \frac{\tan\theta\left(\sigma_{21}^{a} + \nu_A\tan\theta\sigma_{22}^{a}\right) - \left(2+9\tan^2\theta\right)\left(\sigma_{22}^{a} - \tan\theta\sigma_{21}^{a}\right)}{(2-\nu)\tan^2\theta + \left(2+9\tan^2\theta\right)\left[1-2\nu+3(1-\nu)\tan^2\theta\right]} \\ \sigma_{23}^{A*} &= \frac{2\sigma_{23}^{a}}{2+3\tan^2\theta}; \end{aligned} \tag{C2}$$

$0 < \delta a \ll a$, $\overline{y}_1$ moves from (x_1/a) to 1. Above Cauchy-type singular integral equations are inverted in a similar way as in Sect. 2.3.1 to give

$$D_j^{(0)} \cong \alpha_0\alpha_j\frac{1}{\sqrt{1-y_1}};\ (j = 1,\ 2 \text{ and } 3) \tag{C3}$$

$y_1 = x_1/a$, (1-y_1) positive infinitesimal value,

$$\alpha_j(\theta) = \frac{1}{\pi C_1}\left(-\sigma_{21}^{A*}\delta_{j1} + \sigma_{22}^{A*}\delta_{j2} - \delta_{j3}\sigma_{23}^{A*}/(1-\nu)\right) \tag{C4}$$

$(C3)$ is the value of the dislocation j distribution closer to the crack tip. The associated relative displacement of the faces of the crack is

$$\phi_j = 2b\alpha_0\alpha_j a\sqrt{1-y_1}. \tag{C5}$$

The rough conoidal crack extension force per unit length of the crack front $G_{RCo}(P_0)$ at $P_0 \equiv P_D$ (2.4.1) with $R = a$ and $h(a) = a\tan\theta$, reads

$$G_{RCo}(P_0) = \sum_{j=1}^{3}\left(G_{1j}^{(1)(0)} + G_{2j}^{(2)(0)} + G_{3j}^{(3)(0)}\right); \tag{C6a}$$

$$\begin{aligned} G_{11}^{(1)(0)} &= -\frac{g_a\alpha_1}{2}\left(2(1-2\nu)\alpha_2\sin^2\phi W_1^* + \frac{\xi(a,\phi)}{a}\{6W_1^*\right. \\ &\times\left[\alpha_1\sin\phi(1-4\cos^2\phi) - \alpha_3\cos\phi(1-4\sin^2\phi)\right] \end{aligned}$$

$$+ LN_1^*\big[\alpha_1 \sin\phi(3-\cos^2\phi) + \alpha_3\cos\phi(2\nu+\sin^2\phi)\big]\big\}\big),$$

$$G_{12}^{(1)(0)} = g_a\alpha_1\big(\bar{W}\big[-\alpha_1 + \alpha_1\nu\cos^2\phi - \alpha_3\nu\sin 2\phi/2\big]$$

$$+ (\xi(a,\phi)/a)(1-\nu)\alpha_2\sin\phi\overline{LN}\big),$$

$$G_{13}^{(1)(0)} = \frac{g_a\alpha_1}{2}\bigg(\alpha_2(1-2\nu)\sin 2\phi W_2^* + \frac{\xi(a,\phi)}{a}\{6W_2^*$$

$$\times\big[\alpha_1\cos\phi(4\sin^2\phi-1) - \alpha_3\sin\phi(1-4\cos^2\phi)\big]$$

$$+ LN_2^*\big[\alpha_1\cos\phi(1-\nu+\sin^2\phi) + \alpha_3\sin\phi(1-\nu+\cos^2\phi)\big]\big\}\big),$$

$$G_{21}^{(2)(0)} = -g_a\alpha_2\big(W_1^*\big[\alpha_1(-1+\nu\cos^2\phi) - \alpha_3\nu\sin 2\phi/2\big]$$

$$+ (\xi(a,\phi)/a)(1-\nu)\alpha_2\sin\phi LN_1^*\big),$$

$$G_{22}^{(2)(0)} = -\frac{g_a\alpha_2}{2}\bigg(2(1-2\nu)\alpha_2\bar{W} + \frac{\xi(a,\phi)}{a}[\alpha_1\sin\phi + \alpha_3\cos\phi]\overline{LN}\bigg),$$

$$G_{23}^{(2)(0)} = g_a\alpha_2\big(-W_2^*\big[\alpha_1\nu\sin 2\phi/2 + \alpha_3(1-\nu+\nu\cos^2\phi)\big]$$

$$+ (\xi(a,\phi)/a)(1-\nu)\alpha_2\cos\phi LN_2^*\big),$$

$$G_{31}^{(3)(0)} = -\frac{g_a\alpha_3}{2}\bigg((1-2\nu)\alpha_2\sin 2\phi W_1^* + \frac{\xi(a,\phi)}{a}\{6W_1^*$$

$$\times\big[\alpha_1\cos\phi(4\sin^2\phi-1) - \alpha_3\sin\phi(1-4\cos^2\phi)\big]$$

$$+ LN_1^*\big[\alpha_1\cos\phi(1-\nu+\sin^2\phi) + \alpha_3\sin\phi(1-\nu+\cos^2\phi)\big]\big\}\big),$$

$$G_{32}^{(3)(0)} = g_a\alpha_3\big(-\bar{W}\big[\alpha_1\nu\sin 2\phi/2 + \alpha_3\big(1-\nu+\nu\cos^2\phi\big)$$

$$+ \frac{\xi(a,\phi)}{a}(1-\nu)\alpha_2\cos\phi\overline{LN}\bigg]\bigg),$$

$$G_{33}^{(3)(0)} = \frac{g_a\alpha_3}{2}\bigg(\alpha_2 2(1-2\nu)\cos^2\phi W_2^* - \frac{\xi(a,\phi)}{a}\{6W_2^*$$

$$\times\big[\alpha_3\cos\phi(4\sin^2\phi-1) + \alpha_1\sin\phi(1-4\cos^2\phi)\big]$$

$$- LN_2^*\big[\alpha_3\cos\phi(3-\sin^2\phi) + \alpha_1\sin\phi(2\nu+\cos^2\phi)\big]\big\}\big); \tag{C6b}$$

$$
\begin{aligned}
g_a &= b\alpha_0^2(2\ln 2 - 1)aC_1/2,\\
\overline{W} &= \frac{\cos\phi(1+\mathrm{sgn}(\sin\phi)) - \sin\phi(1-\mathrm{sgn}(\cos\phi))}{\sqrt{1+(\partial f/\partial R)^2 + (\partial f/\partial S)^2}},\\
\overline{LN} &= \frac{\ln[\![\{(1+\cos\phi)(1-\sin\phi)\}/\{(1-\cos\phi)(1+\sin\phi)\}]\!]}{\sqrt{1+(\partial f/\partial R)^2 + (\partial f/\partial S)^2}},\\
W_1^* &= \left(\sin\phi\frac{\partial f}{\partial R} + \frac{\cos\phi}{a}\frac{\partial f}{\partial\phi}\right)\overline{W},\\
LN_1^* &= \left(\sin\phi\frac{\partial f}{\partial R} + \frac{\cos\phi}{a}\frac{\partial f}{\partial\phi}\right)\overline{LN},\\
W_2^* &= \left(-\cos\phi\frac{\partial f}{\partial R} + \frac{\sin\phi}{a}\frac{\partial f}{\partial\phi}\right)\overline{W},\\
LN_2^* &= \left(-\cos\phi\frac{\partial f}{\partial R} + \frac{\sin\phi}{a}\frac{\partial f}{\partial\phi}\right)\overline{LN}.
\end{aligned}
\tag{C6c}
$$

In a similar way as for $\tilde{G}_{RC}(P_0)$ (2.4.31), we provide an associated normalized quantity $\tilde{G}_{RCo}(P_0)$ defined as

$$
\tilde{G}_{RCo}(P_0) = G_{RCo}(P_0)/G_C^{(I)}. \tag{C7}
$$

2.5 Conoidal Cracks Growing Uniformly in Cubic Crystals

2.5.1 Conoidal Cracks (Quasistatic Condition)

Model and Crack Analysis

The present focusses on smooth conoidal cracks (vertex O) with elliptical bases in cubic crystals; the elliptic form of the bases is expected in anisotropic media. The model is illustrated in Fig. 2.38.

The medium is infinitely extended in the Ox_j-directions (laboratory Cartesian reference frame). The stresses are uniform, applied at infinity, with tension σ_{22}^a and shears σ_{21}^a and σ_{23}^a in the x_2, x_1 and x_3 directions, respectively. Induced normal Poisson's stresses $-\nu_A(j)\sigma_{22}^a$ along x_1 ($j=1$) and x_3 ($j=3$) are considered. The crack nuclei are arbitrarily oriented with attached Cartesian $(O;x_j')$; $O\,x_2'$ is a symmetrical axis. In $x_1'x_3'$ planes, the bases are elliptical with semiaxes ρ_1 and ρ_2 along x_1' and x_3' such that $a_r = \rho_1/\rho_2 =$ constant about any elevation $x_2' = OO' \equiv h$ along Ox_2'. The running position P_D (x_1',x_2',x_3') at the elevation h along the base is written with respect to $(O;x_j')$ as

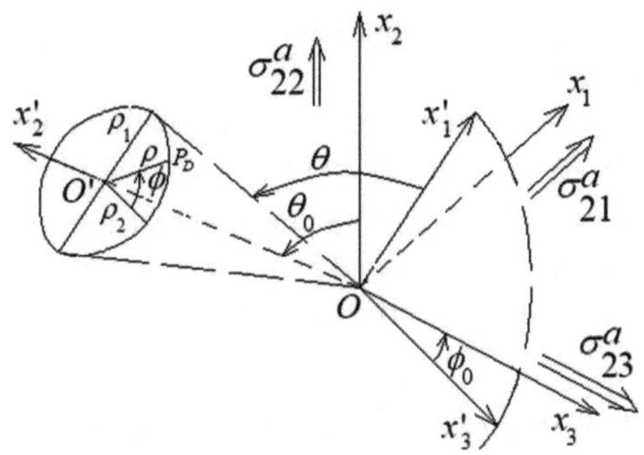

Fig. 2.38 Elliptical base (elevation $x'_2 = OO' \equiv h$) of the conoidal crack with semiaxes ρ_1 and ρ_2 along x'_1 and x'_3. The running point P_D (2.5.1) along the base and angular parameters θ_0 and ϕ_0 that connect $\boldsymbol{x}_j$ and $\vec{x}'_j$ are illustrated. Angle θ is introduced by the relation $\tan\theta = OO'/\rho_1$. The medium suffers uniformly applied tension σ^a_{22} in the vertical x_2-direction and shears σ^a_{21} and σ^a_{23} (parallel to the horizontal x_1x_3-plane) in the x_1 and x_3 directions

$$\overrightarrow{OP}_D = \begin{pmatrix} x'_1 = \rho\sin\phi \\ x'_2 = h = \rho_1\tan\theta \\ x'_3 = \rho\cos\phi \end{pmatrix}; -\pi \le \phi \le \pi, 0 \le \theta < \pi/2;$$

$$\rho^2 = \rho_1^2/(\sin^2\phi + a_r^2\cos^2\phi). \tag{2.5.1}$$

The angle ϕ is between $O'x'_3$ and $O'P_D$ as shown in Fig. 2.38. Angle θ is measured in $O\ x'_1x'_2$ between $\boldsymbol{P_D}$ $(\phi = \pi/2)$ $\boldsymbol{O'}$ and $\boldsymbol{P_D}$ $(\phi = \pi/2)$ $\boldsymbol{O}$ where P_D $(\phi = \pi/2)$ has elevation $x'_2 = h$ from $O\ x'_1x'_3$; its alternate interior angle is shown in Fig. 2.38. Additional angular parameters θ_0 and ϕ_0 (Euler's angles) are introduced that connect $\vec{x}_j$ to $\vec{x}'_j$:

(i) θ_0 (the angle of nutation) defined by unit vectors $\boldsymbol{x}_2$ and $\vec{x}'_2$, the growing sense corresponds to the rotation around $\boldsymbol{x}_3$ (the corkscrew rule applies),
(ii) ϕ_0 (the proper rotation angle) defined by unit vectors $\boldsymbol{x}_3$ and $\vec{x}'_3$, the growing sense corresponds to the rotation around $\vec{x}'_2$.

The crack is represented by a continuous distribution of three dislocation families with Burgers vectors $\boldsymbol{b}_j = b\ \vec{x}'_j$, directed along positive $\vec{x}'_j$, and running point P_D (2.5.1) at the height $x'_2 = OO' \equiv h$ along $O\ x'_2$ (see Fig. 2.38). Distribution functions D_j of the crack dislocations j are defined such that $D_j\ (\rho_1)\ d\rho_1$ represents the number of dislocations j in a small interval $d\rho_1$ located at the position $x'_1 = \rho_1$ on the $O\ \vec{x}'_1$-axis $(-a_1 \le x'_1 = \rho_1 \le a_1)$; $D_j\ (\rho_1) = D_j\ (-\rho_1)$ by symmetry, this restricts ourselves to positive x'_1 values. The elastic fields in the fractured medium read as a superposition of those due to elliptical crack dislocations. In the following, the procedures for determining the elastic fields of the dislocations and crack analysis are explained, first. Then are given the elastic fields (stress and displacement) of the elliptical dislocations j, distribution functions D_j of the crack dislocations and corresponding relative displacement ϕ_j of the faces of the crack, crack-tip stress, and crack extension force G per unit length of the crack front. Discussion and conclusion end the report, respectively.

The three types j ($j = 1, 2$ and 3) of crack dislocation considered have Burgers vectors $\vec{b}_1 = (b, 0, 0)$, $\vec{b}_2 = (0, b, 0)$ and $\vec{b}_3 = (0, 0, b)$ along x'_j; they spread in the $O'x'_1x'_3$-plane in the form (2.5.1). The plastic distortions $\beta_{ij}^{*(l)}$, associated to the elliptical crack dislocations l ($l = 1, 2$ and 3) at the height $h = OO'$ in the distribution (Fig. 2.38), are expressed successively at $\vec{x}' = (x'_1, x'_2, x'_3)$.

$$\beta_{21}^{*(1)} = b\delta(y'_2)\left(H\left[x'_1 + \sqrt{\rho_1^2 - a_r^2 x_3'^2}\right] - H\left[x'_1 - \sqrt{\rho_1^2 - a_r^2 x_3'^2}\right]\right), \ |x'_3| \leq \rho_2$$

$$= \frac{b\rho_1}{(2\pi)^2} \int_{-\infty}^{\infty}\int_{-\infty}^{\infty}\int_{-\infty}^{\infty} \frac{J_1[\eta_2]}{\sqrt{k_3'^2 + a_r^2 k_1'^2}} e^{i\vec{k}'.\vec{x}'_h} d\vec{k}'. \tag{2.5.2}$$

$\eta_2 = (\rho_1/a_r)\sqrt{k_3'^2 + a_r^2 k_1'^2}$, $J_1[\eta_2]$ is the Bessel function of the first kind; $\vec{k}' = (k'_1, k'_2, k'_3)$, $\vec{x}'_h = (x'_1, x'_2 - h \equiv y'_2, x'_3)$, $d\vec{k}' = dk'_1 dk'_2 dk'_3$ and δ and H are the Dirac delta and Heaviside step functions, respectively. $\beta_{21}^{*(1)} = 0$ for $|x'_3| \geq \rho_2$. The other components of the plastic distortion are zero. $\beta_{21}^{*(1)}$ (2.5.2) in its Fourier form is a superposition of single waves. Therefore, associated displacement and stress are similar superpositions, hence,

$$u_m^{(1)}(\vec{x}') = -\int_{-\infty}^{\infty}\int_{-\infty}^{\infty}\int_{-\infty}^{\infty} ik'_l C_{kl12} L_{mk} \overline{\beta}_{21}^{*(1)} e^{i\vec{k}'.\vec{x}'_h} d\vec{k}',$$

$$\sigma_{pq}^{(1)}(\vec{x}') = \int_{-\infty}^{\infty}\int_{-\infty}^{\infty}\int_{-\infty}^{\infty} C_{pqmn}\left(k'_n k'_l C_{kl12} L_{mk} \overline{\beta}_{21}^{*(1)} - \overline{\beta}_{nm}^{*(1)}\right) e^{i\vec{k}'.\vec{x}'_h} d\vec{k}', \tag{2.5.3}$$

$$\overline{\beta}_{21}^{*(1)} = \frac{b\rho_1}{(2\pi)^2} \frac{J_1[\eta_2]}{\sqrt{k_3'^2 + a_r^2 k_1'^2}}$$

$\overline{\beta}_{nm}^{*(1)}$ are zero except $\overline{\beta}_{21}^{*(1)}$. Performing the necessary integrations in (2.5.3) provide the elastic fields of the elliptical dislocations $j = 1$ at the elevation h in the conoidal crack dislocation distribution (Fig. 2.38) in Cartesian coordinates x'_j. We now consider the dislocation $j = 2$ with form (2.5.1) and Burgers vector $\vec{b}_2 = (0, b, 0)$ along x'_2. There are only two non-zero components $\beta_{12}^{*(2)}$ and $\beta_{32}^{*(2)}$ of the plastic distortion.

$$\beta_{12}^{*(2)} = bH(y'_2)\left(\delta\left[x'_1 + \sqrt{\rho_1^2 - a_r^2 x_3'^2}\right] - \delta\left[x'_1 - \sqrt{\rho_1^2 - a_r^2 x_3'^2}\right]\right)$$

$$= \int_{-\infty}^{\infty}\int_{-\infty}^{\infty}\int_{-\infty}^{\infty} \frac{k_1'}{k_2'}\overline{\beta}_{21}^{*(1)} e^{i\vec{k}'.\vec{x}_h'} \mathrm{d}\vec{k}' \equiv \int_{-\infty}^{\infty}\int_{-\infty}^{\infty}\int_{-\infty}^{\infty} \overline{\beta}_{12}^{*(2)} e^{i\vec{k}'.\vec{x}_h'} \mathrm{d}\vec{k}';$$

$$\beta_{32}^{*(2)} = bH(y_2')\left(\delta\left[x_3' + a_r^{-1}\sqrt{\rho_1^2 - x_1'^2}\right] - \delta\left[x_3' - a_r^{-1}\sqrt{\rho_1^2 - x_1'^2}\right]\right)$$

$$= \int_{-\infty}^{\infty}\int_{-\infty}^{\infty}\int_{-\infty}^{\infty} \frac{k_3'}{k_2'}\overline{\beta}_{21}^{*(1)} e^{i\vec{k}'.\vec{x}_h'} \mathrm{d}\vec{k}' \equiv \int_{-\infty}^{\infty}\int_{-\infty}^{\infty}\int_{-\infty}^{\infty} \overline{\beta}_{32}^{*(2)} e^{i\vec{k}'.\vec{x}_h'} \mathrm{d}\vec{k}' \qquad (2.5.4)$$

$$u_m^{(2)} = -\int_{-\infty}^{\infty}\int_{-\infty}^{\infty}\int_{-\infty}^{\infty} ik_l' L_{mk}\left(C_{kl21}\overline{\beta}_{12}^{*(2)} + C_{kl23}\overline{\beta}_{32}^{*(2)}\right) e^{i\vec{k}'.\vec{x}_h'} \mathrm{d}\vec{k}', \qquad (2.5.5)$$

$$\sigma_{pq}^{(2)} = \int_{-\infty}^{\infty}\int_{-\infty}^{\infty}\int_{-\infty}^{\infty} C_{pqmn}\left(k_n' k_l' L_{mk}\left[C_{kl21}\overline{\beta}_{12}^{*(2)} + C_{kl23}\overline{\beta}_{32}^{*(2)}\right] - \overline{\beta}_{nm}^{*(2)}\right) e^{i\vec{k}'.\vec{x}_h'} \mathrm{d}\vec{k}'.$$

We consider the dislocation $j = 3$ with form (2.5.1) and Burgers vector $\vec{b}_3 = (0, 0, b)$ along x_3'. For the plastic distortion, we have

$$\beta_{23}^{*(3)} = b\delta(y_2')H\left(\left[x_1' + \sqrt{\rho_1^2 - a_r^2 x_3'^2}\right]\left[\sqrt{\rho_1^2 - a_r^2 x_3'^2} - x_1'\right]\right), \ |x_3'| \le \rho_2$$

$$= \int_{-\infty}^{\infty}\int_{-\infty}^{\infty}\int_{-\infty}^{\infty} \overline{\beta}_{21}^{*(1)} e^{i\vec{k}'.\vec{x}_h'} \mathrm{d}\vec{k}', \overline{\beta}_{23}^{*(3)} = \overline{\beta}_{21}^{*(1)}; \qquad (2.5.6)$$

$\beta_{23}^{*(3)} = 0$ for $|x_3'| \ge \rho_2$ and the other components of the plastic distortion are zero. Equations (2.5.6) and (2.5.2) are identical although their Cartesian form is written differently.

$$u_m^{(3)} = -\int_{-\infty}^{\infty}\int_{-\infty}^{\infty}\int_{-\infty}^{\infty} ik_l' C_{kl32} L_{mk}\overline{\beta}_{23}^{*(3)} e^{i\vec{k}'.\vec{x}_h'} \mathrm{d}\vec{k}',$$
$$\sigma_{pq}^{(3)} = \int_{-\infty}^{\infty}\int_{-\infty}^{\infty}\int_{-\infty}^{\infty} C_{pqmn}\left(k_n' k_l' C_{kl32} L_{mk}\overline{\beta}_{23}^{*(3)} - \overline{\beta}_{nm}^{*(3)}\right) e^{i\vec{k}'.\vec{x}_h'} \mathrm{d}\vec{k}'. \qquad (2.5.7)$$

At this stage, we can write down the various elastic fields, displacements $\vec{u}^{(j)}$ and $(\sigma)^{(j)}$, associated with the three types j of crack dislocation with form (2.5.1). Our calculation results are displayed below. We shall apply the traction-free condition at any position on the crack boundary:

$$\begin{cases} \overline{\sigma}_{12} - \overline{\sigma}_{11}\partial x_2'/\partial x_1' - \overline{\sigma}_{13}\partial x_2'/\partial x_3' = 0 \\ \overline{\sigma}_{22} - \overline{\sigma}_{21}\partial x_2'/\partial x_1' - \overline{\sigma}_{23}\partial x_2'/\partial x_3' = 0 \\ \overline{\sigma}_{32} - \overline{\sigma}_{31}\partial x_2'/\partial x_1' - \overline{\sigma}_{33}\partial x_2'/\partial x_3' = 0 \end{cases} \tag{2.5.8}$$

$\overline{\sigma}_{ij}$ stands for the total stress at any point P (x_1', x_2', x_3') in the medium and is linked to the crack dislocation distribution D_j. Equation (2.5.8) concerns the positions on the crack faces only. $\overline{\sigma}_{ij}$ is written as (relatively to x_j')

$$\overline{\sigma}_{ij} = \sigma_{ij}'^{A} + \sigma_{ij}^{(C)(1)} + \sigma_{ij}^{(C)(2)} + \sigma_{ij}^{(C)(3)} \tag{2.5.9}$$

$(\sigma')^A$ is the externally applied stress including normal induced stresses from Poisson effect; relatively to x_j', its components are given in *Appendix A* to this section.

$$\sigma_{ij}^{(C)(m)}(\vec{x}') = \int_0^{a_1} \sigma_{ij}^{(m)}(\vec{x}'; \rho_1) D_m(\rho_1) \mathrm{d}\rho_1 (m = 1, 2 \text{ and } 3) \tag{2.5.10}$$

Here we assume that the crack extends from 0 to a_1 along positive values on the O x_1'-axis. $\sigma_{ij}^{(m)}$ is the stress field at $\vec{x}'$ due to a conoidal crack dislocation m with position $x_1' = \rho_1$ along $O\,x_1'$ in the dislocation distribution. Equation (2.5.8) give three integral equations the resolution of which yields the D_m. The relative displacements ϕ_m of the faces of the crack in the x_m'-direction at the position P_D (2.5.1) in the distribution are obtained by integration from the relation $d\phi_m = -bD_m(\rho_1)d\rho_1$:

$$\phi_m = \int_{\rho_1}^{a_1} bD_m(\rho')\mathrm{d}\rho', 0 \le \rho_1 \le a_1 \text{ and } 0 \le \phi \le \pi. \tag{2.5.11}$$

From (2.5.9) to (2.5.10), one can obtain the crack-tip stresses. The crack extension force G per unit length of the crack front is defined in Sects. 2.3.1 and 2.4.1. The front of the crack (Fig. 2.38) with shape (say P_D, (2.5.1) with $\rho_1 = a_1$) is assumed to advance steadily from $\rho_1 = a_1$ (shorter crack) to $\rho_1 = a_1 + \delta a_1$ (lengthened crack). At an arbitrary position P_D (2.5.1) with $a_1 \le \rho_1 \le a_1 + \delta a_1$, is attached a surface element $\overrightarrow{ds}$ that reads

$$\overrightarrow{\mathrm{d}s} = \begin{pmatrix} \mathrm{d}s_1 \\ \mathrm{d}s_2 \\ \mathrm{d}s_3 \end{pmatrix} = -\mathrm{d}\rho_1 \mathrm{d}\phi \frac{\rho_1 \tan\theta}{a_r^3 g_1^{3/2}(\phi)} \begin{pmatrix} \sin\phi \\ -a_r^2 g_1(\phi) \\ a_r^2 \cos\phi \end{pmatrix}_{x_j'} \tag{2.5.12}$$

$$g_1(\phi) = a_r^{-2}\sin^2\phi + \cos^2\phi.$$

The component of the force acting on ds in the x_i'-direction is $\overline{\sigma}_{ij} ds_j$(the summation convention on repeated subscripts applies) where $\overline{\sigma}_{ij}$ are stresses ahead of the shorter or behind the lengthened crack; thus the energy change associated with ds is $\overline{\sigma}_{ij}\mathrm{d}s_j\Delta u^{(i)}/2$(here a summation is also considered over $i = 1, 2$ and 3) where $\Delta u^{(i)}$ is the difference in displacement across the lengthened crack, just behind its tip, in the x_i'-direction. When the crack advances from $\rho_1 = a_1$ to $\rho_1 = a_1 + \delta a_1$, the energy decrease associated with a surface element

$$\Delta s = \int_{\rho_1=a_1}^{a_1+\delta a_1} \mathrm{d}s(P_D) \cong \delta a_1 \mathrm{d}\phi \overline{\Delta s}$$
$$\overline{\Delta s} = a_1 \tan\theta a_r^{-3} g_1^{-3/2}(\phi)\sqrt{\sin^2\phi\cos^2\phi(1-a_r^2)^2 + 2a_r^4 g_1^2(\phi)} \tag{2.5.13}$$

(δa_1 being small and, when used below, will be let to go to zero) is given as

$$-\delta E = \frac{1}{2}\int_{a_1}^{a_1+\delta a_1} \sum_i \sum_j \overline{\sigma}_{ij}\mathrm{d}s_j\Delta u^{(i)}, \tag{2.5.14}$$

the integration being performed with respect to ρ_1; we stress that Δs is the sum of the surface elements ds taken at the various points P_D (2.5.1) as ρ_1 only changes from a_1 to $a_1 + \delta a_1$. The crack extension force G per unit length of the crack front at $P_0 = P_D$ (with $\rho_1 = a_1$, ϕ given) is defined as

$$G(P_0) = \lim_{\delta a_1 \to 0} -\delta E/\Delta s. \tag{2.5.15}$$

Stresses $\sigma_{ij}^{(m)}(\vec{x}'; \rho_1)$ *to Linear Terms with Respect to* $(x_2' - h)$

Physical quantities concerned are the spatial dependence of the crack dislocation distributions D_m about the crack front, the crack-tip stresses $\overline{\sigma}_{ij}$ and the crack extension force G per unit length of the crack front. These require using the stress fields $\sigma_{ij}^{(m)}$ of those dislocations located at the very tip of the crack front. Hence, are displayed first below stresses $\sigma_{ij}^{(m)}$ to linear terms with respect to $(x_2' - h)$. The dislocations m have elevation h in the distribution and semiaxes ρ_1 and ρ_2 ($a_r = \rho_1/\rho_2$) along $O\,x_1'$ and $O\,x_3'$, respectively, under the condition $x_2' < h$ from the crack-tip stress calculations below. Associated stresses are displayed in *Appendix B* to Sect. 2.5.1.

Conoidal Crack Dislocation Distributions D_m

The traction-free conditions (2.5.8) are written for positions $P_C = (x_1', x_2' = x_1' \tan\theta, x_3' = 0)$ that correspond to $\boldsymbol{OP_D}$ ($\phi = \pi/2$) (2.5.1); the crack front is assumed located at $x_1' = a_1$ on the x_1'-axis. This reduces to

$$\begin{cases} \overline{\sigma}_{12} - \overline{\sigma}_{11}\tan\theta = 0 \\ \overline{\sigma}_{22} - \overline{\sigma}_{21}\tan\theta = 0\,. \\ \overline{\sigma}_{32} - \overline{\sigma}_{31}\tan\theta = 0 \end{cases} \tag{2.5.16}$$

Only the spatial dependence of D_m about the crack front is required in the crack analysis. Hence, P_C will be moved closer to the crack-tip *i.e.* $a_1 - \delta a_1 < x_1' \leq a_1$ ($0 < \delta a_1 \leq a_1$). $\sigma_{ij}^{(C)(m)}$ when used in (2.5.16) is identified to

$$\sigma_{ij}^{(C)(m)}(P_C) = \int_{x_1'}^{a_1} \sigma_{ij}^{(m)}(P_C; \overline{x}_1) D_m(\overline{x}_1) \mathrm{d}\overline{x}_1 \tag{2.5.17}$$

where $\sigma_{ij}^{(m)}$ is the stress at P_C due to a conoidal crack dislocation m with position $x_1' = \overline{x}_1$ (height $x_2' = \overline{h} \equiv \overline{x}_1 \tan\theta$) along x_1'; hence, $a_1 - \delta a_1 < x_1' < \overline{x}_1 < a_1$. Essentially, dislocation loops with $\overline{x}_1 < x_1'$ contribute nothing. Under such conditions, (2.5.16) can be managed to provide three singular integral equations with the simple Cauchy type

$$\begin{aligned}
&Term(1) + \frac{b\pi}{(2\pi)^2 C_{11} a_r} \int_{1-\delta a_1/a_1}^{1} \mathrm{d}\overline{z}_1 \frac{D_1(\overline{z}_1)}{1-\overline{z}_1} = 0,\\
&Term(2) + \frac{b\pi}{(2\pi)^2 C_{11} 3 a_r} \int_{1-\delta a_1/a_1}^{1} \mathrm{d}\overline{z}_1 \frac{D_2(\overline{z}_1)}{1-\overline{z}_1} = 0,\\
&\sigma'^{A}_{32} - \tan\theta\sigma'^{A}_{31} - \frac{bC_{44}(C_{11}+4C_{12})}{6\pi C_{11} a_r} \int_{1-\delta a_1/a_1}^{1} \mathrm{d}\overline{z}_1 \frac{D_3(\overline{z}_1)}{1-\overline{z}_1} = 0;
\end{aligned} \tag{2.5.18}$$

$$\begin{aligned}
Term(1) &= \frac{K_{13}^{(2)}\left[\sigma'^{A}_{22} - \tan\theta\sigma'^{A}_{21}\right] + K_{23}^{(2)}\left[\sigma'^{A}_{12} - \tan\theta\sigma'^{A}_{11}\right]}{K_{13}^{(2)}K_{22}^{(1)} + K_{23}^{(2)}K_{12}^{(1)}},\\
Term(2) &= \frac{K_{12}^{(1)}\left[\sigma'^{A}_{22} - \tan\theta\sigma'^{A}_{21}\right] - K_{22}^{(1)}\left[\sigma'^{A}_{12} - \tan\theta\sigma'^{A}_{11}\right]}{K_{13}^{(2)}K_{22}^{(1)} + K_{23}^{(2)}K_{12}^{(1)}};
\end{aligned} \tag{2.5.19}$$

$$\begin{aligned}
K_{22}^{(1)} &= \tan\theta\left(2C_{11}(C_{44}+C_{12}) - (C_{11}^2 + C_{12}C_{44} + 2C_{44}^2)\right),\\
K_{23}^{(2)} &= 4\left[C_{11}^2 + C_{12}C_{44} + C_{11}(C_{44}+C_{12})\right] + 9\tan^2\theta C_{44}^2,\\
K_{12}^{(1)} &= 2C_{44}^2 - \tan^2\theta\left(2(C_{12}C_{44} + C_{11}^2) - C_{11}(C_{12}+C_{44})\right),\\
K_{13}^{(2)} &= \tan\theta\left(9C_{44}^2 + 4\left[C_{11}(C_{44}+C_{12}) + C_{12}C_{44} + C_{11}^2\right]\right);
\end{aligned} \tag{2.5.20}$$

$(\sigma')^A$ is given in *Appendix A* to Sect. 2.5.1. (2.5.18) are inverted in a similar way as in Sect. 2.3.1 to give

$$D_m \cong \alpha_0 \alpha_m \frac{1}{\sqrt{1-z_1}}, \quad m = 1,\ 2 \text{ and } 3, \tag{2.5.21}$$

$z_1 = x_1'/a_1$, $(1 - z_1)$ positive infinitesimal value and α_0 a dimensionless constant (Sect. 2.3.1);

$$\begin{aligned}\alpha_1 &= \frac{4a_r C_{11}}{b} Term(1),\\ \alpha_2 &= \frac{12a_r C_{11}}{b} Term(2),\\ \alpha_3 &= -\frac{6a_r C_{11}}{bC_{44}(C_{11}+4C_{12})}\left[\sigma r_{32}^A - \tan\theta\sigma r_{31}^A\right] \equiv -\frac{6a_r C_{11}}{b} Term(3).\end{aligned} \tag{2.5.22}$$

Equation (2.5.21) is the value of dislocation m distribution closer to the crack tip. The associated relative displacement ϕ_m of the faces of the crack is,

$$\phi_m \cong 2b\alpha_0\alpha_m a_1\sqrt{1-z_1}. \tag{2.5.23}$$

Crack-Tip Stresses

The crack front is assumed moving from the elevation $h(a_1) = a_1 \tan\theta$ to $h(a_1 + \delta a_1) = (a_1 + \delta a_1)\tan\theta$ along x_2' (use Fig. 2.38 for illustration) and consider a position P (x_1', x_2', x_3') on the newly created crack boundary with $x_2' = \overline{x}_1' \tan\theta$, $a_1 \le \overline{x}_1' < a_1 + \delta a_1$. P is given by P_D (2.5.1) with $\rho_1 = \overline{x}_1'$. The stress at P is identified to,

$$\overline{\sigma}_{ij}(P) \equiv \sum_{m=1}^{3} \int_{\overline{x}_1'}^{a_1+\delta a_1} \sigma_{ij}^{(m)}(P;\overline{x}_1) D_m(\overline{x}_1)\mathrm{d}\overline{x}_1 \equiv \sum_{m=1}^{3} \overline{\sigma}_{ij}^{(m)} \tag{2.5.24}$$

$\sigma_{ij}^{(m)}$ are given in *Appendix B* to Sect. 2.5.1. For $D_m(\overline{x}_1)$ we use (2.5.21) with $z_1 = \overline{x}_1/(a_1 + \delta a_1)$. Two types of integration with respect to $\overline{x}_1$ are required in the calculation of $\overline{\sigma}_{ij}(P)$. These are:

$$\begin{aligned}I_t^{(m)}(a) &\equiv \int_{\overline{x}_1'}^{a_1+\delta a_1} \overline{x}_1\left(x_2' - h\right)\mathrm{Re}(I_2) D_m(\overline{x}_1)\mathrm{d}\overline{x}_1\\ &= \frac{\alpha_0\alpha_m 4\tan\theta a_r^3\sqrt{a_1}\cos(\psi-\phi)(a_1+\delta a_1-\overline{x}_1')^{3/2}}{\sqrt{g_1(\phi)g_2^2(\psi)\left(1+\sqrt{\tilde{\omega}}\right)^2\sqrt{1-\tilde{\omega}}\left(a_1+\delta a_1-\overline{x}_1'\sqrt{\tilde{\omega}}\right)^2}},\\ I_t^{(m)}(b) &\equiv \int_{\overline{x}_1'}^{a_1+\delta a_1} \overline{x}_1\mathrm{Re}(Q_1) D_m(\overline{x}_1)\mathrm{d}\overline{x}_1\\ &= \frac{\alpha_0\alpha_m 2a_r^2\sqrt{a_1}(a_1+\delta a_1-\overline{x}_1')^{1/2}}{g_2(\psi)\left(1+\sqrt{\tilde{\omega}}\right)\sqrt{1-\tilde{\omega}}\left(a_1+\delta a_1-\overline{x}_1'\sqrt{\tilde{\omega}}\right)};\end{aligned} \tag{2.5.25}$$

$$\tilde{\omega} = \frac{\cos^2(\psi - \phi)}{g_1(\phi)g_2(\psi)}; \tag{2.5.26}$$

to be introduced in $\overline{\sigma}_{ij}^{(m)}$ to list $\overline{\sigma}_{ij}(P)$ in (2.5.24).

Crack Extension Force

The procedure for calculating G (2.5.15), the crack extension force per unit length of the crack front at P_0, is outlined in above. For $\Delta u^{(i)}$ in (2.5.14), we use ϕ_i (2.5.23) with $z_1 = \overline{x}_1/(a_1 + \delta a_1)$. $G(P_0)$ is written in the form:

$$G(P_0) = \sum_{n,m,i,j} G_{ij}^{(n)}(m), \quad n, m, i, j = 1, 2 \text{ and } 3. \tag{2.5.27}$$

n refer to the relative displacement ϕ_n of the faces of the crack in the x_n'-direction; m corresponds to crack dislocation families; ij point out to the stress component $\overline{\sigma}_{ij}$.

$$G_{11}^{(1)}(1) = -\frac{2^7 a_1 \alpha_0^2 a_r^5 C_{11} \tan\theta \sin\phi Term(1)^2}{3\pi g_1^{1/2}(\phi)\Delta^*}$$

$$\int_{-\pi/2}^{\pi/2} \mathrm{d}\psi \frac{\sin\psi\cos(\psi - \phi)}{g_2^{5/2}(\psi)\tilde{\omega}\sqrt{1-\tilde{\omega}}\left(1+\sqrt{\tilde{\omega}}\right)^2}$$

$$\times \left(2(C_{12}C_{44} + C_{11}^2)\overline{SC}^{1/6} - \frac{F_{11}^{(1)}(a)}{\sqrt{\overline{SC}}} + \frac{4C_{12}C_{11}\cos^2\psi}{\overline{SC}^{1/6}}\right),$$

$$G_{11}^{(1)}(2) = -\frac{4^3 a_1 \alpha_0^2 a_r^4 C_{11} \sin\phi Term(1)Term(2)}{\pi\Delta^*}$$

$$\int_{-\pi/2}^{\pi/2} \frac{\mathrm{d}\psi}{g_2^{3/2}(\psi)\sqrt{\tilde{\omega}}\sqrt{1-\tilde{\omega}}\left(1+\sqrt{\tilde{\omega}}\right)}$$

$$\times \left(\frac{\sin^2\psi F_{11}^{(2)}(a) + C_{12}\cos^2\psi F_{11}^{(2)}(b)}{\overline{SC}^{5/6}}\right.$$

$$\left. + \frac{F_{11}^{(2)}(c)}{\overline{SC}^{1/6}} - \frac{C_{12}(C_{11}+C_{12})\sin^2\psi\cos^2\psi}{\sqrt{\overline{SC}}}\right),$$

$$G_{11}^{(1)}(3) = 0;$$

$$G_{12}^{(1)}(1) = \frac{2^5 a_1 \alpha_0^2 a_r^6 C_{44} C_{11} g_1(\phi) Term(1)^2}{3\pi\Delta^*}$$

$$\int_{-\pi/2}^{\pi/2} \frac{\mathrm{d}\psi}{g_2^{3/2}(\psi)\sqrt{\tilde{\omega}}\sqrt{1-\tilde{\omega}}\left(1+\sqrt{\tilde{\omega}}\right)}$$

$$\times\left(\frac{F_{12}^{(1)}(a)}{\overline{SC}^{5/6}}+\frac{F_{12}^{(1)}(c)}{\overline{SC}^{1/6}}+\frac{F_{12}^{(1)}(b)}{\sqrt{\overline{SC}}}\right),$$

$$G_{12}^{(1)}(2)=-\frac{2^7a_1\alpha_0^2a_r^7C_{44}C_{11}\sqrt{g_1}\tan\theta Term(1)Term(2)}{\pi\Delta^*}$$

$$\times\int_{-\pi/2}^{\pi/2}\mathrm{d}\psi\frac{\sin\psi\cos(\psi-\phi)}{g_2^{5/2}(\psi)\tilde{\omega}\sqrt{1-\tilde{\omega}}\left(1+\sqrt{\tilde{\omega}}\right)^2}\left(\frac{F_{12}^{(2)}(a)}{\overline{SC}^{5/6}}\right.$$

$$\left.+\frac{F_{12}^{(2)}(b)}{\sqrt{\overline{SC}}}+\frac{F_{12}^{(2)}(c)}{\overline{SC}^{1/6}}\right),$$

$$G_{12}^{(1)}(3)=-\frac{4^2a_1\alpha_0^2a_r^6C_{44}C_{11}g_1Term(1)Term(3)}{\pi\Delta^*}$$

$$\times\int_{-\pi/2}^{\pi/2}\frac{\mathrm{d}\psi\sin2\psi}{g_2^{3/2}(\psi)\sqrt{\tilde{\omega}}\sqrt{1-\tilde{\omega}}\left(1+\sqrt{\tilde{\omega}}\right)}\left(\frac{F_{12}^{(3)}(a)}{\overline{SC}^{5/6}}\right.$$

$$\left.+\frac{C_{44}}{\overline{SC}^{1/6}}\right);$$

$$G_{13}^{(1)}(1)=\frac{2^7a_1\alpha_0^2a_r^7C_{44}C_{11}\tan\theta\cos\phi Term(1)^2}{3\pi\sqrt{g_1}\Delta^*}$$

$$\times\int_{-\pi/2}^{\pi/2}\mathrm{d}\psi\frac{\cos\psi\cos(\psi-\phi)}{g_2^{5/2}(\psi)\tilde{\omega}\sqrt{1-\tilde{\omega}}\left(1+\sqrt{\tilde{\omega}}\right)^2}\left(\frac{F_{13}^{(1)}(a)}{\sqrt{\overline{SC}}}\right.$$

$$\left.+2C_{11}\overline{SC}^{1/6}+\frac{4C_{12}\cos^2\psi}{\overline{SC}^{1/6}}\right),$$

$$G_{13}^{(1)}(2)=-\frac{2^5a_1\alpha_0^2a_r^6C_{44}C_{11}\cos\phi Term(1)Term(2)}{\pi\Delta^*}$$

$$\times\int_{-\pi/2}^{\pi/2}\frac{\mathrm{d}\psi\sin2\psi}{g_2^{3/2}(\psi)\sqrt{\tilde{\omega}}\sqrt{1-\tilde{\omega}}\left(1+\sqrt{\tilde{\omega}}\right)}\left(\frac{F_{13}^{(2)}(a)}{\overline{SC}^{5/6}}\right.$$

$$\left.+\frac{2C_{11}}{\overline{SC}^{1/6}}-\frac{C_{12}}{\sqrt{\overline{SC}}}\right),$$

$$G_{13}^{(1)}(3)=0;$$

$$G_{21}^{(2)}(m)=-\frac{\alpha_2\sin\phi}{\alpha_1a_r^2g_1(\phi)}G_{12}^{(1)}(m),\ m=1,2\text{ and }3;$$

$$G_{22}^{(2)}(1)=\frac{2^7a_1\alpha_0^2a_r^7C_{11}\sqrt{g_1}\tan\theta Term(1)Term(2)}{\pi\Delta^*}\int_{-\pi/2}^{\pi/2}\mathrm{d}\psi$$

$$\times \frac{\sin\psi\cos(\psi-\phi)}{g_2^{5/2}(\psi)\tilde{\omega}\sqrt{1-\tilde{\omega}}\left(1+\sqrt{\tilde{\omega}}\right)^2}\left(2C_{11}(C_{44}+C_{12})\overline{SC}^{1/6}\right.$$

$$\left.+\frac{F_{22}^{(1)}(a)}{\sqrt{\overline{SC}}}+\frac{4C_{12}^2\cos^2\psi}{\overline{SC}^{1/6}}\right),$$

$$G_{22}^{(2)}(2)=\frac{2^6a_1\alpha_0^2a_r^6C_{11}g_1\mathit{Term}(1)\mathit{Term}(2)}{\pi\Delta^*}$$

$$\int\limits_{-\pi/2}^{\pi/2}\frac{d\psi}{g_2^{3/2}(\psi)\sqrt{\tilde{\omega}}\sqrt{1-\tilde{\omega}}\left(1+\sqrt{\tilde{\omega}}\right)}\times\left(\frac{F_{22}^{(2)}(a)}{\overline{SC}^{5/6}}\right.$$

$$\left.+\frac{C_{11}(C_{44}+C_{12})}{\overline{SC}^{1/6}}-\frac{2C_{12}^2\sin^2\psi\cos^2\psi}{\sqrt{\overline{SC}}}\right),$$

$$G_{22}^{(2)}(3)=0;$$

$$G_{23}^{(2)}(1)=-\frac{2^5a_1\alpha_0^2a_r^6C_{44}C_{11}\cos\phi\mathit{Term}(1)\mathit{Term}(2)}{\pi\Delta^*}$$

$$\times\int\limits_{-\pi/2}^{\pi/2}\frac{d\psi\sin 2\psi}{g_2^{3/2}(\psi)\sqrt{\tilde{\omega}}\sqrt{1-\tilde{\omega}}\left(1+\sqrt{\tilde{\omega}}\right)}\left(\frac{F_{23}^{(1)}(a)}{\overline{SC}^{5/6}}+\frac{C_{44}}{\overline{SC}^{1/6}}\right),$$

$$G_{23}^{(2)}(3)=\frac{2^43a_1\alpha_0^2a_r^6C_{44}C_{11}\cos\phi\mathit{Term}(2)\mathit{Term}(3)}{\pi\Delta^*}$$

$$\times\int\limits_{-\pi/2}^{\pi/2}\frac{d\psi}{g_2^{3/2}(\psi)\sqrt{\tilde{\omega}}\sqrt{1-\tilde{\omega}}\left(1+\sqrt{\tilde{\omega}}\right)}\left(\frac{F_{23}^{(3)}(a)}{\overline{SC}^{5/6}}\right.$$

$$\left.+\frac{F_{23}^{(3)}(b)}{\overline{SC}^{1/6}}+\frac{F_{23}^{(3)}(c)}{\sqrt{\overline{SC}}}\right);$$

$$G_{31}^{(3)}(m)=\frac{\alpha_3\tan\phi}{\alpha_1a_r^2}G_{13}^{(1)}(m),\quad m=1,2\text{ and }3;$$

$$G_{32}^{(3)}(m)=-\frac{\alpha_3g_1(\phi)}{\alpha_2\cos\phi}G_{23}^{(2)}(m),\quad m=1,2and3;$$

$$G_{33}^{(3)}(1)=\frac{2^6a_1\alpha_0^2a_r^7C_{11}^2\tan\theta\cos\phi\mathit{Term}(1)\mathit{Term}(3)}{\pi\sqrt{g_1}\Delta^*}$$

$$\int\limits_{-\pi/2}^{\pi/2}d\psi\times\frac{\sin\psi\cos(\psi-\phi)}{g_2^{5/2}(\psi)\tilde{\omega}\sqrt{1-\tilde{\omega}}\left(1+\sqrt{\tilde{\omega}}\right)^2}\left(2(C_{11}+C_{44})\overline{SC}^{1/6}\right.$$

$$\left.+\frac{F_{33}^{(1)}(a)}{\sqrt{\overline{SC}}}+\frac{4C_{12}\cos^2\psi}{\overline{SC}^{1/6}}\right),$$

$$G_{33}^{(3)}(2) = \frac{2^5 3 a_1 \alpha_0^2 a_r^6 C_{11} \cos\phi Term(2) Term(3)}{\pi \Delta^*}$$

$$\times \int_{-\pi/2}^{\pi/2} \frac{d\psi}{g_2^{3/2}(\psi)\sqrt{\tilde{\omega}}\sqrt{1-\tilde{\omega}}(1+\sqrt{\tilde{\omega}})} \left(\frac{F_{33}^{(2)}(a)}{SC^{5/6}} \right.$$

$$\left. + \frac{F_{33}^{(2)}(b)}{SC^{1/6}} + \frac{F_{33}^{(2)}(c)}{\sqrt{SC}} \right),$$

$$G_{33}^{(3)}(3) = 0;$$

$$\Delta^* = \sqrt{(1-a_r^2)^2 \sin^2 2\phi + 8a_r^4 g_1^2(\phi)} \tag{2.5.28}$$

Associated average $\langle G \rangle$ value can be defined as

$$\langle G \rangle = \frac{1}{\pi/2} \int_0^{\pi/2} d\phi G(P_0) = \langle G \rangle(\sigma_{ij}^A, \theta_0, \phi_0, a_1, \tan\theta, a_r, C_{nm}), \tag{2.5.29}$$

being function of various mentioned parameters.

DISCUSSION: The crack extension force $G(P_0)$ (2.5.28), per unit length of the crack front, at P_0 invokes an integration with respect to angular variable ψ. In ***k***' spaces involved in the Fourier forms of the dislocation elastic fields (see (2.5.3), (2.5.5) and (2.5.7)), we have used the relations $k_1' = \eta_1 \sin\psi$ and $k_3' = \eta_1 \cos\psi$. The integrations with respect to η_1 have provided functions like $\mathrm{Re}(Q_1)$ and $\mathrm{Re}(I_2)$ in stress expressions (see *Appendix B* to Sect. 2.5.1); remain those with respect to ψ. Various factors and terms in $G(P_0)$ contain ψ, so the integrations with respect to ψ appears laborious and rather unnecessary because the physical parameters (σ_{ij}^A, a_1, θ, a_r and C_{nm}) are revealed in $G(P_0)$ (2.5.28). The average form $\langle G \rangle$ (2.5.29) of $G(P_0)$ is function of various parameters: cubic elastic constants C_{nm} that measure the elastic anisotropy of the medium; $a_r = a_1/a_2$, the ratio of the semiaxes a_1 and a_2 along x_1' and x_3', that is a measure of the departure from the isotropic medium where $a_r = 1$ in steady motion (Sect. 2.4.1); angle θ, evaluated at the height $h = a_1 \tan\theta$, provides a knowledge of the lateral shape of the conoidal crack during its growing; σ_{ij}^A the applied stress matrix with respect to the laboratory reference frame Ox_j; θ_0 and ϕ_0 (Euler's angles) provide the connexion between Ox_j and $O x_j'$ attached to the conoidal crack. Generally, three Euler's angles are needed. Here (Fig. 2.38) only two (θ_0 and ϕ_0) are used because x_3 is contained in the plane $O x_1' x_3'$ with normal x_2'. Consider a material that deforms under loading. The energy E of the system (potential energy of the loading mechanism and the elastic energy of the medium) decreases during its evolution. The material adopts the appropriate possible (depending on external conditions of imposed stress, strain rate and temperature) deformation mode that insures highest energy decrease. The description that follows apply to numerous materials. We refer to copper (Anongba 1989, 1992, 1993; Anongba and Martin 1989, 1991; Anongba et al. 1991, 1993*a* and *b*). Deformation modes observed are:

(1) Plastic deformation by perfect dislocations observable by transmission electron microscopy (TEM). The microstructure after deformation consists of dislocation walls parallel and perpendicular to active {111} slip planes. When the temperature is sufficiently high, subboundaries are formed
that closely satisfy the Frank criterion (subboundary in equilibrium with no long-range stress fields in the medium).
(2) Abundant mechanical twinning which results from the motion of imperfect dislocations. This occurs above 773 K at a well-defined applied stress level τ_V (thermally activated) which defines the beginning of stage V deformation. The twinning develops at various points in the sample and occupies volumes of the order of several mm^3. It seems that this twinning results from the nucleation and propagation of a nucleus that grows in a regular manner over large distance; this may be described in a similar manner as for the crack presented in Fig. 2.38. We thus take up the idea of our predecessors (Bilby and Eshelby 1968; Friedel 1964; for example) who cover the boundary of the nucleus of a mechanical twin by dislocations (see Fig. 6.35 in Friedel 1964). The loaded material develops twins to decrease the energy E of the system in the same way as with cracks. Hence, the present study also entirely applies to twin with the following modification: $(\sigma')^A$ is to be replaced by $(\sigma')^{A-Fr}$ in (2.5.9) and $(\sigma)^A$ (*Appendix A* to Sect. 2.5.1) by $(\sigma)^{A-Fr}$ given by

$$(\sigma)^{A-Fr} = \begin{pmatrix} -\nu_A(1)(\sigma_{22}^a - \sigma_{22}^{fr}) & \sigma_{12}^a - \sigma_{12}^{fr} & 0 \\ \sigma_{21}^a - \sigma_{21}^{fr} & \sigma_{22}^a - \sigma_{22}^{fr} & \sigma_{23}^a - \sigma_{23}^{fr} \\ 0 & \sigma_{32}^a - \sigma_{32}^{fr} & -\nu_A(3)(\sigma_{22}^a - \sigma_{22}^{fr}) \end{pmatrix}_{x_j} \qquad (2.5.30)$$

where σ_{ij}^{fr} is the friction stress opposing the motion of the twin dislocations in the medium. Confrontations with experiments will form *Part II* of this study with attention to the activation of twins in stage V in [112] copper single crystals deformed at constant strain rates in temperature regime 2 (Anongba 1989; Anongba et al. 1993*a* and *b*). The following twinning has been identified (see Fig. 87, page 140, in Anongba (1989)):

(i) $K_1 \equiv (1\bar{1}1) = x_1'x_3'$ twinning plane
(ii) $\eta_1 \equiv [121]$ the shear direction x_1'
(iii) $K_2 \equiv (111)$ second undeformed plane
(iv) $\eta_2 \equiv [\bar{1}2\bar{1}]$ intersection of the plane of shear $(10\bar{1}) = x_1'x_2'$ and plane K_2

In Fig. 2.38, we have

$$\vec{x}_1 = [\bar{2}4\bar{1}]/\sqrt{21}, \vec{x}_2 = [112]/\sqrt{6}, \vec{x}_3 = [31\bar{2}]/\sqrt{14},$$
$$\theta_0 = 62°, \phi_0 = 19°. \qquad (2.5.31)$$

Application to Mechanical Twinning of [112] *Copper Single Crystals*

MECHANICAL BEHAVIOUR AND DEFORMATION STRUCTURES OF [112] COPPER SYNGLE CRYSTALS: [112] copper single crystals have been deformed vertically ($x_2 = [112]$, Fig. 2.38) under constant strain rate tests $\dot{\varepsilon} = m\dot{\gamma}$. The resolved shear stresses τ ($\sigma_{22}^{a} = \tau/m$) and strain γ have been calculated with Schmid factor $m = 0.41$ of the two primary octahedral slip systems $[101](\bar{1}11)$ and $[011](1\bar{1}1)$. A stereographic projection with centre [112] is provided for illustration and identification, Fig. 2.39.

In temperature regime 2 (773 − 1173 K, approximately), the deformation begins with the motion of perfect ½ <011> dislocations. At a well-defined shear stress level denoted τ_V, abundant mechanical twinning set in at different places in the sample. This deformation mode defined what has been named stage V. Twins occupy volumes

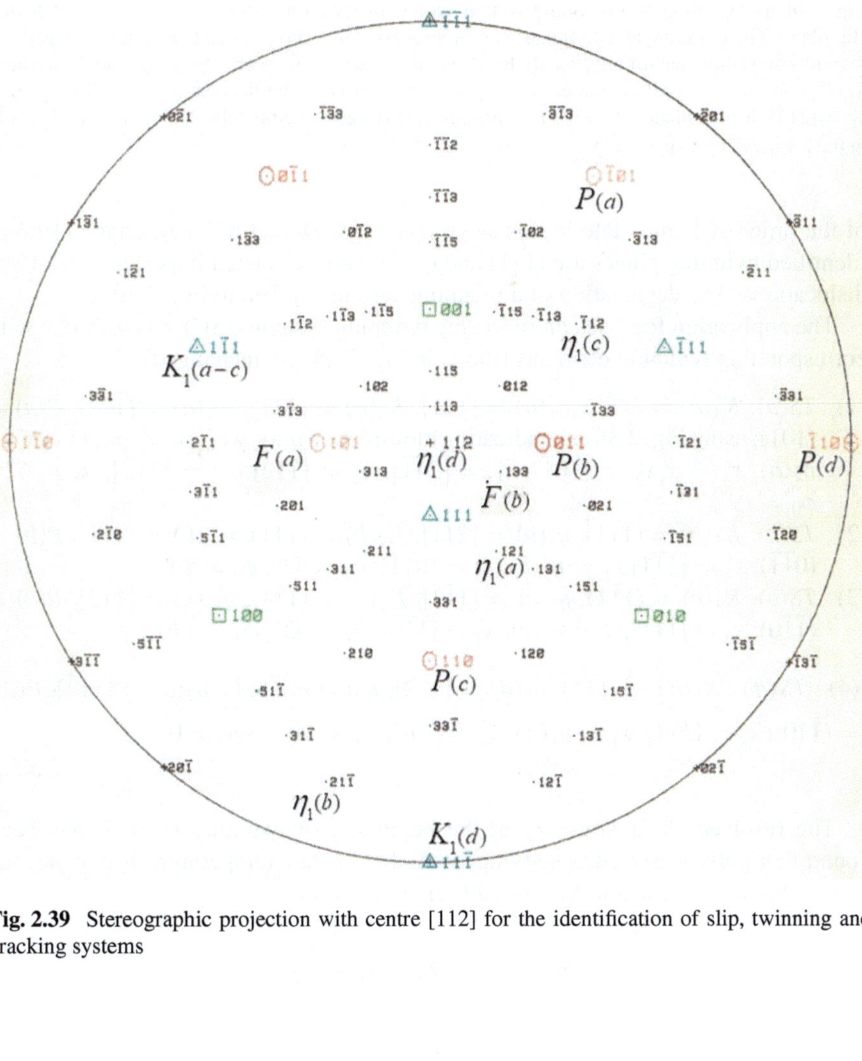

Fig. 2.39 Stereographic projection with centre [112] for the identification of slip, twinning and cracking systems

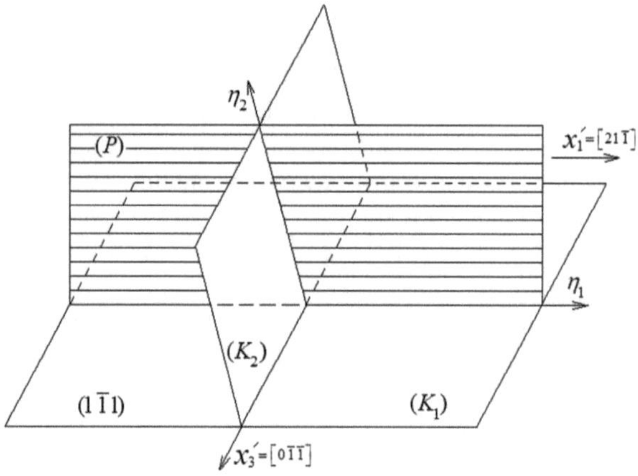

Fig. 2.40 Designation of a twinning system TS and dislocation wall perpendicular to octahedral slip plane. The twinning and conjugate (or reciprocal) planes are (K_1) and (K_2), respectively; the twinning and conjugate (or reciprocal) directions are η_1 and η_2, respectively. η_1, η_2 and the normals to (K_1) and (K_2) are all contained in the plane of shear (P). The dislocations are all parallel to $x_1' = [21\bar{1}]$, form a plane $(P) = (0\bar{1}\bar{1})$, with different Burgers vectors ½[011], ½[110] and ½ [10$\bar{1}$] in the primary slip plane (1$\bar{1}$1)

of the order of 1 mm^3 late in this stage (see Figs. 86 and 87 in Anongba (1989)). Identified twinning planes are ($\bar{1}$11) and (1$\bar{1}$1) with associated imperfect 1/6 <121> dislocations. The designation of a twinning system is given in Fig. 2.40.

The application focusses on following twinning systems *TS*(*i*), *i* = *a, b, c, d* with corresponding twinning elements (the x_j in Fig. 2.38 are unchanged):

(1) *TS*(*a*): $K_1(a) = (1\bar{1}1)$, $\eta_1(a) = [121]$, $K_2(a) = (111)$, $\eta_2\ (a) = [\bar{1}2\bar{1}]$, $P(a) = (10\bar{1})$; using Fig. 2.38 and indicating the direction only, we have $x_2' = [1\bar{1}1]$,$x_1' = \eta_1(a)$, $x_3' = x_1'x_2' = [10\bar{1}]$; $\vec{x}_1 = \left[\bar{2}4\bar{1}\right]$, $\vec{x}_2 = [112]$, $\vec{x}_3 = \left[31\bar{2}\right]$, $\theta_0 = 62°$, $\phi_0 = 19°$;
(2) *TS*(*b*): $K_1(b) = (1\bar{1}1)$, $\eta_1(b) = [21\bar{1}]$, $K_2(b) = (11\bar{1})$, $\eta_2\ (b) = [2\bar{1}1]$, $P(b) = (0\bar{1}\bar{1})$; $x_2' = [1\bar{1}1]$,$x_1' = \eta_1(b)$, $x_3' = [0\bar{1}\bar{1}]$; $\theta_0 = 62°$, $\phi_0 = 80°$;
(3) *TS*(*c*): $K_1(c) = (1\bar{1}1)$, $\eta_1(c) = [1\bar{1}\bar{2}]$, $K_2(c) = (\bar{1}11)$, $\eta_2\ (c) = [1\bar{1}2]$, $P(c) = (\bar{1}\bar{1}0)$; $x_2' = [1\bar{1}1]$,$x_1' = \eta_1(c)$, $x_3' = [\bar{1}\bar{1}0]$; $\theta_0 = 62°$, $\phi_0 = 140°$;

$$\begin{aligned}&(4)\quad TS(d): K_1(d) = (11\bar{1}), \eta_1(d) = [112], K_2(d) = (111), \eta_2(d) = [11\bar{2}], P(d)\\ &= (\bar{1}10); x_2' = [11\bar{1}], x_1' = \eta_1(d), x_3' = [\bar{1}10]; \theta_0 = 90^{\circ},\quad \phi_0 = 0.\end{aligned} \tag{2.5.32}$$

The resolved shear stress τ_V at the beginning of twinning stage *V* has been found thermally activated, satisfying an Arrhenius law temperature *T* dependence (Anongba, Bonneville and Martin 1993*a*), in the form:

$$\dot{\gamma} = A(\tau_V - \tau_\mu)^n \exp -\frac{\Delta H}{kT}$$

or

$$\dot{\varepsilon} = m^{n+1} A(\sigma_{22}^{a} - \sigma_{22}^{fr})^{n} \exp - \frac{\Delta H}{kT}. \tag{2.5.33}$$

The different quantities $\dot{\gamma}$, τ_V, τ_μ, ΔH and constants A, n, and k are given in (Anongba et al. 1993*a*). For the present, σ_{22}^{a} corresponds to τ_V and the friction stress σ_{22}^{fr} to τ_μ (see (2.5.30) for the use of $(\sigma_{22}^{a} - \sigma_{22}^{fr})$ that is called the effective stress). In deformation stage *IV* that preceded stage *V* (twinning), the deformation structure consists of subboundaries. We focus on those perpendicular to primary slip planes $(1\bar{1}1)$ and $(\bar{1}11)$ because some geometries recall the twinning geometry in Fig. 2.40. Subboundary planes are close to $\{011\}$ perpendicular to their dislocation glide planes $\{111\}$; they consist of one, two or three dislocation families, all having their Burgers vectors in the same plane and are parallel to the same direction (<112> have been observed below). They are of pure tilt character because the misorientation axis is parallel to the dislocation direction. These dislocation walls are frequently seen in pairs small distances apart (see subboundaries encircling subgrains C_3 and C_1 in Fig. 7 of Anongba and Martin (1991) and Fig. 6 in Anongba (1993)). A sketch of subboundary *S* of Fig. 6 in Anongba (1993) is depicted in Fig. 2.40: it consists of families 1, 2, and 3 with $\vec{b}_1 = 1/2[011]$, $\vec{b}_2 = 1/2[110]$ and $\vec{b}_3 = 1/2[10\bar{1}]$, respectively. Family 1 is in equal proportion to both families 2 and 3. Their directions are parallel to $[21\bar{1}] = x_1'$ and their common glide plane is $(1\bar{1}1) = (K_1)$. The normal to the subboundary is $[0\bar{1}\bar{1}] = x_3'$. The misorientation angle is 0.6°.

APPLICATION TO MECHANICAL TWINNING IN COPPER: We consider average $\langle G \rangle$ (2.5.29) and present below graphical plots of its normalized value $\langle \tilde{G} \rangle$ defined as

$$\langle \tilde{G} \rangle = \langle G \rangle / G_C^I = \langle \tilde{G} \rangle (\theta, \theta_0, \phi_0, M_{12}, M_{13}, a_r, C_{nm}),$$

$$G_C^I = \frac{2^5 \alpha_0^2 a_r^4}{3\pi^2 C_{44}} (K_I^0)^2; \tag{2.5.34}$$

where $K_I^0 = \sigma_{22}^{a} \sqrt{\pi a_1}$, $M_{12} = \sigma_{21}^{a}/\sigma_{22}^{a}$ and $M_{13} = \sigma_{23}^{a}/\sigma_{22}^{a}$. The applied stresses are viewed as effective stresses acting on the dislocations in the medium. The room temperature average values $C_{11} = 1.691$, $C_{12} = 1.222$ and $C_{44} = 0.7542$ in units of $[10^{11}\ \mathrm{N/m^2}]$ for copper have been used (see Table 2 in Ledbetter and Naimon (1974)). We consider the twinning system *TS*(*a*) (2.5.32) and plot $\langle \tilde{G} \rangle$ as a function of θ, Fig. 2.41.

$\langle \tilde{G} \rangle$ is negative or nearly zero in θ interval [0, 35°] and clearly negative in [79°, 90°] approximately; it becomes clearly positive between [35° 9°], approximately. Negative and positive values indicate forbidden and possible evolution of the system, respectively. Figure 2.42 is a plot of $\langle \tilde{G} \rangle$ as a function of ϕ_0 for $\theta = \pi/3$ and $\theta_0 =$ 62°.

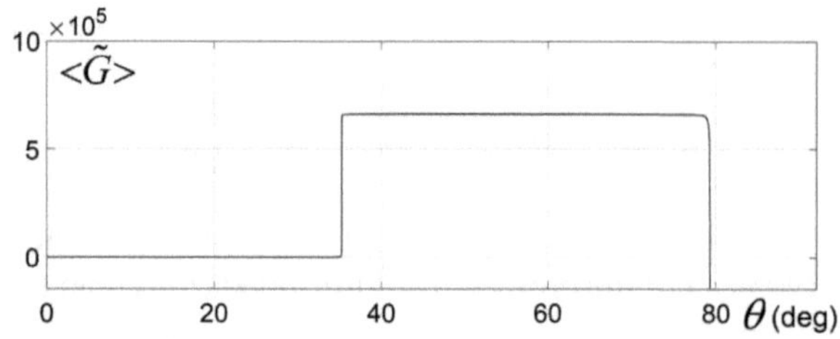

Fig. 2.41 $\tilde{G}$ as a function of θ for the twinning system *TS*(a) (2.5.32); $M_{12} = M_{13} = 10^{-4}$, $\nu_A(1) = \nu_A(3) = 1/3$, $a_r = 4/3$

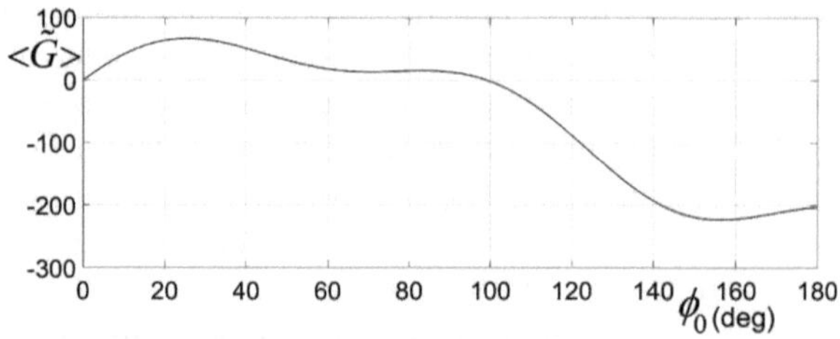

Fig. 2.42 $\langle\tilde{G}\rangle$ as a function of ϕ_0 for $\theta = \pi/3$ and $\theta_0 = 62°$. This corresponds to identical twinning plane $(1\overline{1}1)$ and the twinning systems *TS*(a) ($\phi_0 = 19°$), *TS*(b) ($\phi_0 = 80°$) and *TS*(c) ($\phi_0 = 140°$). $M_{12} = M_{13} = 10^{-4}$, $\nu_A(1) = \nu_A(3) = 1/3$, $a_r = 4/3$

This concerns the identical twinning plane $(1\overline{1}1)$: $\phi_0 = 19°$, $80°$ and $140°$ for the twinning systems *TS*(a), *TS*(b) and *TS*(c), respectively. These have positive $\langle\tilde{G}\rangle$ for *TS*(a) and *TS*(b) indicating that their evolutions are feasible. *TS*(c) displays negative values suggesting that it is forbidden. *TS*(b) is close to $\partial\langle\tilde{G}\rangle/\partial\phi_0 = 0$ (local average $\langle\tilde{G}\rangle$ maximum) as well as *TS*(a) taking account of available twinning system around. Figure 2.43 shows $\langle\tilde{G}\rangle$ as a function of θ_0 for $\theta = \pi/2$ and $\phi_0 = 0$.

DISCUSSION: Fig. 2.40 shows a correspondence of configurations between a dislocation wall $(0\overline{1}\overline{1})$ perpendicular to primary octahedral slip plane $(1\overline{1}1)$ and a twinning system *TS*(b) (2.5.32). The dislocation wall consists of perfect ½<110> dislocations

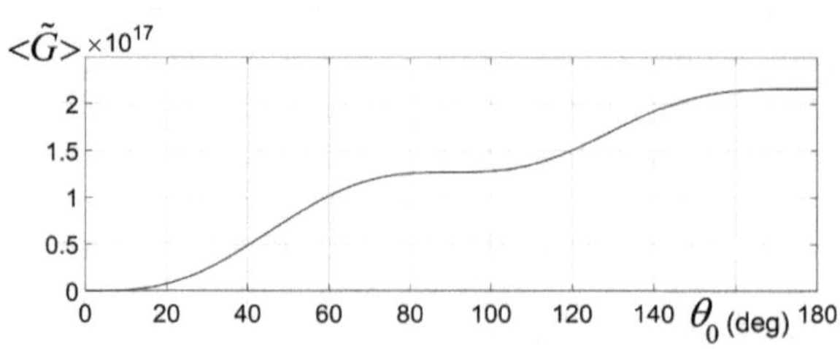

Fig. 2.43 $\langle\tilde{G}\rangle$ (2.5.34) as a function of θ_0 for $\theta = \pi/2$ and $\phi_0 = 0$. For $\theta_0 = 90°$, this corresponds to the twinning system *TS*(d) of the deviation plane $(11\overline{1})$. A local maximum is observed. $M_{12} = M_{13} = 10^{-4}$, $\nu_A(1) = \nu_A(3) = 1/3$, $a_r = 4/3$

with direction $[21\overline{1}]$. To *TS*(b) is associated arrays of parallel imperfects 1/6<112> in plane (P). Both types of walls satisfy the Frank criterion (Hirth and Lothe (1982)) for dislocation arrangements in equilibrium as common property. A pair of subboundary S (Fig. 6 in Anongba (1993)) with opposite Burgers vectors may well be a nucleus of *TS*(b) by the motion in $(1\overline{1}1)$ of partial dislocations $1/6[21\overline{1}]$. In pursuing, we first consider Fig. 2.41. From $\theta = 0$ to 35°, $\langle\tilde{G}\rangle$ is nearly zero or negative indicating that the growing of nuclei of twinning is rejected under such conditions. Hence to be able to grow, a minimum height $OO'_{\rm min} = \rho_1 \tan[35\pi/180]$ for the nuclei is required. In the case of cracks, the fractography of the surfaces of broken specimens in high-cycle fatigue experiments (Sakai et al. 2002; Abdesselam et al. 2018; Stinville et al. 2018) reveals the presence of defects (inclusions, heterogeneities, cavities ...) at the vertex O of observed conoidal cracks. The sizes of these defects may give experimental values for ρ_1 and $OO'_{\rm min}$. We now consider Fig. 2.42. In the situation of cracks, stationary configurations correspond to $\phi_0 = 22°$ and 72°, approximately. These are the configurations that sustain large deformations over time. In the situation of twinning along a <112> direction, both *TS*(a) and *TS*(b) system are stationary. *TS*(c) is forbidden. The plane of deviation $(11\overline{1})$ is also of interest (Fig. 2.43). Indeed *TS*(d) ($\phi_0 = 0$, $\theta_0 = 90°$) is stationary. It has been checked that for $\theta = \pi/3$, the magnitudes of $\langle\tilde{G}\rangle$ observed for *TS*(a), *TS*(b) and *TS*(d) are similar. Configurations of average $\langle\tilde{G}\rangle$ maximum are revealed by the quantity $\langle\tilde{G}\rangle$ (2.5.34). G^I_C (4) contains K^0_I (commonly named stress intensity factor) and carries the Arrhenius law temperature T dependence linking the effective applied stress and the strain rate (2.5.33).

CONCLUSION: This application concerns [112] copper single crystals deformed at constant strain rates between 773 and 1173 K (temperature regime 2). Graphical plots of $\langle\tilde{G}\rangle$ (2.5.34) as a function of (θ_0, ϕ_0) defining crack and twinning systems, have been displayed. Positive $\langle\tilde{G}\rangle$ are observed indicating that these deformation systems can operate under the specified conditions. Configurations of local $\langle\tilde{G}\rangle$ maximum are identified suggesting these correspond to equilibrium states. A minimum size is required for the growing of the nucleus of a given deformation system; this can be determined. A connection between subboundaries of pure tilt character, perpendicular to primary octahedral {111} slip planes and twinning systems has been stressed.

Application to Brittle Fracture Systems of $CoSi_2$ Single Crystals

Figure 2.38 is used to specify a plane of fracture (F). x_2 is vertical, parallel to the applied tension, x_3 is then determined as the intersection between (F) and the laboratory horizontal plane $(x_2) = Ox_1x_3$ allowing x_1 to be known. For definiteness, x_2 is fixed to [112] and we seek possible fracture systems under such conditions. Then graphical plots of $\langle\tilde{G}\rangle$ as a function of ϕ_0 are displayed. The crack propagation directions [T] in (F) are those associated to positive local maxima. It is assumed that the friction stresses are zero. The results for the {111} fractures are those of the twinning systems *TS*(i) mentioned earlier. Additional fracture planes ($F(j)$, $j = a$ and b) considered are:

(1) $F(a) = (101)$: using Figs. 2.38 and 2.39 and indicating the directions only, we have $\theta_0 = 30°, x_2 = [112], x_3 = [11\bar{1}], x_1 = [\bar{1}10], x_2' = [101]$

$$(2)\quad F(b) = (122) : \theta_0 = 18^{\circ}, x_2 = [112], x_3 = [\bar{4}02], x_1 = [1\bar{5}2], x_2' = [122] \tag{2.5.35}$$

For $CoSi_2$, $C_{11} = 2.16$, $C_{12} = 1.35$ and $C_{44} = 0.81$ in units of $[10^{11}\ N/m^2]$, taken from Anongba and Steinemann (1993). $M_{12} = \sigma_{21}^a/\sigma_{22}^a$, $M_{13} = \sigma_{23}^a/\sigma_{22}^a$, $\nu_A(1) = \nu_A(3) = 1/3$, $a_r = 4/3$. Octahedral fracture systems are invariably of the type $< 112 > \{111\}$ as for the twinning systems and examples correspond to $\eta_1[a-d]\ K_1\ [a-d]$ of Fig. 2.39. Fracture plane $F(a) = (101)$ is further considered and Figs. 2.44 and 2.45 are displayed in *CoSi*$_2$ material. Positive maxima of $\langle\tilde{G}\rangle$ are at $\phi_0 = 60°$ and $10°$ approximately, respectively; these correspond to systems $[13\bar{1}](101)$ and $[\bar{1}31](101)$.

Consideration is also given to the fracture plane $(122) = F(b)$ (Figs. 2.46 and 2.47). Positive maxima of $\langle\tilde{G}\rangle$ are at $\phi_0 = 50°$ and $110°$, corresponding to $[\overline{22}3](122)$ and $[\bar{4}11](122)$, respectively.

It is stressed that <110> is not a fracture propagation direction in the fracture planes investigated. Also increasing the shearing stress M_{13} reduces $\langle\tilde{G}\rangle$ to negative values; this means that this stress works against the expansion of the cracks.

CONFRONTATION WITH EXPERIMENTS: The experimental works on *CoSi*$_2$ (Anongba and Steinemann 1993, 1994; Anongba et al. 1994, 1995) will serve for confrontation. Fractures on close-packed octahedral planes $\{111\}$ have invariably a

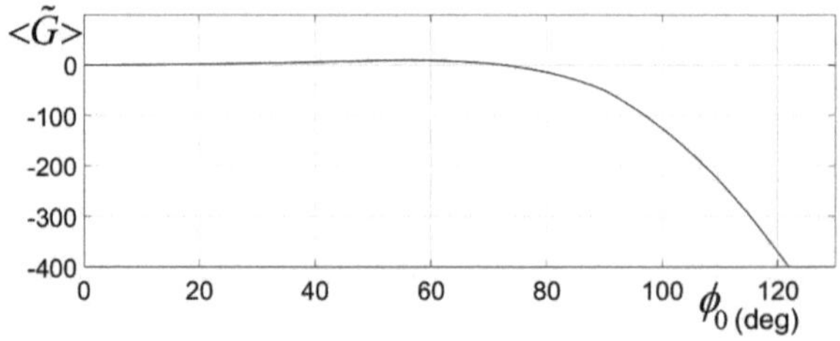

Fig. 2.44 $\langle\tilde{G}\rangle$ (2.5.34) as a function of ϕ_0 for $\theta = \pi/3$ and $\theta_0 = 30°$. This corresponds to the fracture system $[13\bar{1}](101)(\phi_0 = 60°)$. $M_{12} = M_{13} = 10^{-4}$, $\nu_A(1) = \nu_A(3) = 1/3, a_r = 4/3$, $CoSi_2$ material

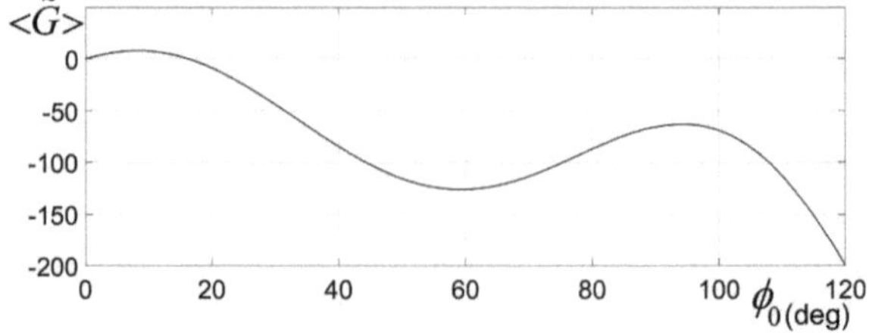

Fig. 2.45 $\langle\tilde{G}\rangle$ as a function of ϕ_0 for $\theta = \pi/3$ and $\theta_0 = 30°$. This corresponds to the fracture system $[\bar{1}31](101)(\phi_0 = 10°)$. $M_{12} = 1$, $M_{13} = 10^{-4}\nu_A(1) = \nu_A(3) = 1/3, a_r = 4/3$, $CoSi_2$ material

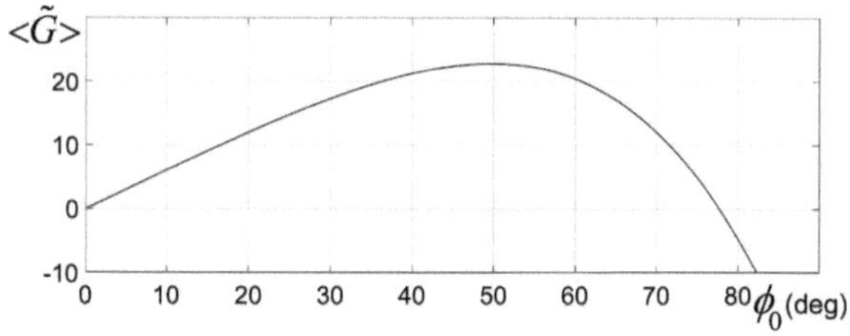

Fig. 2.46 $\langle\tilde{G}\rangle$ as a function of ϕ_0 for $\theta = \pi/3$ and $\theta_0 = 18°$. This corresponds to the fracture system $[\overline{2}\overline{2}\overline{3}](122)$ ($\phi_0 = 50°$). $M_{12} = M_{13} = 10^{-4}$, $\nu_A(1) = \nu_A(3) = 1/3$, $a_r = 4/3$, $CoSi_2$ material

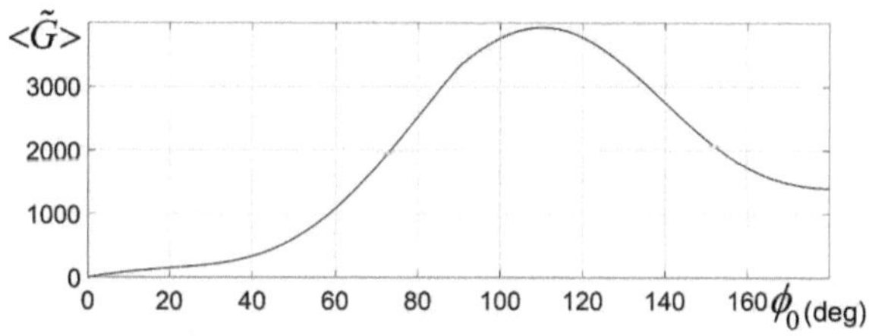

Fig. 2.47 $\langle\tilde{G}\rangle$ as a function of ϕ_0 for $\theta = \pi/3$ and $\theta_0 = 18°$. This corresponds to the fracture system $[\overline{4}11](122)(\phi_0 = 110°)$. $M_{12} = 4$, $M_{13} = 10^{-4}\nu_A(1) = \nu_A(3) = 1/3$, $a_r = 4/3$, $CoSi_2$ material

<112> propagation direction. In [113] $CoSi_2$ single crystals deformed under compression up to brittle fracture, the fracture system is $[112]\,(11\overline{1}) = \eta_1(d)\,K_1(d)$, Fig. 2.39: this is clearly shown on the micrograph in Fig. 2.48. Agreement between theory and experiment is complete. Twinning and fracture systems are the same <112> {111}. The theory is describing twinning and fracture by representing these with continuous distributions of Volterra infinitesimal dislocations. Fractures on (101) (Anongba, Oberli and Steinemann, 1994, 1995) suggest that the direction of fracture propagation is $[13\overline{1}]$ (see Fig. 4 of Anongba, Oberli and Steinemann (1995) and Table 1 there for the numerous numbers of specimens involved). This is confirmed by the present analysis (Fig. 2.44). Increasing the shearing stress M_{12} (Fig. 2.45) provokes a change in the fracture propagation direction to $[\overline{1}31]$. Hence <113> seems the crack propagation direction in {101} planes. Fractures on (212) (Anongba et al. 1994, 1995) clearly indicate that the fracture propagation direction is $[\overline{1}4\overline{1}]$. Figure 2.47 has predicted the cracking system $[\overline{4}\ 11]$ (122) for sufficiently large $M_{12} = 4$. For small shearing stresses, Fig. 2.46 suggests the system $[\overline{2}\overline{2}\overline{3}]$ (122).

Increasing M_{13} works against crack expansion. This is because σ_{23}^a in the modelling (Fig. 2.38) lies parallel to the elliptic crack front. This negative role has been mentioned earlier, Sect. 2.3.1 (Anongba 2021*b*).

SUMMARY: Brittle fracture systems (<112> {111}, <113> {110}, <411> {122}) observed in $CoSi_2$ single crystals have been confronted to the theoretical analysis of Sect. 2.5.1. These do correspond to positive local average $\langle G\rangle$ maxima, where the quantity $\langle G\rangle$ is the crack extension force G per unit length of the crack

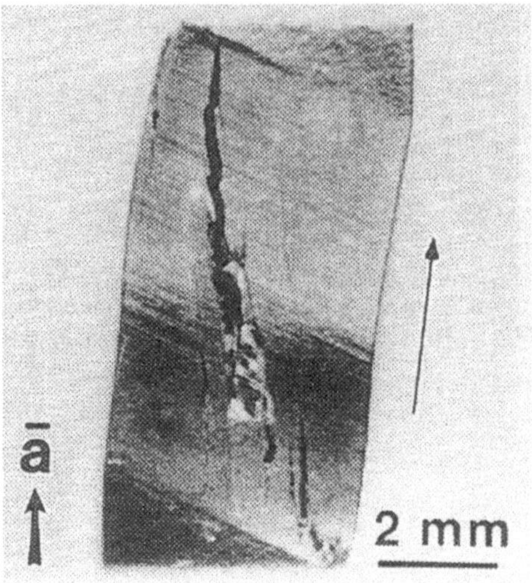

Fig. 2.48 Fractured [113] $CoSi_2$ single crystal under compression testing at 673 K. $(\bar{1}10)$ face: the compression axis at failure is $\bar{a} = [112]$ (perpendicular to the compression faces) and the unlabelled arrow in the deformed region is [113]. The fracture vertical plane is $(11\bar{1})$. After Anongba and Steinemann (1994)

front, averaged other all the positions on the crack front. Hence, equilibrium crack systems provided by the theory are those observed experimentally. Both twinning and cracking systems are of the type <112> {111} as expected. It is understood that above results are achieved within the framework of the theory of linear elasticity incorporating dislocations as crystal defects and basic elements in the mathematical treatment of fracture.

Application to ½ <110> {111} Cross-Slip in FCC Materials

Plastic deformation of [112] copper single crystals (Fig. 2.39) begins in *stage II* with the two symmetrical equally stressed slip systems ½ $[011](1\bar{1}1)$ and ½ $[101](\bar{1}11)$. The hardening rate ($\theta = \mathrm{d}\tau/\mathrm{d}\gamma$; τ and γ, shear stress and strain) in *stage II*, θ_{II}, is constant with stress τ. At a stress level called τ_{III} begins the deformation stage *III* where θ_{III} is decreasing with stress. Abundant cross-slips occur on the deviation plane $(11\bar{1})$ and these are due to screws ½ [011] and ½ [101] gliding initially on $(1\bar{1}1)$ and $(\bar{1}11)$, respectively (Anongba 1989; Anongba et al. 1993*b*). Figure 2.38 shows the conoidal crack geometry under load used to calculate the associated crack extension force *G* per unit length of the crack front (Sect. 2.5.1). Graphical plots are presented of its normalized value $< \tilde{G}_r >$ defined as

$$\langle \tilde{G}_r \rangle = \langle G \rangle / G^I_{Cs} = \langle \tilde{G}_r \rangle (\theta, \theta_0, \phi_0, M_{12}, M_{13}, a_r, C_{nm}),$$

$$G^I_{Cs} = \frac{2^4 \alpha_0^2}{3\pi^2 C_{44}} (K_I^0)^2; \tag{2.5.36}$$

where $K_I^0 = \sigma_{22}^a\sqrt{\pi\Delta}$; Δ is the separation of the partial dislocations. The applied stresses are viewed as effective stresses acting on the dislocations in the medium. The room temperature average values $C_{11} = 1.691$, $C_{12} = 1.222$ and $C_{44} = 0.7542$ in units of $[10^{11}$ N/m$^2]$ for copper have been used (see Table 2 in Ledbetter and Naimon (1974)). Figure 2.38 is used to specify a plane of cross-slip (S). x_2 is vertical, parallel to the applied tension, x_3 is then determined as the intersection between (S) and the laboratory horizontal plane $(x_2) = Ox_1x_3$ allowing x_1 to be known. For definiteness, x_2 is fixed to [112] and we seek possible cross-slip systems under such conditions. Then graphical plots of $\langle \tilde{G}_r \rangle$ as a function of ϕ_0 are displayed. The cross-slip propagation directions [U] in (S) are those associated to positive local maxima. We take (S) = $(11\overline{1})$ (Anongba 1989; Anongba et al. 1993*b*): using Figs. 2.38 and 2.39 and indicating the directions only, we have

$$\theta_0 = \pi/2, x_2 = [112], x_3 = [\overline{1}10], x_1 = [\overline{11}1], x_2' = [11\overline{1}] \tag{2.5.37}$$

Figure 2.49 displays the configuration (at an arbitrary time t) of the screw dislocation ½ [011] during the cross-slip process. The shape is assumed elliptical in $(11\overline{1})$ and the parts in $(1\overline{1}1)$ are hatched. As the ellipse expands in $(11\overline{1})$, the configurations in $(1\overline{1}1)$ remain unchanged; consequently, these latter contribute non-additional value to $\langle G \rangle$ ($d\ \langle G \rangle\ = 0$) there. Hence, $\langle G \rangle$ value is carried by the elliptical shape in $(11\overline{1})$. The quantity $\langle G \rangle$ (see expression (2.5.29)) applies. Positive local maxima of $< \tilde{G}_r >$ correspond to equilibrium states of the cross-slip. The loadings are along x_j as indicated (Fig. 2.49). $a_r = a_1/a_2$, where a_1 and a_2 are the semiaxes along $x_1' = [\overline{1}10]$ and $x_3' = [\overline{112}]$, respectively. In the stationary configuration, we take $2a_2 = \Delta$ the separation of partials (2.5.36). The resolved shear stress τ_{III} at the beginning of *stage III* has been found thermally activated, satisfying an Arrhenius law temperature T dependence. This can be included in the analysis in a similar way as for τ_V the stress at the beginning of twinning. Figure 2.50 is a plot of $\langle \tilde{G}_r \rangle$ (2.5.36) for $(S) = (11\overline{1})$(2.5.37) as a function of ϕ_0. Positive maxima of $\langle \tilde{G}_r \rangle$ are at $\phi_0 = \pi/2$ and π approximately (visual inspection). The former corresponds to the cross-slip system $[011](11\overline{1})$ ($\phi_0 = \pi/2$). The second may be other mechanisms ($[\overline{112}](11\overline{1})$ twinning and fracture, for instance); these are observed at higher stress levels as compared to τ_{III} under similar temperatures. Figure 2.51 shows $\langle \tilde{G}_r \rangle$ for the cross-slip system $[011](11\overline{1})$ ($\theta_0 = \pi/2, \phi_0 = \pi/2$) as a function of a_r. Visual inspection shows non-negative values of $\langle \tilde{G}_r \rangle$ close to zero, from $a_r \cong 0$ up to $a_r \cong 1.5$. Above 1.5, $\langle \tilde{G}_r \rangle$ is clearly negative. A maximum positive $\langle \tilde{G}_r \rangle$ value is expected about $a_r = 1$ which would be the value of a_r at the equilibrium state. $a_1 \cong a_2 \cong \Delta/2$ is expected in the equilibrium configuration of the cross-slip mechanism (Fig. 2.49). A further expansion of the loop from that position would correspond to $d\langle \tilde{G}_r \rangle = 0$. Under such conditions, the cross-slip process is completed allowing motion of the perfect dislocation in the cross-slip plane. We stress that at the beginning of the cross-slip process (Fig. 2.49), θ is expected to be close to $\pi/2$; this is appreciated using Fig. 2.38.

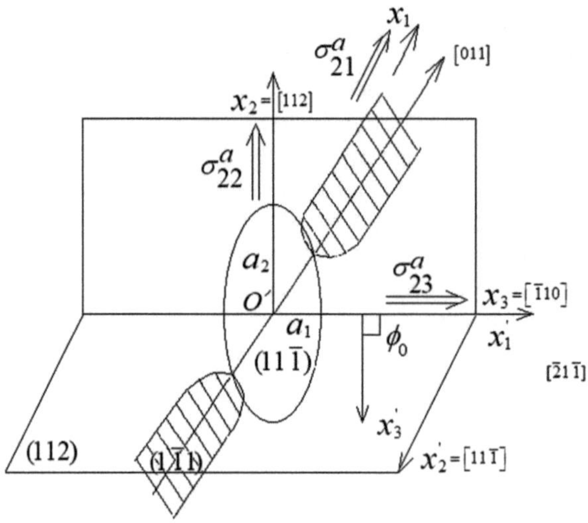

Fig. 2.49 Configuration (at an arbitrary time t) of the screw dislocation 1/2[011] during the cross-slip process. The shape is assumed elliptical in $(11\bar{1})$ and the parts in $(1\bar{1}1)$ are hatched

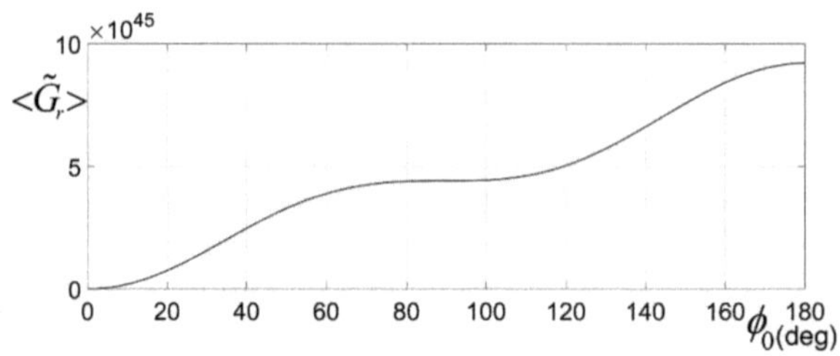

Fig. 2.50 $\langle\tilde{G}_r\rangle$ (2.5.36) as a function of ϕ_0 for $\theta = \pi/2$ and $\theta_0 = \pi/2$. This corresponds to the cross-slip system $[011](11\bar{1})(\phi_0 = \pi/2)$. $M_{12} = M_{13} = 10^{-4}$, $\nu_A(1) = \nu_A(3) = 1/3$, $a_r = 3/4$, copper

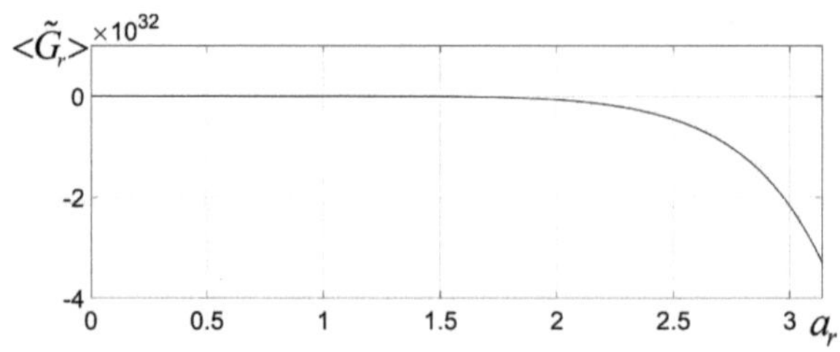

Fig. 2.51 $\langle\tilde{G}_r\rangle$ (2.5.36) as a function of a_r for $\theta = \pi/2$ and the cross-slip system $[011](11\bar{1})(\theta_0 = \phi_0 = \pi/2)$. $M_{12} = M_{13} = 10^{-4}$, $\nu_A(1) = \nu_A(3) = 1/3$, $a_r = 3/4$, copper

DISCUSSION: Assuming the dislocation elliptical in the cross-slip plane at the beginning (Fig. 2.49), $\langle G \rangle$ (2.5.29) applies. Fixing $\theta = \pi/2$ means maintaining a_1 close to zero ($a_1 \simeq 0$) from the definition of θ: we have $\tan\theta = OO'/a_1$ (Fig. 2.38). At $\phi_0 = \pi/2$(Fig. 2.49), the tensile stress along x_3' is $\sigma_{33}^{'A} = \sigma_{22}^{a}$ allowing a_2 to increase. Under such conditions, a_2 can be larger than a_1, meaning that $a_r = a_1/a_2 < 1$. $\langle \tilde{G}_r \rangle$ values in Fig. 2.50 conform to these conditions; positive local maximum of $\langle \tilde{G}_r \rangle$ is at $\phi_0 = \pi/2$. This is the equilibrium state $d\langle \tilde{G}_r \rangle = 0$ of the cross-slip event. Another observed maximum is at $\phi_0 = \pi$. This would correspond to other mechanisms such as twinning and fracture. We can comment about the positive minimum of $\langle \tilde{G}_r \rangle$ observed at $\phi_0 = 0$ (Fig. 2.50): using Fig. 2.49, the tension along x_1' is $\sigma_{11}^{'A} = \sigma_{22}^{a}$ allowing a_1 to increase, but this is hindered by the condition $\theta = \pi/2$ ($a_1 \simeq 0$); Poisson stress is compressive and reads $\sigma_{33}^{'A} = -\nu_A(3)\sigma_{22}^{a}$ hindering also the increase of a_2. Under such conditions $\phi_0 = 0$ is not favoured in Fig. 2.50. In summary, the analysis predicts the observed cross-slip system $[011](11\bar{1})$ of the face-centred-cubic (FCC) structure using the configuration of the screw dislocation, at an arbitrary time t, given by Fig. 2.49. Earliest works (Friedel 1964; Schoeck and Seeger 1955; Wolf 1960; Bonneville et al. 1988) on cross-slip have used configuration like Fig. 2.49. We next make a discussion on cross-slip in close packed hexagonal (CPH) structure. A configuration under load, at a given time t, of the cross-slip event between basal planes is depicted in Fig. 2.52. The dissociation of the dislocation occurs in parallel planes. It is again assumed an elliptical shape for the dislocation on the cross-slip plane at the beginning of the process.

The configuration out of the deviation plane remains unchanged during an increase of time dt; this means that d$\langle G \rangle = 0$ there; hence the change in $\langle G \rangle$ is entirely carried by the elliptical dislocation. This means that the treatment of Sect. 2.5.1 applies for CPH structures. Figure 2.52 is considered in previous works (Duesbery and Hirsch 1968; Escaig 1968).

SUMMARY: Assuming the shape of the dislocation elliptical (Fig. 2.49) in the cross-slip plane at the beginning of the cross-slip process, it is shown that $\langle G \rangle$, a measure of the energy release rate, applies. ½ <110> {111} cross-slip systems, observed in [112] copper single crystals deformed through *stage III* in constant strain rates above

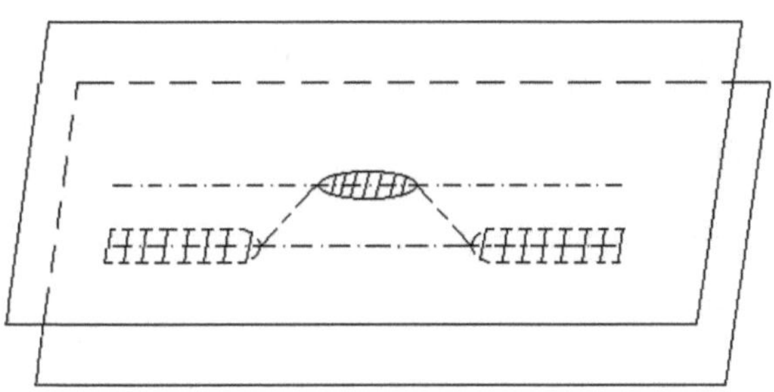

Fig. 2.52 Configuration, under load, of the screw dislocation in the cross-slip process between basal planes of CPH structure

room temperature, do correspond to positive local $\langle G \rangle$ maxima. It is also suggested that the same treatment applies to the cross-slip between parallel basal planes in close packed hexagonal (CPH) structures.

2.5.2 *Conoidal Cracks Growing Uniformly*

A conoidal crack with elliptical bases within cubic crystals, under arbitrarily applied loadings, has been studied in the quasistatic case (Sect. 2.5.1). Referring to this for definitions and notations, we assume that at initial time $t = 0$, the semi-axe of its elliptical crack-front is $a_1 = a_1(0)$ along x_1'-axis of the attached Cartesian system x_j'. At increasing time $t > 0$, its front starts moving uniformly with $a_1 = a_1(0) + v\,t$ (v is constant). $\langle G \rangle$ (at $t = 0$) is given by (2.5.29) (Sect. 2.5.1) and written in the form (2.5.34)

$$\langle \tilde{G} \rangle \equiv \langle G \rangle(t = 0)/G_C^I = \langle \tilde{G} \rangle(\theta, \theta_0, \phi_0, M_{12}, M_{13}, a_r, C_{nm}),$$

$$G_C^I = \frac{2^5 \alpha_0^2 a_r^4}{3\pi^2 C_{44}} (K_I^0)^2;$$

where $K_I^0 = \sigma_{22}^a \sqrt{\pi a_1(0)}$. To obtain its $\langle G \rangle$ (t) from (2.5.34), we just replace $(K_I^0)^2/C_{44}$ by $G^{(I)}$ taken from (2.1.28). Hence, we have

$$\begin{aligned} \langle G \rangle(t) &= G_C^I \langle \tilde{G} \rangle(\theta, \theta_0, \phi_0, M_{12}, M_{13}, a_r, C_{nm}), \\ G_C^I &= \frac{2^5 \alpha_0^2 a_r^4}{3\pi^2} G^{(I)}; \end{aligned} \tag{2.5.38}$$

where in $G^{(I)}$ (2.1.28), $K_\text{I} = \sigma_{22}^a \sqrt{\pi c}$, $c \equiv a_1 = a_1(0) + v\,t$. At $t = 0$, the conoidal crack starts growing uniformly (see Sect. 2.1.2).

2.5.3 *Appendix to Conoidal Cracks (Sect. 2.5.1)*

APPENDIX A: APPLIED STRESSES $(\sigma')^A$

$$\begin{aligned} \sigma_{11}^{'A} = {} & \cos^2\phi_0 \left(\cos^2\theta_0 \sigma_{11}^A + \sin^2\theta_0 \sigma_{22}^A\right) + \sin^2\phi_0 \sigma_{33}^A + \sin 2\theta_0 \cos^2\phi_0 \sigma_{12}^A \\ & + \sin 2\phi_0 \sin\theta_0 \sigma_{23}^A \end{aligned}$$

$$\sigma_{22}^{'A} = \sin^2\theta_0 \sigma_{11}^A - \sin 2\theta_0 \sigma_{12}^A + \cos^2\theta_0 \sigma_{22}^A$$

$$\sigma_{33}^{'A} = \sin^2\phi_0\cos^2\theta_0\sigma_{11}^A + \sin 2\theta_0\sin^2\phi_0\sigma_{12}^A - \sin 2\phi_0\cos\theta_0\sigma_{31}^A + \sin^2\phi_0\sin^2\theta_0\sigma_{22}^A - \sin 2\phi_0\sin\theta_0\sigma_{23}^A + \cos^2\phi_0\sigma_{33}^A$$

$$\sigma_{12}^{'A} = \sin 2\theta_0\cos\phi_0(\sigma_{22}^A - \sigma_{11}^A)/2 + \cos\phi_0\cos 2\theta_0\sigma_{21}^A - \sin\phi_0\sin\theta_0\sigma_{31}^A + \sin\phi_0\cos\theta_0\sigma_{23}^A$$

$$\sigma_{13}^{'A} = -\sin 2\phi_0(\cos^2\theta_0\sigma_{11}^A + \sin^2\theta_0\sigma_{22}^A - \sigma_{33}^A)/2 - \sin 2\phi_0\sin 2\theta_0\sigma_{21}^A/2 + \cos 2\phi_0(\cos\theta_0\sigma_{31}^A + \sin\theta_0\sigma_{23}^A)$$

$$\sigma_{23}^{'A} = -\sin 2\theta_0\sin\phi_0(\sigma_{22}^A - \sigma_{11}^A)/2 + \cos\phi_0\cos\theta_0\sigma_{23}^A - \sin\phi_0\cos 2\theta_0\sigma_{21}^A - \sin\theta_0\cos\phi_0\sigma_{31}^A;$$

$$(\sigma)^A = \begin{pmatrix} -\nu_A(1)\sigma_{22}^a & \sigma_{12}^a & 0 \\ \sigma_{21}^a & \sigma_{22}^a & \sigma_{23}^a \\ 0 & \sigma_{23}^a & -\nu_A(3)\sigma_{22}^a \end{pmatrix}_{x_j}.$$

$\nu_A(m)$ ($m = 1$ and 3) is Poisson's ratio in the direction x_m perpendicular to the applied tension x_2.

APPENDIX B: CRACK DISLOCATION (m) STRESSES $\sigma_{ij}^{(m)}(\vec{x}'; \rho_1)(\vec{x}'; \rho_1)$

$$\sigma_{11}^{(1)}(\vec{x}'; \rho_1) = \frac{b\rho_1}{2\pi 3C_{11}} \int_{-\pi/2}^{\pi/2} d\psi \frac{\sin\psi}{\sqrt{g_2(\psi)}}(x_2' - h)\mathrm{Re}(I_2) \times \left(2(C_{12}C_{44} + C_{11}^2)\overline{SC}^{1/6} - \frac{F_{11}^{(1)}(a)}{\sqrt{\overline{SC}}} + \frac{4C_{12}C_{11}\cos^2\psi}{\overline{SC}^{1/6}}\right),$$

$$\overline{SC} = \sin^6\psi + \cos^6\psi,\ g_2(\psi) = \cos^2\psi + a_r^2\sin^2\psi,$$

$$F_{11}^{(1)}(a) = \sin^4\psi C_{11}(C_{12} + C_{44}) + \cos^4\psi C_{11}(C_{12} + C_{11}) - 2C_{12}^2\sin^2\psi\cos^2\psi;$$

$$\sigma_{11}^{(2)} = \frac{b\rho_1}{\pi 3C_{11}} \int_{-\pi/2}^{\pi/2} \frac{d\psi}{\sqrt{g_2(\psi)}}\mathrm{Re}(Q_1)\left(\frac{\sin^2\psi F_{11}^{(2)}(a) + C_{12}\cos^2\psi F_{11}^{(2)}(b)}{\overline{SC}^{5/6}} + \frac{F_{11}^{(2)}(c)}{\overline{SC}^{1/6}} - \frac{C_{12}(C_{11} + C_{12})\sin^2\psi\cos^2\psi}{\sqrt{\overline{SC}}}\right),$$

$$F_{11}^{(2)}(a) = F_{11}^{(1)}(a),\ F_{11}^{(2)}(c) = \sin^2\psi(C_{12}C_{44} + C_{11}^2) + \cos^2\psi C_{12}(C_{11} + C_{44})$$

$$F_{11}^{(2)}(b) = 2\sin^4\psi C_{11} + \cos^4\psi(C_{11} + C_{44}) - 2C_{12}\sin^2\psi\cos^2\psi;$$

$$\sigma_{11}^{(3)}(\vec{x}'; \rho_1) = 0;$$

$$\sigma_{22}^{(1)}(\vec{x}'; \rho_1) = \frac{b\rho_1}{2\pi 3C_{11}} \int_{-\pi/2}^{\pi/2} \mathrm{d}\psi \frac{\sin\psi}{\sqrt{g_2(\psi)}}(x_2' - h)\mathrm{Re}(I_2) \times \left(2C_{11}(C_{44} + C_{12})\overline{SC}^{1/6} + \frac{F_{22}^{(1)}(a)}{\sqrt{\overline{SC}}} + \frac{4C_{12}^2\cos^2\psi}{\overline{SC}^{1/6}}\right),$$

$$F_{22}^{(1)}(a) = -\big(\sin^4\psi(C_{11}^2 + C_{12}C_{44}) + \cos^4\psi C_{11}(C_{12} + C_{11}) - 2C_{12}C_{11}\sin^2\psi\cos^2\psi\big);$$

$$\sigma_{22}^{(2)} = \frac{b\rho_1}{\pi 3C_{11}} \int_{-\pi/2}^{\pi/2} \frac{\mathrm{d}\psi \mathrm{Re}(Q_1)}{\sqrt{g_2(\psi)}} \left(\frac{F_{22}^{(2)}(a)}{\overline{SC}^{5/6}} + \frac{C_{11}(C_{44} + C_{12})}{\overline{SC}^{1/6}} - \frac{2C_{12}^2\sin^2\psi\cos^2\psi}{\sqrt{\overline{SC}}}\right),$$

$$F_{22}^{(2)}(a) = C_{11}^2(\sin^4\psi + \cos^4\psi) - C_{11}C_{12}\sin^2\psi\cos^2\psi + C_{12}C_{44}\overline{SC};$$

$$\sigma_{22}^{(3)}(\vec{x}'; \rho_1) = 0;$$

$$\sigma_{33}^{(1)}(\vec{x}'; \rho_1) = \frac{b\rho_1}{6\pi} \int_{-\pi/2}^{\pi/2} \mathrm{d}\psi \frac{\sin\psi}{\sqrt{g_2(\psi)}}(x_2' - h)\mathrm{Re}(I_2) \times \left(2(C_{11} + C_{44})\overline{SC}^{1/6} + \frac{F_{33}^{(1)}(a)}{\sqrt{\overline{SC}}} + \frac{4C_{12}\cos^2\psi}{\overline{SC}^{1/6}}\right),$$

$$F_{33}^{(1)}(a) = -\big(2\cos^4\psi C_{11} + \sin^4\psi(C_{11} + C_{44}) - 2C_{12}\sin^2\psi\cos^2\psi\big);$$

$$\sigma_{33}^{(2)} = \frac{b\rho_1}{\pi 3C_{11}} \int_{-\pi/2}^{\pi/2} \frac{\mathrm{d}\psi \mathrm{Re}(Q_1)}{\sqrt{g_2(\psi)}} \left(\frac{F_{33}^{(2)}(a)}{\overline{SC}^{5/6}} + \frac{F_{33}^{(2)}(b)}{\overline{SC}^{1/6}} + \frac{F_{33}^{(2)}(c)}{\sqrt{\overline{SC}}}\right),$$

$$F_{33}^{(2)}(a) = C_{11}C_{12}(\sin^4\psi + \cos^4\psi) - 2C_{12}^2\sin^2\psi\cos^2\psi + C_{12}C_{44}\sin^6\psi + C_{11}C_{12}\sin^2\psi\cos^4\psi + C_{11}C_{44}\cos^6\psi + C_{11}^2\sin^4\psi\cos^2\psi,$$

$$F_{33}^{(2)}(b) = C_{44}C_{12} + C_{11}C_{12}\sin^2\psi + C_{11}^2\cos^2\psi,$$

$$F_{33}^{(2)}(c) = -C_{12}(C_{11} + C_{12})\sin^2\psi\cos^2\psi;$$

$$\sigma_{33}^{(3)}(\vec{x}'; \rho_1) = 0; \tag{B1}$$

$$\sigma_{12}^{(1)} = \frac{C_{44}b\rho_1}{C_{11}6\pi}\int_{-\pi/2}^{\pi/2}\frac{d\psi\,\mathrm{Re}(Q_1)}{\sqrt{g_2(\psi)}}\left(\frac{F_{12}^{(1)}(a)}{\overline{SC}^{5/6}} + \frac{F_{12}^{(1)}(c)}{\overline{SC}^{1/6}} + \frac{F_{12}^{(1)}(b)}{\sqrt{\overline{SC}}}\right),$$

$$F_{12}^{(1)}(a) = 2\big(C_{11}\sin^2\psi\cos^4\psi - 2C_{12}\sin^4\psi\cos^2\psi - C_{11}\cos^6\psi\big),$$

$$F_{12}^{(1)}(b) = C_{44}\sin^4\psi + C_{11}\cos^4\psi,\ F_{12}^{(1)}(c) = 2\big(C_{44}\sin^2\psi - 2C_{12}\cos^2\psi\big);$$

$$\sigma_{12}^{(2)}(\vec{x}'; \rho_1) = -\frac{C_{44}b\rho_1}{C_{11}6\pi}\int_{-\pi/2}^{\pi/2} d\psi\frac{\sin\psi}{\sqrt{g_2(\psi)}}(x_2' - h)\mathrm{Re}(I_2)$$
$$\times\left(\frac{F_{12}^{(2)}(a)}{\overline{SC}^{5/6}} + \frac{F_{12}^{(2)}(b)}{\sqrt{\overline{SC}}} + \frac{F_{12}^{(2)}(c)}{\overline{SC}^{1/6}}\right),$$

$$F_{12}^{(2)}(a) = 2\big(C_{11}(\sin^4\psi + \cos^4\psi) - 2C_{12}\sin^2\psi\cos^2\psi - C_{11}\overline{SC}\big),$$

$$F_{12}^{(2)}(c) = 2\big(C_{44} - 2C_{12}\cos^2\psi\big),\ F_{12}^{(2)}(b) = C_{44}\sin^4\psi + C_{11}\cos^4\psi;$$

$$\sigma_{12}^{(3)} = \frac{C_{44}b\rho_1}{C_{11}6\pi}\int_{-\pi/2}^{\pi/2} d\psi\frac{\sin 2\psi}{\sqrt{g_2(\psi)}}\mathrm{Re}(Q_1)\left(\frac{F_{12}^{(3)}(a)}{\overline{SC}^{5/6}} + \frac{C_{44}}{\overline{SC}^{1/6}}\right),$$

$$F_{12}^{(3)}(a) = C_{11}(\sin^4\psi + \cos^4\psi) - 2C_{12}\sin^2\psi\cos^2\psi;$$

$$\sigma_{13}^{(1)}(\vec{x}'; \rho_1) = -\frac{C_{44}b\rho_1}{C_{11}6\pi}\int_{-\pi/2}^{\pi/2} d\psi\frac{\cos\psi}{\sqrt{g_2(\psi)}}(x_2' - h)\mathrm{Re}(I_2)$$
$$\times\left(\frac{F_{13}^{(1)}(a)}{\sqrt{\overline{SC}}} + 2C_{11}\overline{SC}^{1/6} + \frac{4C_{12}\cos^2\psi}{\overline{SC}^{1/6}}\right),$$

$$F_{13}^{(1)}(a) = -\big(C_{44}\sin^4\psi + C_{11}\cos^4\psi\big);$$

$$\sigma_{13}^{(2)} = \frac{C_{44}b\rho_1}{C_{11}6\pi}\int\limits_{-\pi/2}^{\pi/2} \mathrm{d}\psi \frac{\sin 2\psi}{\sqrt{g_2(\psi)}}\mathrm{Re}(Q_1)\left(\frac{F_{13}^{(2)}(a)}{\overline{SC}^{5/6}} + \frac{2C_{11}}{\overline{SC}^{1/6}} - \frac{C_{12}}{\sqrt{\overline{SC}}}\right),$$

$$F_{13}^{(2)}(a) = (C_{11} + C_{44})(\sin^4\psi + \cos^4\psi);$$

$$\sigma_{13}^{(3)}(\vec{x}'; \rho_1) = 0;$$

$$\sigma_{23}^{(1)} = \frac{C_{44}b\rho_1}{C_{11}6\pi}\int\limits_{-\pi/2}^{\pi/2} \mathrm{d}\psi \frac{\sin 2\psi}{\sqrt{g_2(\psi)}}\mathrm{Re}(Q_1)\left(\frac{F_{23}^{(1)}(a)}{\overline{SC}^{5/6}} + \frac{C_{44}}{\overline{SC}^{1/6}}\right),$$

$$F_{23}^{(1)}(a) = C_{11}(\sin^4\psi + \cos^4\psi) - 2C_{12}\sin^2\psi\cos^2\psi;$$

$$\sigma_{23}^{(2)}(\vec{x}'; \rho_1) = -\frac{C_{44}b\rho_1}{C_{11}6\pi}\int\limits_{-\pi/2}^{\pi/2} \mathrm{d}\psi \frac{\cos\psi}{\sqrt{g_2(\psi)}}(x_2' - h)\mathrm{Re}(I_2)$$
$$\times\left(\frac{F_{23}^{(2)}(a)}{\overline{SC}^{5/6}} + \frac{F_{23}^{(2)}(c)}{\sqrt{\overline{SC}}} + \frac{F_{23}^{(2)}(b)}{\overline{SC}^{1/6}}\right),$$

$$F_{23}^{(2)}(a) = F_{12}^{(2)}(a),\ F_{23}^{(2)}(b) = 2\big(C_{44} - 2C_{12}\sin^2\psi\big),$$

$$F_{23}^{(2)}(c) = C_{11}\sin^4\psi + C_{44}\cos^4\psi;$$

$$\sigma_{23}^{(3)} = \frac{C_{44}b\rho_1}{C_{11}6\pi}\int\limits_{-\pi/2}^{\pi/2} \mathrm{d}\psi \frac{\mathrm{Re}(Q_1)}{\sqrt{g_2(\psi)}}\left(\frac{F_{23}^{(3)}(a)}{\overline{SC}^{5/6}} + \frac{F_{23}^{(3)}(b)}{\overline{SC}^{1/6}} + \frac{F_{23}^{(3)}(c)}{\sqrt{\overline{SC}}}\right),$$

$$F_{23}^{(3)}(a) = 2\big(C_{11}\cos^2\psi(\sin^4\psi + \cos^4\psi) - 2C_{12}\sin^2\psi\cos^4\psi - C_{11}\overline{SC}\big),$$

$$F_{23}^{(3)}(b) = 2\big(C_{44}\cos^2\psi - 2C_{12}\sin^2\psi\big),\ F_{23}^{(3)}(c) = F_{23}^{(2)}(c); \tag{B2}$$

$\mathrm{Re}(Q_1) = \frac{\overline{y}_1}{(\overline{y}_1^2 - \Omega^2)^{3/2}}$ and $\mathrm{Re}(I_2) = -\frac{3\overline{y}_1\Omega}{(\overline{y}_1^2 - \Omega^2)^{5/2}}$, $\Omega < \overline{y}_1$ for both, where $\overline{y}_1 = \frac{\rho_1}{a_r}\sqrt{g_2(\psi)}$, $g_2(\psi) = \cos^2\psi + a_r^2\sin^2\psi$ and $\Omega = \sin\psi x_1' + \cos\psi x_3'$.

2.6 Electromagnetic Fields Due to Steadily Propagating Cracks

Electromagnetic phenomena generated in materials by dislocations and cracks are known for more than half a century; a review of works on electrical effects of dislocations in various materials (ionic crystals, metals, semiconductors) has been given by Nabarro (1967) who also provides us with the following simple result valid for metals: the inhomogeneous dilatation $\Delta_D(\vec{x})$ ($\vec{x}$ position vector) of the lattice due to the dislocation leads to a redistribution of the conduction electrons. As a result, an electric potential $V_D(\vec{x})$ and electric field $\overrightarrow{E}_D(\vec{x})$ are built round a dislocation; furthermore the relation linking Δ_D and V_D is linear. The proportionality relation between V_D and Δ_D revealed in metallic materials is helpful. This suggests that a similar relation between the electric potential $V_C(\vec{x})$(resp. V_{CD}) and the corresponding lattice dilatation $\Delta_C(\vec{x})$(resp. Δ_{CD}) exists in presence of a crack (resp. elastic–plastic crack). This is because a crack is equivalent to a continuous distribution of dislocations with infinitesimal Burgers vectors. This study aims at providing analytical expressions for V_C,V_{CD} and corresponding electric fields $\overrightarrow{E}_C$ and $\overrightarrow{E}_{CD}$ for simple crack geometries. These expressions depend implicitly on time through parameters such as applied stress, crack length and size of plastic zone ahead of the crack front. We can then link the electric fields to their induced magnetic fields through Maxwell equations and interpret in this way the corresponding electromagnetic phenomena. The results apply first to metals for which several works on the emission of electromagnetic radiation exist (Misra 1975; Misra and Kumar 2004; Jagasivamani and Iyer 1988, among others). This work also finds application in any material where proportionality exists between electric potential and lattice dilatation such as in ice (Petrenko 1993, 1996; Evtushenko et al. 1987). The present analysis demonstrates once again the conceptual advantage of representing the crack by a continuous distribution of infinitesimal dislocations. This allows a simplified treatment (calculation of electric polarization vector, electric potential and electric and magnetic fields, for instance) based on the dislocation only. Additional experiments of concern are numerous and involve several different materials (see for example: Golovin et al. 1985; Cress et al. 1987; Frid et al. 2003; Mori et al. 2004; Hadjicontis et al. 2004). In what follows, an elastic–plastic crack model under mode I loading is considered from which an expression for the electric potential is calculated; the isolated crack is equally treated. The electric field induced magnetic field and Poynting's vector are derived. Graphical representations of the various electromagnetic quantities are performed for the isolated crack that propagates steadily. A discussion is made of the results, followed by a conclusion.

The Model

The model we shall refer to is shown in Fig. 2.53. This is a linear isotropic elastic medium to which we attach a Cartesian coordinate system x_i. It contains a planar crack in the Ox_1x_3-plane that extends from $x_1 = -c$ to c and runs indefinitely in the

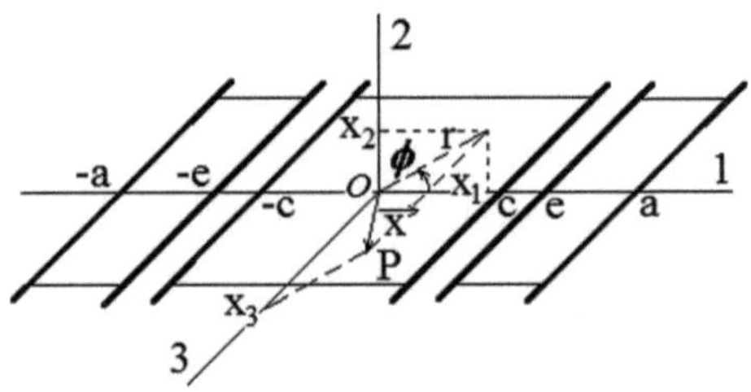

Fig. 2.53 An elastic–plastic crack system located in the Ox_1x_3-plane. The crack extends from $x_1 = -c$ to c with a straight front running indefinitely in the x_3-direction. On both sides of the crack, a dislocation-free zone $c < |x_1| < e$ and plastic domain $e \leq |x_1| \leq a$ are present. The system is subjected to uniform applied tension σ_a in the x_2-direction (see text)

x_3-direction. Coplanar to the crack, and on both sides of it, a dislocation-free zone ($c < |x_1| < e$) and a localized crack-tip plastic zone ($e \leq |x_1| \leq a$) are present.

This model of elastic–plastic crack explains quite well the relationship between applied stress, yield stress and parameters c, e and a (Bilby et al. 1963; Anongba and Vitek 2003). We can therefore expect this modelling to give appropriate values of the electromagnetic fields in presence of crack and localized plasticity at crack tip. In many experiments, dislocation generation is observed from the tip of the crack prior to crack motion. Anongba and Vitek (2003) have shown that the condition $G = 2\gamma$ holds when the plastic zone is fixed at the initiation stage of crack propagation, and they provide an appropriate expression for G. We shall restrict ourselves to mode I loading only (uniform applied stress σ_a in the x_2-direction). Both the crack and the plastic domain are mathematically represented by planar continuous arrays of infinitely long straight edge dislocations (Burgers vector $(0, b, 0)$,$b > 0$) lying parallel to the x_3-axis. The dislocations are in equilibrium, can climb freely in the crack but must overcome a friction stress $\sigma_y(>\sigma_a)$ representing the yield stress in the plastic region.

The Electric Potential

To calculate the electric potential, we assume as in metals that it is proportional to the dilatation of the lattice. We begin with the contribution due to a single dislocation and then, by superposition, display the results for the isolated crack (i.e. a crack with no plastic zone) and the elastic–plastic crack of Fig. 2.53. Consider an edge dislocation (Burgers vector $(0, b, 0)$) at the origin that lies indefinitely in the x_3-direction. The dilatation Δ_D at a point P with coordinates (x_1, x_2, x_3) is proportional to $\sum_i \sigma_{ii}$ where (σ_{ij}) is the stress tensor. In the present geometry, (σ_{ij}) may be taken from Anongba (2009, 2010) or Anongba et al. 2011; this gives

$$\Delta_D = \frac{b}{2\pi}\left(\frac{1-2\nu}{1-\nu}\right)\frac{\cos\phi}{r}; \tag{2.6.1}$$

ν is Poisson's ratio, r and ϕ are the associated cylindrical coordinates (see Fig. 2.53) with $r^2 = x_1^2 + x_2^2$. In metallic materials, the electric potential $V_D(\vec{x})$, $\vec{x} = \overrightarrow{OP}$, round the dislocation is proportional to Δ_D as (Landauer 1951; Dexter 1952; Nabarro 1967; Brown 1966):

$$V_D(\vec{x}) = c_0 \frac{E_F}{q} \frac{b}{2\pi} \left(\frac{1-2\nu}{1-\nu} \right) \frac{\cos\phi}{r} \equiv A \frac{\cos\phi}{r}. \tag{2.6.2}$$

Here $E_F = \hbar^2 k_F^2 / 2m$ where k_F is the conduction electron wave number on the Fermi surface, m and q the electron mass and charge, $\hbar$ has its usual meaning and c_0 is a constant, given by Nabarro (1967) as $c_0 = 4/15$. By superposition, the electric potential V_{CD} round the crack system in Fig. 2.53 can be written as

$$V_{CD}(\vec{x}) = A \left(\int_{-a}^{-e} + \int_{-c}^{c} + \int_{e}^{a} \right) \frac{x_1 - x_1'}{\left(x_1 - x_1'\right)^2 - x_2^2} f_{CD}\left(x_1'\right) \mathrm{d}x_1' \tag{2.6.3}$$

where the dislocation distribution function $f_{CD}(x_1)$ is such that $f_{CD}(x_1)\mathrm{d}x_1$ gives the number of dislocations in the interval dx_1 at the position x_1 ($|x_1| < c$ and $e < |x_1| < a$) and $f_{CD}(\pm e) = f_{CD}(\pm a) = 0$. Analytical expressions for f_{CD} have been given in different forms by Chang and Ohr (1983) and Anongba and Vitek (2003); the latter expression appears simpler to deal with and will be adopted in the following. Before displaying the result of the integration in (2.6.3), we shall first write the electric potential V_C for the isolated crack (i.e. the crack of Fig. 2.53 only, without the plastic regions $e < |x_1| < a$).

Bilby and Eshelby (1968) have given the distribution function of crack dislocations for the isolated crack to be $f_C(x_1) = \sigma_a x_1 / \pi C_1 \sqrt{c^2 - x_1^2}$ ($C_1 = \mu b / 2\pi(1-\nu)$, μ the shear modulus), that we use and, by integration, we obtain

$$V_C(\vec{x}) = (g_C(z) + g_C(\overline{z}))/2 = \mathrm{Re}[g_C(z)] \tag{2.6.4}$$

with

$$g_C(z) = \overline{A}\sigma_a \left(-1 + \sqrt{\frac{z^2}{z^2 - c^2}} \right) \tag{2.6.5}$$

where $\overline{A} = A/C_1 = c_0(1-2\nu)E_F/\mu q$, $z = x_1 + ix_2$ is a complex variable and Re[...] denotes the real part of the complex quantity inside the brackets [...].

For the elastic–plastic crack of Fig. 2.53, Anongba and Vitek (2003) have given the dislocation distribution function f_{CD} in the form: for $|x_1| < c$,

$$f_{CD}(x_1) = \frac{2\sigma_y F(\pi/2,k)}{C_1\pi^2}\left(\frac{(e^2-c^2)x_1}{e\sqrt{a^2-c^2}}\sqrt{\frac{a^2-x_1^2}{(c^2-x_1^2)(e^2-x_1^2)}}\right.$$
$$\left. + \operatorname{sgn}(x_1)Z(\beta_1(x_1),k)\right)$$

and

$$f_{CD}(x_1) = \frac{2\sigma_y F(\pi/2,k)}{C_1\pi^2}\operatorname{sgn}(x_1)Z(\beta_2(x_1),k)\text{for} e \le |x_1| \le a; \tag{2.6.6}$$

here

$k^2 = c^2(a^2-e^2)/e^2(a^2-c^2)$, $\operatorname{sgn}(x_1) = x_1/|x_1|$, $\beta_1(x_1) = \sin^{-1}(1/\sqrt{n(x_1)})$, $\beta_2(x_1) = \sin^{-1}(\sqrt{n(x_1)}/k)$,$n(x_1) = c^2(e^2-x_1^2)/e^2(c^2-x_1^2)$.

and F and Z are the elliptic integral of first kind and Jacobi's zeta function, respectively. With the help of this expression for f_{CD}, we can perform the necessary integrations in (2.6.3) and write V_{CD} in the following form:

$$V_{CD}(\vec{x}) = (g_{CD}(z) + g_{CD}(\bar{z}))/2 = \operatorname{Re}[g_{CD}(z)] \tag{2.6.7}$$

with

$$g_{CD}(z) = 2(\overline{A}/\pi^2)\sigma_y F(\pi/2,k)\left(\frac{e^2-c^2}{e^2(a^2-c^2)}\left[-2a^2\prod(\pi/2,-\frac{c^2}{a^2-c^2},k)\right.\right.$$
$$\left. + \frac{2(a^2-e^2)F(\pi/2,k)z^2}{z^2-e^2} + 2(c^2-e^2)\frac{z^2(a^2-z^2)}{(z^2-e^2)(z^2-c^2)}\prod(\pi/2,n(z),k)\right]$$
$$+Zln(\pi/2,m(z),k) - Zln(\pi/2,n(z),k) + Zln(\pi/2,c^2/e^2,k)$$
$$\left.-Zln(\pi/2,\frac{a^2-e^2}{a^2-c^2},k)\right); \tag{2.6.8}$$

where

$m(z) = (a^2-e^2)(z^2-c^2)/(a^2-c^2)(z^2-e^2)$, $n(z) = c^2(e^2-z^2)/e^2(c^2-z^2)$.

and Π and Zln are the elliptic integral of third kind and a "special" function defined as (here ϕ, n and k are variables)

$$Z\ln(\phi,n,k) = \int_0^\phi \ln(1-n\sin^2\beta)\mathrm{d}Z(\beta,k).$$

In this definition of $Z\ln$, ln is the natural logarithm, dZ is the differentiation of the Jacobi's zeta function $Z(\beta,k)$ with respect to variable β only (i.e. $\mathrm{d}Z \equiv (\partial Z/\partial\beta)\mathrm{d}\beta$).

Electric and Magnetic Fields

The electric potential $V(\vec{x})$ and electric field $\vec{E}(\vec{x})$ are related by the relation $\vec{E} = -\overrightarrow{grad}V$; when, as in the present study, V can be written as $V(\vec{x}) = \text{Re}[g(z)]$ with $z = x_1 + ix_2$, we arrive at

$$\vec{E}(\vec{x}) = \begin{pmatrix} -\text{Re}[\partial g/\partial z] \\ \text{Im}[\partial g/\partial z] \\ 0 \end{pmatrix} \tag{2.6.9}$$

in the basis $(\vec{e}_1, \vec{e}_2, \vec{e}_3)$ of the Cartesian system $Ox_1x_2x_3$(Fig. 2.53). Im[...] denotes the imaginary part of the function inside the brackets [...]. $\vec{E}_C$ and $\vec{E}_{CD}$ are obtained from (2.6.9) with g taking the values g_C and g_{CD} respectively. Because the electric field $\vec{E}$ depends on parameters c, e, a, and σ_u, it depends implicitly on time through the variation with time of these parameters. Consequently, an associated induced magnetic field $\vec{B}$ exists that can be found from Maxwell's equations. In a linear isotropic medium, this induced $\vec{B}$ is linked to $\vec{E}$ by

$$\overrightarrow{rot}\left(\frac{\vec{B}}{\mu_0\mu_1}\right) = \frac{\partial}{\partial t}\left(\varepsilon_0\varepsilon_1\vec{E}\right); \tag{2.6.10}$$

ε_0 and μ_0 are the electric and magnetic constants; ε_1 and μ_1 are the relative permittivity and permeability, respectively. We should add to the electric field an additional term $\vec{E}'$ corresponding to the temporal variation of the magnetic field through the relation

$$\overrightarrow{rot}\,\vec{E}' = -\partial\vec{B}/\partial t. \tag{2.6.11}$$

Now assuming in a similar way as $\vec{E}$(2.6.9), $\vec{E}'$ to be independent of x_3 with no component along the x_3-direction, (2.6.11) leads to $\partial B_i/\partial t = 0$, $i = 1$ and 2, where B_i is the component along the x_i-axis of $\vec{B}$. Consequently, B_i ($i = 1$ and 2) are constant with time and can be set equal to zero if initially there is no current. In the following, we shall restrict ourselves to the calculation of $\vec{B}$(2.6.10) assuming $\vec{B}$ to be parallel to x_3 only. Also, $\vec{E}'$ is viewed as of second order in magnitude, and will no more be considered. We thus obtain

$$\vec{B} = B_3\vec{e}_3 = -\varepsilon_0\varepsilon_1\mu_0\mu_1(\partial\text{Im}[g(z)]/\partial t)\vec{e}_3. \tag{2.6.12}$$

Again, $\vec{B}_C$(isolated crack) and $\vec{B}_{CD}$(elastic–plastic crack, Fig. 2.53) are obtained from (2.6.12) with g taking the values g_C and g_{CD}. We stress that we assume no current for simplicity. This applies to pure metals. Foreign ions, introduced into the lattice, could flow under the electric field and generate electric currents. The Poynting's vector $\overrightarrow{S}$ is the instantaneous energy that the electromagnetic wave transmits per unit time through a unit surface perpendicular to the direction of wave propagation. It reads

$$\begin{aligned}\overrightarrow{S} &= \overrightarrow{E} \wedge \frac{\overrightarrow{B}}{\mu_0\mu_1} \\ &= \frac{B_3}{\mu_0\mu_1}\begin{pmatrix}\mathrm{Im}[\partial g/\partial z] \\ \mathrm{Re}[\partial g/\partial z] \\ 0\end{pmatrix}\end{aligned} \tag{2.6.13}$$

with $\vec{S}_C$,for the isolated crack, and $\overrightarrow{S}_{CD}$, for the elastic–plastic crack, both defined in a manner similar as previously.

Steady Propagation of Isolated Crack

We consider here that the crack is isolated, and we present graphical representations of various electromagnetic quantities. We restrict ourselves to the first quadrant (i.e. both x_1 and $x_2 \geq 0$) as justified into two steps: (1) g_C and $\partial g_C/\partial z$ are even and odd respectively, this reduces the domain to $-\pi/2 \leq \arg z \leq \pi/2$(i.e.$x_1 \geq 0$). (2) Changing the sign of x_2 in g_C and $\partial g_C/\partial z$ is equivalent to taking their complex conjugates because z changes into $\bar{z}$. Hence, in the expression of the electric field, $\mathrm{Re}[\partial g_C/\partial z]$ remains unchanged while $\mathrm{Im}[\partial g_C/\partial z]$ changes sign only. We can conveniently ascribe x_2 to positive values only.

Electric Potential

The electric potential $V_C(\vec{x})$, shown in Fig. 2.54a, is a surface that consists of various coloured bands. To each strip is associated a small range of values reported for convenience on a vertical scale (see in Fig. 2.54a). In the Ox_1x_3-plane ($x_2 = 0$), the electric potential is constant for $x_1 < c$ (zone occupied by the crack), blue colour at the bottom of the vertical scale to which is attached the value (− 1). For $x_1 > c$, the potential, from zero (x_1 large), increases indefinitely when x_1 tends towards the tip of the crack ($x_1 = c$). In the Ox_2x_3-plane ($x_1 = 0$), the electric potential increases in the form of a parabola, from a value (− 1) ($x_2 = 0$) (blue, at the bottom of the vertical scale), up to a value close to zero (brown) for x_2 large. In x_1x_2 planes ($x_3 =$ constant), positive constant $V_C/\overline{A}\sigma_a$ contours have concentric shapes that shrink about the crack tip. The associated $V_C/\overline{A}\sigma_a$ values increase indefinitely.

Electric Field

The electric field $\overrightarrow{E}_C(\vec{x})$ is given by (2.6.9) with

$$\frac{\partial g_C}{\partial z} = -\overline{A}\sigma_a \frac{c^2}{\left(z^2 - c^2\right)^{3/2}}. \tag{2.6.14}$$

At $P(r, \phi, x_3)$, the following points can be quoted:

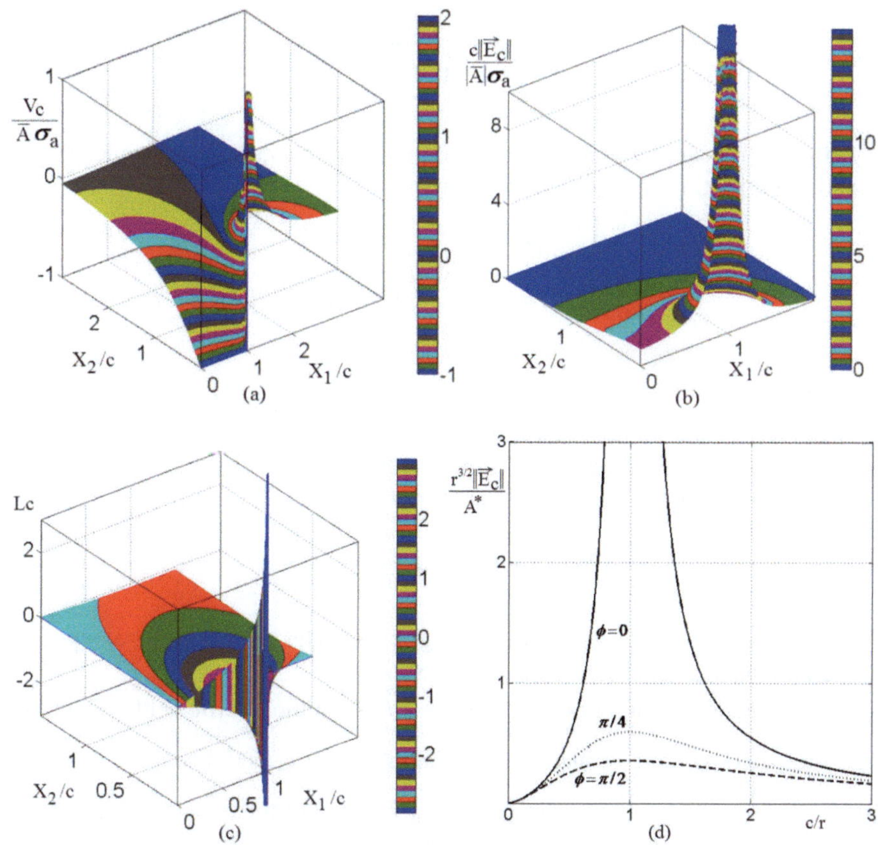

Fig. 2.54 **a** Electric potential V_C (2.6.4). **b** Spatial distribution of the magnitude $\left\|\overrightarrow{E}_C\right\|$ of electric field (2.6.9) (2.6.14) at different (x_1, x_2) and (d) $\left\|\overrightarrow{E}_C\right\|$ values (2.6.17) calculated at three fixed positions $P(r, \psi, x_3)$, $\psi = 0, \pi/4$ and $\pi/2$, as a function of crack half-length c. We assume the Griffith relation (2.6.16) to hold at any time. **c** Lines of electric field $Lc =$ constant (2.6.15)

- In the Ox_1x_3-plane ($\phi = 0$), the x_1-component of $\overrightarrow{E}_C$,$E_C(1)$, is equal to zero for $r < c$ because $\partial g_C/\partial z$ is pure imaginary. The reduced x_2-component of $\overrightarrow{E}_C$,$E_C(2)/\overline{A}\sigma_a$, is then negative and given by

$$\frac{E_C(2)}{\overline{A}\sigma_a} = -c^2\sqrt{\frac{1}{\left(c^2 - r^2\right)^3}}.$$

For $r > c$, $\partial g_C/\partial z$ is pure real and $E_C(2)$ is zero while $E_C(1)/\overline{A}\sigma_a$ is positive and reads

$$\frac{E_C(1)}{\overline{A}\sigma_a} = c^2\sqrt{\frac{1}{\left(r^2 - c^2\right)^3}}.$$

- In the Ox_2x_3-plane ($\phi = \pi/2$),$\partial g_C/\partial z$ is pure imaginary, $E_C(1)$ is identically zero and $E_C(2)/\overline{A}\sigma_a$ is negative and given by

$$\frac{E_C(2)}{\overline{A}\sigma_a} = -c^2\sqrt{\frac{1}{\left(c^2 + r^2\right)^3}}.$$

- For $0 < \phi < \pi/2$, the electric field $\overrightarrow{E}_C$ is tangent to the lines of electric field at any point P, by definition; this corresponds to the condition $d\vec{x}(P) \wedge \overrightarrow{E}_C(P) = 0$ where $d\vec{x}(P)$ is an infinitesimal displacement vector. Using this condition, it is easy to show that the lines of field are given by

$$L_C(\vec{x}) = \text{Im}[g_C(z)] = \text{constant}. \tag{2.6.15}$$

Constant L_C contours are shown on Fig. 2.54c. The sense of $\overrightarrow{E}_C$ along these curves may be deduced from (2.6.9) and (2.6.14).

The spatial distribution of the magnitude $\left\|\overrightarrow{E}_C\right\|$ of the electric field is shown on Fig. 2.54b. In Ox_1x_3, $\left\|\overrightarrow{E}_C\right\|$ increases indefinitely when x_1 tends towards c from both sides. $\left\|\overrightarrow{E}_C\right\|$ takes a non-zero value (value 1.12, purple) for $x_1 = 0$; it decreases to zero for x_1 large (above c). In Ox_2x_3, $\left\|\overrightarrow{E}_C\right\|$ decreases, from the value(1.12, purple) ($x_2 = 0$), towards zero when x_2 increases. In x_1x_2(x_3 = constant), spatial positions corresponding to $\left\|\overrightarrow{E}_C\right\|$ = constant, form circles that shrink about $x_1 = c$ with associated $\left\|\overrightarrow{E}_C\right\|$ increasing indefinitely. We would like to express quantity $\left\|\overrightarrow{E}_C(P)\right\|$ at a given position $P(x_1, x_2, x_3)$ with cylindrical coordinates (r, ϕ, x_3) as a function of the crack half-length c. Because $\left\|\overrightarrow{E}_C\right\|$ depends on the applied stress σ_a, it is necessary to specify the dependence of σ_a with respect to c. In the present study, we shall assume that the Griffith relation $G = 2\gamma$ is satisfied at any time. This leads to

$$c = \frac{2\gamma E}{\pi\left(1 - \nu^2\right)}\frac{1}{\sigma_a^2} \tag{2.6.16}$$

where E is Young's modulus. We can write, making use of (2.6.9), (2.6.14) and (2.6.16), that

$$\left\|\overrightarrow{E}_C(\vec{x})\right\| = A^*\left(\frac{c}{\left|z^2 - c^2\right|}\right)^{3/2} \tag{2.6.17}$$

with

$$A^* = \frac{c_0(1-2\nu)E_F}{\mu|q|}\sqrt{\frac{2\gamma E}{\pi(1-\nu^2)}}.$$

Figure 2.54d displays $\left\|\vec{E}_C\right\|$ for three different points $P(r, \phi, x_3)$,$\phi = 0$,$\pi/4$ and $\pi/2$, as a function of c. When $\phi = 0$, the point P lies in Ox_1x_3, the plane of the crack. From the value $c = 0$, $\left\|\vec{E}_C\right\|$ increases (from the value zero) indefinitely when c tends towards P ($c \to r$). When $c > r$, $\left\|\vec{E}_C\right\|$ decreases gradually when c increases. For the two other positions of P, $\phi = \pi/4$ and $\pi/2$, $\left\|\vec{E}_C\right\|$ increases from zero ($c = 0$) up to a maximum for c about r; then $\left\|\vec{E}_C\right\|$ decreases continually when c further increases in the range of large values.

Magnetic Field

$B_3(\vec{x})$, the x_3-component of the magnetic field (2.6.12), is represented in Fig. 2.55a, b. We assume that g_C depends on time through σ_a and c; we link σ_a to c by (2.6.16), this gives

$$B_3(\vec{x}) = \varepsilon_0\varepsilon_1\mu_0\mu_1 A^*(\mathrm{d}c/\mathrm{d}t)c^{-3/2}\mathrm{Im}\left[-\frac{1}{2}\sqrt{\frac{z^2}{z^2-c^2}} + c^2\sqrt{\frac{z^2}{\left(z^2-c^2\right)^3}}\right]. \tag{2.6.18}$$

In $Ox_1x_3(x_2 = 0)$, B_3 (Fig. 2.55a) is zero for $x_1 > c$; for $x_1 < c$, B_3 has the shape of a parabola; it increases, from the value zero for $x_1 = 0$(small strip of purple limited by the parabola), towards infinity when x_1 tends towards c. The corresponding stripes are displayed vertically and are limited by a parabola. Out of $Ox_1x_3(0 < x_1 < c)$, B_3(Fig. 2.55a) takes values close to zero (negative for both x_2 small and x_1 sufficiently close to c) and then increases slowly to reach a maximum when x_2 increases; further increase of x_2 leads to a decrease of B_3. Out of $Ox_1x_3(x_1 > c)$, B_3 (Fig. 2.55a) has negative values close to zero for x_2 small; when x_2 increases, B_3 first decreases up to a negative minimum (the magnitude of which increases as x_1 tends towards c); then B_3 increases from that minimum with x_2. A boundary curve, that is visible on Fig. 2.55a, separates region 1 (with B_3 positive) from region 2 (with B_3 negative). We have represented B_3(Fig. 2.55b) at three different positions $P(r, \phi, x_3)(\phi = 0$, $\pi/4$ and $\pi/8)$, as a function of crack half-length c. When $\phi = 0$ (i.e. P on the Ox_1x_3 plane) B_3 is zero for $c < r$. In the region $c > r$, B_3 increases indefinitely when c decreases towards r. For the two other positions of P, $\phi = \pi/4$ and $\pi/8$,B_3 (from the value zero for $c = 0$) displays a single oscillation with a negative minimum and positive maximum for $c < r$ and $c > r$ respectively. B_3 tends towards zero for large

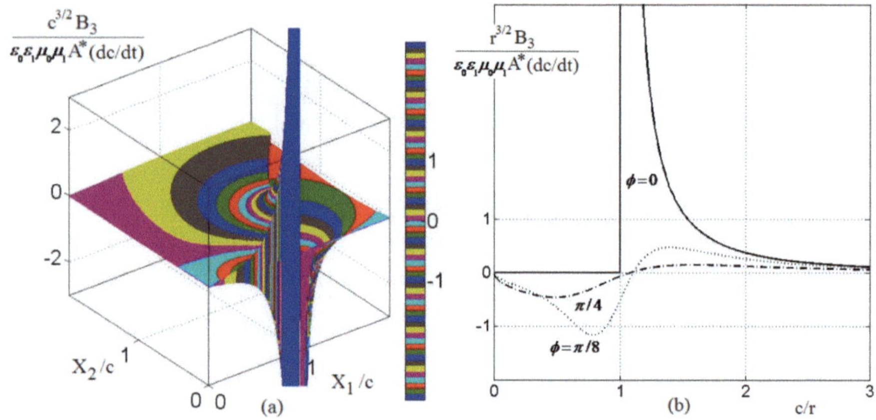

Fig. 2.55 x_3-component B_3 (2.6.18) of the magnetic field $\overrightarrow{B}_C$ (2.6.12) for the isolated crack: **a** spatial distribution at different (x_1, x_2) for fixed crack half-length c and **b** B_3 values calculated at three fixed positions $P(r, \phi, x_3)$, $\phi = 0, \pi/4$ and $\pi/8$, as a function of crack half-length c assuming the Griffith relation (2.6.16) to hold

values of c. Note that for $\phi = \pi/2(Ox_2x_3$-plane, $x_1 = 0)$, B_3(2.6.18) is identically zero.

Poynting's Vector

To illustrate the Poynting vector $\overrightarrow{S}$ (2.6.13), we can make use of (2.6.18) for B_3 and (2.6.14) for $\partial g_C/\partial z$. In the Ox_1x_3-plane ($x_2 = 0$), $\overrightarrow{S}$ is equal to zero for $x_1 > c$ because B_3 is zero; for $x_1 < c$, $\partial g_C/\partial z$(2.6.14) is pure imaginary, therefore the x_2-component of $\overrightarrow{S}$ is equal to zero because it is proportional to $\text{Re}[\partial g_C/\partial z] = 0$. Consequently, $\overrightarrow{S}$ is parallel to the x_1-direction and its sense is directed towards the tip of the crack. Observe that in the Ox_2x_3-plane $\overrightarrow{S} = 0$ because B_3 is identically zero. Out of the Ox_1x_3 and Ox_2x_3 planes, it is convenient to make use of the lines of Poynting vector. We define a line of Poynting vector as a curve whose tangent at running point P is parallel to $\overrightarrow{S}$ (P). In a similar way as for the lines of electric field, we find that the lines of Poynting vector are given by the condition

$$Ls(\vec{x}) = \text{Re}\left[\sqrt{\frac{z^2}{z^2 - c^2}}\right] = \text{constant}. \tag{2.6.19}$$

Constant Ls contours are displayed on Fig. 2.56b. The Poynting vector is tangent to these contours. The intensity of the electromagnetic wave at position P is defined as the temporal average $\langle \left\| \overrightarrow{S} \right\| \rangle$ of the magnitude of the Poynting vector:

$$\langle \left\| \vec{S} \right\| \rangle = \frac{1}{T} \int_0^T \left\| \vec{S} \right\| \mathrm{d}t \tag{2.6.20}$$

where T is a time interval for the integration. Assuming the crack to expand with time, with half lengths c_1 and c_2 at $t = 0$ and $t = T$ respectively, we can write (with the help of (2.6.12) and (2.6.13))

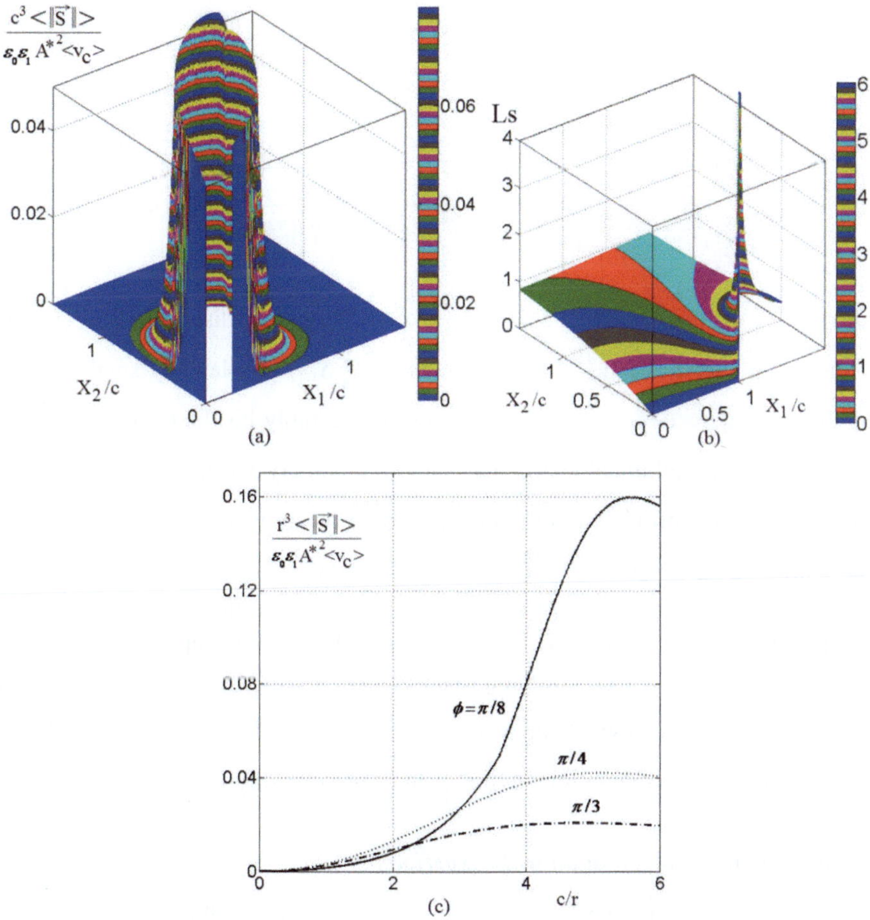

Fig. 2.56 Intensity $\langle \left\| \vec{S} \right\| \rangle$ (2.6.22) of the electromagnetic wave: **a** spatial distribution at different (x_1, x_2) and **c** $\langle \left\| \vec{S} \right\| \rangle$ values calculated at three fixed positions $P(r, \phi, x_3)$, $\phi = \pi/8, \pi/4$ and $\pi/3$, as a function of crack half-length c assuming the Griffith relation (2.6.16) to hold. **b** Lines of Poynting vector corresponding to Ls = constant (2.6.19)

$$\langle\left\|\vec{S}\right\|\rangle = \varepsilon_0\varepsilon_1\langle v_c\rangle\frac{1}{c_2 - c_1}\int\limits_{c_1}^{c_2}\left|\frac{\partial \mathrm{Im}[g_C(z)]}{\partial c}\right|\left|\frac{\partial g_C(z)}{\partial z}\right|\mathrm{d}c \tag{2.6.21}$$

where $\langle v_c\rangle = (c_2 - c_1)/T$ is the distance travelled by the crack divided by the travelling time T. $\langle v_c\rangle$ may be viewed as an average crack expansion velocity. With the help of (2.6.16) to (2.6.18), we can further manage $\langle\left\|\vec{S}\right\|\rangle$ to read

$$\langle\left\|\vec{S}\right\|\rangle = \frac{\varepsilon_0\varepsilon_1 A^{*2}\langle v_c\rangle}{(c_2 - c_1)^3}\langle\tilde{s}\rangle \tag{2.6.22}$$

with

$$\langle\tilde{s}\rangle = \int\limits_{\tilde{c}_1}^{\tilde{c}_1+1}\frac{1}{\left|\tilde{z}^2 - \tilde{c}^2\right|^{3/2}}\left|\mathrm{Im}\left[-\frac{1}{2}\sqrt{\frac{\tilde{z}^2}{\tilde{z}^2 - \tilde{c}^2}} + \tilde{c}^2\sqrt{\frac{\tilde{z}^2}{\left(\tilde{z}^2 - \tilde{c}^2\right)^3}}\right]\right|\mathrm{d}\tilde{c} \tag{2.6.23}$$

where $\tilde{c}_1 = c_1/(c_2 - c_1)$ and $\tilde{z} = z/(c_2 - c_1)$. The spatial distribution of $\langle\left\|\vec{S}\right\|\rangle$ is shown on Fig. 2.56a ($c_1 = 0$ and $c_2 \equiv c$). $\langle\left\|\vec{S}\right\|\rangle$ is relatively small outside the domain $[x_1 < c, x_2 < c]$ and increases considerably inside. Constant $\langle\left\|\vec{S}\right\|\rangle$ contours have an elliptical shape that shrinks gradually when $\langle\left\|\vec{S}\right\|\rangle$ increases. Figure 2.56c may provide a better understanding of the behaviour of $\langle\left\|\vec{S}\right\|\rangle$. On Fig. 2.56c, we have reported $\langle\left\|\vec{S}\right\|\rangle$ (2.6.22), ($c_1 = 0$ and $c_2 \equiv c$) at three different points $P(r, \phi, x_3)$($\phi = \pi/8$,$\pi/4$ and $\pi/3$), as a function of the crack half-length c. The position with $\phi = \pi/8$ is closest to Ox_1x_3, the plane occupied by the crack. The behaviour of $\langle\left\|\vec{S}\right\|\rangle$ is common to all points. Starting from the value zero for $c = 0$, $\langle\left\|\vec{S}\right\|\rangle$ increases with c up to a maximum at $c/r \cong 5.5$ for $\phi = \pi/8$ and then decreases slowly, the maximum of $\langle\left\|\vec{S}\right\|\rangle$ increasing rapidly when ϕ tends towards zero.

Discussion and Confrontation with Experiments

In absence of crack-tip plasticity (isolated crack), we have assumed the crack half-length c to depend implicitly on time and have linked σ_a to c through the Griffith relation $G = 2\gamma$. By doing so, we have introduced a quantity $\mathrm{d}c/\mathrm{d}t = v_c$, the velocity of the tip of the crack. The equilibrium condition requires that v_c be constant (steady crack propagation). A similar procedure can be adopted when the crack is elastic plastic (Fig. 2.53) but we have different possible scenarios (A and B, for example) with respect to the variation with time of the various parameters c, e, a and σ_a:

A. We may assume that c is constant and a varies with time; then σ_a is linked to a through Chang and Ohr (1983) condition that imposes a finite stress at the plastic zone boundaries (see also Anongba and Vitek (2003) for the form below):

$$\frac{\pi\sigma_a}{2\sigma_y} = \frac{e^2 - c^2}{e\sqrt{a^2 - c^2}} \prod(\pi/2, e^2k^2/c^2, k) \tag{2.6.24}$$

(for the definition of the various parameters, see (2.6.6)). The experimental situation corresponds to a spreading of the plastic zone under loading while the crack length is fixed. This treatment introduces a new parameter $da/dt = v_a$.

B. This scenario corresponds to c variable, while the plastic zone is fixed at the initiation stage of fracture propagation. We can then proceed by making use of the Griffith relation $G = 2\gamma$ with the appropriate value of the crack extension force G given by Anongba and Vitek 2003.

The resulting temporal dependences (so far described) for the electromagnetic fields correspond to quasi static changes of the configuration of the crack systems. Regarding the application of the results of the present study, it is appropriate to refer first to ice. Because Evtushenko et al. (1987) and Petrenko (1993, 1996) attribute some importance to lattice dilatation as a possible cause for the emission of electromagnetic radiation observed during the fracture of ice under loading. They propose expressions for the electric potential and electric field that involve lattice dilatation. The dilatation Δ_{CD} is related to V_{CD} (2.6.7) by

$$\Delta_{CD}(\vec{x}) = (q/c_0E_F)V_{CD}(\vec{x}). \tag{2.6.25}$$

Introducing Δ_{CD} (2.6.25) into previous results for ice (see for example Petrenko (1993, 1996) allows a treatment of the emission of electromagnetic radiation in ice when both cracks and plastic zones are present. Note that the yield stress σ_y is considered by the modelling. In absence of plasticity, the dilatation Δ_C and the electric potential V_C are also related by (2.6.25). Petrenko (1996) has performed the following experiment in ice: fracture specimens with edge notch are subjected to tensile tests in which the applied stress is oscillatory; the associated electric potential is measured at different positions round the notch. Agreement between measured and calculated values of the potential is obtained. The electric potential is proportional to the dilatation of the lattice. Petrenko (1993, 1996) did not discuss quasi static crack propagation on the line expounded above that uses the Griffith $G = 2\gamma$ criterion to link the crack length to the applied stress. This treatment is of interest because physical quantities are revealed: crack-front velocity dc/dt, average crack extension velocity $\langle v_c \rangle$ and surface energy γ. The parameter A^* is proportional to $\sqrt{\gamma}$. Hence $\left\|\vec{E}_C\right\|$ (2.6.17) and $\left\|\vec{B}_C\right\|$ (i.e. $|B_3|$ (2.6.18)) are proportional to $\sqrt{\gamma}$ (Figs. 2.54d and 2.55a and b, respectively); the corresponding intensity $\langle\left\|\vec{S}\right\|\rangle$ (2.6.22) of the electromagnetic wave is proportional to γ (Fig. 2.56a, c). $\left\|\vec{B}_C\right\|$ is proportional to

dc/dt (Fig. 2.55a, b) and $\langle \| \vec{S} \| \rangle$ to $\langle v_c \rangle$ (Fig. 2.56a, c). As the crack is expanding with time (i.e. c), electromagnetic quantities increases in magnitude up to a maximum and then decreases slowly at large distances far away from position P (see Fig. 2.55d for $\| \vec{E}_C \|$ and Fig. 2.56c for $\langle \| \vec{S} \| \rangle$). The component $|B_3|$ of the magnetic field on the over hand displays a single oscillation (Fig. 2.55b). Petrenko (1996) reports two types of signals (recorded by antenna) captured from a growing crack. Type 1 signal increases during crack propagation up to a maximum and then relaxes after crack arrest. Type 2 signal displays oscillations. Both types of signal have been attributed to pre-existing intrinsic electric fields in ice although type 1 signal could correspond to quasi static propagation of the crack. This suggests re-examination of measured signals. Regarding the dependence of the electromagnetic fields upon surface energy, we may refer to the experimental work of Misra and Kumar (2004) with metallic materials. These authors have measured a linear variation of the electromagnetic radiation peak voltage with bond energy; this agrees with the proportionality between $\langle \| \vec{S} \| \rangle$ and γ given by the present analysis. Similarly, attention should be paid to experimental evidence of the dependence of the electromagnetic fields upon crack extension velocity. This brings us, for example, to the work by Frid et al. 2003: these authors have found an empirical equation that fits quite well electromagnetic radiation signals displaying oscillations in the form of a pulse. The signals are recorded during compression fracture of rocks (carbonate and igneous) and transparent materials (glass, PMMA and glass ceramics). They propose a derivation of the envelope of the pulse that is found to be proportional to the average crack velocity. In their modelling, they assume the envelope of the pulse to be associated with the fracture process itself and signal oscillations to vibration of charge carriers about equilibrium positions. This agrees with the result that the intensity of the electromagnetic wave $\langle \| \vec{S} \| \rangle$ is proportional to the average crack velocity $\langle v_c \rangle$. The results of the present study find application if the dilatation of the lattice caused by dislocations and cracks produces an electric potential. This is believed to be the case in metals (Nabarro 1967; Molotskii 1980; Misra et al. 2007) in addition to ice (Evtushenko et al. 1987; Petrenko 1993, 1996).The emission of electromagnetic radiation in metals has been observed in a number of different situations: at the onset of plastic flow, at different deformation stages on the stress–strain curves, at the initiation stage of crack propagation (for a review, see Misra et al. 2007). For these various situations, one must first find an appropriate model of the arrangement and dynamics of the dislocations and then proceed to calculate the associated lattice dilatation; in general, this is a formidable task. The model in Fig. 2.53 is presumably suitable in the situation of localized plasticity at the tip of the crack. Detailed confrontation with experiment is missing at present. Because available experimental works do not connect the emission of electromagnetic radiation with the size and location of crack-tip plastic zones. Recently, Misra et al. (2007) have proposed a theoretical model for the emission of electromagnetic radiation associated with plasticity and fracture in metallic materials. This work is based on a proposal made earlier by Molotskii (1980) that imposes the dislocation dipole moment to be proportional to the whole length of the dislocation. We first stress the weakness of this assumption and then indicate some difficulties with the modelling by Misra et al. (2007). Molotskii (1980) working hypothesis can be described as follows. Assume a straight edge dislocation at the origin lying along the x_3-direction (use Fig. 2.53 for the x_i-directions) with Burgers vector $(b, 0, 0)$ in the x_1-direction. Under action of an applied stress, at a given time, its shape in the glide plane Ox_1x_3 is given at position $(0, 0, x_3)$ by $\xi(x_3, t)$. He then considers that the dislocation, at any time, has a dipole moment p per unit length constant, independent of position x_3. By doing so, Molotskii finds that the dipole moment of the dislocation is proportional to the dislocation length and increases for small deviation from the straight edge position. Molotskii's result is difficult to accept because the dipole moment per unit length of a dislocation is associated with the dilatation of the lattice (Nabarro 1967) and therefore depends on dislocation character; it decreases to zero for screws in linear elasticity (the lattice dilatation for screws is zero). To see the weakness of Molotskii's assumption that the dislocation dipole moment increases with its length, consider a straight edge dislocation segment with length $L = AB$ along Ox_3 that bows out in its glide plane from its initial straight edge position. The result is an increase of the dislocation length. Now assume its curved shape to consist of edge and screw portions only. Because all the screw portions have zero dipole moment, we discover that the curved dislocation has a dipole moment essentially identical to that of the segment AB along Ox_3, contrary to Molotskii's working assumption. Misra et al. (2007) investigates electromagnetic radiation as a function of deformation up to fracture propagation using specimens with edge notch tested in tension. Plastic yielding from the notch precedes crack propagation. A specimen geometry with edges (see Figs. 1 to 3 in Misra et al. 2007) is used to elaborate a theoretical analysis that presents serious difficulties (*A* and *B*):

A. According to Misra et al. (2007) the straight edge dislocation is parallel to *Oz*, has Burgers vector along *Ox* and is associated with a supplementary half-plane of atoms *Oyz* ($y > 0$). The dipole moment of the dislocation is parallel to *Oy*. As the dislocation bows out with time from the straight edge position in the *Oxz* slip plane, its whole curved shape can be approximated by small straight edge and screw portions. The edge portions all have a dipole moment parallel to *Oy* and the screws contribute nothing. Consequently, the variation with time of the dipole moment of the dislocation remains parallel to *Oy*. It is impossible for the considered gliding edge dislocation to have the first temporal derivative of the dipole moment in the z-direction as assumed without any demonstration by the authors. We stress that a variation of the dislocation dipole moment lying in the z-direction would imply necessarily a spreading of the dislocation in the *Oyz* plane which contains both the dipole moment and the dislocation line direction *Oz*; this corresponds to a climbing motion of the edge dislocation, a mechanism that is not considered by the authors.
B. Misra et al. (2007) also uses the incorrect assumption of Molotskii that stipulates that the dislocation dipole moment increases with its length.

In view of the difficulties indicated above, treatments by Molotskii (1980) and Misra et al. (2007) appear to be incorrect.

SUMMARY: This work provides formulas for the lattice dilatation, electric potential and electric field corresponding to an elastic–plastic crack and a brittle crack without crack-tip plasticity, under mode I loading. Taking account of an implicit variation with time of these quantities, this work gives expressions for the induced magnetic field and the associated Poynting vector. The results are applied to the quasi-static propagation of an isolated crack assuming the Griffith condition $G = 2\gamma$ to hold; this reveals the relationships between the electromagnetic fields and various parameters such as crack-tip velocity and surface energy. It is indicated how a similar application can be performed when the crack is associated with a plastic zone; this would provide a better interpretation of experimental findings. The present study applies first to pure metals but could be adapted to any material where lattice dilatation produces an electric field: ice is discussed as one example. Finally, the inadequacy of several models, proposed to explain the emission of electromagnetic radiation in metallic materials, is demonstrated.

2.7 Crack-Front Vibration

In thermal equilibrium, crack fronts vibrate freely; this is a direct consequence of the vibration of atoms at the very tip of the crack as illustrated in Fig. 2.57. The crack front would deviate more and more from straight line as temperature increases. Crack-front deviation can generally be decomposed into components parallel and perpendicular to the average crack plane. The former component, at most of the order of nearest neighbour distance between atoms, is generally too small compared to the crack length which is an important parameter in the Griffith crack-propagation condition; consequently, we may expect this component to have an unnoticeable effect. We shall only be concerned with the crack-front oscillation out of the average crack plane as measured for instance in planes perpendicular to the crack propagation direction. Let ξ (x, t) be such an oscillation (x and t are space and time variables, respectively). We expect ξ to have the form of a standing wave because of the superposition of transverse waves travelling along the crack front. ξ satisfies the d'Alembert wave equation for long wavelengths and consequently can be assumed to be a sum of standing sine wave modes of the form $\xi_n \cos \kappa_n x \cos \omega_n t$ (where n a positive integer refers to the mode and ξ_n, κ_n and ω_n are amplitude, wave number and frequency, respectively). We shall make use of the concept that a crack is equivalent to a continuous of dislocations so that the vibration of the crack would be described by the vibration of the associated crack dislocations. The model of the crack system is an infinitely extended elastic medium containing a vibrating crack. We take the crack to be in the x_1x_3-plane, extending from $x_1 = -a_1$ to a_1 and running indefinitely in the x_3-direction. The crack front located at $x_1 = \pm a_1$ vibrates freely in the form

$$\xi = \sum_n \xi_n \cos \kappa_n x_3 \cos \omega_n t \tag{2.7.1}$$

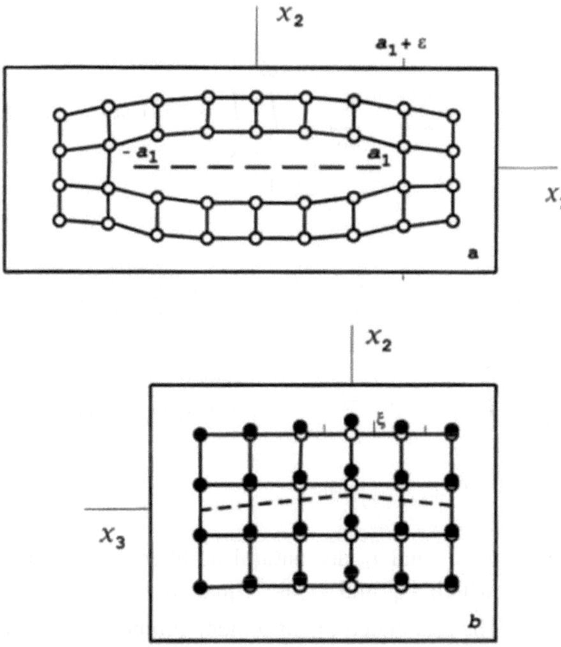

Fig. 2.57 Schematic illustration of the vibration of a static crack front in a simple cubic lattice. **a** Atoms in equilibrium (arbitrary) are indicated by open circles. The crack (dashed line) extends from $x_1 = -a_1$ to a_1 in the x_1x_3-plane of a Cartesian system x_i. The atomic plane perpendicular to the crack just ahead of the crack tip is assumed to be located at $x_1 = a_1 + \varepsilon$. **b** Drawing of a plane of atoms, normal to the x_1 axis ahead of the crack tip (the $x_1 = a_1 + \varepsilon$ plane for instance). Atoms displaced during passage of a transverse wave are indicated in solid. The crack front (dashed) oscillates accordingly. Displacements ξ are indicated

in the x_2x_3-plane. The geometry of the crack corresponds to Fig. 2.57. The system is subjected to a uniform applied stress σ_{22}^a along x_2. The Cartesian system at rest is attached to the loading mechanism. An analysis of this model is given by Anongba (2003).

In what follows are successively given, displacement and stress fields due to a dislocation vibrating in form (2.7.1), vibrating crack analysis, mean square gradient of a vibrating crack front, discussion and conclusions.

Displacement and Stress Fields Due to a Vibrating Edge Dislocation

We consider a dislocation with Burgers vector (0, b, 0) lying in the x_3-direction and oscillating in the x_2x_3-plane in the form of a standing wave (2.7.1). The plastic distortion, on the linear approximation of the present study, is given by $\beta_{12}^*(\boldsymbol{x},t) = b\delta(x_1)H(x_2) - b\xi\,\delta(x_1)\delta(x_2)$ where δ and H are the Dirac delta function and the Heaviside step function, respectively. Here, the first term is due to the undisturbed dislocation. The corresponding displacement is known (Anongba, 1995). We shall therefore concentrate on the second term denoted $\beta_{12}^{*\xi}$. Its Fourier form may be written as

$$\beta_{12}^{*\xi}(\vec{x},t) = \sum_{n} \sum_{\tau_n, k_{3n'}=-\infty}^{\infty} \int_{-\infty}^{\infty} \int_{-\infty}^{\infty} \bar{\beta}_{12}^{*\xi_n}(k_1, k_2, k_{3n'}, \tau_n)$$
$$e^{i(k_1x_1+k_2x_2+k_{3n'}\kappa_n x_3-\tau_n\omega_n t)}\mathrm{d}k_1\mathrm{d}k_2 \tag{2.7.2}$$

$$\bar{\beta}_{12}^{*\xi_n}(k_1, k_2, k_{3n'}, \tau_n) = \frac{\omega_n}{2\pi} \int_{-\pi/\omega_n}^{\pi/\omega_n} e^{i\tau_n\omega_n t}\mathrm{d}t \frac{\kappa_n}{2\pi} \int_{-\pi/\kappa_n}^{\pi/\kappa_n} e^{-ik_{3n'}\kappa_n x_3}\mathrm{d}x_3$$
$$\times \frac{1}{(2\pi)^2} \int_{-\infty}^{\infty} \int_{-\infty}^{\infty} e^{-i(k_1x_1+k_2x_2)}(-b\xi_n \cos\kappa_n x_3 \cos\omega_n t)$$
$$\delta(x_1)\delta(x_2)\mathrm{d}x_1\mathrm{d}x_2 \tag{2.7.3}$$

k_1 and k_2 are real and k'_{3n} and τ_n are natural numbers. $\overline{\beta}_{12}^{*\xi_n}$ is non-zero only for couples (k'_{3n}, τ_n) for which k'_{3n} and τ_n are equal to 1 or $-$ 1; for these couples, $\overline{\beta}_{12}^{*\xi_n} = -b\xi_n/(4\pi)^2$. The associated displacement components u_i^{ξ} may be obtained, using relation (8.9) in the book by Mura (1987), as

$$u_i^{\xi}(\vec{x},t) = -\sum_{n,\tau_n,k_{3n'}} \int_{-\infty}^{\infty} \int_{-\infty}^{\infty} i\bar{\beta}_{12}^{*\xi_n} c_{jl12} k_l^{(n)} N_{ij}(\vec{k}^{(n)}, \varpi_n) D^{-1}(\vec{k}^{(n)}, \varpi_n)$$
$$\times e^{i(k_1x_1+k_2x_2+k_3^{(n)}x_3-\varpi_n t)}\mathrm{d}k_1\mathrm{d}k_2 \tag{2.7.4}$$

in which $\vec{k}^{(n)} = (k_1^{(n)} = k_1, k_2^{(n)} = k_2, k_3^{(n)} = k'_{3n}\kappa_n)$ and $\varpi_n = \tau_n\omega_n$. Explicit expressions for N_{ij}, D and c_{jl12} are Eqs. (9.7), (9.25) and (A2.1) by Mura (1987). Direct calculations give:

$$u_i^{\xi}(\vec{x},t) = \sum_n \frac{b\xi_n}{2\pi} \cos\kappa_n x_3 \cos\omega_n t \times \left(\frac{\partial}{\partial(x_1\delta_{i2}+x_2\delta_{i1})} I_t^{(n)} \right.$$
$$\left. + 2\frac{c_t^2}{\omega_n^2} \frac{\partial^3}{\partial(x_1\delta_{i1}+x_2\delta_{i2})^2 \partial(x_1\delta_{i2}+x_2\delta_{i1})} \left(I_t^{(n)} - I_l^{(n)} \right) \right) \tag{2.7.5}$$

$$u_3^{\xi}(\vec{x},t) = \sum_n \frac{b\xi_n}{2\pi} (-\kappa_n sin\kappa_n x_3) \cos\omega_n t 2\frac{c_t^2}{\omega_n^2} \frac{\partial^2}{\partial x_1 \partial x_2} \left(I_t^{(n)} - I_l^{(n)} \right) \tag{2.7.6}$$

$$I_z^{(n)} = \frac{1}{2\pi} \int_{-\infty}^{\infty} \int_{-\infty}^{\infty} \frac{1}{k_1^2 + k_2^2 + P_z^{(n)^2}} e^{i(k_1x_1+k_2x_2)}\mathrm{d}k_1\mathrm{d}k_2 = K_0\left[P_z^{(n)} r\right] \tag{2.7.7}$$

the subscript z taking the values t or l. The subscript i takes the values 1 or 2 in (2.7.5) as also in (2.7.8) below; $r^2 = x_1^2 + x_2^2$ and $P_z^{(n)^2} = \kappa_n^2 - \omega_n^2/c_z^2$. c_t and c_l (i.e., c_z with $z = t$ and $z = l$) are the velocities of transverse and longitudinal sound waves respectively. $K_n[x]$ is the nth-order modified Bessel function usually so denoted and δ_{ij} is the Kronecker delta. Finally, the total displacement takes the form:

$$\begin{aligned} u_i(\vec{x}, t) = u_i^0(\vec{x}) + \sum_n \frac{b\xi_n}{2\pi} \cos\kappa_n x_3 \cos\omega_n t \Bigg(2\frac{c_t^2}{\omega_n^2} \Bigg\{ \left(\frac{x_1}{r^2}\delta_{i2} + \frac{x_2}{r^2}\delta_{i1} \right) \\ \times \left(P_t^{(n)^2} K_2[P_t^{(n)} r] - P_l^{(n)^2} K_2[P_l^{(n)} r] \right) \\ - \frac{x_1 x_2 x_i}{r^3} \left(P_t^{(n)^3} K_3[P_t^{(n)} r] - P_l^{(n)^3} K_3[P_l^{(n)} r] \right) \Bigg\} \\ - P_t^{(n)} K_1[P_t^{(n)} r] \left(\frac{x_1}{r}\delta_{i2} + \frac{x_2}{r}\delta_{i1} \right) \Bigg) \end{aligned} \tag{2.7.8}$$

$$u_3(\vec{x}, t) = \sum_n \frac{b\xi_n}{\pi} (-\kappa_n \sin\kappa_n x_3) \cos\omega_n t \frac{c_t^2}{\omega_n^2} \frac{x_1 x_2}{r^2} \left(P_t^{(n)^2} K_2[P_t^{(n)} r] - P_l^{(n)^2} K_2[P_l^{(n)} r] \right). \tag{2.7.9}$$

u_i^0 is the displacement due to the undisturbed dislocation. It is easy to show that the displacement, given by (2.7.8) and (2.7.9), satisfies the equation of motion; this may be understood from the fact that its derivation makes use of a dynamic solution that also satisfies the equation of motion (Mura 1987). When only one vibration mode is assumed, (2.7.8) and (2.7.9) correspond to the displacement due to an edge dislocation oscillating in the form of a standing sine wave (i.e., $\xi = \xi_n$ *cos* $\kappa_n x_3$ *cos* $\omega_n t$). This is the case considered earlier by Laub and Eshelby (1966) using a different method; however, their relation (2), as compared to the present results, contains some errors. The stresses σ_{ij} $(\boldsymbol{x}, t)$ and elastic strains e_{ij} $(\boldsymbol{x}, t)$ can be obtained by differentiating (2.7.8) and (2.7.9). Restricting ourselves to σ_{22}, σ_{23} and e_{12}, since they are involved in the vibrating crack analysis (see below), we find

$$\begin{aligned} \sigma_{22}(\vec{x}, t) = \sigma_{22}^0(\vec{x}) + \frac{2\mu\nu}{1-2\nu}\frac{\partial}{\partial x_1} u_1^\xi(\vec{x}, t) + \frac{2\mu(1-\nu)}{1-2\nu}\frac{\partial}{\partial x_2} u_2^\xi(\vec{x}, t) \\ + \frac{2\mu\nu}{1-2\nu}\frac{\partial}{\partial x_3} u_3(\vec{x}, t) \end{aligned} \tag{2.7.10}$$

$$\begin{aligned} \sigma_{23}(\vec{x}, t) = \sum_n \frac{\mu b}{2\pi} \xi_n (-\kappa_n \sin\kappa_n x_3) \cos\omega_n t \Bigg(4\frac{c_t^2}{\omega_n^2} \Bigg\{ \frac{x_1}{r^2} \left(P_t^{(n)^2} K_2[P_t^{(n)} r] \right. \\ \left. - P_l^{(n)^2} K_2[P_l^{(n)} r] \right) - \frac{x_1 x_2^2}{r^3} \left(P_t^{(n)^3} K_3[P_t^{(n)} r] - P_l^{(n)^3} K_3[P_l^{(n)} r] \right) \Bigg\} \\ - \frac{x_1}{r} P_t^{(n)} K_1[P_t^{(n)} r] \Bigg) \end{aligned} \tag{2.7.11}$$

and

$$e_{12}(\vec{x},t) = \left\| 1 - \xi \frac{\partial}{\partial x_2} \right\| e_{12}^0(\vec{x}) + e_{12}^R(\vec{x},t) \tag{2.7.12}$$

$$\begin{aligned} e_{12}^R(\vec{x},t) = \sum_n \frac{b\xi_n}{2\pi} \cos\kappa_n x_3 \cos\omega_n t \Bigg(\frac{c_t^2}{\omega_n^2} \Bigg\{ -\frac{1}{r^2}\left(P_t^{(n)^2} - P_t^{(n)^2}\right)\left(1 - \frac{8x_1^2x_2^2}{r^4}\right). \\ - \frac{2}{r}\left(P_t^{(n)^3} K_1[P_t^{(n)} r] - P_l^{(n)^3} K_1[P_l^{(n)} r]\right) \\ - \frac{6}{r^2}\left(P_t^{(n)^2} K_2[P_t^{(n)} r] - P_l^{(n)^2} K_2[P_l^{(n)} r]\right) \\ + \frac{2x_1^2x_2^2}{r^4}\left(P_t^{(n)^4} K_4[P_t^{(n)} r] - P_l^{(n)^4} K_4[P_l^{(n)} r]\right) \Bigg\} + \frac{1}{2} P_t^{(n)^2} K_0[P_t^{(n)} r] \Bigg) \end{aligned} \tag{2.7.13}$$

u_i^ξ corresponds to the oscillating part of the displacement field. σ_{22}^0 and e_{12}^0 are field components due to the straight edge dislocation. The term in ‖ ‖ is an operator that acts on e_{12}^0. μ is the shear modulus and ν is Poisson ratio.

Analysis of the Vibrating Crack

The crack may be represented by a continuous distribution $D(x_1)$ of edge dislocations vibrating perpendicular to the x_1-axis such that $D(x_1)dx_1$ is the number of dislocations in dx_1 about x_1. Each of the climbing dislocations has the form (2.7.1), Burgers vector (0, b, 0), and displacement field given by (2.7.8) and (2.7.9). Therefore, the displacement field of our crack system, which is a superposition of individual dislocation displacements, satisfies the equation of motion. Next, we shall consider the total force exerted at any point on our dislocation distribution; let $\overline{F}_i$ be its component along the x_i axis. $\overline{F}_1$ is set equal to zero, this leads to the governing equation. We demand $\overline{F}_2$ to correspond to uniform motion only, this leads to the conditions for free vibration. $\overline{F}_3$ carries no additional condition; on average over x_3 it is zero.

GOVERNING EQUATION. The distribution function D is given by the condition that each of the climbing dislocations is acted on, at any point P, by zero total force along x_1. Consider the dislocation located at x_1: to it are associated a running point $P = (x_1, \xi(x_3, t), x_3)$ and corresponding line sense $\boldsymbol{L} = \{1/[1 + (\partial\xi(x_3, t)/\partial x_3)^2]^{1/2}\}$ $(0, \partial\xi(x_3, t)/\partial x_3, 1)$. The condition that the total force at P be zero is

$$\sigma_{22}^a + \int_{-a_1}^{a_1} (\sigma_{22} - \sigma_{23}L_2/L_3)D(x_1')\mathrm{d}x_1' = 0 \tag{2.7.14}$$

where σ_{22} and σ_{23} are the stresses at P due to the dislocation located at x_1' and L_i (i = 2 and 3) are the components at P of the dislocation line sense. For convenience we choose P to be located on the average crack plane; this means that $P = (x_1, 0, x_3^*)$, where x_3^* corresponds to the nodes of $\xi(x_3,t)$. Consequently, we may write

$$\sigma_{22} = \frac{C}{x_1 - x_1'} \tag{2.7.15}$$

$$\sigma_{23} = \frac{\nu C}{x_1 - x_1'} \frac{\partial \xi}{\partial x_3}(x_3^*, t) + \nu C \sum_n \xi_n(-\kappa_n \sin \kappa_n x_3^*) \cos \omega_n t R^{(n)}(x_1 - x_1') \tag{2.7.16}$$

$$\begin{aligned} R^{(n)}(x_1 - x_1') = \frac{1}{x_1 - x_1'} \Bigg(& \frac{4(1-\nu)}{\nu} \frac{c_t^2}{\omega_n^2} \Big(P_t^{(n)^2} K_2[P_t^{(n)}(x_1 - x_1')] \\ & - P_l^{(n)^2} K_2[P_l^{(n)}(x_1 - x_1')] \Big) - 1 \Bigg) \\ & - \frac{1-\nu}{\nu} P_t^{(n)} K_1[P_t^{(n)}(x_1 - x_1')] \mathrm{sgn}(x_1 - x_1') \end{aligned} \tag{2.7.17}$$

$C = \mu b/2\pi(1 - \nu)$ and $\mathrm{sgn}(x_1 - x_1') = |x_1 - x_1'|/(x_1 - x_1')$. Defining the temporal average of any function of time $f(t)$ with period T to be $\langle f \rangle = (1/T) \int_0^T f(t)\mathrm{d}t$ the temporal average of (2.7.14) reads

$$\begin{aligned} \sigma_{22}^a + \int_{-a_1}^{a_1} C\Bigg(& \Big[1 - \nu\langle(\partial\xi(x_3^*, t)/\partial x_3)^2\rangle\Big]/(x_1 - x_1') \\ & - \frac{\nu}{2} \sum_n \xi_n^2 \kappa_n^2 R^{(n)}(x_1 - x_1') \Bigg) D(x_1')\mathrm{d}x_1' = 0 \end{aligned} \tag{2.7.18}$$

The Cauchy principal value of the integral is to be taken. The integral Eq (2.7.18) determines the average arrays of continuously distributed vibrating edge dislocations; it is like the corresponding one obtained in the analysis of a wavy crack with a rigid sinusoidal front (e.g., Eq. (9) by Anongba, 1996). When the parameter $<(\partial\xi(x_3^*, t)/\partial x_3)^2>$ is set equal to zero, (2.7.18) corresponds to a planar crack with a rigid straight front in mode I loading. Considering the vibration of the crack front reveals new properties of the crack system. We shall show below that $<(\partial\xi(x_3^*, t)/\partial x_3)^2>$ should be lower than $1/\nu$ which removes the singularity from (2.7.18). We shall refer to this parameter in the following as the mean-square gradient of a vibrating crack front. On the average fracture plane at the nodes x_3^* (i.e., on the x_1x_3-plane)

$$G(x_3^*, t) = \frac{1 - \nu\left(\partial\xi(x_3^*, t)/\partial x_3\right)^2}{\sqrt{1 + \left(\partial\xi(x_3^*, t)/\partial x_3\right)^2}} G_0 \tag{2.7.19}$$

where $G_0 = K_I^2(1 - \nu^2)/E$ is the crack opening force per unit edge length of the rigid planar crack (K_I is the stress intensity factor of mode I loading and E is Young's modulus). When $\partial\xi(x_3^*, t)/\partial x_3$ is small in magnitude, we can take the temporal average of G at the nodes to be

$$G(x_3^*, t) = \frac{1 - \nu\langle\left(\partial\xi(x_3^*, t)/\partial x_3\right)^2\rangle}{\sqrt{1 + \langle\left(\partial\xi(x_3^*, t)/\partial x_3\right)^2\rangle}} G_0. \tag{2.7.20}$$

Here we quote the important fact that G, when averaged over time, is different from G_0; this corresponds to a crack front that is non straight on average over time. Its average form (2.7.20) decreases to zero as $<(\partial\xi(x_3^*, t)/\partial x_3)^2>$ increases to the limiting value $1/\nu$. No extension along x_1 of the existing crack is expected when G is identically zero.

FREE VIBRATION CONDITIONS: DISPERSION LAW AND WAVE VELOCITY. The x_2-component of the total force exerted on an oscillating edge dislocation of our vibrating crack system at point $P = (x_1, \xi(x_3, t), x_3)$ is obtained from the well-known Peach and Koehler (1950) formula to be

$$\overline{F}_2 = -bL_3\overline{\sigma}_{12} = -2\mu bL_3\overline{e}_{12} \tag{2.7.21}$$

where $\overline{\sigma}_{12}$ and $\overline{e}_{12}$ are field components that refer to the total stress and strain at P, respectively. $\overline{e}_{12}$ may be considered to originate from the continuously distributed vibrating edge dislocations. It may be written as

$$\begin{aligned}\bar{e}_{12} &= \int_{-a_1}^{a_1} e_{12}(x_1 - x_1', \xi, x_3, t)D(x_1')\mathrm{d}x_1' \\ &= \int_{-a_1}^{a_1} \left\| 1 - \xi\frac{\partial}{\partial x_2} \right\| e_{12}^0(x_1 - x_1', \xi)D(x_1')\mathrm{d}x_1' \\ &+ \int_{-a_1}^{a_1} e_{12}^R(x_1 - x_1', \xi, x_3, t)D(x_1')\mathrm{d}x_1'.\end{aligned} \tag{2.7.22}$$

$e_{12}(\boldsymbol{x}, t)$ is given by (2.7.12) and (2.7.13). The first expression with ‖ ‖ represents the strain due to a planar crack displaced statically by (2.7.1) (i.e., by ξ) and coinciding locally with the actual vibrating crack. In the linear approximation it equally represents the field of a planar crack coinciding with and moving uniformly with the same instantaneous velocity as the actual crack. The motion of the crack in dynamic conditions will also be possible if we apply an external field that annuls the second expression with e_{12}^R. In free vibration conditions no external force is needed and the second term must vanish (i.e., only the field corresponding to uniform motion is admissible for free motion). This gives the condition for free vibration

$$\int_{-a_1}^{a_1} e_{12}^R(x_1 - x_1', \xi, x_3, t)D(x_1')\mathrm{d}x_1' = 0 \tag{2.7.23}$$

We note here that Laub and Eshelby (1966) first used a similar argument to find the condition for free motion of a vibrating dislocation; the present analysis is a direct extension to continuous arrays of dislocations. We shall deduce a simple dispersion law relating the frequency ω_n to the wave number κ_n, which may conveniently be written as $\omega_n = c_n\, \kappa_n$. Introducing this relation in (2.7.23) in which e^R_{12} is given by (2.7.13), we get

$$\sum_n \frac{b\xi_n}{2\pi} \cos\kappa_n x_3 \cos\omega_n t \kappa_n^2$$
$$\int_{-a_1}^{a_1} \left(\frac{c_t^2}{c_n^2} \left\{ -\frac{a_t^{(n)^2} - a_l^{(n)^2}}{r_n'^2} - \frac{2}{r_n'} \left(a_t^{(n)^3} K_1[a_t^{(n)} r_n'] - a_l^{(n)^3} K_1[a_l^{(n)} r_n'] \right) \right.\right.$$
$$\left. - \frac{6}{r_n'^2} \left(a_t^{(n)^2} K_2[a_t^{(n)} r_n'] - a_l^{(n)^2} K_2[a_l^{(n)} r_n'] \right) \right\}$$
$$\left. + \frac{1}{2} a_t^{(n)^2} K_0[a_t^{(n)} r_n'] \right) D(x_1') dx_1' = 0 \tag{2.7.24}$$

$$r_n'^2 = \kappa_n^2((x_1 - x_1')^2 + \xi^2),\ a_t^{(n)2} = 1 - \frac{c_n^2}{c_t^2} \text{and} a_l^{(n)2} = 1 - \frac{c_n^2}{c_l^2}. \tag{2.7.25}$$

The terms in $x_1^2 x_2^2$ in e^R_{12} with $x_2 = \xi$ give a term of order ξ^2 and have thus been discarded in our linear approximation. For $r_n' \ll 1$ (this would be the case for relatively small κ_n values), (2.7.24) corresponds to

$$\sum_n \frac{b\xi_n}{2\pi} \cos\kappa_n x_3 \cos\omega_n t \kappa_n^2 \int_{-a_1}^{a_1} \left(\frac{c_t^2}{c_n^2} \left\{ -\frac{1}{4} \left(a_t^{(n)^4} \ln(\frac{1}{2} a_t^{(n)} r_n') \right.\right.\right.$$
$$\left. \left. - a_l^{(n)^4} \ln(\frac{1}{2} a_l^{(n)} r_n') \right) - \frac{1}{4} \left(a_t^{(n)^4} - a_l^{(n)^4} \right)(\gamma + 1/4) \right\}$$
$$\left. - \frac{1}{2} a_t^{(n)^2} \left(\ln(a_t^{(n)} r_n'/2) + \gamma \right) \right) D(x_1') \mathrm{d}x_1' = 0 \tag{2.7.26}$$

In our vibrating crack model, we assume that there is no resultant dislocation content i.e., $\int_{-a_1}^{a_1} D(x_1)\mathrm{d}x_1 = 0$. Therefore, terms independent of x_1 in the various brackets in (2.7.26) contribute nothing. Equation (2.7.26) may consequently be rearranged to read

$$\sum_n \frac{b\xi_n}{2\pi} \cos\kappa_n x_3 \cos\omega_n t \left(-\frac{\kappa_n^2}{4} \right)$$
$$\left\{ \frac{2c_l^2 c_t^4 - c_n^2 (c_t^4 + c_l^4)}{c_t^2 c_l^4} \right\} \int_{-a_1}^{a_1} D(x_1') \ln(r_n'/2) dx_1' = 0 \tag{2.7.27}$$

The solution of (2.7.27) corresponds to the constant c_0 value of c_n that cancels the term in curly brackets:

$$c_n = \frac{\sqrt{2}c_l c_t^2}{\sqrt{c_t^4 + c_l^4}} \equiv c_0. \tag{2.7.28}$$

Thus, free vibration would imply a dispersion law that reads

$$\omega_n = c_0 \kappa_n \tag{2.7.29}$$

in the range of small wave numbers. c_0 may be viewed as the velocity of travelling sine waves along the crack front. It is equal to the velocity of a travelling sine wave along an edge dislocation found by Laub and Eshelby (1966) for long waves (i.e., their relation (21) with $\theta = \pi/2$). It is easy to show that $c_0 < c_t$, consequently $\omega_n/c_t < \kappa_n$; under this condition, these authors have also shown that no elastic energy is radiated from a single vibrating dislocation. This holds also for a continuous array of vibrating dislocations by using the superposition principle. In free vibration conditions the motion of our vibrating crack is self-maintaining, and no elastic energy is radiated.

Mean-Square Gradient of a Vibrating Crack Front

In the following, a calculation is made of the temperature dependence of the mean-square gradient of a vibrating crack front. We assume long waves. This assumption will be seen to hold below. We consider a vibrating crack in mode *I* loading in an isotropic elastic material. We take the crack to be on average in the x_1x_3-plane extending from $x_1 = -a_1$ to a_1 and running in the positive x_3-direction over a finite edge length *L*. The spacing between crack-tip atoms in the crack-front direction is *a*. We shall adopt the series form (2.7.1) for the crack front with periodic boundary conditions (i.e., $\xi(0,t) = \xi(L, t)$) and the dispersion law for free vibration (2.7.29). These give

$$\xi(x_3, t) = \sum_{n=1}^{N} \xi_n \cos \kappa_n x_3 \cos \omega_n t \tag{2.7.30}$$

where $\kappa_n = 2n\pi/L$, $\omega_n = c_0\kappa_n$ and $\kappa N = 2N\pi/L = \pi/a$ (κ_n is limited to the first Brillouin zone). The mean-square gradient of a vibrating crack front, that we denote by $\langle(\partial\xi/\partial x_3)^2\rangle^*$, has been defined as the temporal average of $(\partial\xi/\partial x_3)^2$ measured at the nodes of the vibrating crack front (i.e., on the average crack plane).

Requiring (without loss of generality) ξ to have nodes $x_3^* = (2p + 1)L/4$ (with $p = 0$ and 1), it reads

$$\langle(\partial\xi/\partial x_3)^2\rangle^* = \sum_{n=1}^{N} \xi_n^2\kappa_n^2/2 \tag{2.7.31}$$

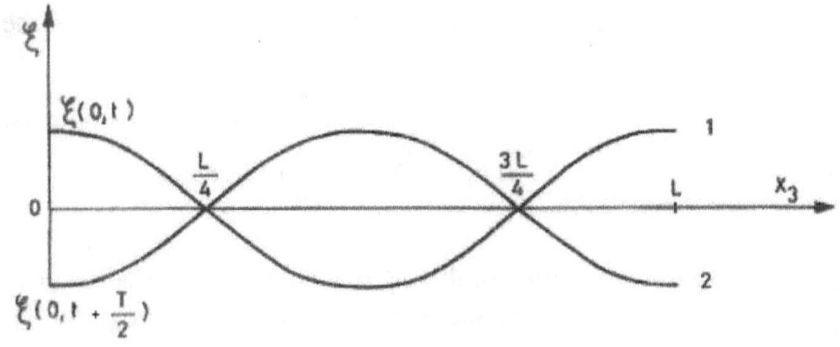

Fig. 2.58 Shape of ξ (arbitrary) to illustrate various defined parameters ($x_1 = \pm a_1$). ξ is assumed to be L and T-periodical with respect to space and time respectively, with nodes located at points $x_3^* = (2p+1)L/4$ where p is a natural number. Curves 1 and 2 correspond to two different times separated by $T/2$. The series form of ξ would be $\xi(x_3,t) = \sum_n \xi_n \cos \kappa_n x_3 \cos \omega_n t$ under the following conditions: $\kappa_n = 2n\pi/L$, $\omega_n = 2n\pi/T$ and $\xi_n = 0$ for n even. Note that $\partial\xi(0,t)/\partial x_3 = 0$ at any time as required

A schematic illustration of ξ with the various defined parameters is shown in Fig. 2.58. It is easy to show that the temporal average ΔE_c^n of the kinetic energy of a crack front with length L vibrating in the standing sine wave mode n of amplitude $\xi_n \cos \kappa_n x_3 \cos \omega_n t$ is

$$\Delta E_c^n = \frac{L}{8} m_n \xi_n^2 \omega_n^2 = \frac{Lc_0^2}{8} m_n \xi_n^2 \kappa_n^2 \tag{2.7.32}$$

where m_n is the effective mass per unit length of the crack front in mode n vibration. Considering that the energy ΔE_n in the mode, as in any harmonic oscillator, is half kinetic energy and half potential energy, when averaged over time, we may write

$$\langle(\partial\xi/\partial x_3)^2\rangle^* = \frac{2}{Lc_0^2} \sum_{n=1}^{N} \frac{1}{m_n} \Delta E_n \tag{2.7.33}$$

The crack front is regarded as an edge dislocation. Eshelby (1962) has shown that m_n is proportional to $\ln(\kappa_N/\kappa_n)$. For small κ_n, this coefficient varies slowly and may be approximated by $\ln(\kappa_N/\kappa_1)$. This corresponds to a constant $\overline{m}$ value of m_n given by

$$\overline{m} = \frac{\rho b^2}{4\pi}\left(1 + \frac{c_t^4}{c_l^4}\right) \ln \frac{L}{2a} \tag{2.7.34}$$

where ρ is the mass density of the elastic medium and b the Burgers vector associated with the crack front. Equation (2.7.33) becomes

$$\langle(\partial\xi/\partial x_3)^2\rangle^* = \frac{2}{L\beta_0} \Delta E \tag{2.7.35}$$

where $\beta_0 = \overline{m}c_0^2$ is the line tension associated with the crack front and ΔE the vibration energy is taken to be

$$\Delta E = \sum_{n=1}^{N} \Delta E_n = \sum_{n=1}^{N} (j_n + 1/2)\hbar\omega_n \tag{2.7.36}$$

when mode n is excited to quantum number j_n; that is, when the mode is occupied by j_n phonons (Kittel 1996). $\hbar$ has its usual meaning. ΔE consists of two parts. The first part $\Delta E' = \sum_n \hbar\omega_n/2$ is the zero-point energy and the second part $\Delta E'' = \sum_n \hbar j_n\omega_n$ is the energy associated with thermally excited phonons. We calculate both separately, starting with $\Delta E''$. j_n is given by the Planck distribution function (Kittel 1996)

$$j_n = \frac{1}{e^{\hbar\omega_n/k_B T} - 1} \tag{2.7.37}$$

where k_B is Boltzmann's constant. We have

$$\Delta E'' = \sum_{n=1}^{N} \frac{\hbar\omega_n}{e^{\hbar\omega_n/k_B T} - 1} \tag{2.7.38}$$

It is convenient to replace the sum by an integral. The number of modes per unit range of κn is $L/2\pi$ (periodic boundary conditions), this gives

$$\Delta E'' = \int_{\kappa_1}^{\kappa_N} \frac{L}{2\pi} \frac{\hbar\omega_n}{e^{\hbar\omega_n/k_B T} - 1} \mathrm{d}\kappa_n = \int_{\omega_1}^{\omega_N} \frac{L}{2\pi c_0} \frac{\hbar\omega}{e^{\hbar\omega/k_B T} - 1} \mathrm{d}\omega \tag{2.7.39}$$

where $\omega_N \equiv c_0\,\kappa_N$. Let $x = \hbar\omega/k_B T$ and ω_1 goes to zero for L large, (2.7.39) becomes

$$\Delta E'' = \frac{L k_B^2}{2\pi\hbar c_0} T^2 f\left(\frac{T_N}{T}\right) \tag{2.7.40}$$

$$f\left(\frac{T_N}{T}\right) = \int_0^{T_N/T} \frac{x\mathrm{d}x}{e^x - 1} \tag{2.7.41}$$

and $T_N \equiv \hbar\omega_N/k_B = \pi\hbar c_0/a\,k_B$. Similarly, we calculate $\Delta E'$ to be

$$\Delta E' = \frac{\pi\hbar L c_0}{8a^2} \tag{2.7.42}$$

Equation (2.7.35) becomes

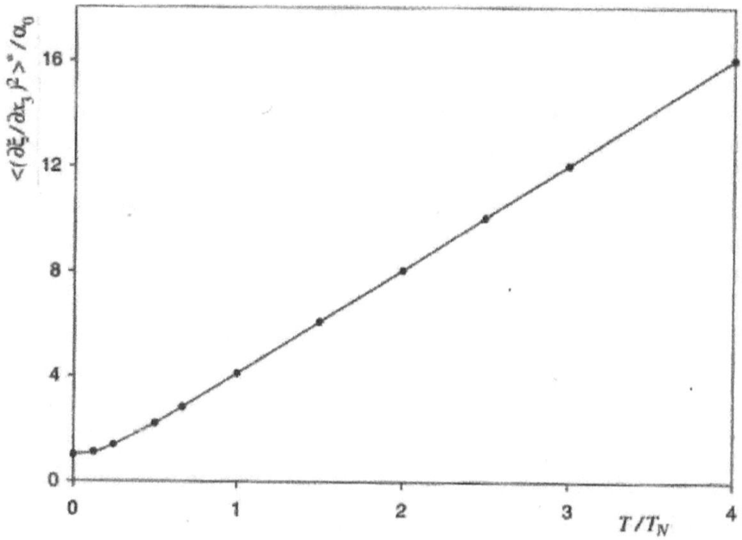

Fig. 2.59 Mean-square gradient (normalized by $\alpha_0 = k_B\,T_N/4\beta_0 a$) of a vibrating crack front in a solid as a function of temperature (reduced by $T_N = \pi\hbar c_0/a\,k_B$). $\alpha_0 \cong 2.8 \times 10^{-4}$ and $T_N \cong 118$ K in tungsten for instance; see the text

$$\langle(\partial\xi/\partial x_3)^2\rangle^* = \alpha_0\left(1 + 4\left(\frac{T}{T_N}\right)^2 f\left(\frac{T_N}{T}\right)\right) \tag{2.7.43}$$

where $\alpha_0 = k_B\,T_N/4\beta_0\,a$. For arbitrary $(T_N/T), f\,(T_N/T)$ (i.e. (2.7.41)) can only be evaluated numerically. Hence the mean-square gradient of a vibrating crack front at different temperatures must be calculated numerically using (2.7.41) and (2.7.43). This is given in Fig. 2.59. In two limiting cases analytical expressions of (2.7.43) can be obtained as follows:

(i) When $T \ll T_N, f\,(T_N/T) \cong f\,(\infty) = \pi^2/6$; this gives

$$\langle(\partial\xi/\partial x_3)^2\rangle^* \cong \alpha_0\left(1 + \frac{2\pi^2}{3}\left(\frac{T}{T_N}\right)^2\right) \tag{2.7.44}$$

(ii) When $T \gg T_N, f\,(T_N/T) \cong T_N/T$; this leads to

$$\langle(\partial\xi/\partial x_3)^2\rangle^* \cong \alpha_0\left(1 + 4\frac{T}{T_N}\right) \tag{2.7.45}$$

At absolute zero ($T = 0$) there exists a non-zero value of the mean-square gradient of a vibrating crack front given by α_0. It is proportional to T^2 at very low temperature (2.7.44) and to T at higher temperatures (2.7.45). The derivation of $\langle(\partial\xi/\partial x_3)^2\rangle^*$ has been performed under the assumption that we are concerned with long waves.

This would correspond to $\langle(\partial\xi/\partial x_3)^2\rangle^*$ much lower than unity. The corresponding temperature range would satisfy the condition $k_B T \ll \beta_0 a$ by using (2.7.45). $\beta_0 a/k_B$ is generally very high; its value is 1070×10^2 K in tungsten, which we consider in the following, and largely exceeds the melting point 3695 K. To estimate the magnitude of the mean-square gradient of a vibrating crack front, we apply our development to tungsten, a nearly isotropic material. We take the average crack plane to be (001) and the crack front to lie along $\langle 110\rangle$ with a length $L = 10^{-2}$ m. The atomic spacing along the crack front is $a = a_0\sqrt{2}$ where $a_0 = 3.16 \times 10^{-10}$ m is the lattice constant. The Burgers vector b associated with the crack front is taken to be $b = a_0/2$ which corresponds to the closest distance between two parallel (001) atomic planes. We also need the magnitudes of c_0 and $\overline{m}$. They may be obtained by making use of the following physical constants at absolute zero: density ρ = 19,317 kg m^{-3}, elastic constants $c_{11} = 532.6$, $c_{12} = 204.9$ and $c_{44} = 163.1$ GPa, taken from Kittel (1972). These serve first to calculate $c_t \cong \sqrt{c_{44}/\rho}$ and $c_l \cong \sqrt{c_{11}/\rho}$; then we obtain $c_0 =$ 2180 m s^{-1} (using (2.7.28)) and $\overline{m} = 68.3 \times 10^{-17}$ kg m^{-1} (using (2.7.34)). With the appropriate physical parameters above, we get $T_N = 118$ K and $\alpha_0 = 2.8 \times 10^{-4}$. We define the mean crack front inclination angle θ by

$$\tan^2\theta = \langle(\partial\xi/\partial x_3)^2\rangle^* \tag{2.7.46}$$

This leads to an angle θ of about 1° at absolute zero, 3.2° at room temperature, and 10.6° at the melting point. These values are low, justifying the long wave assumption. Let us now emphasize similar results obtained here and in surface physics. At temperatures higher than T_N, the temperature dependent part of $\langle(\partial\xi/\partial x_3)^2\rangle^*$ is $k_B T/\beta_0 a$ (see (2.7.45)) where β_0 is the line tension of the crack front and a is the atomic spacing in the crack-front direction. In surface physics, in the theory of the roughening transition, the influence of thermal fluctuations on the shape of atomic steps on surfaces is a subject of investigation. Using classical statistics, the deviation from a straight line of the equilibrium profile $y(x)$ of a step edge (represented by a continuous line) as a function of temperature has been studied (Desjonquères and Spanjaard 1993). The equilibrium fluctuations of y'^2 obtained there (e.g., their relation (2.4.6)) is $k_B T/\beta_0\xi$ where β_0 is a linear internal energy per unit length and ξ (in their notation) is the line thickness with a similar meaning as a above.

Discussion

The mean-square gradient of a vibrating crack front is a strictly positive parameter (see for instance (2.7.43)). The average form $\langle G\rangle^*$ of the crack opening force (2.7.20) that depends on this parameter is therefore different from G_0. We can express $\langle G\rangle^*$ as a function of the mean crack-front inclination angle defined by (2.7.46); this gives

$$\langle G\rangle^* = (1 - \nu\tan^2\theta)\cos\theta G_0. \tag{2.7.47}$$

In a recent study of a wavy crack with a rigid sinusoidal front (Anongba 1996), the crack extension force calculated at positions (on the crack front) located on the mean fracture plane is identical in form to (2.7.47), θ in this study being the acute angle between the rigid crack front and the mean fracture plane. The relations for $\langle G\rangle^*$ ((2.7.20) or (2.7.47)) lead to view the vibrating crack front as being non straight on average over time. The present study therefore appears to belong to a more general one that aims at determining the exact contribution of the abundantly observed non-planar character of cleavage fracture surfaces, on μm or sub-μm levels, to the conditions for crack motion. Consider two static crack fronts under thermal vibration at two different temperatures forced to move at a given time. A question arises whether we can expect differences in subsequent fractured surfaces due to an increase of θ with increasing temperature. Differences indeed may occur, for instance through an increase in the roughness of fractured surfaces as T increases. The roughness of broken surfaces is observed from very low temperatures in brittle crystalline materials when there are no dislocation activities (e.g., in silicon at room temperature (George and Michot 1993), for instance). We can expect part of the observed roughness to be the result of crack-front vibration. Let the local surface energy of the crack at the nodes of our vibrating crack front be γ. The condition for the initiation of crack motion corresponds locally to $\langle G\rangle^* = 2\gamma$. Referring to (2.7.20) and (2.7.43) and taking account of the fact that $\langle(\partial\xi/\partial x_3)^2\rangle^*$ is much smaller than unity, the stress intensity factor in mode I loading at the beginning of fracture K_{IC} may approximately be expressed as

$$K_{IC} \cong \left\{1 + \frac{(1+2\nu)\alpha_0}{4} + (1+2\nu)\alpha_0\left(\frac{T}{T_N}\right)^2 f\left(\frac{T_N}{T}\right)\right\}\left(\frac{2\gamma E}{1-\nu^2}\right)^{1/2} \tag{2.7.48}$$

Thus, when $T \ll T_N$

$$K_{IC} \cong \left\{1 + \frac{(1+2\nu)\alpha_0}{4} + \frac{\pi^2(1+2\nu)\alpha_0}{6}\left(\frac{T}{T_N}\right)^2\right\}\left(\frac{2\gamma E}{1-\nu^2}\right)^{1/2} \tag{2.7.49}$$

And when $T \gg T_N$

$$K_{IC} \cong \left\{1 + \frac{(1+2\nu)\alpha_0}{4} + (1+2\nu)\alpha_0\frac{T}{T_N}\right\}\left(\frac{2\gamma E}{1-\nu^2}\right)^{1/2} \tag{2.7.50}$$

Here in these expressions of K_{IC}, additional terms to $\sqrt{2\gamma E/(1-\nu^2)}$ are due to crack-front vibration, second term corresponding to zero-point vibration. Thus, by changing the shape of the crack front, thermal vibration contributes to the increase of K_{IC} with temperature. However, the values of α_0 and T_N in tungsten indicate that this contribution is small.

SUMMARY: The crack opening force G per unit length of the crack front, on the average fracture plane, is found to be different from that of the rigid planar crack, when averaged over time, but the magnitude of the difference is relatively small. G decreases with increasing temperature T and is found to be proportional to T^2 at very low temperature and to T at higher temperatures. The relation for G leads to view the vibrating crack front as being non straight on average over time. In tungsten, a nearly isotropic material, the theory yields a mean crack front inclination angle of about 1° at absolute zero, 3.2° at room temperature and 10° at the melting point. The mean crack-front inclination angle θ is an acute angle, measured in the plane perpendicular to the crack propagation direction, between the crack-front and the average crack plane.

Chapter 3
Brittle Cracks Under Compression and Bending

3.1 Cracks Under Compression: Introducing Poisson Effect

Consider as an illustration in Fig. 3.1 a rectangular fracture specimen with large dimensions, isotropic and elastic. Initially the specimen is dashed. With respect to a Cartesian coordinate system x_i and under the action of uniform compressive stress $(-\sigma_a)$ along the x_1-direction, the specimen shrinks in the x_1-direction but extends according to Poisson in the x_2 and x_3 directions; the shape then taken by the specimen is in solid. To the specimen is attached a planar crack with finite dimensions along x_1 and x_2 but infinite in the calculations (below) in the x_3-direction. The crack is symmetrical with respect to Ox_3, inclined by an angle θ with respect to Ox_1x_3 and has a straight front along x_3. When the crack is in Ox_1x_3 ($\theta = 0$), considering the applied compression $(-\sigma_a)$ only, we show in the following that the crack extension force G per unit length of the crack front is zero. However, this result conflicts with numerous experimental observations revealing that the crack propagates axially along x_1. We shall come back to the experimental evidence latter. At this point, we only stress that in addition to an extension along x_1, there is an opening of the crack along x_2. Things happen as if the crack was subjected to a tension in the x_2-direction. A possible origin of an induced tension is the well-known Poisson effect which, in the framework of linear elasticity obeying Hooke's law, leads to a stress $(\nu\sigma_a)$ directed along x_2 (Fig. 3.1) where ν is Poisson ratio. The goal of the present study is to give the conditions for the propagation of the crack depicted schematically in Fig. 3.1 taking account of Poisson effect. The physical quantity to be provided by the analysis is the crack extension force G as a function of the inclination angle θ.

The crack is represented by a continuous distribution of infinitesimal dislocations. The stress field induced by the crack in the surrounding medium will be given by the dislocations of the distribution. This method for describing the crack is well understood from the general work by Bilby and Eshelby (1968). We stress below that two distinct couples of edges lead to the same value of the crack extension force, a value that is in full agreement with stress intensity factors given in the literature

P. N.B. Anongba, *Mechanics of Steadily Propagating Cracks*,
https://doi.org/10.1007/978-981-96-5703-2_3

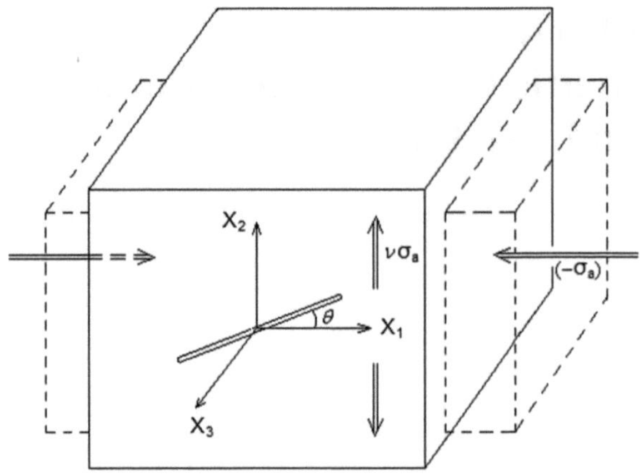

Fig. 3.1 Fracture specimen under uniform applied compression $(-\sigma_a)$, $\sigma_a > 0$, in the x_1-direction. The specimen is dashed before compression and solid under loading. An inclined planar crack with a straight front parallel to x_3 is associated to the body as well as an assumed uniform internal tension $(\nu\sigma_a)$ in the x_2-direction corresponding to the Poisson effect

(Sih et al. 1962; Sih and Liebowitz 1968) in the case of an inclined planar crack with the same geometry as in Fig. 3.1 but under tension along x_1. Next, we first report the stresses of the edge dislocations used in the crack analysis as well as the assumed stress field of Poisson effect. Then, crack analysis is performed, and a discussion is made of the contribution of Poisson effect in the understanding of brittle fracture in compression.

Poisson Effect and Dislocation Stress Fields

In the situation of Fig. 3.1 where the specimen is loaded in compression in the x_1-direction, Poisson effect corresponds to a strain $\varepsilon_{22}^{(P)}$ in the x_2-direction and $\varepsilon_{33}^{(P)}$ in the x_3-direction. The effect of $\varepsilon_{22}^{(P)}$ corresponds to an opening of the crack when $\theta = 0$. Poisson law reads

$$\varepsilon_{22}^{(P)} = -\nu\varepsilon_{11}^{a} \tag{3.1}$$

where ε_{11}^{a} is the strain along x_1 that results from the compression $(-\sigma_a)$. $(-\sigma_a) = E\varepsilon_{11}^{a}$ from Hooke's law and E is Young's modulus. To $\varepsilon_{22}^{(P)}$ is associated an internal stress $\sigma_{22}^{(P)}$ using Hooke's law, this leads to

$$\sigma_{22}^{(P)} = \nu\sigma_a. \tag{3.2}$$

We assume an internal uniform Poisson stress. Similarly, Poisson tension $\sigma_{33}^{(P)}$ in the x_3-direction reads $\sigma_{33}^{(P)} = \nu\sigma_a$ so that the stress tensor corresponding to Poisson in the basis $(\vec{e}_1, \vec{e}_2, \vec{e}_3)$ associated with the directions x_i is given as

$$\left(\sigma^{(P)}\right) = \begin{pmatrix} 0 & 0 & 0 \\ 0 & \nu\sigma_a & 0 \\ 0 & 0 & \nu\sigma_a \end{pmatrix}. \tag{3.3}$$

We shall deal with two types of edges parallel to x_3 arranged on an inclined crack plane. The crack dislocations would then have an elevation h with respect to the horizontal Ox_1x_3 plane. For a dislocation with Burgers vector $(b, 0, 0)$ lying indefinitely in the x_3-direction and displaced by $x_2 = h$ from the origin, the stress field is given at $\vec{x} = (x_1, x_2, x_3)$ by

$$\begin{aligned}
\sigma_{12}^{(1)}(\vec{x}) &= C\frac{x_1\left(x_1^2 - (x_2 - h)^2\right)}{r^4},\\
\sigma_{11}^{(1)}(\vec{x}) &= -C\frac{(x_2 - h)\left(3x_1^2 + (x_2 - h)^2\right)}{r^4},\\
\sigma_{22}^{(1)}(\vec{x}) &= C\frac{(x_2 - h)\left(x_1^2 - (x_2 - h)^2\right)}{r^4},\\
\sigma_{33}^{(1)}(\vec{x}) &= -2C\nu\frac{(x_2 - h)}{r^2},\\
\sigma_{13}^{(1)} &= \sigma_{23}^{(1)} = 0;
\end{aligned} \tag{3.4}$$

$r^2 = x_1^2 + (x_2 - h)^2$ and $C = \mu b/2\pi(1 - \nu)$ where μ is the shear modulus. For a dislocation with Burgers vector $(0, b, 0)$ lying along x_3 and displaced by $x_2 = h$ from the origin, the stress is

$$\begin{aligned}
&\sigma_{12}^{(2)}(\vec{x}) = C\frac{(x_2 - h)(x_1^2 - (x_2 - h)^2)}{r^4};\\
&\sigma_{ii}^{(2)}(\vec{x}) = C\frac{x_1}{r^2}\left(\frac{x_1^2(\delta_{i1} + \delta_{i2}) + (x_2 - h)^2(-\delta_{i1} + 3\delta_{i2})}{r^2} + 2\nu\delta_{i3}\right), i = 1,\ 2 \text{ and } 3;\\
&\sigma_{j3}^{(2)}(\vec{x}) = 0, j = 1 \text{ and } 2;
\end{aligned} \tag{3.5}$$

δ_{ij} is the Kronecker delta.

Analysis of Cracks Under Compression

The crack analysis will be performed with two different dislocation arrangements corresponding to Figs. 3.2 and 3.3. For each arrangement we shall successively give the equilibrium dislocation distributions, the crack-tip stresses and the crack extension force. Consider first the crack dislocations description given in Fig. 3.2. The medium is infinitely extended with a planar crack (straight front) in the Ox_1x_3 plane extending from $x_1 = -a$ to a and running indefinitely in the x_3-direction. We consider two families of straight edge dislocations in the crack. Families 1 and 2 are both parallel to x_3 with Burgers vectors $\vec{b}_1 = (b, 0, 0)$ and $\vec{b}_2 = (0, b, 0)$ in the x_1 and x_2 directions respectively. The two dislocation families are assumed to be continuously distributed over the crack area between $x_1 = -a$ to a.

The system is subjected to uniform applied compression $(-\sigma_a)$ at infinity in the $\vec{e}_1'$ direction (Fig. 3.2). The basis $(\vec{e}_1', \vec{e}_2')$ is obtained from $(\vec{e}_1, \vec{e}_2)$ after a rotation about Ox_3 by an angle θ. Furthermore, we assume uniform Poisson tension $\nu\sigma_a$ in the $\vec{e}_2'$

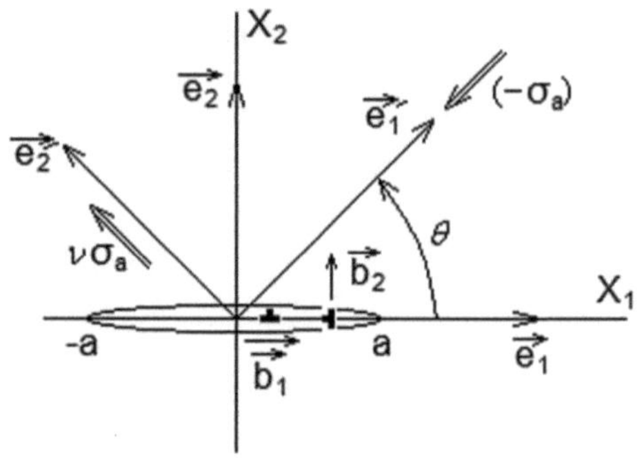

Fig. 3.2 Crack dislocations geometry corresponding to straight edges parallel to x_3 with $\vec{b}_1$ along x_1 in the crack plane and $\vec{b}_2$ along x_2 perpendicular to crack. Uniform applied compression $(-\sigma_a)$ acts along $\vec{e}_1'$ and internal Poisson $(\nu\sigma_a)$ along $\vec{e}_2'$. Note that in this arrangement $\vec{b}_1$ and $\vec{b}_2$ are not directed along the acting forces

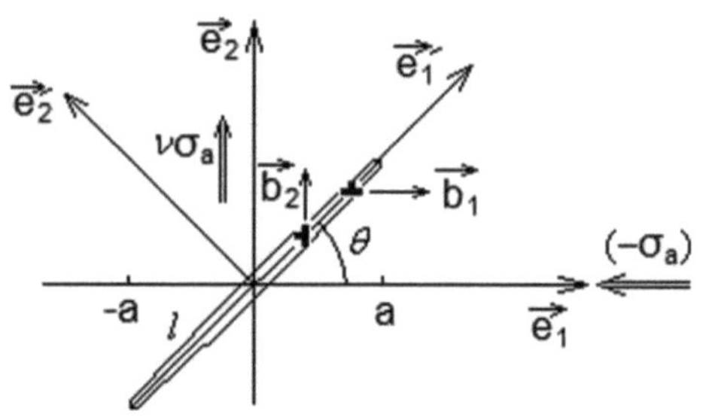

Fig. 3.3 Crack dislocations arrangement corresponding to straight edges parallel to x_3 with $\vec{b}_1$ along x_1 out of the crack plane and $\vec{b}_2$ along x_2 non perpendicular to crack. Uniform applied compression $(-\sigma_a)$ acts along $\vec{e}_1$ and internal Poisson $(\nu\sigma_a)$ along $\vec{e}_2$. The projected half length of the crack along x_1 is a. In this geometry $\vec{b}_1$ and $\vec{b}_2$ are directed along the acting forces. We also have $a = l\cos\theta$ where l is the half length of the inclined crack

direction. The dislocation distribution function $D_i(x_1)$ ($i = 1$ and 2 for edges 1 and 2) gives the number of dislocations i in a small interval dx_1 about x_1 as $D_i(x_1)dx_1$. We are concerned with the problem of finding the equilibrium distributions D_i of the dislocations under the combined action of their mutual repulsions and the force exerted on them by $(-\sigma_a)$ and $\nu\sigma_a$. We ask the crack faces to be traction free, this gives

$$\begin{cases} \overline{\sigma}_{12} = 0 \\ \overline{\sigma}_{22} = 0\,. \\ \overline{\sigma}_{23} = 0 \end{cases} \tag{3.6}$$

$\overline{\sigma}_{ij}$ stands for the total stress at any point (x_1, x_2, x_3) in the medium and is linked to D_i; in (3.6), we are only concerned with the points of the crack faces. $\overline{\sigma}_{ij}$ is written as

$$\overline{\sigma}_{ij} = \sigma_{ij}^{a} + \sigma_{ij}^{(P)} + \overline{\sigma}_{ij}^{(1)} + \overline{\sigma}_{ij}^{(2)} \tag{3.7}$$

where σ_{ij}^{a} corresponds to the applied compression, $\sigma_{ij}^{(P)}$ to Poisson and

$$\overline{\sigma}_{ij}^{(n)}(x_1, x_2, x_3) = \int_{-a}^{a} \sigma_{ij}^{(n)}(x_1 - x_1', x_2, x_3) D_n(x_1')\mathrm{d}x_1' \; (n = 1 \text{ and } 2); \tag{3.8}$$

here $\sigma_{ij}^{(n)}(n = 1 \text{or } 2)$ is the stress field produced by a dislocation displaced by $(x_2 - h)$ $(h = 0$ in the case of Fig. 3.2) from the origin with Burgers vector $(b, 0, 0)$ or $(0, b, 0)$ [(3.4) and (3.5)]. With respect to the system $(O, \vec{e}_1, \vec{e}_2, \vec{e}_3)$ we have

$$\left(\sigma^{a}\right) = (-\sigma_a)\begin{pmatrix} \cos^2\theta & \sin 2\theta/2 & 0 \\ \sin 2\theta/2 & \sin^2\theta & 0 \\ 0 & 0 & 0 \end{pmatrix},$$

$$\left(\sigma^{(P)}\right) = \nu\sigma_a\begin{pmatrix} \sin^2\theta & -\sin 2\theta/2 & 0 \\ -\sin 2\theta/2 & \cos^2\theta & 0 \\ 0 & 0 & 1 \end{pmatrix}; \tag{3.9}$$

$\sigma_{ij}^{(n)}$ in (3.8) may be taken from (3.4) and (3.5) with $h = 0$. The traction free boundary condition (3.6) then leads to the following system of integral equations

$$\begin{cases} (-\sigma_a)\dfrac{(1+\nu)p}{1+p^2} + C\displaystyle\int_{-a}^{a} \dfrac{D_1(x_1^{;})\mathrm{d}x_1^{;}}{x_1 - x_1^{;}} = 0 \\ (-\sigma_a)\left(\dfrac{p^2 - \nu}{1+p^2}\right) + C\displaystyle\int_{-a}^{a} \dfrac{D_2(x_1^{;})}{x_1 - x_1^{;}}\mathrm{d}x_1^{;} = 0 \end{cases} \tag{3.10}$$

where $p = \tan\theta$. The Cauchy principal values of the integrals are to be taken. The type of solution is well known (Bilby and Eshelby 1968):

$$D_1(x_1) = \frac{(1+\nu)p}{1+p^2}\frac{(-\sigma_a)}{\pi C}\frac{x_1}{\sqrt{a^2 - x_1^2}},$$

$$D_2(x_1) = \left(\frac{p^2 - \nu}{1+p^2}\right)\frac{(-\sigma_a)}{\pi C}\frac{x_1}{\sqrt{a^2 - x_1^2}}. \tag{3.11}$$

The corresponding relative displacement φ_i of the crack faces, in the x_1 $(i = 1)$ and x_2 $(i = 2)$ directions, are:

$$\varphi_1(x_1) = \frac{(1+\nu)p}{1+p^2}(-\sigma_a b/\pi C)(a^2 - x_1^2)^{1/2},$$

$$\varphi_2(x_1) = \left(\frac{p^2 - \nu}{1 + p^2}\right)(-\sigma_a b/\pi C)(a^2 - x_1^2)^{1/2}. \tag{3.12}$$

D_i is unbounded at $x_1 = \pm a$ and the φ_i curve vertical at these end points. We are now interested in stress values in the neighbourhood of the crack tip located at $x_1 = a$ at point P with coordinates $(x_1, x_2 = 0, x_3)$. Substituting $x_1 = a + s$, $0 < s << a$, $\overline{\sigma}_{ij}$(3.7) is given by the following formula:

$$\overline{\sigma}_{ij}(s) = \sum_{n=1}^{2} \int_{a-\delta a}^{a} \sigma_{ij}^{(n)}(a + s - x_1', x_2 = 0, x_3) D_n(x_1') \mathrm{d}x_1' \tag{3.13}$$

with $\delta a << a$. This stress expression means that only those dislocations located about the crack front in x_1-interval $[a - \delta a, a]$ will contribute significantly to the stress at $x_1 = a+s$ ahead of the crack tip as s tends to zero; any other contribution will be negligible for a sufficiently small value of s. We observe that this formula is precise with no place for any other kind of additional stress term. Restricting ourselves to $\overline{\sigma}_{12}$ and $\overline{\sigma}_{22}$ since they are involved in the calculation of the crack extension force G (see in the following) we obtain

$$\overline{\sigma}_{12}(s) = \frac{K_1}{\sqrt{2\pi}} \frac{1}{\sqrt{s}}, \quad \overline{\sigma}_{22}(s) = \frac{K_2}{\sqrt{2\pi}} \frac{1}{\sqrt{s}} \tag{3.14}$$

$$K_1 = \frac{(1 + \nu)p}{1 + p^2}(-\sigma_a)\sqrt{a\pi}, \quad K_2 = \frac{p^2 - \nu}{1 + p^2}(-\sigma_a)\sqrt{a\pi}. \tag{3.15}$$

By removing ν in (3.15), we recover the stress intensity factors displaced by Sih et al. (1962) (see also Sih and Liebowitz 1968) for an inclined planar crack loaded in tension. We proceed to calculate the crack extension force following Bilby and Eshelby (1968). This procedure is detailed since it is also used in the analysis of the crack of Fig. 3.3. Allow the right-hand tip of the crack in Fig. 3.2 to advance (say rigidly for simplicity) from $x_1 = a$ to $a + \delta a$, but apply forces to the freshly formed surfaces to prevent relative displacement; the energy of the system is unaltered. Now allow these forces to relax to zero so that the crack extends effectively from a to $a+\delta a$. The work done by these forces corresponds to a decrease of the energy of the system which we shall estimate (the energy of the system consists of the elastic energy of the medium and the energy of the loading mechanism). The element $\mathrm{d}s = \mathrm{d}x_3\mathrm{d}x_1$ of the fracture plane ahead of the crack front, at a point $P = (x_1, x_2 = 0, x_3)$, may be defined by $\mathrm{d}\vec{s} = \vec{\gamma}\mathrm{d}s$ where $\vec{\gamma}$ is the unit vector perpendicular to ds pointing to the positive x_2-direction. We obtain $\mathrm{d}\vec{s} = (0, 1, 0)\mathrm{d}x_1\mathrm{d}x_3$. The relevant component of the force acting on ds in the x_i-direction is $\overline{\sigma}_{ij}\mathrm{d}s_j$ (the summation convention on repeated subscripts applies) where $\overline{\sigma}_{ij}$ are stresses ahead of the shorter crack; thus the energy change associated with ds is $\overline{\sigma}_{ij}\mathrm{d}s_j\Delta u^{(i)}/2$ (here a summation is also considered over $i = 1$ and 2) where $\Delta u^{(i)}$ is the difference in displacement across the lengthened crack, just behind its tip, in the x_i-direction. When the crack advances from $x_1 = a$

to $a + \delta a$, the energy decrease associated with a surface element $\mathrm{d}x_3\delta a$ is

$$-\delta E = \mathrm{d}x_3 \left(\frac{1}{2} \int_a^{a+\delta a} \overline{\sigma}_{12} \Delta u^{(1)} \mathrm{d}x_1 + \frac{1}{2} \int_a^{a+\delta a} \overline{\sigma}_{22} \Delta u^{(2)} \mathrm{d}x_1 \right). \tag{3.16}$$

Let G be a derivative of the energy of the system with respect to crack area. G corresponds to the limiting value taken by $-\delta E/\mathrm{d}x_3\delta a$ as δa decreases to zero: $G = \lim_{\delta a \to 0} -\delta E/\mathrm{d}x_3\delta a$. Stresses $\overline{\sigma}_{ij}$ generally consist of terms that are either bounded or unbounded as x_1 tends to a; only those stress terms that are singular may contribute a non-zero value to G; the bounded terms all contribute nothing. Hence, we can use (3.14) for $\overline{\sigma}_{12}$ and $\overline{\sigma}_{22}$. $\Delta u^{(i)}$ may be obtained from the solution of (3.10) modified to allow for the fact that the crack extends from $x_1 = -a$ to $a + \delta a$ instead of from $-a$ to a. We may use (3.12) for $\Delta u^{(i)}$ following Bilby and Eshelby (1968); this leads to

$$\begin{aligned} G(P_0) &= \left(K_1^2 + K_2^2\right)\left(1 - \nu^2\right)/E \\ &= \left(\frac{\nu^2 + p^2}{1 + p^2}\right) a\pi\sigma_a^2\left(1 - \nu^2\right)/E. \end{aligned} \tag{3.17}$$

Expression (3.17) gives the value of G at an arbitrary point $P_0(a, x_2 = 0, x_3)$ along the front of the planar crack with half-length a. G is defined as the crack extension force per unit edge length of the crack front. Consider again the inclined planar crack problem under uniform compression but with a different type of crack dislocations, Fig. 3.3. The main difference with Fig. 3.2 is that the Burgers vectors of the dislocations are now directed along the applied compression $(-\sigma_a)$ x_1-direction and Poisson $(\nu\sigma_a)$ x_2-direction, but are not linked to the plane of the crack. Here a designates the projected half-length l of the crack along x_1. The traction free boundary condition at an arbitrary point $P = (x_1, x_2 = px_1, x_3)$, $|x_1| < a$, of the faces of the crack reads

$$\begin{cases} \overline{\sigma}_{12} - p\overline{\sigma}_{11} = 0 \\ \overline{\sigma}_{22} - p\overline{\sigma}_{12} = 0. \\ \overline{\sigma}_{23} - p\overline{\sigma}_{13} = 0 \end{cases} \tag{3.18}$$

Again for $\overline{\sigma}_{ij}$ we use (3.7): σ_{ij}^a and $\sigma_{ij}^{(P)}$ are obvious; for $\overline{\sigma}_{ij}^{(n)}$ we need to express $\sigma_{ij}^{(n)}$ [(3.4) and (3.5)] paying attention to the fact that h in (3.4) and (3.5) is position dependent in the dislocation distribution. Introducing the expressions for $\overline{\sigma}_{ij}$ in (3.18) we obtain

$$\begin{cases} p\sigma_a + C \int\limits_{-a}^{a} \dfrac{D_1(x_1^{`})\mathrm{d}x_1^{`}}{x_1 - x_1^{`}} = 0 \\ \nu\sigma_a + C \int\limits_{-a}^{a} \dfrac{D_2(x_1^{`})}{x_1 - x_1^{`}}\mathrm{d}x_1^{`} = 0 \end{cases}. \tag{3.19}$$

The corresponding solutions are

$$D_1^*(x_1) = \frac{p\sigma_a}{\pi C}\frac{x_1}{\sqrt{a^2 - x_1^2}},$$
$$D_2^*(x_1) = \frac{\nu\sigma_a}{\pi C}\frac{x_1}{\sqrt{a^2 - x_1^2}} \tag{3.20}$$

with relative displacement φ_i^* of the faces of the crack, in the x_1 ($i = 1$) and x_2 ($i = 2$) directions

$$\varphi_1^*(x_1) = (p\sigma_a b/\pi C)(a^2 - x_1^2)^{1/2}$$
$$\varphi_2^*(x_1) = (\nu\sigma_a b/\pi C)(a^2 - x_1^2)^{1/2}. \tag{3.21}$$

It is seen from (3.20) that in absence of Poisson effect $D_2^* = 0$ and in absence of the applied compression $D_1^* = 0$. Hence dislocation family 1 responds to the applied compression and family 2 to Poisson. Using (3.13) and paying attention to put $x_2 = px_1$ in $\sigma_{ij}^{(n)}$, the stress $\overline{\sigma}_{ij}$ at a point $P = (x_1, x_2 = px_1, x_3)$ ahead of the crack front ($x_1 = a + s, 0 < s \ll a$) may be calculated. We only give $\overline{\sigma}_{11}$, $\overline{\sigma}_{12}$ and $\overline{\sigma}_{22}$ (to be used for the calculation of G):

$$\overline{\sigma}_{11} = \frac{1}{(1+p^2)^2}\left(-p(3+p^2)K_1^* + (1-p^2)K_2^*\right)\frac{1}{\sqrt{2\pi}\sqrt{s}},$$
$$\overline{\sigma}_{12} = \frac{1-p^2}{(1+p^2)^2}\left(K_1^* + pK_2^*\right)\frac{1}{\sqrt{2\pi}\sqrt{s}},$$
$$\overline{\sigma}_{22} = \frac{1}{(1+p^2)^2}\left(p(1-p^2)K_1^* + (1+3p^2)K_2^*\right)\frac{1}{\sqrt{2\pi}\sqrt{s}} \tag{3.22}$$

where $K_1^* = p\sigma_a\sqrt{a\pi}$ and $K_2^* = \nu\sigma_a\sqrt{a\pi}$; terms with K_i^* are due to dislocation family i ($i = 1$ and 2). To calculate G, consider that the crack tip at $x_1 = a$ and elevation $x_2 = pa$ (see Fig. 3.3) advances from $x_1 = a$ to $a + \delta a$ on the inclined crack plane. We must attach to a surface element Δs ahead of the crack front an energy decrease $(-\delta E)$. We take $\Delta s = \int_a^{a+\delta a} \mathrm{d}s$ with $\mathrm{d}s = \sqrt{1+p^2}\mathrm{d}x_1\mathrm{d}x_3$; this gives $\Delta s = \sqrt{1+p^2}\delta a\mathrm{d}x_3$. The unit vector $\vec{\gamma}$ perpendicular to $\mathrm{d}s$ is $\vec{\gamma} = 1/\sqrt{1+p^2}(-p, 1, 0)$ so that $d\vec{s} = \vec{\gamma}\mathrm{d}s = (-p, 1, 0)\mathrm{d}x_1\mathrm{d}x_3$. We have $(-\delta E) = \frac{1}{2}\int_a^{a+\delta a}\sum_i(\overline{\sigma}_{i1}ds_1 + \overline{\sigma}_{i2}ds_2)\Delta u^{(i)}$, the integration being performed with respect to x_1. Making use of (3.22) for $\overline{\sigma}_{ij}$ and (3.21) for $\Delta u^{(i)}$ we arrive at

$$G(P_0) = \lim_{\delta a \to 0} -\delta E/\Delta s$$
$$= \frac{1}{\sqrt{1+p^2}}\left(K_1^{*2} + K_2^{*2}\right)(1-\nu^2)/E$$
$$= \left(\frac{\nu^2 + p^2}{1+p^2}\right) l\pi\sigma_a^2(1-\nu^2)/E \tag{3.23}$$

that is identical to (3.17); $P_0 = (a, x_2 = pa, x_3)$. The two different ways for describing the crack dislocations (Figs. 3.2 and 3.3) lead to the same expression of the crack extension force. The factor $(\nu^2+p^2)/(1+p^2)$ in (3.23) increases continuously with p from the value ν^2 ($p = 0$) to a value limited by 1 (p large). It is interesting to mention from (3.23) that G consists of two separate terms, with K_i^* (i = 1 and 2) associated with the dislocation families 1 and 2 (Fig. 3.3) respectively. The arrangement of the crack dislocations given in Fig. 3.3 is convenient. Consider a fracture specimen that breaks under general loading along a fracture surface (this may be a non-planar surface) corresponding to the loading conditions. One may decompose the applied loading into a tension σ_{22}^a(mode I) along x_2 (use Fig. 3.1 to illustrate), a shear σ_{12}^a (mode II) along x_1 and a shear σ_{23}^a (mode III) along x_3. Hence it will be possible to represent the crack by families i of dislocations with Burgers vectors $\vec{b}_i$ along x_i ($i = 1, 2$ and 3) responding to the associated mode i (II, I and III respectively); see for example, Anongba (2009, 2010).

Discussion

We first observe that Poisson effect plays no role during the propagation process when the crack is loaded in tension. Indeed, if the crack of Fig. 3.1 is assumed to be located in Ox_1x_3 ($p = 0$) and loaded in tension along x_2, Poisson effect would correspond to a lateral contraction of the medium parallel to x_1; under such conditions there is no relative displacement of the faces of the crack associated with that contraction. We conclude that Poisson effect plays no role in tension. In contrast, under applied compression ($-\sigma_a$) along x_1, Fig. 3.1, Poisson tension ($\nu\sigma_a$) along x_2 opens a crack located in Ox_1x_3 thus improving the conditions for crack motion. Again, consider the specimen in Fig. 3.1 with $p = 0$. In absence of the Poisson effect, we obtain $G = 0$ from (3.17) or (3.23): this result conflicts with experimental observations (see below). When the fracture specimen breaks at a stress level σ_c, using the condition $G = 2\gamma$ where γ is the surface energy, we obtain

$$\sigma_c = \frac{1}{\nu}\sqrt{\frac{2\gamma E}{\pi(1-\nu^2)l}}. \tag{3.24}$$

The same specimen with half-length l loaded in tension along x_2 (Fig. 3.1, $p = 0$) would fracture at a stress level σ_t given by

$$\sigma_t = \sqrt{\frac{2\gamma E}{\pi(1-\nu^2)l}}. \tag{3.25}$$

We arrive at

$$\sigma_c = \frac{1}{\nu}\sigma_t. \tag{3.26}$$

Assuming $\nu = 1/3$ (isotropic medium), this gives $\sigma_c = 3\sigma_t$ suggesting that 3 times larger stress is needed to fracture a specimen in compression as compared to tension. There is a considerable number of experimental observations showing cracks propagating axially along x_1 under compression. We may refer, among others, to the following works: Fig. 14.1 inPaul (1968); Figs. 6.9.1a and 6.9.2a in Nemat-Nasser and Hori (1999). These cracks display the following features (refer to Fig. 3.1 with $\theta = 0$ for the x_i-directions):

- Crack propagation occurs in absence of plasticity.
- The crack length may be of the order of ½ mm and even larger.
- Larger cracks are more opened. In other words, there is a gradual opening of the faces of the crack during crack extension. The opening occurs in the x_2-direction perpendicular to the crack Ox_1x_3 plane.

It appears necessary to assume a tension in the x_2-irection to be able to explain axial propagation of the crack along x_1 under compression. The present study gives the contribution of the Poisson effect to the conditions for crack propagation under compression in the case of an isotropic homogeneous elastic body. It is interesting to discuss Poisson effect in connection with experimental measurements of the stress to fracture in compression. In most available works (see, for example, Brace and Bombolakis 1963; Hoek and Bieniawski 1965; Horii and Nemat-Nasser 1985), specimens contain a starter crack in the form of a slit oriented at an acute angle $\theta \neq 0$ with respect to the x_1-direction of compression as illustrated in Fig. 3.1. Unfortunately fracture of the specimen under load does not correspond to an extension of the starter crack in its initial inclination direction, a necessary condition for result (3.23) to apply. In contrast, a new crack initiates from the tip of the slit and propagates in a completely different direction to become gradually parallel to the x_1-axis of compression. Under such conditions it appears necessary that the starter crack be parallel to the x_1-axis ($\theta = 0$), a situation that is uncommonly reported. In the experiments by Ashby and Hallam (1986) in PMMA plates, the smallest reported slit inclination angle is $\theta = 15°$. This is because as θ decreases to zero, the stress to fracture the specimen increases and falls in a range corresponding to crack propagation in presence of plasticity. The conditions we shall take from their work (see Fig. 4 inAshby and Hallam 1986) are: $\theta = 15°$ in our notation; $l = 8$ mm starter crack half length; σ_c(stress to fracture) given by them in the form of a ratio $R = \sigma_c\sqrt{\pi l}/K_{IC} = 6.5$ with $K_{IC} = 1.0$ MPa $m^{1/2}$; K_{IC} is the fracture toughness of the PMMA for slow crack growth taken from Marshall and Williams (1973). Introducing Poisson effect using (3.23) with $p = 0$ approximately, $\nu\sigma_c\sqrt{\pi l}$ is the appropriate quantity to be compared with K_{IC}, this gives a ratio $R^{(P)} = \nu\sigma_c\sqrt{\pi l}/K_{IC} = 2.5$ ($\nu = 0.38$, PMMA). We would expect $R^{(P)} = 1$ if Poisson effect operates exclusively in the crack propagation process along x_1. The observed discrepancy may have different possible origins. We may first indicate that the starter crack is $\theta = 15°$ inclined from the compression x_1-axis and should be perfectly aligned along x_1. We may also invoke the facts that the slit is not perfectly sharp, or a localized plasticity exists at the tip of the crack. In summary, under uniform applied compression ($-\sigma_a$) (Fig. 3.1), an axial x_1—propagation of a planar straight-fronted crack in Ox_1x_3 requires a non-zero relative displacement of the faces of the crack in the x_2-direction. This may be provided by Poisson effect and requires ($1/\nu$) times larger stress when compared with tensile fracture experiments.

SUMMARY: An estimate is made of the contribution of Poisson effect to the conditions for the propagation of an inclined planar crack subjected to uniform compression in the framework of linear isotropic elasticity. An expression of the crack extension force G per unit length of the crack front is provided as a function of the inclination slope p of the crack. In terms of stress and when comparison is made with tension, it is $(1/\nu)$ times more difficult to break a fracture specimen in compression. Our treatment uses the representation of the inclined crack by a continuous distribution of edge dislocations with infinitesimal Burgers vectors. It is shown that two different arrangements of the crack dislocations lead to the same value of the crack extension force.

3.2 Cracked Bars Bent by Terminal Transverse Load and Couple

Applied Stress Tensor

Consider a rectangular bar (Fig. 3.4) of length L_0, embedded in a wall at one end in a horizontal position and subjected to a downward vertical force F on the other end. We use a Cartesian coordinate system x_i where origin O is the centre of the sample cross section at the wall, the direction x_1 is the axis of the sample, x_2 is the upward vertical and $Ox_1x_2x_3$ forms a direct orthogonal axis system, i.e. the rotation around Ox_3 from Ox_1 to Ox_2 advances the corkscrew in the positive direction on Ox_3. The objective of this study is to investigate the conditions of crack propagation in the sample under load. The specimen is assumed large, homogeneous, isotropic, deforming elastically and fracturing without plasticity.

The elastic deformation in the bar under load at equilibrium has been well studied (Love 1927) and the subject is presented in courses on the elastic properties of solids (bachelor cycle at the university) (see Buttet 1981) for example). The upper part (1) in the bar lengthens, the lower portion (2) shrinks in length. There is no length variation in the median plane (plane separating parts (1) and (2) before loading); this plane is called the "neutral plane". As a result, region (1) is in tension and (2) in compression. Consider a portion of the specimen located between the points P_0' and P_0 on the neutral line. Its shape (E_0) is dashed before loading with a length l. In the deformed state, its solid-line shape (E) is a portion of a circular ring whose faces at points P_0' and P_0 meet at a point C, centre of the ring. We define a Cartesian

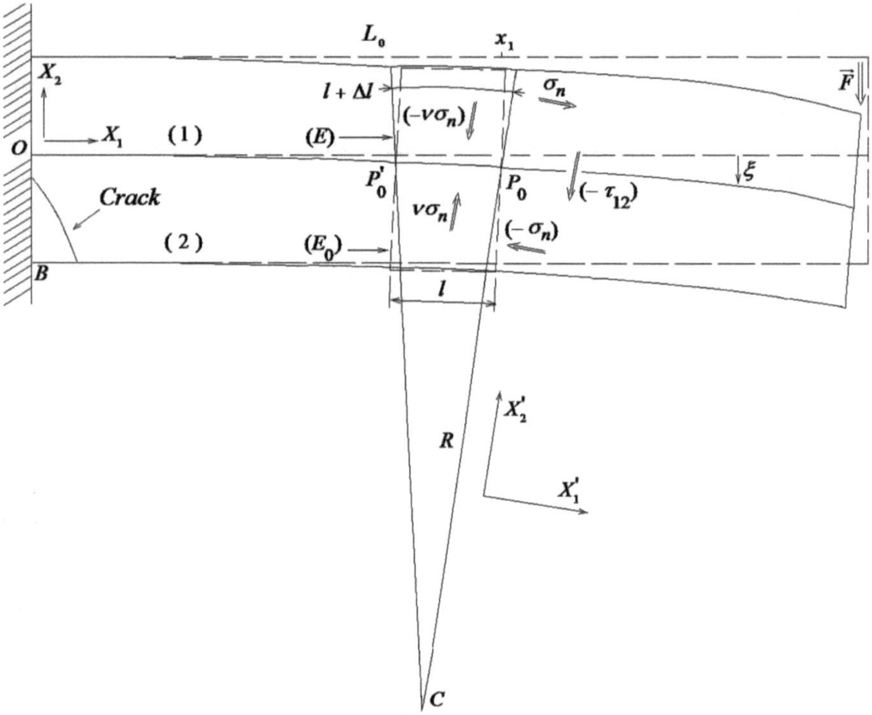

Fig. 3.4 Rectangular bar (length L_0, thickness e_0 and width l_0, measured along x_1, x_2 and x_3, respectively) under bending by a terminal transverse force F. Regions (1) and (2) are in tension and compression, respectively; these are separated by the neutral line. The specimen is dashed before bending and solid under load. Internal stresses are represented about a point P_0 on the neutral line with coordinate x_1: these are the tension and compression ($\pm\sigma_n$), The associated induced Poisson ($\pm\nu\sigma_n$), and the shear ($-\tau_{12}$). The other symbols are defined in the text

system $P_0x'_1x'_2x_3$ where $P_0x'_1$ is perpendicular to CP_0 with a positive direction directed towards the end of the bar under load, $P_0x'_2$ is the axis CP_0 directed upward and P_0x_3 is parallel to Ox_3 through P_0. The length $(l + \Delta l)$ of (E) along x'_1 depends on the value x'_2 of the position where the measurement is performed; we have

$$\frac{\Delta l}{l} = \frac{x'_2}{R} \tag{3.27}$$

where $R = CP_0$. The corresponding internal stress is

$$\sigma'_{11} = E\frac{x'_2}{R} \tag{3.28}$$

where E is Young's modulus. The magnitude of σ'_{11} is denoted by σ_n ($|\sigma'_{11}| \equiv \sigma_n$). σ'_{11} and x'_2 have the equal sign, positive in part (1) of (E) and negative in part (2);

hence (E) suffers tension and compression forces in regions (1) and (2), respectively, both parallel to P_0x_1'. The force associated to the stresses on (E) about P_0 is equal to F. This means that tangential forces are present that are directed along P_0x_2' (in the negative direction), to which corresponds a shear stress σ_{12}' sensibly equal to (Buttet 1981)

$$\sigma_{12}' = -F/S_0 \equiv -\tau_{12} \tag{3.29}$$

The magnitude of the resultant moment M of the forces at P_0 is given by

$$M = \int_{Section} x_2'\sigma_n \mathrm{d}s = \frac{E}{R}\int_{Section} x_2'^2 \mathrm{d}s \equiv \frac{EI}{R} \tag{3.30}$$

where $I = \int x_2'^2 \mathrm{d}s$ is the geometric moment of inertia of sample section at P_0 about P_0x_3. Writing that M is equal to the moment (at P_0) $F(L_0 - x_1)$ of the forces acting on the complementary part $(L_0 - x_1)$ of specimen, we obtain

$$\frac{1}{R} = \frac{F(L_0 - x_1)}{EI}; \tag{3.31}$$

we can then write

$$\sigma_{11}' = \frac{F(L_0 - x_1)}{I}x_2' = \tau_{12}\frac{S_0(L_0 - x_1)}{I}x_2'. \tag{3.32}$$

The equilibrium shape of the bar is extracted from

$$\frac{1}{R} = \frac{\partial^2\xi/\partial x_1^2}{\left[1 + (\partial\xi/\partial x_1)^2\right]^{3/2}} \cong \frac{\partial^2\xi}{\partial x_1^2} \tag{3.33}$$

for small displacements ξ (Fig. 3.4) of points of the bar along x_2. Integrating (3.33) gives

$$\xi(x_1) = \frac{F}{EI}\left(\frac{L_0x_1^2}{2} - \frac{x_1^3}{6}\right) \tag{3.34}$$

with boundary conditions $\xi(x_1 = 0) = 0$ and $\partial\xi/\partial x_1(x_1 = 0) = 0$.

Model and Crack Analysis

How crack behaves under load in the specimen? We refer to Anongba (2015). Assume that a crack expands from an arbitrary x_1 position inside the material. From the stresses about $P_0(x_1, \xi(x_1), x_3)$ (Fig. 3.4), one can distinguish two distinct types of crack propagation behaviour, in tension region (1) and compression region (2). In the former zone (1), the crack is essentially loaded in tension σ_n. From experiments, it is well-known that a crack loaded in tension, after fracture propagation over large distances, is directed perpendicularly to the applied tension. In this position, Poisson's contraction $(-\nu\sigma_n)$ produces no relative displacement of the faces of the crack, hence does not contribute to crack motion (as commented earlier by Anongba et al. 2011). However, considering the shear $(-\tau_{12})$ (much smaller than σ_n), the crack may deviate from its natural position by a small angle. In the latter region (2), the crack suffers mainly a compression $(-\sigma_n)$(Fig. 3.4). It is also well-known that a crack, experiencing a compressive stress, aligns itself parallel to the compression direction (see Anongba et al. 2011 and references therein). In this orientation, an opening of the crack is performed by Poisson's tension $\nu\sigma_n$. A deviation of the crack (due to $(-\tau_{12})$) from the compression direction is also to be considered. The crack extension force (per unit length of the crack front) G_1 at the tip of the crack that moves in region (1) is much larger than G_2 the corresponding quantity in part (2) of the specimen. Hence, this is under the condition $G_2 = 2\gamma$(where γ is the surface energy) that complete failure will take place. The crack model used to calculate G_2 is described below. Physical quantities associated with that modelling (stresses about the crack tip, crack extension force and natural orientation of the crack tip, for instance) are given. The properties of fracture (crack path and failure stresses) are described and discussed. It is planned to calculate a crack extension force G_2 pertaining to the tip a crack that is moving downward (region (2) in Fig. 3.4). The considered model is presented in Fig. 3.5. This is an infinitely extended elastic medium. With respect to a Cartesian system $O'x_1'x_2'x_3'$, the crack with half-length l extends along x_1' from $x_1' = -a$ to a; it is inclined by an acute angle θ_2 with respect to $O'x_1'x_3'$ with a straight front running indefinitely in the x_3'-direction. The acting stresses are: (1) a uniform compression $(-\sigma_n)$ acting in the x_1'-direction, (2) a uniform Poisson's tension $\nu\sigma_n$ in the x_2'-direction and (3) a uniform shear $(-\tau_{12})$ lying in $x_2'x_3'$ and parallel to x_2'. The crack is represented.

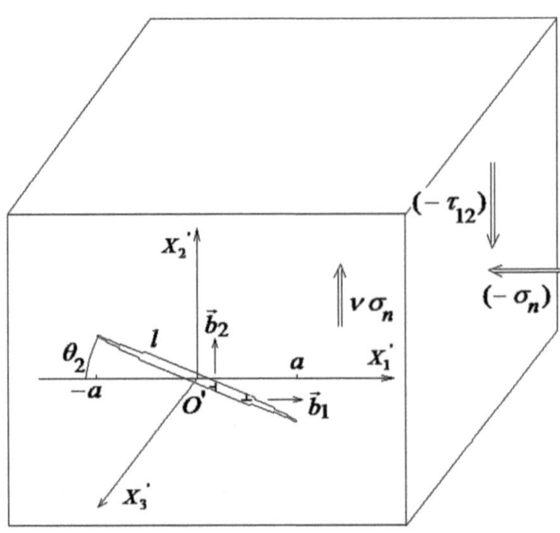

Fig. 3.5 Inclined crack by angle θ_2 with respect to $O'x_1'x_3'$, with a straight front parallel to $O'x_3'$. Crack dislocations (1 and 2) are straight edges parallel to x_3' with $\vec{b}_1$ along the applied compression x_1'-direction and $\vec{b}_2$ along the induced Poisson's tension x_2'-direction. A shear $(-\tau_{12})$ acts in $x_2'x_3'$ planes in the x_2'-direction

by two families (1) and (2) of straight edge dislocations parallel to x'_3 with infinitesimal Burgers vectors $\vec{b}_1 = (b, 0, 0)$ and $\vec{b}_2 = (0, b, 0)$, respectively. The two families are assumed to be continuously distributed over the crack area from $x'_1 = -a$ to a. The dislocation distribution function $D_i(x'_1)$ ($i = 1$ and 2) gives the number of dislocations i in a small interval dx'_1 about x'_1 as $D_i(x'_1)dx'_1$. The correspondence with Fig. 3.4 is that O' can be viewed as located in region (2) or about P_0 and the directions x'_i have the same meaning; O' and the stresses depend on the position x_1 along the bar.

The tensor (σ^A) of the applied stresses in $O'x'_1x'_2x'_3$ reads

$$(\sigma^A) = \begin{pmatrix} -\sigma_n & -\tau_{12} & 0 \\ -\tau_{12} & \nu\sigma_n & 0 \\ 0 & 0 & \nu\sigma_n \end{pmatrix}. \tag{3.35}$$

For a dislocation with Burgers vector $(b, 0, 0)$ lying indefinitely in the x'_3-direction and displaced by $x'_2 = h$ from the origin, the stress field is given at $\vec{x}' = (x'_1, x'_2, x'_3)$ by

$$\begin{aligned}
\sigma^{(1)}_{12}(\vec{x}') &= C\frac{x'_1\left(x'^2_1 - (x'_2 - h)^2\right)}{r^4}, \\
\sigma^{(1)}_{11}(\vec{x}') &= -C\frac{(x'_2 - h)\left(3x'^2_1 + (x'_2 - h)^2\right)}{r^4}, \\
\sigma^{(1)}_{22}(\vec{x}') &= C\frac{(x'_2 - h)\left(x'^2_1 - (x'_2 - h)^2\right)}{r^4}, \\
\sigma^{(1)}_{33}(\vec{x}') &= -2C\nu\frac{(x'_2 - h)}{r^2}, \\
\sigma^{(1)}_{13} &= \sigma^{(1)}_{23} = 0;
\end{aligned} \tag{3.36}$$

$r^2 = x'^2_1 + (x'_2 - h)^2$ and $C = \mu b/2\pi(1 - \nu)$ where μ is the shear modulus. For a dislocation with Burgers vector $(0, b, 0)$ lying along x'_3 and displaced by $x'_2 = h$ from the origin, the stress is

$$\begin{aligned}
\sigma^{(2)}_{12}(\vec{x}') &= C\frac{(x'_2 - h)(x'^2_1 - (x'_2 - h)^2)}{r^4}; \\
\sigma^{(2)}_{ii}(\vec{x}') &= C\frac{x'_1}{r^2}\left(\frac{x'^2_1(\delta_{i1} + \delta_{i2}) + (x'_2 - h)^2(-\delta_{i1} + 3\delta_{i2})}{r^2} + 2\nu\delta_{i3}\right), \ i = 1, 2 \text{ and } 3 \\
\sigma^{(2)}_{j3}(\vec{x}') &= 0, \ j = 1 \text{ and } 2;
\end{aligned} \tag{3.37}$$

δ_{ij} is the Kronecker delta. The crack analysis requires the determination of the equilibrium crack dislocation distribution D_i, relative displacement of the faces of the crack, crack-tip stress and crack extension force. Any point on the crack is given by $P_s = (x'_1, x'_2 = -px'_1, x'_3)$ with $|x'_1| < a$ and $p = \tan\theta_2 > 0$. We ask the crack faces to be free of traction, this gives

$$\begin{cases} \overline{\sigma}_{12} + p\overline{\sigma}_{11} = 0 \\ \overline{\sigma}_{22} + p\overline{\sigma}_{21} = 0. \\ \overline{\sigma}_{32} + p\overline{\sigma}_{31} = 0 \end{cases} \tag{3.38}$$

$\overline{\sigma}_{ij}$ stands for the total stress at any point (x_1', x_2', x_3') in the medium and is linked to D_i; in (3.38), we are only concerned with the points of the crack faces. $\overline{\sigma}_{ij}$ is written as

$$\overline{\sigma}_{ij} = \sigma_{ij}^A + \overline{\sigma}_{ij}^{(1)} + \overline{\sigma}_{ij}^{(2)} \tag{3.39}$$

where σ_{ij}^A are the *ij*- elements of matrix (σ^A)(3.35) and

$$\overline{\sigma}_{ij}^{(n)}(x_1', x_2', x_3') = \int_{-a}^{a} \sigma_{ij}^{(n)}(x_1' - x, x_2', x_3') D_n(x) \mathrm{d}x \quad (n = 1 \text{ and } 2); \tag{3.40}$$

here $\sigma_{ij}^{(n)}(n = 1 \text{ or } 2)$ is the stress field produced by a dislocation displaced by $x_2' = h = -px$ from the origin with Burgers vector $(b, 0, 0)$ or $(0, b, 0)$ [(3.36) and (3.37)]. Introducing the expressions for $\overline{\sigma}_{ij}$ in (3.38), we obtain

$$\begin{cases} -(\tau_{12} + p\sigma_n) + C \int_{-a}^{a} \dfrac{D_1(x)\mathrm{d}x}{x_1' - x} = 0 \\ \nu\sigma_n - p\tau_{12} + C \int_{-a}^{a} \dfrac{D_2(x)\mathrm{d}x}{x_1' - x} = 0 \end{cases}. \tag{3.41}$$

(3.41) forms the governing equations of our modelling. The corresponding solutions are well-known

$$D_1(x_1') = -\frac{\tau_{12} + p\sigma_n}{\pi C} \frac{x_1'}{\sqrt{a^2 - x_1'^2}},$$
$$D_2(x_1') = \frac{\nu\sigma_n - p\tau_{12}}{\pi C} \frac{x_1'}{\sqrt{a^2 - x_1'^2}} \tag{3.42}$$

with relative displacement φ_i of the faces of the crack in the x_1' $(i = 1)$ and x_2' $(i = 2)$ directions

$$\varphi_1(x_1') = -\frac{b(\tau_{12} + p\sigma_n)}{\pi C} \sqrt{a^2 - x_1'^2},$$
$$\varphi_2(x_1') = \frac{b(\nu\sigma_n - p\tau_{12})}{\pi C} \sqrt{a^2 - x_1'^2}. \tag{3.43}$$

Below are presented expressions for the stress about the crack tip and crack extension force G_2. We shall look for crack configuration at which G_2 is maximum.

This configuration would correspond to the natural configuration of the propagating crack; this is under this configuration that we shall apply the condition for crack propagation $G_2 = 2\gamma$. The stresses ahead of the crack tip at point $P_s = (x_1', x_2' = -px_1', x_3')$, $x_1' = a + s$, $0 < s \ll a$, are given, for sufficiently small values of s, by the following relation:

$$\overline{\sigma}_{ij}(s) = \sum_{n=1}^{2} \int_{a-\delta a}^{a} \sigma_{ij}^{(n)}(x_1' - x, -px_1', x_3')D_n(x)\mathrm{d}x;\ x_1' = a + s \tag{3.44}$$

with $\delta a \ll a$. Restricting ourselves to the stresses involved in the calculation of the crack extension force, we obtain

$$\begin{aligned}
\overline{\sigma}_{11}(s) &= \frac{p(3+p^2)K_1^* + (1-p^2)K_2^*}{(1+p^2)^2}\frac{1}{\sqrt{2\pi}\sqrt{s}},\\
\overline{\sigma}_{12}(s) &= \frac{(1-p^2)(K_1^* - pK_2^*)}{(1+p^2)^2}\frac{1}{\sqrt{2\pi}\sqrt{s}},\\
\overline{\sigma}_{11}(s) &= \frac{-p(1-p^2)K_1^* + (1+3p^2)K_2^*}{(1+p^2)^2}\frac{1}{\sqrt{2\pi}\sqrt{s}}
\end{aligned} \tag{3.45}$$

where $K_1^* = -(\tau_{12} + p\sigma_n)\sqrt{a\pi}$ and $K_2^* = (\nu\sigma_n - p\tau_{12})\sqrt{a\pi}$; terms with K_i^* are due to dislocation family i ($i = 1$ and 2). The crack extension force can be calculated following a procedure described by Bilby and Eshelby (1968); we have also referred to this in several works (Anongba et al. 2011; Anongba, 2010, among many others). This gives G_2 at $P_a = (x_1' = a, x_2' = -pa, x_3')$ as

$$G_2(P_a) = \frac{(1-\nu^2)l\pi}{E(1+p^2)}\left((\tau_{12} + p\sigma_n)^2 + (\nu\sigma_n - p\tau_{12})^2\right) \equiv G_a. \tag{3.46}$$

The behaviour of G_a as a function of p is as follows. G_a is minimum at $p = 0$, increases up to a maximum at $p = p_+$ and then decreases asymptotically to a value $G_a(\infty)$ as p tends to infinity; we have

$$\begin{aligned}
p_+ &= \frac{1}{2}\left((1+\nu)\overline{\sigma}_n + \sqrt{(1+\nu)^2\overline{\sigma}_n^2 + 4}\right),\\
G_a(p=0) &= \frac{(1-\nu^2)l\pi\tau_{12}^2}{E}\left(1 + \nu^2\overline{\sigma}_n^2\right),\\
G_a(p=p_+) &= \frac{(1-\nu^2)l\pi\tau_{12}^2}{E}\left(1 + \overline{\sigma}_n^2 + \frac{2(1-\nu)\overline{\sigma}_n}{(1+\nu)\overline{\sigma}_n + \sqrt{(1+\nu)^2\overline{\sigma}_n^2 + 4}}\right) \equiv G_a^{\max},\\
G_a(\infty) &= \frac{(1-\nu^2)l\pi\tau_{12}^2}{E}\left(1 + \overline{\sigma}_n^2\right),
\end{aligned} \tag{3.47}$$

where $\overline{\sigma}_n = \sigma_n/\tau_{12}$. The fracture stress $\tau_{12} \equiv \sigma_f$ is obtained from $G_a^{\max} = 2\gamma$ as

$$\sigma_f = \sigma_t \sqrt{\frac{(1+\nu)\overline{\sigma}_n + \sqrt{(1+\nu)^2\overline{\sigma}_n^2 + 4}}{(1+\nu)\overline{\sigma}_n^3 + (3-\nu)\overline{\sigma}_n + (1+\overline{\sigma}_n^2)\sqrt{(1+\nu)^2\overline{\sigma}_n^2 + 4}}} \quad (3.48)$$

where

$$\sigma_t = \sqrt{\frac{2\gamma E}{\pi(1-\nu^2)l}}. \quad (3.49)$$

Fracture Properties

We shall assume that σ_n in our modelling (Fig. 3.5) is given by the magnitude of σ'_{11} (3.32) that depends on the x_1-position along the specimen and value of x'_2 along the axis CP_0(Fig. 3.4). Therefore, the various derived physical quantities depend on x_1 and x'_2. σ'_{11}(3.32) (hence σ_n) is maximum in magnitude for $x_1 = 0$ and $x'_2 = -e_0/2$.This corresponds to position $B = (x_1 = 0, x_2 = -e_0/2)$ in Fig. 3.4. We have $\overline{\sigma}_n(B) = 6L_0/e_0$ (for $I = l_0 e_0^3/12$) that is generally much larger than unity. We first discuss the behaviour of the crack about B. The crack inclination angle θ_2^+ with respect to the x'_1-direction (Fig. 3.5) (given by $p_+ = \tan\theta_2^+$ (3.47) takes about B for $\overline{\sigma}_n \gg 1$, the simple form $\tan\theta_2^+(B) = (1+\nu)\overline{\sigma}_n(B) = 6(1+\nu)L_0/e_0$. With $L_0/e_0 = 10$, this gives $\theta_2^+(B) = 89°$ that corresponds to a crack path not far from the x_2-direction. Using (3.32) with $x_1 = 0, x'_2 = x_2$ and writing

$$dx_2/dx_1 = -(1+\nu)\overline{\sigma}_n = -(1+\nu)S_0L_0|x_2|/I,$$

we obtain by integration

$$x_2 = x_2(0)e^{(1+\nu)V_0x_1/I} \quad (3.50)$$

where $x_2(0) < 0$ in compression region (2) and $V_0 = L_0S_0$ is the volume of the specimen. (3.50) indicates that the crack path is in the form of an exponential about the wall in region (2). A complete crack shape about the wall may be a path parallel to Ox_2 for $x_2 > 0$ and in the form of an exponential for $x_2 < 0$ with a beginning point somewhere below O (as shown schematically in Fig. 3.4). A pre-existing flaw may promote failure out of the wall; again, the exponential path of the crack would be preserved there. The fracture stress σ_f (3.48) in bending by terminal load F (Fig. 3.4) is proportional to the quantity σ_t (3.49). This latter expression is nothing else than the stress at failure when a specimen with a crack of half-length l is loaded in tension. When $\overline{\sigma}_n >> 1$, (3.48) reduces to

$$\frac{\sigma_f}{\sigma_t} = \frac{1}{\overline{\sigma}_n} = \frac{e_0}{6L_0} \quad (3.51)$$

and in term of the bending moment $M \equiv M_f$,

$$M_f = \frac{2I\sigma_t}{e_0}. \tag{3.52}$$

It is important to stress that these relations are intimately related to the existence of a shear stress τ_{12}. In presence of a shear stress, G_a (3.46) has a maximum from which is deduced σ_f (3.48); however, in absence of a shear stress ($\tau_{12} = 0$) as we shall discuss, G_a has no maximum with p and (3.48) (hence 3.51, 3.52) does not exist. This brings us to discuss the case of bending without shear stress. An example corresponds to the bending of a bar by terminal couple (Love 1927); the bending moment $M \equiv M_c$ (3.30) is constant along the bar so that $\sigma'_{11} = M_c x'_2/I$ (3.32). The crack extension force G_{2c} in this situation is given by (3.46) with $\tau_{12} = 0$. We have

$$G_{2c}(P_a) = \frac{(1-\nu^2)\sigma_n^2 l\pi}{E}\left(\frac{\nu^2 + p^2}{1+p^2}\right). \tag{3.53}$$

This is the expression of the crack extension force at the tip of an inclined crack under compression and internal Poisson tension discussed by Anongba, Bonneville and Joulain (2011). The factor $(\nu^2 + p^2)/(1 + p^2)$ in (3.53) increases continuously with p from ν^2 ($p = 0$) to a value limited by 1 (p large). In absence of shear stress, the inclined crack (Fig. 3.5) tends to align parallel to the applied compression ($-\sigma_n$) direction (see Anongba et al. 2011 and references therein). The natural configuration of the crack system corresponds to $p = 0$. Under such conditions, the relation $G_{2c} = 2\gamma$ yields

$$\sigma_n = \frac{1}{\nu}\sigma_t. \tag{3.54}$$

Taking $\sigma_n = M_c e_0/2I$, we have

$$M_c = \frac{2I\sigma_t}{\nu e_0}. \tag{3.55}$$

To compare with (3.52), we take $\nu = 1/3$ for isotropic materials and obtain $M_c = 3M_f$; this indicates that three times larger bending moment is required to fracture a specimen in bending by terminal couple. The present modelling assumes a crack path perpendicular to the applied tension direction in tension regions of the fractured specimen and parallel to the applied compression in compression regions. This means that fracture specimens must have large dimensions in order that crack propagation over macroscopic distance is the controlling mechanism of failure. Crack nucleation would be internal. Fracture under boundary effects so frequently associated to small specimens may not be described by the present study.

SUMMARY: In the present study, an analysis has been performed of the conditions of failure in a rectangular bar, horizontally in a wall at one end and under load about the other end, either by terminal transverse load or terminal couple. The specimen consists of two distinct parts (1) and (2) separated by a non-deformed (neutral) plane. Region (1) suffers tension and (2) compression. It is known from experiment that a crack propagating over macroscopic distances under tension, tends to align perpendicularly to the applied tension direction, whilst under a compressive stress, the equal crack would be directed along the compression direction. In the last orientation, this is the induced Poisson's tension, much smaller than the applied stress that is responsible for the opening of the crack faces. We are thus led to consider two distinct crack extension forces G_1 and G_2 pertaining to both regions. Because G_1 is much larger than G_2, this is under the condition $G_2 = 2\gamma$ that complete failure would take place. We then proceed to a calculation of G_2, using a model of a finite crack in an infinitely extended homogeneous medium, that permits us to describe a natural configuration of the crack under load and a relation between the failure stress and the dimensions of the specimen. This study allows to distinguish two types of applied loading: (1) flexion associated with shear (case of terminal transverse load) and (2) pure bending without shear (case of terminal couple). Three times larger bending moment is required to fracture the specimen in the latter applied loading condition.

3.3 DCB Specimen Bent by Opposite Terminal Transverse Loads

Consider a rectangular bar $BAA'B'$ (Fig. 3.6) of thickness $BB' = 2\ e_0$, embedded in wall at one end BB' in a horizontal position; at the other end AA', a pair of opposite terminal transverse vertical loads $\boldsymbol{F}$ and $-\boldsymbol{F}$ separate into two equal parts $(J)\ BAA_mB_m$ and $(J)'\ B_mA_mA'B'$ the bar $BAA'B'$. These are drawn dashed before separation and full under loads. Such specimen under loads is named "Double-cantilever beam (*DCB*) specimen" (see Fig. 2.13 in Lawn 1993) with beams (J) and $(J)'$. To the initial dashed form of the specimen is attached a Cartesian coordinate system x_i where the direction x_1 (left-to-right) is horizontal (axis of the sample), x_2 is the upward vertical and $x_1x_2x_3$ forms a direct orthogonal basis. An origin for this coordinate system may be located at the center of the sample cross section at the wall. The objective is to investigate the conditions for the separation of the bar under loads into two equal parts (J) and $(J)'$. The specimen is assumed large, homogeneous, isotropic, deforming elastically, and fracturing without plasticity. Gravitational forces are neglected. The elastic stress fields in the *DCB* sample can be understood from Sect. 3.2. The bottom part (J) of the specimen can be viewed as a rectangular bar bent by terminal transverse load $\boldsymbol{F}$. Hence, various notations and definitions reported in part (J) (Fig. 3.6) are identical to those of Fig. 3.4; remains to specify what is L_0 in the present. It is assumed that the value of F corresponds to a position O_m^0 such that $O_m^0A_m = L_0$. Moving on the central line from A_m to B_m, O_m^0 is the first spatial position for which the two arms (J) and (J)' have horizontal tangents parallel to B_mA_m (*i.e.* to x_1). Under such conditions, the tensile stresses $\sigma_{11} \equiv \sigma_n(O_m^0)$ at O_m^0 in both (J) and $(J)'$ are parallel to x_1, and read (Sect. 3.2):

$$\sigma_n(O_m^0) = \frac{FL_0e_0}{2I},$$
$$I = \frac{l_0e_0^3}{12}. \tag{3.56}$$

Considering tensile forces $\sigma_n(O_m^0)$ only, there is no relative displacement of faces of the crack in the x_1 and x_2 directions. Consequently, the crack would stop about O_m^0; this corresponds to $G(O_m^0) = 0$. However there exists tensile stresses parallel to x_2 (Sect. 3.2) that open the crack faces about O_m^0. These are $\sigma_{22}(O_m^0)' \equiv \nu\sigma_n(O_m^0)$ and $\sigma_{22}(O_m^0) \equiv -\nu\sigma_n(O_m^0)$ in arms (*J*)' and (*J*) (ν is Poisson's modulus). In x_1-region between B_m and O_m^0, the tip of the crack is under mode *I* loading. Assuming a planar straight-fronted crack in uniform motion v = constant, the crack extension force *G* can be taken from Anongba (2020a). Hence in the subsonic velocity regime ($v < c_t$ the velocity of transverse sound wave), we may write:

$$G^{(I)}(O_m(t)) = \frac{bK_I^2}{4\pi C_1^{(I)}},$$

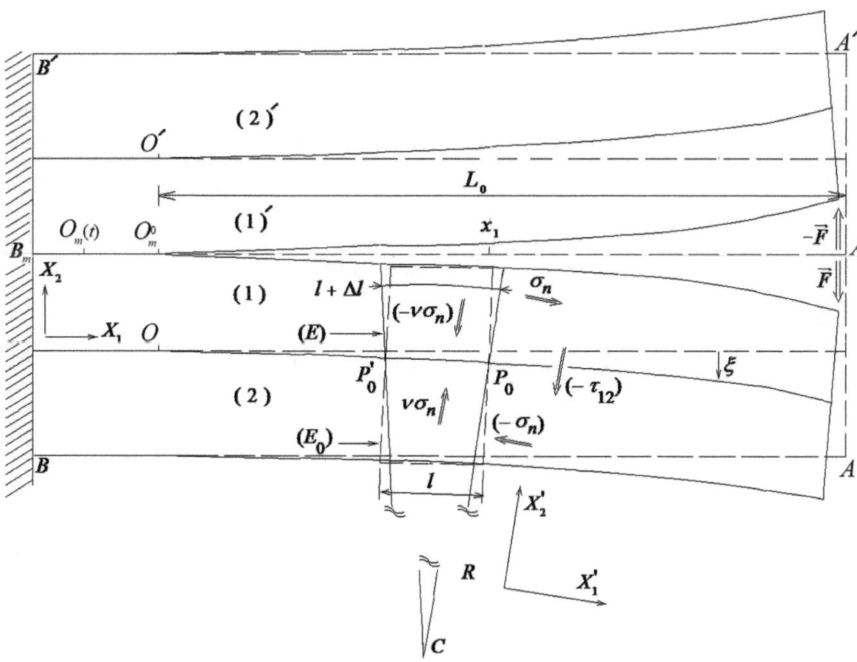

Fig. 3.6 Rectangular bar BAA′B′ (length BA, thickness BB′ = 2 e_0 and width l_0, measured along x_1, x_2 and x_3, respectively) embedded in a wall at one end BB′ in horizontal position and separated into two arms at the other end AA′ by a pair of opposite vertical loads **F** and − **F**. It is assumed that no shearing stresses exist in the sample region located between vertical lines BB_mB' and OO_m^0O'; see text for details

$$K_I = \nu \sigma_n(O_m^0)\sqrt{\pi c},$$
$$c = vt + L_0; \qquad (3.57)$$

$C_1^{(I)}$ is given by (2.1.5) (Sect. 2.1). It is stressed that O_m^0 position along $B_m A_m$ (Fig. 3.6) may be determined experimentally by observing the shape of the bar $BAA'B'$ under loads; BB' is allowed locally to contract and extend freely along x_2 in the wall. In summary, considering a *DCB* specimen bent by pairs of opposite terminal transverse loads $\boldsymbol{F}$ and $-\boldsymbol{F}$ (Fig. 3.6), two different situations emerge: (i) When F is small, the separation into two beams would end about O_m^0 before reaching the wall; this corresponds to $G(O_m^0) = 0$. (ii) For sufficiently large F, a crack would develop from O_m^0 towards B_m. Assuming its motion uniform with constant crack-front velocity v, the crack extension force G per unit length of the crack-front is finite. Its value $G^{(I)}$ is given by (3.57). We may therefore equal $G^{(I)}$ to 2γ, where the surface energy is γ, under stationary conditions.

3.4 Failure Stresses in Tension, Compression and Bending

Based on the analyses of Sects. 3.1 and 3.2, the present description emphasizes the fact that relations between failure stresses in brittle specimens tested in tension, compression and bending are attainable in isotropic materials if crack propagation is the controlling mechanism of failure, and not the initiation of crack (starter crack) motion or the nucleation of crack. Consider a fracture specimen with large dimensions under load. We assume that an internal crack is nucleated and propagates over large distances; if a starter crack is initially present, the equal configuration corresponds to fracture initiation and propagation over macroscopic distance. Then we have:

TENSION

The crack, with half-length l, is perpendicular to the applied tension direction. The failure stress σ_t is given by (3.25)

$$\sigma_t = \sqrt{\frac{2\gamma E}{\pi(1-\nu^2)l}}$$

COMPRESSION

The identical crack is aligned parallel to the compression direction. The failure stress σ_c is given by (3.26)

$$\sigma_c = \frac{1}{\nu}\sigma_t,$$

indicating that three times larger applied stress is required to bring to failure an isotropic specimen in compression as compared to tension.

BENDING

The identical crack will have one tip moving through the compression region (2). We assume that this is the condition $G_2 = 2\gamma$ that controls the complete failure of the fractured specimen, where G_2 is the crack extension force (per unit length of the crack front) pertaining to the tip of the crack that moves through the compression part (2) of the specimen; the other part (1) of the specimen is in tension. Consequently, we have.

(a) **bending with shear** (case of bending by terminal transverse load)

$$\frac{\sigma_f}{\sigma_t} = \frac{1}{\overline{\sigma}_n} = \frac{e_0}{6L_0} \tag{3.51}$$

(under the condition: $6L_0/e_0 \gg 1$) where σ_f is the failure stress and e_0 and L_0 are the thickness and length of the bending specimen measured along the applied load and specimen axis directions, respectively; in term of the bending moment M_{f} this relation becomes

$$M_f = \frac{2I\sigma_t}{e_0} \tag{3.52}$$

where I is the geometric moment of inertia of sample section about width axis through middle of the section;

(b) **bending without shear** (case of bending by terminal couple only)

$$M_{\mathrm{c}} = M_{\mathrm{f}}/\nu$$

where M_{c} is the constant bending moment; this suggests that three times larger bending moments are required to fracture isotropic samples in the latter applied loading.

The present modelling assumes fracture propagation over large distance. Why this is important?

Because:

(1) The propagating cracks have unique (single) orientations with respect to the respective applied loadings
(2) The crack lengths have similar magnitude
(3) the mechanism of failure is crack propagation and not initiation of crack motion or crack nucleation

Chapter 4
Cracks in Half-Space with Gravitational Forces

4.1 Manifestation of Gravitational Forces in Crack Propagation

The present aims at introducing gravitational forces in fracture mechanics. The motivation comes from the following common observation. Blocks of stone with large dimensions (say of the order of three meters or larger) can be easily fractured into two pieces (Fig. 4.1). First, cylindrical holes are introduced at the top surface using drills. Second, fracture is initiated from the holes with the help of sledgehammers and wedges. Without any additional action, the crack will move with time downward other very large distance and separate the block of stone into two parts. The fracture surface is perfectly flat. This sentence can be posed: a crack with a critical size in a solid can begin an expansion under gravitational forces only to split completely into two pieces a large solid. The explanation is as follows. The weight $\sigma_{22}^{AW} = \rho g x_2$ increases downward vertically (along the x_2-direction), with the position x_2 of the crack front (ρ is the density of the stone and g the acceleration due to gravity); in the two other directions x_1 and x_3 exist induced normal Poisson's stresses $-\nu_A \sigma_{22}^{AW}$ (ν_A is Poisson's ratio). These are Poisson's stresses that produce induced macroscopic tensile forces perpendicular to cracking planes. These promote failure of the stone over large distance.

4.2 Crack Model

To do so, we shall refer to the crack model used in previous works (Anongba 2019, 2023) (Fig. 4.2). This is a three-dimensional elastic half-space on the flat boundary of which is posed, in the x_3-direction, a long cylinder with radius R. A planar straight-fronted crack with length $AB = l$, inclined around x_3 by angle θ from the surface,

P. N.B. Anongba, *Mechanics of Steadily Propagating Cracks*,
https://doi.org/10.1007/978-981-96-5703-2_4

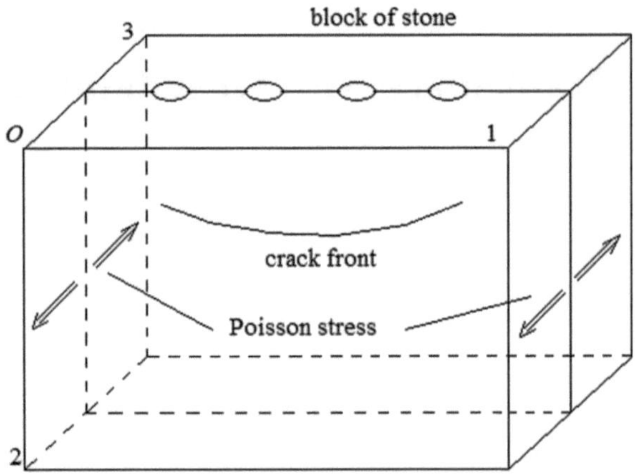

Fig. 4.1 Block of stone with four holes at top surface to introduce fracture with the help of sledge-hammer and wedge. The crack initiates and moves downward. These are Poisson's stresses which produce macroscopic tensile forces perpendicular to the crack plane; these induced tensile forces produce failure of the stone over large distance. The weight of the stone is the externally applied force and Poisson's forces are induced normal forces

is present; $\boldsymbol{OA} = (a_1, a_2, 0)$ and $\boldsymbol{OB} = (b_1, b_2, 0)$. This is an expected crack configuration at a large distance from the cylinder under contact pressure. The relevance of this modelling may be understood as follows. A slab of cylinder with thickness dx_3' at spatial position O' $(0,0, x_3')$ exerts elastic fields (displacement and stress) proportional to those of a point load at *O'* (proportionality coefficient dx_3'). Physically, this corresponds to the action of a spherical indenter to which is associated a conoidal fracture surface for sufficiently large load (Roesler 1956 as quoted by Frank and Lawn 1967; see also Lawn 1993). The coalescence of conoidal cracks from different slabs of cylinder along Ox_3 would produce planar fracture surface envelopes parallel to x_3 at large crack lengths. Therefore, we expect this modelling to provide the experimentally observed fracture surface inclination angle θ and crack length l as a function of critical load P by both a spherical indenter and cylinder. A symmetrical crack with respect to Ox_2x_3 is expected to develop between *A'* and B'. The crack under load is represented by a continuous distribution of dislocations with infinitesimal Burgers vectors. The stress induced by the crack is equivalent to that produced by the dislocations. Two straight edge dislocation families J ($J = //$ and $\perp$) parallel to x_3 with Burgers vectors $\vec{b}_{//} = (b, 0, 0)$ and $\vec{b}_{\perp} = (0, b, 0)$ parallel and perpendicular to the solid flat surface are considered. To a crack dislocation J located at x_1 is associated an elevation h from Ox_1x_3 (Fig. 4.2) with distribution function D_J such that $D_J(x_1')dx_1'$ represent the number of crack dislocations J in small x_1-interval dx_1' about x_1'. It is required to find the equilibrium dislocation distributions under the combined actions of the cylinder $(\sigma)^A$, gravitational forces $(\sigma)^{AW}$ and crack dislocation stresses $(\sigma)^{(J)}$. The applied stress tensor $(\sigma)^A$ will include the stress $(\sigma)^a$ due to the cylinder itself and the induced normal stresses due to the Poisson effect, namely that, to a normal stress σ_{ii}^a acting in the x_i-direction also correspond normal stresses

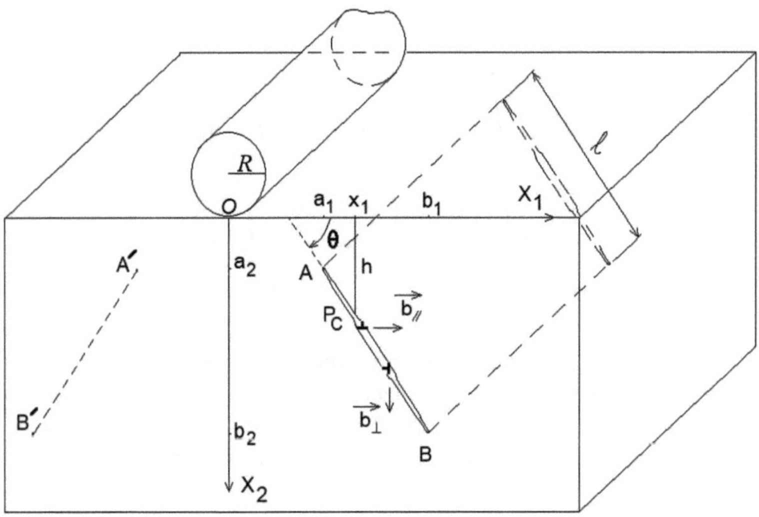

Fig. 4.2 Brittle elastic half-space under load by a cylinder (P is the load per unit length of the cylinder) posed along the x_3-direction on its flat boundary surface; a planar crack of finite extension l is present, located between positions A and B in Ox_1x_2-plane and inclined by angle θ from the planar boundary. The crack front is straight, parallel to the cylinder axis. The modelling assumes the half-space to be infinitely extended and the cylinder (radius R) and crack to run indefinitely along the x_3-direction

$(-\ \nu\sigma^a_{ii})$ in the two other associated x_j-directions. It is important to mention that considering the induced normal stresses due to Poisson effect is in fact necessary for the x_2-component of the applied force exerted on a boundary surface element ds, at any spatial position P_S of the three-dimensional elastic half-space flat boundary, to be zero except on Ox_3 (see below). Stress fields due to straight edges in half-space are available (Head 1953; Mura 1968). We provide below dislocation J elastic fields from a different method involving Galerkin vectors with biharmonic functions in Fourier forms; a similar procedure has been used to investigate the elastic fields of interfacial dislocations (straight and sinusoidal) (Anongba 2015, 2017a). In what follows, $(\sigma)^A$, $(\sigma)^{AW}$, $(\sigma)^{(J)}$, equilibrium crack dislocation distributions, crack-tip stresses and crack extension force are given.

4.3 Applied Elastic Fields

The displacement corresponding to a pressure at a point on a plane boundary may be taken from Love (1927); we assume a slab of cylinder of thickness dx'_3, located at O' $(0, 0, x'_3)$ and acting in the x_2-direction, to produce a displacement corresponding to that of a pressure point multiplied by dx'_3. This gives by superposition the displacement $\vec{u}^a$ produced by the cylinder, at arbitrary position (x_1, x_2, x_3), in the form

$$
\begin{aligned}
u_1^a &= \frac{Px_1x_2}{4\pi\mu}\int\limits_{-\infty}^{\infty}\frac{dx_3'}{r'^3} - \frac{Px_1}{4\pi(\lambda+\mu)}\int\limits_{-\infty}^{\infty}\frac{dx_3'}{r'(x_2+r')},\\
u_2^a &= \frac{Px_2^2}{4\pi\mu}\int\limits_{-\infty}^{\infty}\frac{dx_3'}{r'^3} + \frac{P(\lambda+2\mu)}{4\pi\mu(\lambda+\mu)}\int\limits_{-\infty}^{\infty}\frac{dx_3'}{r'},\\
u_3^a &= 0;
\end{aligned}
\tag{4.1}
$$

$r'^2 = x_1^2 + x_2^2 + (x_3 - x_3')^2$ and λ and μ are Lamé's constants. The associated stress fields $(\sigma)^a$ are obtained from the displacement $\vec{u}^a$ by partial differentiation with respect to coordinates x_i. Induced normal stresses originating from Poisson effect are considered; hence, the applied stress field $(\sigma)^A$ has the form

$$
(\sigma)^A = \begin{pmatrix} \sigma_{11}^a - \nu(\sigma_{22}^a + \sigma_{33}^a) & \sigma_{12}^a & 0\\ \sigma_{12}^a & \sigma_{22}^a - \nu(\sigma_{11}^a + \sigma_{33}^a) & 0\\ 0 & 0 & \sigma_{33}^a - \nu(\sigma_{11}^a + \sigma_{22}^a) \end{pmatrix}. \tag{4.2}
$$

The displacement $\vec{u}^a$ ($r^2 = x_1^2 + x_2^2$, $r \neq 0$) obtained from (4.1) is:

$$
\begin{aligned}
u_1^a &= \frac{P}{2\pi\mu}\left(2(1-2\nu)\frac{x_1\ln r}{x_2} + \frac{x_1x_2}{r^2} - 2(1-2\nu)\tan^{-1}\frac{x_1}{x_2+r}\right),\\
u_2^a &= \frac{P}{2\pi\mu}\left(-2(1-\nu)\ln r + \frac{x_2^2}{r^2}\right)\\
u_3^a &= 0
\end{aligned}
\tag{4.3}
$$

Constants with infinite values are omitted. We use (4.3) for $x_2 \neq 0$. Arguments in the logarithm are dimensionless. These conditions also apply for the corresponding stress $(\sigma)^a$ that follows:

$$
\begin{aligned}
(\sigma)^a &= \begin{pmatrix} \sigma_{11}^a & \sigma_{12}^a & 0\\ \sigma_{12}^a & \sigma_{22}^a & 0\\ 0 & 0 & \sigma_{33}^a \end{pmatrix};\\
\sigma_{11}^a &= \frac{2P}{\pi x_2}\left((1-\nu)\ln r + \frac{(1-\nu)x_1^2}{r^2} - \frac{x_1^2x_2^2}{r^4}\right),\\
\sigma_{22}^a &= \frac{2P}{\pi x_2}\left(\nu\ln r + \frac{\nu x_1^2}{r^2} - \frac{x_2^4}{r^4}\right),\\
\sigma_{33}^a &= \frac{2\nu P}{\pi x_2}\left(\ln r + \frac{x_1^2 - x_2^2}{r^2}\right),
\end{aligned}
$$

$$\sigma_{12}^{a} = -\frac{Px_1}{\pi x_2^2}\left((1-2\nu)\ln r - \frac{(1-2\nu)x_2^2}{r^2} + \frac{2x_2^4}{r^4}\right). \tag{4.4}$$

The applied stress field $(\sigma)^A$ (4.2) becomes

$$(\sigma)^A = \begin{pmatrix} \sigma_{11}^A & \sigma_{12}^a & 0 \\ \sigma_{12}^a & \sigma_{22}^A & 0 \\ 0 & 0 & 0 \end{pmatrix};$$

$$\sigma_{11}^A = \frac{2(1+\nu)P}{\pi x_2}\left((1-2\nu)\ln r + \frac{(1-2\nu)x_1^2 - (1-\nu)x_2^2}{r^2} + \frac{x_2^4}{r^4}\right),$$

$$\sigma_{22}^A = \frac{2Px_2}{\pi r^2}\left(\nu^2 + \frac{\nu x_1^2 - x_2^2}{r^2}\right) \tag{4.5}$$

The traction $\mathrm{d}\vec{F}_H$ at an arbitrary point on any surface element ds parallel to x_1x_3 is given by

$$\frac{\mathrm{d}\vec{F}_H}{\mathrm{d}s} = \begin{pmatrix} \sigma_{12}^a \\ \sigma_{22}^A \\ 0 \end{pmatrix}. \tag{4.6}$$

At P_s $(x_1, 0, x_3)$, the component σ_{22}^A perpendicular to the boundary plane Ox_1x_3 vanishes when $x_2 \to 0$, except at $r = 0$, as expected. $\sigma_{33}^A = 0$ is equally wanted. Hence, $(\sigma)^A$ (4.5) behaves adequately.

$$(\sigma)^{AW} = \begin{pmatrix} \sigma_{11}^{AW} & 0 & 0 \\ 0 & \sigma_{22}^{AW} & 0 \\ 0 & 0 & \sigma_{33}^{AW} \end{pmatrix}; \tag{4.7}$$

$$\sigma_{22}^{AW} = -\rho g x_2, \sigma_{11}^{AW} = \sigma_{33}^{AW} - \nu_A \sigma_{22}^{AW}.$$

4.4 Elastic Fields of Straight Dislocations in a Half-Space

A three-dimensional half-space (infinitely extended elastic solid) is considered with shear modulus μ and Poisson's ratio ν (Anongba 2019). The solid occupies the region $x_2 > 0$ and its flat surface is Ox_1x_3; it contains a straight dislocation parallel to x_3 and displaced by $x_2 = h$ from the origin (Fig. 4.3). The dislocations concerned are edges with Burgers vectors $\vec{b}_{//} = (b, 0, 0)$ parallel to x_1 and $\vec{b}_{\perp} = (0, b, 0)$ parallel to x_2 (Fig. 4.3). We present below a methodology for determining their elastic fields that is equally valid for a screw dislocation with Burgers vector $\vec{b}_{III} = (0, 0, b)$. Let $\vec{u}^{(J)}$

and $(\sigma)^{(J)}$ be the displacement and stress fields in the medium due to the dislocation J (J = //, ⊥ and III). Very generally, the following description is expected to apply:

- Ox_1x_3 is free from traction; at P_s $(x_1, 0, x_3)$, this gives

$$\sigma_{12}^{(J)} = 0, \sigma_{22}^{(J)} = 0 \text{ and } \sigma_{23}^{(J)} = 0. \tag{4.8}$$

- Far from the dislocation and free surface, the elastic fields correspond to those $(\vec{u}^{(J)\infty}, (\sigma)^{(J)\infty})$ of a straight dislocation displaced by $x_2 = h$ from the origin with Burgers vector $\vec{b}_J$ in the whole three-dimensional space:

$$\begin{aligned} \vec{u}^{(J)} &\to \vec{u}^{(J)\infty} \\ (\sigma)^{(J)} &\to (\sigma)^{(J)\infty} \end{aligned} \tag{4.9}$$

  when one moves far away in the x_2- direction ($|x_2 - h| \to \infty$).
- The elastic fields may be expressed in the form

$$\begin{aligned} \vec{u}^{(J)} &= \vec{u}^{(J)\infty} - \vec{u}^{(J)W} \\ (\sigma)^{(J)} &= (\sigma)^{(J)\infty} - (\sigma)^{(J)W} \end{aligned} \tag{4.10}$$

  where $\vec{u}^{(J)W}$ and $(\sigma)^{(J)W}$ satisfy the equations of equilibrium and possess the following properties.

$$\begin{aligned} \vec{u}^{(J)\infty}(P_S) &= \vec{u}^{(J)W}(P_S) \\ (\sigma)^{(J)\infty}(P_S) &= (\sigma)^{(J)W}(P_S); \end{aligned} \tag{4.11}$$

  this ensures the traction-free boundary condition (4.8).
- $\vec{u}^{(J)W}$ and $(\sigma)^{(J)W}$ cancel far from free surface and dislocation; this means that

$$\begin{aligned} \vec{u}^{(J)W} &\to 0 \\ (\sigma)^{(J)W} &\to 0 \end{aligned} \tag{4.12}$$

  when $|x_2 - h| \to \infty$; this ensures the veracity of (4.9) above.

The elastic fields $\vec{u}^{(J)}$ and $(\sigma)^{(J)}$ thus obtained are valuable representations to Figure 4.3. The associated $\vec{u}^{(J)W}$ and $(\sigma)^{(J)W}$ are investigated with the help of Galerkin vectors in a similar way as in our previous works (Anongba 2015, 2017a). We take Galerkin vectors $\vec{V}^{(J)}$ with only one non-zero, x_1-component $V_1^{(//)}$ for J = //, x_2-component $V_2^{(\perp)}$ for $J = \perp$, and x_3-component $V_3^{(III)}$ for $J = III$, with same form

$$V_j^{(J)}(\vec{x}) = \overline{\alpha}^{(J)}(\vec{k})e^{i\vec{k}.\vec{x}} + \overline{\beta}^{(J)}(\vec{k})x_2 e^{i\vec{k}.\vec{x}}, \tag{4.13}$$

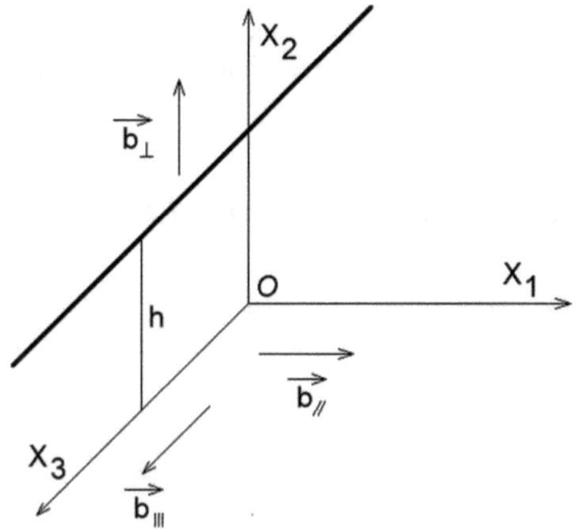

Fig. 4.3 Elastic half-space $x_2 > 0$ containing an infinitely long straight dislocation J parallel to x_3 and displaced by $x_2 = h$ from the origin with Burgers vector $\vec{b}_J$ directed along x_1 ($J = //$), x_2 ($J = \perp$) or x_3 ($J = III$)

$j = 1$ to 3 for $J = //, \perp, III$ respectively, under the condition $\vec{k}^2 = k_1^2 + k_2^2 + k_3^2 = 0$ that ensures the biharmonicity of $V_j^{(J)}$. For $V_j^{(J)}$ to cancel far from the dislocation and free surface, we take $k_2 = i\sqrt{k_1^2 + k_3^2}$. The elastic fields under consideration are x_3-independent; hence k_3 is set equal to zero. This leads to

$$k_2 = i|k_1|. \tag{4.14}$$

The elastic fields corresponding to $V_j^{(J)}$ (4.13) may be first calculated; then more general forms $\vec{u}^{(J)W}$ and $(\sigma)^{(J)W}$ are constructed from the previous ones by superposition over k_1. To calculate $\overline{\alpha}^{(J)}$ and $\overline{\beta}^{(J)}$ (4.13), we use (4.11) for stress and restrict ourselves to stress components involved in the traction-free boundary condition (4.8) only. $(\sigma)^{(J)\infty}$ is taken from Anongba (2010) and written in Fourier forms at P_S. We obtain

$$\begin{aligned}
\overline{\alpha}^{(//)} &= \frac{iC_1}{2k_1^3}(-2 + 2\nu + (5 - 4\nu)h|k_1|)e^{-h|k_1|},\\
\overline{\beta}^{(//)} &= \frac{iC_1}{2k_1^2}(-1 + 2h|k_1|)\mathrm{sgn}(k_1)e^{-h|k_1|};\\
\overline{\alpha}^{(\perp)} &= \frac{iC_1}{2k_1^3}(-2\nu + (1 - 4\nu)h|k_1|)e^{-h|k_1|};\\
\overline{\beta}^{(\perp)} &= -\frac{iC_1}{2k_1^2}(1 + 2h|k_1|)\mathrm{sgn}(k_1)e^{-h|k_1|};\\
\overline{\alpha}^{(III)} &= 0,\\
\overline{\beta}^{(III)} &= -\frac{iD_1}{4(1-\nu)}\frac{\mathrm{sgn}(k_1)}{k_1^2}e^{-h|k_1|}
\end{aligned} \tag{4.15}$$

$C_1 = \mu b/2\pi(1-\nu)$, $D_1 = \mu b/2\pi$. . Again with (4.15), $\vec{u}^{(J)W}$ and $(\sigma)^{(J)W}$ are known; the associated dislocation J elastic fields are given by (4.10).

4.5 Dislocation Stress Fields

Galerkin vectors $\vec{V}^{(J)}$ (4.13, 4.15), $J = //, \perp$andIII, have been used to estimate the dislocation elastic fields. We display below the stresses only because they are involved in the crack analysis.

$$\begin{aligned}
\sigma_{ii}^{(\perp)} &= C_1 x_1 \left(\frac{\delta_{i1}+\delta_{i2}+2\nu\delta_{i3}}{r_{-h}^2} + \frac{2(x_2-h)^2(\delta_{i2}-\delta_{i1})}{r_{-h}^4} - \frac{\delta_{i1}+\delta_{i2}+2\nu\delta_{i3}}{r_{+h}^2} \right. \\
&\quad - \frac{2(x_2+h)(\delta_{i2}(x_2+h)-\delta_{i1}(x_2-3h)+\delta_{i3}4\nu h)}{r_{+h}^4} \\
&\quad \left. + \frac{4hx_2\big(3(x_2+h)^2-x_1^2\big)(\delta_{i1}-\delta_{i2})}{r_{+h}^6} \right), \\
\sigma_{12}^{(\perp)} &= C_1 \left(\frac{(x_2-h)\big(x_1^2-(x_2-h)^2\big)}{r_{-h}^4} + \frac{(x_2-h)\big((x_2+h)^2-x_1^2\big)}{r_{+h}^4} \right. \\
&\quad \left. + \frac{4hx_2(x_2+h)\big((x_2+h)^2-3x_1^2\big)}{r_{+h}^6} \right), \\
\sigma_{j3}^{(\perp)} &= 0;
\end{aligned} \tag{4.16}$$

$$\begin{aligned}
\sigma_{ii}^{(//)} &= C_1 \left(-\frac{(x_2-h)(\delta_{i1}-\delta_{i2}+2\nu\delta_{i3})}{r_{-h}^2} - \frac{2(x_2-h)\big(\delta_{i1}x_1^2+\delta_{i2}(x_2-h)^2\big)}{r_{-h}^4} \right. \\
&\quad + \frac{x_2(3\delta_{i1}-\delta_{i2}+2\nu\delta_{i3})+h(5\delta_{i1}+\delta_{i2}+2\nu\delta_{i3})}{r_{+h}^2} + \frac{16hx_2(x_2+h)^3(\delta_{i1}-\delta_{i2})}{r_{+h}^6} \\
&\quad + \frac{\delta_{i3}4h\nu\big(x_1^2-(x_2+h)^2\big)}{r_{+h}^4} \\
&\quad \left. + \frac{2(x_2+h)((x_2+h)[x_2(\delta_{i2}-\delta_{i1})-h(3\delta_{i1}+\delta_{i2})]+6hx_2(\delta_{i2}-\delta_{i1}))}{r_{+h}^4} \right), \\
\sigma_{12}^{(//)} &= C_1 x_1 \left(\frac{x_1^2-(x_2-h)^2}{r_{-h}^4} - \frac{x_1^2-(x_2+h)^2}{r_{+h}^4} - \frac{4hx_2\big(3(x_2+h)^2-x_1^2\big)}{r_{+h}^6} \right), \\
\sigma_{j3}^{(//)} &= 0
\end{aligned} \tag{4.17}$$

$$\sigma_{j3}^{(III)} = D_1\left(\frac{\delta_{j1}(h - x_2) + \delta_{j2}x_1}{r_{-h}^2} + \frac{\delta_{j1}(h + x_2) - \delta_{j2}x_1}{r_{+h}^2}\right),$$

$$\sigma_{ii}^{(III)} = 0, \sigma_{12}^{(III)} = 0. \tag{4.18}$$

In (4.16–4.18), δ_{ij} is the Kronecker delta, subscripts i and j take values (1, 2 and 3) and (1 and 2), respectively; $r_{-h}^2 = x_1^2 + (x_2 - h)^2$, $r_{+h}^2 = x_{,1}^2 + (x_2 + h)^2$, $C_1 = \mu b/2\pi(1 - \nu)$, $D_1 = \mu b/2\pi$.. These results are in complete agreement with previous works (Head 1953; Mura 1968).

4.6 Crack Dislocation Distributions

We consider one crack in the half-space, located between A and B in the Ox_1x_2-plane (Fig. 4.2). The crack system is completely defined when the dislocation distributions D_J ($J = //$ and $\perp$) are known. For this purpose, we ask the crack faces to be free from traction:

$$\begin{cases} \overline{\sigma}_{12} - \partial h/\partial x_1 \overline{\sigma}_{11} = 0 \\ \overline{\sigma}_{22} - \partial h/\partial x_1 \overline{\sigma}_{12} = 0 \end{cases}. \tag{4.19}$$

$(\overline{\sigma})$ is the stress at any position $P(x_1, x_2, x_3)$ in the medium and is linked to D_J. In (4.19), we are concerned with the point $P_C(x_1, x_2 = h(x_1), x_3)$ of the crack faces only.

$$(\overline{\sigma}) = (\sigma)^A + (\sigma)^{AW} + (\overline{\sigma})^{(//)} + (\overline{\sigma})^{(\perp)} \tag{4.20}$$

$$\overline{\sigma}_{ij}^{(J)}(P) = \int_{a_1}^{b_1} \sigma_{ij}^{(J)}(x_1 - x_1', x_2, x_3) D_J(x_1') \mathrm{d}x_1' \quad (J = // \text{ and } \perp). \tag{4.21}$$

$\sigma_{ij}^{(J)}$ is the stress produced by a dislocation J located at an elevation $h(x_1')$ from the half-space boundary. (4.19) provides two integral equations with Cauchy-type singular kernels that determine the D_J. We just need to use the spatial dependence of D_J about the crack tip B. The condition (4.19) can be managed to read:

$$\begin{cases} f_{//}^{A-W}(B) + C_1 \displaystyle\int_{1-\delta b_1/b_1}^{1} \mathrm{d}y_1' D_{//}(y_1') \frac{1}{1 - y_1'} = 0 \\ f_{\perp}^{A-W}(B) + C_1 \displaystyle\int_{1-\delta b_1/b_1}^{1} \mathrm{d}y_1' D_{\perp}(y_1') \frac{1}{1 - y_1'} = 0 \end{cases}; \tag{4.22}$$

$f_{//}^{A-W}(B) = \sigma_{12}^{A} + \sigma_{12}^{AW} - p_0(\sigma_{11}^{A} + \sigma_{11}^{AW}), f_{\perp}^{A-W}(B) = \sigma_{22}^{A} + \sigma_{22}^{AW} - p_0(\sigma_{12}^{A} + \sigma_{12}^{AW})$ $p_0 = \tan\theta, C_1 = \mu b/2\pi(1-\nu)$. Here ν is Poisson's ratio and μ is the shear modulus. Inverting (4.22) in a similar way as in Sect. 2.3.1, we arrive at

$$D_J \cong \alpha_0 \alpha_J \frac{1}{\sqrt{1-y_1}}, \phi_J \cong 2b\alpha_0\alpha_J b_1\sqrt{1-y_1}; \tag{4.23}$$

$\alpha_J = f_J^{A-W}/\pi C_1$, $y_1 = x_1/b_1$, $(1 - y_1)$ positive infinitesimal value and α_0 dimensionless constant.

4.7 Crack-Tip Stresses

In the crack plane and ahead of the crack-tip at spatial position $P_C(x_1 = b_1 + s, x_2 = h(x_1), x_3), 0 < s \ll b_1$, the total stress $\overline{\sigma}_{ij}(P_C)$ is identified to the following formula.

$$\overline{\sigma}_{ij}(s) = \sum_{J=//\text{and}\perp} \int_{b_1-\delta b_1}^{b_1} \sigma_{ij}^{(J)}(b_1 + s - x_1')D_J(x_1')\mathrm{d}x_1', \delta b_1 \ll b_1. \tag{4.24}$$

This formula means that only those dislocations located about the crack front in x_1-interval $[b_1 - \delta b_1, b_1]$ will contribute significantly to the stress at $x_1 = b_1 + s$ ahead of the crack-tip as s tends to zero; any other contribution will become negligible for a sufficiently small value of s. Using (4.23) for D_J and integrating (4.24), we obtain

$$\begin{aligned}
\overline{\sigma}_{12}(s) &= \frac{(1-p_0^2)}{(1+p_0^2)^2}(\alpha_{//} + p_0\alpha_{\perp})C_1\alpha_0\frac{\pi\sqrt{b_1}}{2\sqrt{s}},\\
\overline{\sigma}_{11}(s) &= \frac{1}{(1+p_0^2)^2}[(1-p_0^2)\alpha_{\perp} - p_0(3+p_0^2)\alpha_{//}]C_1\alpha_0\frac{\pi\sqrt{b_1}}{2\sqrt{s}},\\
\overline{\sigma}_{22}(s) &= \frac{1}{(1+p_0^2)^2}[p_0(1-p_0^2)\alpha_{//} + (1+3p_0^2)\alpha_{\perp}]C_1\alpha_0\frac{\pi\sqrt{b_1}}{2\sqrt{s}},\\
\overline{\sigma}_{33}(s) &= \frac{2\nu}{1+p_0^2}(\alpha_{\perp} - p_0\alpha_{//})C_1\alpha_0\frac{\pi\sqrt{b_1}}{2\sqrt{s}}\\
\overline{\sigma}_{j3} &= 0 \quad j = 1 \text{ and } 2.
\end{aligned} \tag{4.25}$$

4.8 Crack Extension Force

The expression of the crack extension force also applies in the presence of gravitational forces. We obtain at B $(b_1, h(b_1), x_3)$

$$G(B) = \frac{2(1-\nu)b_1\alpha_0^2}{\mu\sqrt{1+p_0^2}}\left(\left[f_{//}^{A-W}(B)\right]^2 + \left[f_{\perp}^{A-W}(B)\right]^2\right). \tag{4.26}$$

For numerical application, we take A to coincide with the origin O (Fig. 4.2) and both the medium and cylinder to have equal density ρ. We define $\tilde{G}$ a reduced value of G as

$$\tilde{G} = \frac{G(B)}{G_0}, \quad G_0 = 2(1-\nu)(\alpha_0\rho g)^2R^3/\mu. \tag{4.27}$$

When applied to the solid posed on the surface of earth, the specimen suffers compression, so we write $\sigma_{22}^{AW} = -\rho g x_2$. $\tilde{G}$ is written as a function of $\tilde{b}_2 = b_2/R$ for three different values of θ in Fig. 4.4 and as a function of θ for three values of $\tilde{b}_2$ in Fig. 4.5.

DISCUSSION: The modelling is considered applicable for large cracks with B far from the cylinder. To estimate the appropriate height $\tilde{b}_2(B_C)$ of B at which gravitational forces only will break into two pieces the medium, from Fig. 4.4, this would occur at $B = B_C$ for which $G(B_c) = 2\gamma$ in steady motion, where γ is the surface energy. α_0 may be provided from experiment as an adjustable parameter.

Figure 4.1 corresponds to $\theta = \pi/2$ with the cylinder (see Fig. 4.2) on the four holes of the top surface. From Fig. 4.5, it is seen that $\tilde{G}$ is larger for small θ. The

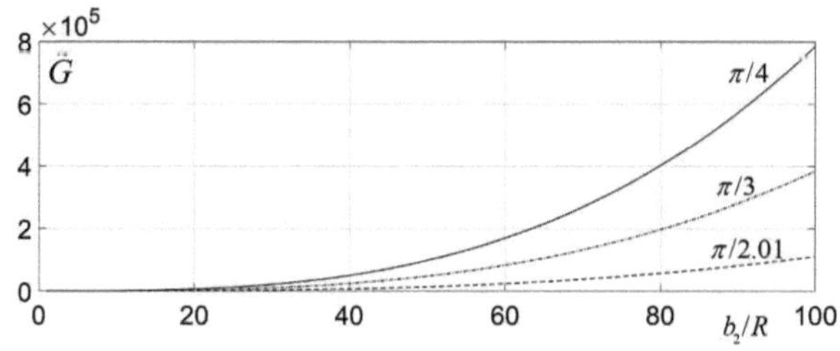

Fig. 4.4 $\tilde{G}$ (4.27) as a function of $\tilde{b}_2 = b_2/R$ for three different values of θ. It increases continuously with $\tilde{b}_2$

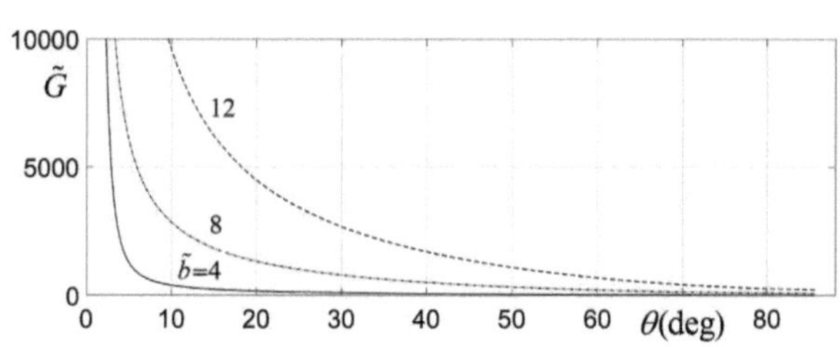

Fig. 4.5 $\tilde{G}$ (4.27) as a function of θ for three different values of $\tilde{b}_2$. It decreases continuously with θ

depth of the crack or equivalently the length of the crack can be evaluated in steady motion as follows. For smaller θ values (use Fig. 4.5), G is larger than 2γ for the crack is observed; then it decreases with depth and should stop inside the medium when $G \leq 2\gamma$. Under gravitational forces, solids suffer compression. Hence, fracture surfaces over large distance will be aligned parallel to $\boldsymbol{g}$, the acceleration due to gravity, that becomes a zone axis for families of fractured planes.

SUMMARY: We have investigated fracture propagation in a three-dimensional elastic half-space subject to the rectilinear contact pressure of a cylinder lying on the flat Ox_1x_3-plane boundary and in presence of gravitational forces. Analytical expressions have been obtained for G, the crack extension force per unit length of the crack front. That is a merit of the work. This makes it possible to account for the commonly observed fact that cracks can break into two very large blocks of stone in absence of externally applied forces from the experimenter.

Chapter 5
Cracks at the Interface of Bi-Materials

5.1 Planar Interface Straight-Fronted Cracks

When two welded dissimilar elastic materials are loaded in tension, perpendicularly to their common interface, an internal shear stress develops. This is the result of a difference in the magnitudes of the Poisson's contractions that are generated in both media. This has been recognized and incorporated in the mathematical analysis of the propagation of an interface crack loaded in tension (Anongba 2014); Poisson's effect increases significantly the crack extension force. Consequently, it becomes of interest to discuss Poisson's effect in connection with experiments. We may consider (among others) Liechti and Chai (1991, 1992) which use edge-cracked bi-material (epoxy/glass) strip specimens in biaxial loading experiments; the pair epoxy/glass possesses a large difference in the values of Poisson's contractions on crossing the interface. An edge-cracked (epoxy/glass) strip specimen under applied tension normal to the interface clearly displays an internal shear stress due to Poisson's effect directed in one direction in the main part of the uncracked region of the specimen ahead of the tip of the starter crack. Hence, applying a shear in the same direction would increase much more the magnitude of the crack extension force whereas in the opposite direction, this would decrease. In the present, we shall extend our interface crack analysis to bi-axial applied loading (mixed mode $I + II$) and confront the theoretical findings to experiments with attention to Liechti and Chai.

Internal Shear Stress on the Interface Due to Poisson's Contraction

Consider two isotropic elastic media (1) and (2) (Fig. 5.1), isotropic and elastic, with notation ν_i, μ_i, E_i ($i = 1$ and 2) to designate Poisson's ratio, shear modulus and Young's modulus, respectively. We assume that the two solids are firmly linked along a planar interface, such that: mediums 1 and 2 occupy the regions $x_2 > 0$ and $x_2 < 0$ respectively; Ox_1x_3 represents the interface with origin O at the centre. An interface crack of finite length $2a$ is present at the origin, extending from $x_1 = -a$ to a with a straight front running indefinitely in the x_3-direction. Under the action of a uniform

P. N.B. Anongba, *Mechanics of Steadily Propagating Cracks*,
https://doi.org/10.1007/978-981-96-5703-2_5

tension σ_{22}^a, applied at infinity, the solids (1) and (2) suffer a Poisson's contraction of different magnitude; this induces an internal shear stress in the surrounding media about the interface. For illustration purpose only of the displacement, we construct on the continuum on both sides of the interface an identical cubic lattice. This is not a lattice in the sense of crystallography (i.e. this is not associated with atomic arrangement). We assume that Hooke's law is valid throughout. These lattices are illustrated in dashed lines under the assumption $\nu_2\sigma_{22}^a/E_2 > \nu_1\sigma_{22}^a/E_1$ before the welding of the solids. Consider two nodes A (medium 1) and B (medium 2) on either side of the interface which have the equal coordinate x_1 (before loading) and a separation distance d (along x_1) under load when the two materials are not welded. In the welded state, A is moved horizontally by a distance of $u_1 < 0$ and B by $u_2 > 0$. We ignore vertical displacement along x_2. In Fig. 5.1, the displacements u_1 and u_2 at different places are represented by short horizontal arrows and the resultant welded lattices are partially drawn in solid lines. The induced internal shear stress τ_{12}^a is represented about the node A and is given, at an arbitrary point $P(x_1, 0, x_3)$ on the interface, the following expression (Anongba 2014):

$$\tau_{12}^a(P) = -\frac{\nu_s\mu_s}{a_s}\left(\frac{\nu_1}{E_1} - \frac{\nu_2}{E_2}\right)\sigma_{22}^a x_1 \equiv -\overline{\alpha}x_1 \tag{5.1.1}$$

where μ_s corresponds to the shear modulus about the interface and a_s is the distance between A and B along x_2 (Fig. 5.1); $\nu_s = \varphi/d$ is the ratio of φ (distance along x_1 between A and B in the welded state under load) by d (the corresponding

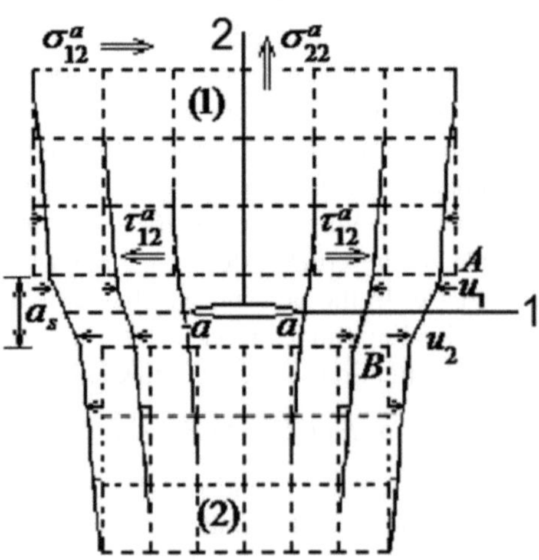

Fig. 5.1 Two elastic mediums (1) and (2) subjected to a uniform tension σ_{22}^a at infinity. Far from the interface (Ox_1x_3), the mediums suffer uniform Poisson's contractions $(-\nu_i\sigma_{22}^a/E_i)$ ($i = 1$ and 2) in the x_1-direction. The induced shear stress τ_{12}^a is indicated about the interface. A crack of finite length $2a$ located at the origin is present. This sketch corresponds to $(\nu_2\sigma_{22}^a/E_2) > (\nu_1\sigma_{22}^a/E_1)$. For mixed mode $I + II$ loading, an additional applied shear stress σ_{12}^a is present. In the regions $x_1 > 0$ and $x_1 < 0$, positive σ_{12}^a works with and against τ_{12}^a respectively

distance in the non-welded state). τ_{12}^a is odd function of x_1 changing sign in the region $x_1 < 0$. When an additional uniform shear stress σ_{12}^a is applied externally, the associated displacements in the media amplify or reduce those due to τ_{12}^a depending on the sign of σ_{12}^a. In the situation of Fig. 5.1 where $\overline{\alpha} < 0$, positive σ_{12}^a amplifies and reduces τ_{12}^a in $x_1 > 0$ and $x_1 < 0$, respectively.

Interface Crack Model and Singular Integral Equations

Our method of analysis consists in representing the crack by a continuous distribution of dislocation families with infinitesimal Burgers vectors. The crack in Fig. 2.38 is assumed to be filled with two families (1 and 2) of straight edge dislocations parallel to x_3 with Burgers vectors $\vec{b}_1 = (b, 0, 0)$ and $\vec{b}_2 = (0, b, 0)$ in the x_1 and x_2 directions, respectively. The system is subjected to uniform applied shear σ_{12}^a and tension σ_{22}^a at infinity; in addition we consider internal shear stresses transmitted by the interface as the result of the existence of internal uniform Poisson stresses $(-\nu_1\sigma_{22}^a)$ and $(-\nu_2\sigma_{22}^a)$ in the x_1- direction in materials 1 and 2, respectively. The dislocation distribution function $D_i(x_1)$ ($i = 1$ and 2) gives the number of dislocations of family i in a small interval dx_1 about x_1 as $D_i(x_1)dx_1$. To find the equilibrium dislocation distributions, we may ask for zero total force on the crack faces; on our modelling, this gives

$$\begin{cases} \overline{\sigma}_{12} = 0 \\ \overline{\sigma}_{22} = 0 \end{cases}; \tag{5.1.2}$$

$\overline{\sigma}_{ij}$ stands for the total stress at any point $P(x_1, x_2, x_3)$ in the surrounding medium and is linked to D_i; in (5.1.2), we are only concerned with the points of the crack faces. $\overline{\sigma}_{ij}(P)$ is written as

$$\overline{\sigma}_{ij} = \sigma_{ij}^a + \sigma_{ij}^{(s)} + \overline{\sigma}_{ij}^{(1)} + \overline{\sigma}_{ij}^{(2)} \tag{5.1.3}$$

where σ_{ij}^a corresponds to both the applied stresses and the assumed uniform internal Poisson stress, $\sigma_{ij}^{(s)}$ to the stress induced by the interface under load in absence of the crack and

$$\overline{\sigma}_{ij}^{(n)} = \int_{-a}^{a} \sigma_{ij}^{(n)}(x_1 - x_1', x_2, x_3) D_n(x_1') dx_1' \quad (n = 1 \text{ and } 2); \tag{5.1.4}$$

here $\sigma_{ij}^{(n)}$ is the stress field produced by a dislocation of family n at the origin.

To write explicitly (5.1.2), we display the following results on the interface (points $P(x_1, x_2 = 0, x_3)$):

- $\sigma_{12}^{(s)}$ is given by τ_{12}^a (5.1.1) and $\sigma_{22}^{(s)} \equiv 0$;
- Edges with $\vec{b}_1 = (b, 0, 0)$ (Family 1).

$$\sigma_{12}^{(1)}(x_1 - x_1', x_2 = 0, x_3) = \frac{bC}{\pi} \frac{1}{x_1 - x_1'},$$
$$\sigma_{22}^{(1)}(x_1 - x_1', x_2 = 0, x_3) = -b\beta C\delta(x_1 - x_1'); \tag{5.1.5}$$

- Edges with $\vec{b}_2 = (0, b, 0)$ (Family 2)

$$\sigma_{12}^{(2)}(x_1 - x_1', x_2 = 0, x_3) = b\beta C\delta(x_1 - x_1'),$$
$$\sigma_{22}^{(2)}(x_1 - x_1', x_2 = 0, x_3) = \frac{bC}{\pi} \frac{1}{x_1 - x_1'}. \tag{5.1.6}$$

In (5.1.5) and (5.1.6), taken from Comninou and Dundurs (1980), δ is the Dirac delta function. Moreover,

$$C = \frac{2\mu_1(1-\alpha)}{(\kappa_1 + 1)(1 - \beta^2)} = \frac{2\mu_2(1+\alpha)}{(\kappa_2 + 1)(1 - \beta^2)} \tag{5.1.7}$$

with

$$\alpha = \frac{\mu_1(\kappa_2 + 1) - \mu_2(\kappa_1 + 1)}{\mu_1(\kappa_2 + 1) + \mu_2(\kappa_1 + 1)}; \quad -1 \le \alpha \le 1 \tag{5.1.8}$$

$$\beta = \frac{\mu_1(\kappa_2 - 1) - \mu_2(\kappa_1 - 1)}{\mu_1(\kappa_2 + 1) + \mu_2(\kappa_1 + 1)}; \quad -\frac{1}{2} \le \beta \le \frac{1}{2} \tag{5.1.9}$$

where $\kappa_i = 3 - 4\nu_i$. The traction free boundary condition (5.1.2) yields

$$\begin{aligned} &(\sigma_{12}^a - \overline{\alpha} x_1)\delta_{i1} + \sigma_{22}^a \delta_{i2} \\ &+ b\beta C(D_2(x_1)\delta_{i1} - D_1(x_1)\delta_{i2}) \\ &+ \frac{bC}{\pi} \int_{-a}^{a} \frac{D_i(x_1')}{x_1 - x_1'} \mathrm{d}x_1' = 0, \end{aligned} \tag{5.1.10}$$

$i = 1$ and 2, $|x_1| < a$, and δ_{ij} is the Kronecker delta. We arrive at a system of two integral equations with Cauchy-type singular kernels with unknown functions D_1 and D_2. The Cauchy principal values of the integrals must be taken. Next, an analytical solution is given to the governing Eqs. (5.1.10).

Analytical Solution to Singular Integral Equations

From the second of (5.1.10) and using the book by Muskhelishvili (1992) (see Chap. 11 "Inversion Formulae for Arcs"), we write

$$D_2(x_1) = \frac{1}{\pi bC} \int_{-a}^{a} \sqrt{\frac{a^2 - x_1'^2}{a^2 - x_1^2}} \frac{(\sigma_{22}^a - b\beta C D_1(x_1'))}{x_1 - x_1'} \mathrm{d}x_1'. \tag{5.1.11}$$

Introducing (5.1.11) in the first equation of (5.1.10), we obtain

$$\int_{-a}^{a}\left(\sqrt{a^2-x_1^2}-\beta^2\sqrt{a^2-x_1'^2 ieq414}\right)\frac{D_1(x_1')}{x_1'-x_1}\mathrm{d}x_1'=\frac{\pi}{bC}f(x_1) \tag{5.1.12}$$

where

$$f(x_1)=(\sigma_{12}^a-\overline{\alpha}x_1)\sqrt{a^2-x_1^2}+\beta\sigma_{22}^a x_1. \tag{5.1.13}$$

We temporary modify our notation and write (5.1.12) as

$$\frac{\sqrt{1-s^2}}{\pi}\int_{-1}^{1}\frac{g(t)}{t-s}\mathrm{d}t+\frac{\lambda^2}{\pi}\int_{-1}^{1}\frac{\sqrt{1-t^2}g(t)}{t-s}\mathrm{d}t=\frac{1}{abC}\overline{f}(s) \tag{5.1.14}$$

where

$$\begin{aligned} g(t)&=D_1(at)\\ \overline{f}(s)&=f(as)\\ \lambda&=i\beta \end{aligned} \tag{5.1.15}$$

(i. e. $\lambda^2=-\beta^2$). Equation (5.1.14) takes the form of "Example 43" (page 158) in the book by Estrada and Kanwal 2000; by a convincing operational approach, it is shown there that the solution of (5.1.14) can be written as

$$g(s)=\frac{\overline{\psi}_1(s)-\overline{\psi}_2(s)}{2\lambda\sqrt{1-s^2}} \tag{5.1.16}$$

where $\overline{\psi}_1$ and $\overline{\psi}_2$ are solutions of the pair of Cauchy type integral equations

$$\begin{aligned} &\overline{\psi}_n(s)+(\delta_{n1}-\delta_{n2})\frac{\lambda}{\pi}\int_{-1}^{1}\frac{\overline{\psi}_n(t)}{t-s}\mathrm{d}t\\ &=\frac{\overline{f}(s)}{abC}+A\delta_{n1}+B\delta_{n2};\ n=1 \text{ and } 2 \end{aligned} \tag{5.1.17}$$

that satisfy the additional requirement

$$\frac{1}{\pi}\int_{-1}^{1}g(t)\ln\left(\frac{1+t}{1-t}\right)\mathrm{d}t=\int_{-1}^{1}\frac{\overline{f}(t)}{abC\sqrt{1-t^2}}\mathrm{d}t; \tag{5.1.18}$$

A and B are arbitrary constants.

The solutions of (5.1.17) ($\overline{\psi}_1$ and $\overline{\psi}_2$) can be worked out by using the treatment of Muskhelishvili 1992 (see Chap. 14 "The Case of Continuous Coefficients"); we have

$$\overline{\psi}_n(s) = \frac{1}{1-\beta^2}\left(\frac{\overline{f}(s)}{abC} + A\delta_{n1} + B\delta_{n2}\right)$$
$$+ (\delta_{n1} - \delta_{n2})\frac{1}{\pi i}\left(\frac{\beta}{1-\beta^2}\right)\left(\frac{1-s}{1+s}\right)^{i\delta(-\delta_{n1}+\delta_{n2})}$$
$$\times\left(const_1\delta_{n1} + const_2\delta_{n2} + \int_{-1}^{1}\frac{\left(\overline{f}(t)/abC + A\delta_{n1} + B\delta_{n2}\right)\mathrm{d}t}{(1-t/1+t)^{i\delta(-\delta_{n1}+\delta_{n2})}(t-s)}\right) \quad (5.1.19)$$

where

$$\delta = \frac{1}{2\pi}\ln\left(\frac{1+\beta}{1-\beta}\right); \quad (5.1.20)$$

$n = 1$ and 2; $const_1$ and $const_2$ are arbitrary constants. Using (5.1.18) and some considerations on the nature (odd and even) of $\overline{f}$ and g, we find that (i) $const_1 = const_2, A = -B$ or (ii) $const_1 = -const_2, A = B$. Then, a simple form for g follows from which is deduced the following expression for D_1 the distribution function of the edge dislocation family 1:

$$D_1(x_1) = \frac{1}{\pi\sqrt{a^2-x_1^2}}\left(const_1\mathrm{Re}\left[\left(\frac{a-x_1}{a+x_1}\right)^{i\delta}\right] + const_2\mathrm{Im}\left[\left(\frac{a-x_1}{a+x_1}\right)^{i\delta}\right]\right)$$
$$-\frac{1}{\pi bC(1-\beta^2)\sqrt{a^2-x_1^2}}\mathrm{Re}\left[\left(\frac{a-x_1}{a+x_1}\right)^{i\delta}\int_{-a}^{a}\left(\frac{a+t}{a-t}\right)^{i\delta}\frac{f(t)}{t-x_1}\mathrm{d}t\right], \quad (5.1.21)$$

$-a < x_1, t < a$, Re[...] and Im[...] denote the real and imaginary parts of the complex quantity inside the brackets [.], and here also, $const_1$ and $const_2$ are arbitrary constants. From D_1 (5.1.21), we can reach D_2 using (5.1.11). Closed form solutions are obtained after performing the various integrations in (5.1.21) using (5.1.13) for f. These are given below in the next section assuming zero dislocation content after unloading.

Dislocation Distributions; Relative Displacements of the Faces of the Crack

Closed-form expressions have been obtained for the distribution functions D_1 and D_2 of the edge crack dislocations. These are

$$D_n(x_1) = \frac{ch(\pi\delta)}{2bC\sqrt{a^2 - x_1^2}} \mathrm{Re}\big[(i\delta_{n1} + \delta_{n2})\{2(\sigma_{22}^a - i\sigma_{12}^a)(x_1 + 2ia\delta) - \big(ia^2(1+4\delta^2) + 4a\delta x_1 - 2ix_1^2\big)\overline{\alpha}\}\left(\frac{a - x_1}{a + x_1}\right)^{i\delta}\Big], \quad (5.1.22)$$

$n = 1$ and 2. D_1 and D_2 only differ by a factor i in front of the curly brackets $\{\}$. ch is the hyperbolic cosine function. The relative displacement φ_n ($n = 1$ and 2) of the faces of the crack in the x_n-direction is

$$\varphi_n(x_1) = \int_{x_1}^{a} bD_n(x_1')\mathrm{d}x_1', \ |x_1| \le a, \quad (5.1.23)$$

and after integration

$$\varphi_n = \frac{ch(\pi\delta)}{C} \mathrm{Re}\left[(i\delta_{n1} + \delta_{n2})\frac{(a - x_1)^{i\delta+1/2}}{(2a)^{i\delta-1/2}}\left\{(\sigma_{22}^a - i\sigma_{12}^a)\left(F(X) - \frac{(a - x_1)F(Y)}{2a(i\delta + 3/2)}\right) + 2ia\overline{\alpha}\left(\frac{1 + 2i\delta}{4}F(X) - \frac{a - x_1}{2a(i\delta + 3/2)}(i\delta F(Y) + F(Z))\right)\right\}\right], \quad (5.1.24)$$

$n = 1$ and 2; F is Gauss's hypergeometric function and X, Y and Z are arguments for F with values

$$X = (i\delta + 1/2, i\delta + 1/2; i\delta + 3/2; (a - x_1)/2a)$$
$$Y = (i\delta + 1/2, i\delta + 3/2; i\delta + 5/2; (a - x_1)/2a)$$
$$Z = (i\delta - 1/2, i\delta + 3/2; i\delta + 5/2; (a - x_1)/2a).$$

Other relations of interest (in the calculation of the crack extension force below, for instance) are expressions for φ_n, in the vicinity of the crack tip at $x_1 = a$. At the distance $s = a - x_1, 0 < s \ll a$, we get using (5.1.24)

$$\varphi_n(s) = \frac{ch(\pi\delta)\sqrt{2as}}{C} \mathrm{Re}\left[(i\delta_{n1} + \delta_{n2})(\sigma_{22}^a - i\sigma_{12}^a - a\delta\overline{\alpha} + ia\overline{\alpha}/2)\left(\frac{s}{2a}\right)^{i\delta}\right], \quad (5.1.25)$$

$n = 1$ and 2.

Stresses at the Crack Tip and Crack Extension Force

We would like to express the total stresses $\overline{\sigma}_{12}$ and $\overline{\sigma}_{22}$ in the plane of the crack in the neighbourhood of the crack tip at $x_1 = a$. These are given by the dominant terms of $(\overline{\sigma}_{ij}^{(1)} + \overline{\sigma}_{ij}^{(2)})$ in (5.1.3) at point $P = (x_1, x_2 = 0, x_3)$ with $x_1 = a + s_1, 0 < s_1 << a$. Using (5.1.3–5.1.6) and (5.1.22) we obtain

$$\overline{\sigma}_{n2}(s_1) = \sqrt{\frac{a}{2s_1}} \\ \mathrm{Re}\left[(i\delta_{n1} + \delta_{n2})(1 + 2i\delta)\big(\sigma_{22}^a - i\sigma_{12}^a - a\delta\overline{\alpha} + ia\overline{\alpha}/2\big)\Big(\frac{s_1}{2a}\Big)^{i\delta} \right], \quad (5.1.26)$$

$n = 1$ and 2. The crack extension force G is defined as

$$G = \lim_{\Delta a \to 0} \frac{1}{\Delta a} \left(\frac{1}{2} \int_a^{a+\Delta a} \overline{\sigma}_{12}(s_1)\varphi_1(s_2)\mathrm{d}x_1 + \frac{1}{2} \int_a^{a+\Delta a} \overline{\sigma}_{22}(s_1)\varphi_2(s_2)\mathrm{d}x_1 \right); \quad (5.1.27)$$

for $\overline{\sigma}_{12}$ and $\overline{\sigma}_{22}$, we use (5.1.26) with $s_1 = x_1 - a$, and for φ_1 and φ_2 (5.1.25) with $s_2 = a + \Delta a - x_1$. We obtain

$$G = \frac{a\pi(1 + 4\delta^2)}{4C} \left(\big(\sigma_{22}^a - a\delta\overline{\alpha}\big)^2 + \left(\frac{a\overline{\alpha}}{2} - \sigma_{12}^a \right)^2 \right). \quad (5.1.28)$$

When G is defined as

$$G = G_1 + G_2 \quad (5.1.29)$$

with

$$G_n = \lim_{\Delta a \to 0} \frac{1}{\Delta a} \left(\frac{1}{2} \int_a^{a+\Delta a} \overline{\sigma}_{n2}(s_1)\phi_n(s_2)\mathrm{d}x_1 \right), n = 1 \text{ and } 2, \quad (5.1.30)$$

it follows

$$G_n = \frac{G}{2} + (-\delta_{n1} + \delta_{n2}) \frac{ach(\pi\delta)}{4C} \\ \times \lim_{\Delta a \to 0} \mathrm{Re}\left[(1 + 2i\delta)\eta^2 B(i\delta + 3/2, i\delta + 1/2) \left(\frac{\Delta a}{2a} \right)^{2i\delta} \right], \\ n = 1 \text{ and } 2, \quad (5.1.31)$$

where

$$\eta = \sigma_{22}^a - i\sigma_{12}^a - a\delta\overline{\alpha} + ia\overline{\alpha}/2; \tag{5.1.32}$$

$B(x, y)$ is the beta function (Euler's integral of the first kind). It thus appears that G (5.1.28) is well defined, but G_n, associated with individual loading modes *I* and *II*, carries an oscillatory term with Δa indefinitely. The crack extension force possesses the following aspects:

- G (5.1.28) the total energy release rate is well defined,
- G_1 and G_2 (5.1.30) the individual energy release rates associated with modes *II* and *I* carry an oscillatory term with Δa indefinitely,
- If the oscillatory term is neglected, then $G_1 = G_2 = G/2$ (5.1.31).

These behaviours of the crack extension force have been described earlier by Sun and Jih (1987) and Raju et al. (1988).

Confrontation with Experiments

Experimental works provide the crack extension force (i.e. fracture toughness) as a function of phase angle, also termed mode mix or mixity (Liechti and Chai 1991, 1992, among others). The phase angle ψ at position $P = (x_1, x_2 = 0, x_3)$, on the interface ahead of the crack front, with $x_1 = a + s_1, o < s_1 \ll a$, is defined in terms of crack-tip stresses as

$$\psi = \tan^{-1}(\overline{\sigma}_{12}(s_1)/\overline{\sigma}_{22}(s_1)). \tag{5.1.33}$$

Making use of (5.1.26), we write

$$\tan\psi = -\frac{x_0 \tan\theta_1 + y_0}{x_0 - y_0 \tan\theta_1} \tag{5.1.34}$$

where

$$\begin{aligned} \theta_1 &= \delta \ln(s_1/2a), \\ x_0 &= \sigma_{22}^a(1 - 2\delta\kappa_s\tilde{a} + 2\delta M), \\ y_0 &= \sigma_{22}^a\big(2\delta + (1 - 4\delta^2)\kappa_s\tilde{a}/2 - M\big) \end{aligned} \tag{5.1.35}$$

with $\kappa_s = \nu_s\mu_s(\nu_1/E_1 - \nu_2/E_2)$, $\tilde{a} = a/a_s$ and $M = \sigma_{12}^a/\sigma_{22}^a$. The crack extension force G (5.1.28) can be expressed in terms of the phase angle as

$$G = (1 - \delta\kappa_s\tilde{a})^2 \\ G(M = 0; \kappa_s = 0)\frac{(1 + 4\delta^2)(1 + \tan^2\theta_1)(1 + \tan^2\psi)}{(1 - 2\delta\tan\theta_1 - (2\delta + \tan\theta_1)\tan\psi)^2} \tag{5.1.36}$$

where

$$G(M = 0; \kappa_s = 0) = \frac{a\pi(1 + 4\delta^2)\sigma_{22}^{a2}}{4C}. \tag{5.1.37}$$

$G(M = 0; \kappa_s = 0)$ is the value of the crack extension force when both M (the applied shear by tension ratio) and κ_s are zero. From (5.1.1), it may be seen that $\kappa_s = 0$ corresponds to $\tau_{12}^a = 0$ (absence of Poisson's effect). Poisson's effect is considered only when $\kappa_s \neq 0$. Other relations are of interest. We first mention the relation between M and the phase angle ψ; this is derived from (5.1.34) as

$$M = \frac{A_1 + B_1 \tan\psi}{1 - 2\delta\tan\theta_1 - (2\delta + \tan\theta_1)\tan\psi} \tag{5.1.38}$$

where

$$A_1 = 2\delta + (1 - 4\delta^2)\kappa_s\tilde{a}/2 + (1 - 2\delta\kappa_s\tilde{a})\tan\theta_1,$$
$$B_1 = 1 - 2\delta\kappa_s\tilde{a} - (2\delta + (1 - 4\delta^2)\kappa_s\tilde{a}/2)\tan\theta_1.$$

It is also interesting to mention the following result on the value of the phase angle at the minimums of the crack extension force. When $\kappa_s = 0$, G (5.1.28) is minimum for $M = 0$ (i.e. $\sigma_{12}^a = 0$); this minimum corresponds to $G(M = 0; \kappa_s = 0)$ (5.1.37). When $\kappa_s \neq 0$, G (5.1.28) is minimum for $M = \kappa_s\tilde{a}/2 \equiv M_E$ with value

$$G_E = (1 - \delta\kappa_s\tilde{a})^2 G(M = 0; \kappa_s = 0). \tag{5.1.39}$$

It is noted that both minimums of G correspond to the equal phase angle ψ_E [use (5.1.34) for instance] given by

$$\tan\psi_E = -\frac{2\delta + \tan\theta_1}{1 - 2\delta\tan\theta_1}. \tag{5.1.40}$$

ψ_E depends on $\theta_1 = \delta\ln(s_1/2a)$ and therefore on the values given to s_1 and a.

For illustration purpose and to confront theory with experiments, we consider the work of Liechti and Chai (1992) with the bi-material system (1)/(2) of our modelling corresponding to the pair epoxy/glass. The parameters taken from their work are: $\delta = -0.0605$; $E_1 = 2.03$ GPa, $\nu_1 = 0.37$; $E_2 = 68.95$ GPa, $\nu_2 = 0.20$. We also use $E_i = 2\mu_i(1 + \nu_i)$ and take $\mu_s = (\mu_1 + \mu_2)/2$, $\nu_s = 0.5$.

Figure 5.2a is a graphical representation of $\tilde{G}$, the crack extension force G (5.1.28) reduced by $G(M = 0; \kappa_s = 0)$ (5.1.37), as a function of $M = \sigma_{12}^a/\sigma_{22}^a$. When $\kappa_s = 0$, $\tilde{G}$ is minimum for $M = 0$; when Poisson's effect is taken into account ($\kappa_s \neq 0$), this minimum is shifted to $M = \kappa_s\tilde{a}/2 \equiv M_E$ (see below (5.1.35) for $\tilde{a}$). The reduced $\tilde{G}$ increases indefinitely when M moves away from M_E along the opposite directions. If one takes $2a = 7.65$ cm [the length of the edge starter crack of the specimens (Liechti and Chai 1992)] and $a_s = 0.5$ cm ($\tilde{a} = a/a_s = 7.65$), this gives $M_E = 5.04$. Figure 5.2b represents the phase angle ψ as a function of M [using (5.1.38)]. Here also, dashed and solid curves correspond to $\kappa_s = 0$ and $\kappa_s \neq 0$

respectively. We take values close to those of Liechti and Chai 1992: $2a = 7.65$ cm and $s_1 = 1.27$ cm; this gives $\psi_E = 0.67°$, $M_E = 5.04$. As mentioned earlier ψ_E corresponds to $M = 0$ on the $\kappa_s = 0$ curve and $M = M_E$ on the $\kappa_s \neq 0$ curve. For the system (1)/(2) corresponding to the epoxy/glass, τ_{12}^a (5.1.1) is negative in the $x_1 > 0$ region of the fracture specimen; this means that positive M reduces the internal strain caused by τ_{12}^a while negative M amplifies this strain in that region. For $\kappa_s \neq 0$ (solid curve in Fig. 5.2b) and $M = 0$, the phase angle is negative with a large magnitude (about $-70°$); with increasing positive applied M, the phase angle ψ increases drastically: M interval of [0, 10] corresponds to ψ interval of [$-70°$, $70°$] approximately. The corresponding decrease of ψ for negative M is much less pronounced: M interval of [-20, 0] for ψ interval of [$-85°$, $-72°$] approximately. For the experimentalist, starting from zero applied M, this is the manifestation of an apparent asymmetry. Edge-cracked bi-material strip specimens in bi-axial loading experiments would display an asymmetrical behaviour with respect to the direction of the applied shear stress σ_{12}^a as may be noted in Liechti and Chai (1992). We further defined from (5.1.36) $\hat{G}(\psi)$ as

$$\hat{G} = \frac{G}{(1-\delta\kappa_s\tilde{a})^2 G(M=0;\kappa_s=0)} = \frac{(1+4\delta^2)(1+\tan^2\theta_1)(1+\tan^2\psi)}{(1-2\delta\tan\theta_1-(2\delta+\tan\theta_1)\tan\psi)^2}. \tag{5.1.41}$$

Figure 5.3 represents $\hat{G}(\psi)$ for $2a = 7.65$ cm and $s_1 = 1.27$ cm. Also reported are experimental values (symbol *) of the fracture toughness versus phase angle ψ given by Liechti and Chai (their Fig. 6). What we do first is to identify (by simple inspection) the minimum value: we find it to be located at $\psi_m = 20°$, approximately. We then report the experimental data so that ψ_m be read ψ_E (5.1.40); with the values given to a and s_1 above, $\psi_E = 0.67°$. Hence for each value of the phase angle provided by Liechti and Chai, we add the negative value $\psi_E - \psi_m$; then, the experimental fracture toughness is reduced by 3.75 J/m^2(taken as the associated minimum). Comparison between $\hat{G}$ (5.1.41) and experimental points (symbol *) seems quite satisfactory. We use $\hat{G}$ instead of $\tilde{G} = G/G(M = 0; \kappa_s = 0)$ because Liechti and Chai didn't take any account of Poisson's effect in their considerations with theory. Considering Poisson's effect would require multiplying the fracture toughness values in Fig. 5.3 by the quantity $(1-\delta\kappa_s\tilde{a})^2$ [see (5.1.36)]; with $a_s = 1/2$ cm and $2a = 7.65$ cm ($\tilde{a} = 7.65$) and epoxy/glass, this gives the value 2.6 approximately. In other words, we would consider $\tilde{G}$ instead of $\hat{G}$. We also note that our crack model of Fig. 5.1 is different from the edge-cracked bi-material system in Liechti and Chai 1992 (see their Fig. 1); however, this difference seems not too serious in the graphical representation of fracture toughness versus phase angle, on view of Fig. 5.3.

SUMMARY: The mathematical analysis of a bi-material interface crack of finite length $2a$ under applied mixed mode $I + II$ loading, incorporating an internal shear stress that originates from a difference in Poisson's contractions on crossing the interface, has been performed. Various physical quantities, involved in the discussion

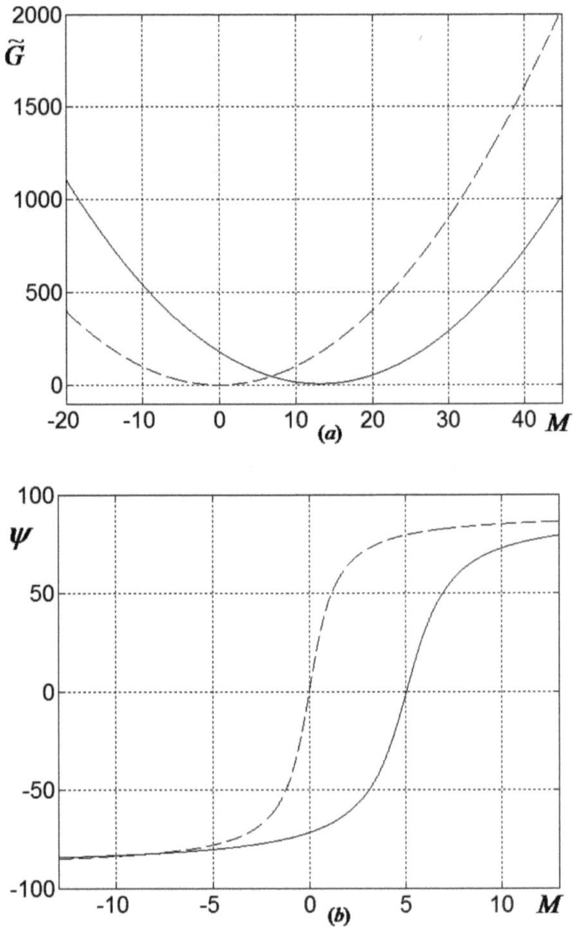

Fig. 5.2 **a** Reduced crack extension force $\tilde{G} = G/G(M = 0; \kappa_s = 0)$ as a function of $M = \sigma_{12}^a/\sigma_{22}^a$, $\tilde{a} = 20$. **b** Phase angle ψ(degrees) as a function of M. The identical $\psi_E(= 0.67°)$ corresponds to both $M = 0$ $(\kappa_s = 0)$ and $M = M_E = 5.04$ $(\kappa_s \neq 0)$; $2a = 7.65\,\text{cm}$, $s_1 = 1.27\,\text{cm}$, $a_s = 0.5\,\text{cm}$. In both (**a**) and (**b**), dashed and solid curves correspond to $\kappa_s = 0$ and $\kappa_s \neq 0$ respectively; material parameters for the system (1)/(2) correspond to epoxy/glass (see text)

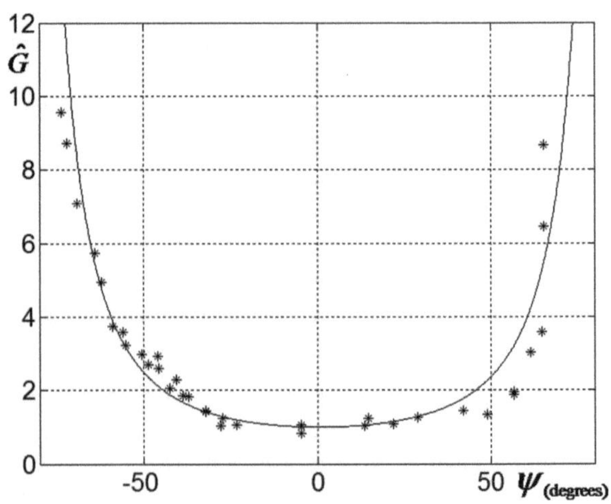

Fig. 5.3 Reduced crack extension force $\hat{G}$ (5.1.41) (solid curve) as a function of phase angle ψ (degrees). Experimental points (symbol *) are those of Liechti and Chai (1992) (their Fig. 6); $2a = 7.65\,\text{cm}$, $s_1 = 1.27\,\text{cm}$ and material parameters correspond to epoxy/glass [system(1)/(2)] (see text)

of the conditions for crack motion, are given in closed-form expressions. These are the relative displacement of the faces of the crack, the crack-tip stress and the crack extension force. This would help to a better interpretation of experimental findings in several situations. Confrontation with experiments has been done with attention to the work of Liechti and Chai (1992) with edge-cracked bi-material epoxy/glass strip specimens in bi-axial loading experiments.

5.2 Interfacial Dislocation Elastic Fields

Consider a pair of different solids $S1$ and $S2$, of infinite sizes, welded along a non-planar sinusoidal surface S defined by the surfing point $P_S(x_1, x_2 = \xi_n \sin \kappa_n x_3, x_3)$, in such a way that $S1$ and $S2$ occupy the regions $x_2 > \xi_n \sin \kappa_n x_3$ and $x_2 < \xi_n \sin \kappa_n x_3$, respectively. The situation is shown in the Fig. 5.4 where $S1$ and $S2$ are confined for illustration purpose in a parallelepiped of finite sizes.

The present aims at providing expressions for the displacement and stress fields of sinusoidal dislocations, lying on that interface at the origin, extending indefinitely in the x_3-direction and spreading in the Ox_2x_3-plane in the sinusoidal form $A_n = \xi_n \sin \kappa_n x_3$. The dislocation is edge on average for Burgers vectors in the x_1 and x_2 directions; for the former direction, the Burgers vector is perpendicular to the plane of location of the dislocation, hence this is a climb-type edge dislocation; for the latter, the Burgers vector is in the plane of the dislocation, consequently, this is a glide-type edge dislocation. Here, we restrict ourselves to a Burgers vector $\vec{b} = (0, b, 0)$ in the x_2-direction. Using the results of such a study is at several levels: (1) the elastic fields due to an arbitrary form of interface dislocation in x_2x_3 planes with the same Burgers vector can be derived by superposition (Fourier series expansion); (2) a non-planar large interface crack loaded in tension in the x_2-direction and propagating in the x_1-direction may be represented mathematically by a continuous array of long non-straight dislocations with infinitesimal Burgers vectors $\vec{b} = (0, b, 0)$. Denote by $\vec{u}^{(m)}$ and $(\sigma)^{(m)}$ ($m = 1$ and 2) the displacement and stress fields in the solid (m)

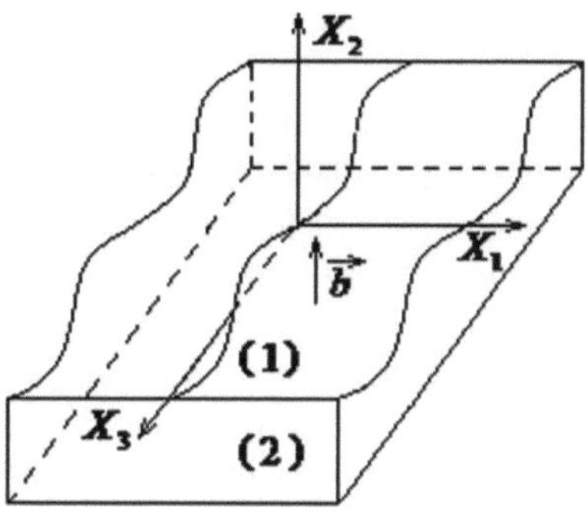

Fig. 5.4 Two elastic mediums (1) and (2) welded along a non-planar sinusoidal surface and containing an interface sinusoidal dislocation at the origin. The dislocation lies in the Ox_2x_3-plane and runs indefinitely in the x_3-direction. The Burgers vector $\vec{b}$ is present

due to the interface sinusoidal dislocation. We assume that the following description applies:

(1) $\vec{u}^{(m)}$ and $(\sigma)^{(m)}$ are continuous at the crossing of the interface

$$\vec{u}^{(1)}(P_S) = \vec{u}^{(2)}(P_S) \text{ and } (\sigma)^{(1)}(P_S) = (\sigma)^{(2)}(P_S) \tag{5.2.1}$$

(2) Far from the dislocation and the interface, the elastic fields in the medium (m) correspond to those of a sinusoidal dislocation in an infinitely extended homogeneous solid with the equal elastic constants, that we denote by $\vec{u}^{(m)\infty}$ and $(\sigma)^{(m)\infty}$, hence

$$\begin{aligned}\vec{u}^{(m)} &\to \vec{u}^{(m)\infty}\\ (\sigma)^{(m)} &\to (\sigma)^{(m)\infty}\end{aligned} \tag{5.2.2}$$

when one moves far away in the x_2-direction.

(3) The elastic fields may be expressed in the form

$$\begin{aligned}\vec{u}^{(m)} &= \vec{u}^{(m)\infty} - \vec{u}^{(m)W}\\ (\sigma)^{(m)} &= (\sigma)^{(m)\infty} - (\sigma)^{(m)W}\end{aligned} \tag{5.2.3}$$

where $\vec{u}^{(m)W}$ and $(\sigma)^{(m)W}$ satisfy the equations of equilibrium and possess the properties that follow.

(4) $\Delta\vec{u}^{\infty}(P_S) \equiv \vec{u}^{(2)\infty} - \vec{u}^{(1)\infty} = \vec{u}^{(2)W} - \vec{u}^{(1)W} \equiv \Delta\vec{u}^{W}(P_S)$

$$(\Delta\sigma)^{\infty}(P_S) \equiv (\sigma)^{(2)\infty} - (\sigma)^{(1)\infty} = (\sigma)^{(2)W} - (\sigma)^{(1)W} \equiv (\Delta\sigma)^{W}(P_S), \tag{5.2.4}$$

these conditions ensure the continuity of the elastic fields across the interface.

(5) $\vec{u}^{(m)W}$ and $(\sigma)^{(m)W}$ cancel far from the dislocation and interface; this means that

$$\begin{aligned}\vec{u}^{(m)W} &\to 0\\ (\sigma)^{(m)W} &\to 0\end{aligned} \tag{5.2.5}$$

when $|x_2| \to \infty$; this ensures the veracity of condition (5.2.2) above.

The elastic fields $\vec{u}^{(m)}$ and $(\sigma)^{(m)}$ thus obtained are expected to be valuable representations of the physical situation illustrated in the Fig. 5.4. The associated $\vec{u}^{(m)W}$ and $(\sigma)^{(m)W}$ are investigated with the help of Galerkin vectors and corresponding equations of equilibrium. In what follows: in Sect. 5.2.1, $\Delta\vec{u}^{\infty}(P_S)$ and $(\Delta\sigma)^{\infty}(P_S)$ are expressed in a Fourier series form that involves terms with $\exp(i\vec{k}.\vec{x})$, with $\vec{k} = (k_1, k_2, k_3)$ (k_i real numbers) and $\vec{x} = (x_1, x_2, x_3)$ vector position; in Sect. 5.2.2, a Galerkin vector with components involving $\exp(i\vec{k}.\vec{x})$ is considered. The associated elastic fields also consist of terms proportional to $\exp(i\vec{k}.\vec{x})$. These are managed in the equal Fourier series form. Then equations of the type (5.2.4) can be posed.

In Sect. 5.2.3, the search for the appropriate elastic fields $\vec{u}^{(m)W}$ and $(\sigma)^{(m)W}$ is performed. In Sect. 5.2.4, calculation results are discussed and confronted to previous studies. The beginning part of this study has been reported (Anongba 2015).

5.2.1 Interface Boundary Values Carried by the Elastic Fields of a Sinusoidal Dislocation in a Homogeneous Solid

The elastic fields due to a sinusoidal edge dislocation ($\vec{b} = (0, b, 0)$) lying in the Ox_2x_3-plane in the sinusoidal form $A_n(x_3) = \xi_n \sin \kappa_n x_3$ have been provided in infinite series forms by Anongba (1995). In a similar way as in previous studies (Anongba 1996, 2003, 2009, 2010; Anongba and Vitek 2003), we shall assume ξ_n small and limit the elastic solutions up to terms of first order with respect to ξ_n. In this way, the elastic fields consist of two terms:

$$\vec{u}^{(m)\infty} = \vec{u}^{(0)(m)\infty} + \vec{u}^{A_n(m)\infty}$$
$$(\sigma)^{(m)\infty} = (\sigma)^{(0)(m)\infty} + (\sigma)^{A_n(m)\infty} \tag{5.2.6}$$

where $\vec{u}^{(0)(m)\infty}$ and $(\sigma)^{(0)(m)\infty}$ are of zero order with respect to ξ_n corresponding to the fields of a straight edge dislocation; $\vec{u}^{A_n(m)\infty}$ and $(\sigma)^{A_n(m)\infty}$ are proportional to the sinusoid $A_n(x_3)$ or to its spatial derivative $\partial A_n/\partial x_3$. The purpose here is to write down the differences $\Delta\vec{u}^{\infty}$ and $(\Delta\sigma)^{\infty}$ (5.2.4) on crossing the interface at arbitrary point $P_S(x_1, x_2 = \xi_n \sin \kappa_n x_3, x_3)$. We use the notation $x_2 = \xi$ ($\xi = \xi_n \sin \kappa_n x_3$ small) and take the MacLaurin series expansions of the elastic fields up to terms of first order with respect to ξ; this means that

$$\Delta\vec{u}^{\infty}(x_1, x_2 = \xi, x_3) = \Delta\vec{u}^{\infty}(x_1, 0, x_3) + \frac{\partial \Delta\vec{u}^{\infty}}{\partial x_2}(x_1, 0, x_3)\xi,$$
$$(\Delta\sigma)^{\infty}(x_1, x_2 = \xi, x_3) = (\Delta\sigma)^{\infty}(x_1, 0, x_3) + \frac{\partial (\Delta\sigma)^{\infty}}{\partial x_2}(x_1, 0, x_3)\xi. \tag{5.2.7}$$

$\vec{u}^{(m)\infty}$ and $(\sigma)^{(m)\infty}$ are taken from previous works; we obtain (u_i is the i-component of vector $\vec{u}$ and σ_{ij} the ij-element of the stress matrix (σ); $i, j = 1$ to 3)

$$\Delta u_i^{\infty}(x_1, x_2 = \xi, x_3) = \Delta u_i^{(0)\infty} + \Delta u_i^{A_n\infty}$$
$$\Delta \sigma_{ij}^{\infty}(x_1, x_2 = \xi, x_3) = \Delta \sigma_{ij}^{(0)\infty} + \Delta \sigma_{ij}^{A_n\infty}$$

as

$$\Delta u_1^{(0)\infty} = \frac{bC_\nu}{4\pi} \ln|x_1| = -\frac{bC_\nu}{8\pi} \int_{-\infty}^{\infty} \frac{1}{|k_1|} e^{ik_1x_1} dk_1$$

$$\begin{aligned}\Delta u_1^{A_n\infty} &= \frac{bC_\nu\kappa_n A_n}{4\pi|x_1|}(\kappa_n|x_1|K_0 + K_1)\xi \\ &= -\frac{bC_\nu A_n}{8\pi}\int\limits_{-\infty}^{\infty}\frac{k_1^2}{\sqrt{k_1^2+\kappa_n^2}}e^{ik_1x_1}\mathrm{d}k_1\xi\end{aligned}$$

$$\Delta u_2^{(0)\infty} = \frac{bC_\nu}{4\pi x_1}\xi = -\frac{ibC_\nu}{8\pi}\int\limits_{-\infty}^{\infty}\mathrm{sgn}(k_1)e^{ik_1x_1}\mathrm{d}k_1\xi$$

$$\Delta u_2^{A_n\infty} = -\frac{bC_\nu\kappa_n A_n}{4\pi}\mathrm{sgn}(x_1)K_1 = \frac{ibC_\nu A_n}{8\pi}\int\limits_{-\infty}^{\infty}\frac{k_1}{\sqrt{k_1^2+\kappa_n^2}}e^{ik_1x_1}\mathrm{d}k_1$$

$$\begin{aligned}\Delta u_3^{(0)\infty} &= 0 \\ \Delta u_3^{A_n\infty} &= -\frac{bC_\nu\kappa_n}{4\pi}\frac{\partial A_n}{\partial x_3}\mathrm{sgn}(x_1)K_1\xi \\ &= \frac{ibC_\nu}{8\pi}\frac{\partial A_n}{\partial x_3}\int\limits_{-\infty}^{\infty}\frac{k_1}{\sqrt{k_1^2+\kappa_n^2}}e^{ik_1x_1}\mathrm{d}k_1\xi\end{aligned}$$

$$\begin{aligned}\Delta\sigma_{11}^{(0)\infty} &= \frac{C_2-C_1}{x_1} = -2Q_b\int\limits_{-\infty}^{\infty}\mathrm{sgn}(k_1)e^{ik_1x_1}\mathrm{d}k_1 \\ \Delta\sigma_{11}^{A_n\infty} &= \frac{(C_2-C_1)\kappa_n A_n}{x_1}\left(3\kappa_n K_0 + \frac{(6+\kappa_n^2x_1^2)K_1}{|x_1|}\right)\xi \\ &= 2Q_bA_n\int\limits_{-\infty}^{\infty}\frac{k_1(3k_1^2+2\kappa_n^2)}{\sqrt{k_1^2+\kappa_n^2}}e^{ik_1x_1}\mathrm{d}k_1\xi\end{aligned}$$

$$\begin{aligned}\Delta\sigma_{22}^{(0)\infty} &= \frac{C_2-C_1}{x_1} = -2Q_b\int\limits_{-\infty}^{\infty}\mathrm{sgn}(k_1)e^{ik_1x_1}\mathrm{d}k_1 \\ \Delta\sigma_{22}^{A_n\infty} &= -\frac{(C_2-C_1)\kappa_n A_n}{x_1}\left(\kappa_n K_0 + \frac{2K_1}{|x_1|}\right)\xi \\ &= -2Q_bA_n\int\limits_{-\infty}^{\infty}k_1\sqrt{k_1^2+\kappa_n^2}e^{ik_1x_1}\mathrm{d}k_1\xi\end{aligned}$$

$$\begin{aligned}\Delta\sigma_{33}^{(0)\infty} &= -\frac{8iQ_c}{x_1} = -4Q_c\int\limits_{-\infty}^{\infty}\mathrm{sgn}(k_1)e^{ik_1x_1}\mathrm{d}k_1 \\ \Delta\sigma_{33}^{A_n\infty} &= -\frac{\kappa_n A_n}{x_1}\left(8iQ_c\kappa_n K_0 + \left(\frac{16iQ_c}{|x_1|} + \kappa_n^2(C_2-C_1)|x_1|\right)K_1\right)\xi\end{aligned}$$

$$= 2A_n \int_{-\infty}^{\infty} \frac{k_1(2Q_c k_1^2 + (2Q_c + Q_b)\kappa_n^2)}{\sqrt{k_1^2 + \kappa_n^2}} e^{ik_1x_1} \mathrm{d}k_1 \xi$$

$$\Delta\sigma_{12}^{(0)\infty} = \frac{C_2 - C_1}{x_1^2}\xi = 2iQ_b \int_{-\infty}^{\infty} |k_1| e^{ik_1x_1} dk_1 \xi$$

$$\Delta\sigma_{12}^{A_n\infty} = \kappa_n A_n \left(4iQ_c\kappa_n K_0 - \frac{(C_2 - C_1)K_1}{|x_1|} \right)$$

$$= -2iA_n \int_{-\infty}^{\infty} \frac{Q_b k_1^2 + (Q_b - Q_c)\kappa_n^2}{\sqrt{k_1^2 + \kappa_n^2}} e^{ik_1x_1} \mathrm{d}k_1$$

$$\Delta\sigma_{13}^{(0)\infty} = 0$$

$$\Delta\sigma_{13}^{A_n\infty} = 4i\kappa_n \frac{\partial A_n}{\partial x_3} \left(Q_b\kappa_n K_0 + \frac{(2Q_b - Q_c)K_1}{|x_1|} \right)\xi$$

$$= 2i \frac{\partial A_n}{\partial x_3} \int_{-\infty}^{\infty} \frac{(Q_c - 2Q_b)k_1^2 + (Q_c - Q_b)\kappa_n^2}{\sqrt{k_1^2 + \kappa_n^2}} e^{ik_1x_1} \mathrm{d}k_1 \xi$$

$$\Delta\sigma_{23}^{(0)\infty} = 0$$

$$\Delta\sigma_{23}^{A_n\infty} = -4iQ_c\kappa_n \frac{\partial A_n}{\partial x_3} \mathrm{sgn}(x_1) K_1 = -2Q_c \frac{\partial A_n}{\partial x_3} \int_{-\infty}^{\infty} \frac{k_1}{\sqrt{k_1^2 + \kappa_n^2}} e^{ik_1x_1} \mathrm{d}k_1 \quad (5.2.8)$$

where

$$C_v = [1/(1 - \nu_1) - 1/(1 - \nu_2)], \quad Q_b = i(C_2 - C_1)/4,$$
$$Q_c = i(\nu_2 C_2 - \nu_1 C_1)/4, \quad C_m = b\mu_m/2\pi(1 - \nu_m);$$

K_i is the ith-order modified Bessel function with argument $\kappa_n|x_1|$ and $\mathrm{sgn}(k_1) = k_1/|k_1|$; μ_m and ν_m are shear modulus and Poisson's ratio. In the various expressions in (5.2.8), constant terms are omitted.

5.2.2 Galerkin Vectors and Interface Boundary Conditions

A Galerkin vector $\vec{V}(\vec{x})$ is a vector whose components are biharmonic spatial functions ($\Delta\Delta\vec{V} = 0$; Δ Laplace operator) in order to satisfy the equilibrium equations with zero body forces. Then the associated displacement $\vec{u}$ is expressed as

$$2\mu\vec{u} = 2(1-\nu)\Delta\vec{V} - \vec{\nabla}(\vec{\nabla}.\vec{V}) \tag{5.2.9}$$

where μ and ν are shear modulus and Poisson's ratio respectively; $\vec{\nabla}$ is the operator *nabla*, $\vec{\nabla} = (\partial/\partial x_1, \partial/\partial x_2, \partial/\partial x_3)$. The stress field (σ) is obtained from the displacement $\vec{u}$ by partial differentiation with respect to coordinates x_i. The matter is treated in several books (see Fung 1965; Soutas-Little 1973; Barber 2002; among many others). For the present problem, we arrive at Galerkin vectors with only one non-zero x_2-component, arranged in the form

$$V_2(\vec{x}) = \overline{\alpha}_2(\vec{k})e^{i\vec{k}.\vec{x}} + \overline{\beta}_2(\vec{k})x_2 e^{i\vec{k}.\vec{x}} \tag{5.2.10}$$

under the condition $\vec{k}^2 = k_1^2 + k_2^2 + k_3^2 = 0$ that ensures the biharmonicity of V_2. For V_2 to cancel far from the interface, we write

$$k_2 = k_2^{(m)} \equiv (-1)^{m-1} i\sqrt{k_1^2 + k_3^2} \tag{5.2.11}$$

with $m = 1$ when $x_2 > \xi_n \sin\kappa_n x_3$ (half-space 1) and $m = 2$ when $x_2 < \xi_n \sin\kappa_n x_3$ (half-space 2). We use the notations

$$\vec{k}^{(m)} \equiv \left(k_1, k_2^{(m)}, k_3\right), \overline{\alpha}_2^{(m)} \equiv \overline{\alpha}_2\left(\vec{k}^{(m)}\right), \overline{\beta}_2^{(m)} \equiv \overline{\beta}_2\left(\vec{k}^{(m)}\right);$$

hence for half-space 1 ($x_2 > \xi_n \sin\kappa_n x_3$), solid (1)

$$V_2(\vec{x}) \equiv V_2^{(1)}(\vec{x}) = \overline{\alpha}_2^{(1)} e^{i\vec{k}^{(1)}.\vec{x}} + \overline{\beta}_2^{(1)} x_2 e^{i\vec{k}^{(1)}.\vec{x}}$$

and for half-space 2 ($x_2 < \xi_n \sin\kappa_n x_3$), solid (2)

$$V_2(\vec{x}) \equiv V_2^{(2)}(\vec{x}) = \overline{\alpha}_2^{(2)} e^{i\vec{k}^{(2)}.\vec{x}} + \overline{\beta}_2^{(2)} x_2 e^{i\vec{k}^{(2)}.\vec{x}}.$$

The elastic fields corresponding to V_2 (5.2.10) may be first calculated starting with (5.2.9); then, more general forms $\vec{u}^{(m)V}$ and $(\sigma)^{(m)V}$ are constructed from the previous ones by superposition over k_1 and k_3 (here the superscript V is just a notation, not to be confused with $\left\|\vec{V}\right\|$). For $\vec{u}^{(m)V}$ and $(\sigma)^{(m)V}$ to conform with $\vec{u}^{(m)\infty}$ and $(\sigma)^{(m)\infty}$ (5.2.6), the summation over k_1 is continuous and that over k_3 is discrete. k_3 takes three values: $-\kappa_n$, 0, κ_n. The fields corresponding to $k_3 = 0$ are denoted $\vec{u}^{(0)(m)V}$ and $(\sigma)^{(0)(m)V}$ and terms associated with $k_3 = -\kappa_n$ and κ_n are merged to form expressions denoted $\vec{u}^{A_n(m)V}$ and $(\sigma)^{A_n(m)V}$; this is made possible by requiring that

$$\overline{\alpha}_2^{(m)}(\kappa_n) \equiv -\overline{\alpha}_2^{(m)}(-\kappa_n),\ \overline{\beta}_2^{(m)}(\kappa_n) \equiv -\overline{\beta}_2^{(m)}(-\kappa_n). \tag{5.2.12}$$

In (5.2.12), $\overline{\alpha}_2^{(m)}(\kappa_n)$ stands for $\overline{\alpha}_2(k_1, k_2^{(m)}, \kappa_n)$. We write

$$\begin{aligned} \vec{u}^{(m)V} &= \vec{u}^{(0)(m)V} + \vec{u}^{A_n(m)V} \\ (\sigma)^{(m)V} &= (\sigma)^{(0)(m)V} + (\sigma)^{A_n(m)V}. \end{aligned} \tag{5.2.13}$$

$\vec{u}^{(0)(m)V}$ and $(\sigma)^{(0)(m)V}$ are x_3-independent; $\vec{u}^{A_n(m)V}$ and $(\sigma)^{A_n(m)V}$ are proportional to the sinusoid $A_n(x_3)$ or to its spatial derivative $\partial A_n/\partial x_3$. Here also, for points P_S on the interface, $\Delta\vec{u}^V$ and $(\Delta\sigma)^V$ are expanded up to terms of first order with respect to $x_2 = \xi$ in a similar manner as in (5.2.8) for $\Delta\vec{u}^\infty$ and $(\Delta\sigma)^\infty$. Requiring $\Delta\vec{u}^V = \Delta\vec{u}^\infty$ and $(\Delta\sigma)^V = (\Delta\sigma)^\infty$ lead to the following equations, writing first the conditions corresponding to $k_3 = 0$ (i.e. $\Delta u_i^{(0)V} = \Delta u_i^{(0)\infty}$ and $\Delta\sigma_{ij}^{(0)V} = \Delta\sigma_{ij}^{(0)\infty}$).

$$\Delta u_1^{(0)V} = \Delta u_1^{(0)\infty} \Rightarrow$$

$$|k_1|\left(\frac{\overline{\alpha}_2^{(2)}}{\mu_2} + \frac{\overline{\alpha}_2^{(1)}}{\mu_1}\right) + \left(\frac{\overline{\beta}_2^{(2)}}{\mu_2} - \frac{\overline{\beta}_2^{(1)}}{\mu_1}\right) = -\frac{ibC_\nu}{4\pi}\frac{\operatorname{sgn}(k_1)}{k_1^2} \tag{a}$$

$$|k_1|\left(\frac{\overline{\alpha}_2^{(2)}}{\mu_2} - \frac{\overline{\alpha}_2^{(1)}}{\mu_1}\right) + 2\left(\frac{\overline{\beta}_2^{(2)}}{\mu_2} + \frac{\overline{\beta}_2^{(1)}}{\mu_1}\right) = 0 \tag{b}$$

$$\Delta u_2^{(0)V} = \Delta u_2^{(0)\infty} \Rightarrow$$

$$|k_1|\left(\frac{\overline{\alpha}_2^{(2)}}{\mu_2} - \frac{\overline{\alpha}_2^{(1)}}{\mu_1}\right) - 2\left(\frac{(1-2\nu_2)\overline{\beta}_2^{(2)}}{\mu_2} + \frac{(1-2\nu_1)\overline{\beta}_2^{(1)}}{\mu_1}\right) = 0 \tag{c}$$

$$|k_1|\left(\frac{\overline{\alpha}_2^{(2)}}{\mu_2} + \frac{\overline{\alpha}_2^{(1)}}{\mu_1}\right) - \left(\frac{(1-4\nu_2)\overline{\beta}_2^{(2)}}{\mu_2} - \frac{(1-4\nu_1)\overline{\beta}_2^{(1)}}{\mu_1}\right) = \frac{ibC_\nu}{4\pi}\frac{\operatorname{sgn}(k_1)}{k_1^2} \tag{d}$$

$$\Delta\sigma_{11}^{(0)V} = \Delta\sigma_{11}^{(0)\infty} \Rightarrow$$

$$|k_1|\left(\overline{\alpha}_2^{(2)} + \overline{\alpha}_2^{(1)}\right) + (1+2\nu_2)\overline{\beta}_2^{(2)} - (1+2\nu_1)\overline{\beta}_2^{(1)} = -2Q_b\frac{\operatorname{sgn}(k_1)}{k_1^2} \tag{e}$$

$$|k_1|\left(\overline{\alpha}_2^{(2)} - \overline{\alpha}_2^{(1)}\right) + 2(1+\nu_2)\overline{\beta}_2^{(2)} + 2(1+\nu_1)\overline{\beta}_2^{(1)} = 0 \tag{f}$$

$$\Delta\sigma_{22}^{(0)V} = \Delta\sigma_{22}^{(0)\infty} \Rightarrow$$

$$|k_1|\left(\overline{\alpha}_2^{(2)} + \overline{\alpha}_2^{(1)}\right) - (1-2\nu_2)\overline{\beta}_2^{(2)} + (1-2\nu_1)\overline{\beta}_2^{(1)} = 2Q_b\frac{\operatorname{sgn}(k_1)}{k_1^2} \tag{g}$$

$$|k_1|\left(\overline{\alpha}_2^{(2)} - \overline{\alpha}_2^{(1)}\right) + 2\nu_2\overline{\beta}_2^{(2)} + 2\nu_1\overline{\beta}_2^{(1)} = 0 \tag{h}$$

$$\Delta\sigma_{33}^{(0)V} = \Delta\sigma_{33}^{(0)\infty} \Rightarrow$$

$$\nu_2\overline{\beta}_2^{(2)} - \nu_1\overline{\beta}_2^{(1)} = -2Q_c\frac{\operatorname{sgn}(k_1)}{k_1^2} \tag{i}$$

$$\nu_2\overline{\beta}_2^{(2)} + \nu_1\overline{\beta}_2^{(1)} = 0 \tag{j}$$

$$\Delta\sigma_{12}^{(0)V} = \Delta\sigma_{12}^{(0)\infty} \Rightarrow (h) \text{ and } (e) \text{ above} \tag{5.2.14}$$

In Eqs. (5.2.14a–j) above, $\overline{\alpha}_2^{(m)}$ stands for $\overline{\alpha}_2(k_1, k_2^{(m)}, k_3 = 0)$. The conditions corresponding to $\Delta u_i^{A_nV} = \Delta u_i^{A_n\infty}$ and $\Delta\sigma_{ij}^{A_nV} = \Delta\sigma_{ij}^{A_n\infty}$ are now listed as:

$$\Delta u_1^{A_nV} = \Delta u_1^{A_n\infty} \Rightarrow$$

$$\sqrt{k_1^2+\kappa_n^2}\left(\frac{\overline{\alpha}_2^{(2)}}{\mu_2} + \frac{\overline{\alpha}_2^{(1)}}{\mu_1}\right) + \frac{\overline{\beta}_2^{(2)}}{\mu_2} - \frac{\overline{\beta}_2^{(1)}}{\mu_1} = 0 \tag{a}$$

$$\sqrt{k_1^2+\kappa_n^2}\left(\frac{\overline{\alpha}_2^{(2)}}{\mu_2} - \frac{\overline{\alpha}_2^{(1)}}{\mu_1}\right) + 2\left(\frac{\overline{\beta}_2^{(2)}}{\mu_2} + \frac{\overline{\beta}_2^{(1)}}{\mu_1}\right) = -\frac{bC_\upsilon\xi_n}{8\pi}\frac{k_1}{k_1^2+\kappa_n^2} \tag{b}$$

$$\Delta u_2^{A_nV} = \Delta u_2^{A_n\infty} \Rightarrow \text{(c) and (d) (displayed below)}$$

$$\sqrt{k_1^2+\kappa_n^2}\left(-\frac{\overline{\alpha}_2^{(2)}}{\mu_2} + \frac{\overline{\alpha}_2^{(1)}}{\mu_1}\right) + 2\left(\frac{(1-2\nu_2)\overline{\beta}_2^{(2)}}{\mu_2} + \frac{(1-2\nu_1)\overline{\beta}_2^{(1)}}{\mu_1}\right) = \frac{bC_\upsilon\xi_n}{8\pi}\frac{k_1}{k_1^2+\kappa_n^2} \tag{c}$$

$$\sqrt{k_1^2+\kappa_n^2}\left(\frac{\overline{\alpha}_2^{(2)}}{\mu_2} + \frac{\overline{\alpha}_2^{(1)}}{\mu_1}\right) - \frac{(1-4\nu_2)\overline{\beta}_2^{(2)}}{\mu_2} + \frac{(1-4\nu_1)\overline{\beta}_2^{(1)}}{\mu_1} = 0 \tag{d}$$

$$\Delta u_3^{A_nV} = \Delta u_3^{A_n\infty} \Rightarrow \text{(a) and (b) above}$$

$$\begin{aligned}\Delta\sigma_{11}^{A_nV} = \Delta\sigma_{11}^{A_n\infty} \Rightarrow\ & k_1^2\sqrt{k_1^2+\kappa_n^2}\left(\overline{\alpha}_2^{(2)} + \overline{\alpha}_2^{(1)}\right) \\ & + [(1+2\nu_2)k_1^2 + 2\nu_2\kappa_n^2]\overline{\beta}_2^{(2)} \\ & - [(1+2\nu_1)k_1^2 + 2\nu_1\kappa_n^2]\overline{\beta}_2^{(1)} = 0\end{aligned} \tag{e}$$

$$k_1^2\sqrt{k_1^2+\kappa_n^2}\left(\overline{\alpha}_2^{(2)}-\overline{\alpha}_2^{(1)}\right)+2[(1+\nu_2)k_1^2+\nu_2\kappa_n^2]\overline{\beta}_2^{(2)}$$
$$+2[(1+\nu_1)k_1^2+\nu_1\kappa_n^2]\overline{\beta}_2^{(1)}=-\frac{iQ_b\xi_n k_1(3k_1^2+2\kappa_n^2)}{k_1^2+\kappa_n^2} \tag{f}$$

$$\Delta\sigma_{22}^{A_nV}=\Delta\sigma_{22}^{A_n\infty}\Rightarrow\sqrt{k_1^2+\kappa_n^2}\left(\overline{\alpha}_2^{(2)}+\overline{\alpha}_2^{(1)}\right)$$
$$-(1-2\nu_2)\overline{\beta}_2^{(2)}+(1-2\nu_1)\overline{\beta}_2^{(1)}=0 \tag{g}$$

$$\sqrt{k_1^2+\kappa_n^2}\left(\overline{\alpha}_2^{(2)}-\overline{\alpha}_2^{(1)}\right)$$
$$+2(\nu_2\overline{\beta}_2^{(2)}+\nu_1\overline{\beta}_2^{(1)})=-\frac{iQ_b\xi_n k_1}{k_1^2+\kappa_n^2} \tag{h}$$

$$\Delta\sigma_{33}^{A_nV}=\Delta\sigma_{33}^{A_n\infty}\Rightarrow\kappa_n^2\sqrt{k_1^2+\kappa_n^2}\left(\overline{\alpha}_2^{(2)}+\overline{\alpha}_2^{(1)}\right)$$
$$+[(1+2\nu_2)\kappa_n^2+2\nu_2k_1^2]\overline{\beta}_2^{(2)}$$
$$-[(1+2\nu_1)\kappa_n^2+2\nu_1k_1^2]\overline{\beta}_2^{(1)}=0 \tag{i}$$

$$\kappa_n^2\sqrt{k_1^2+\kappa_n^2}\left(\overline{\alpha}_2^{(2)}-\overline{\alpha}_2^{(1)}\right)+2[(1+\nu_2)\kappa_n^2+\nu_2k_1^2]\overline{\beta}_2^{(2)}$$
$$+2[(1+\nu_1)\kappa_n^2+\nu_1k_1^2]\overline{\beta}_2^{(1)}=-\frac{i\xi_n k_1[2Q_ck_1^2+(Q_b+2Q_c)\kappa_n^2]}{k_1^2+\kappa_n^2} \tag{j}$$

$$\Delta\sigma_{12}^{A_nV}=\Delta\sigma_{12}^{A_n\infty}\Rightarrow$$

$$\sqrt{k_1^2+\kappa_n^2}\left(\overline{\alpha}_2^{(2)}-\overline{\alpha}_2^{(1)}\right)+2(\nu_2\overline{\beta}_2^{(2)}+\nu_1\overline{\beta}_2^{(1)})$$
$$=-\frac{i\xi_n[Q_bk_1^2+(Q_b-Q_c)\kappa_n^2]}{k_1(k_1^2+\kappa_n^2)} \tag{k}$$

$$\sqrt{k_1^2+\kappa_n^2}\left(\overline{\alpha}_2^{(2)}+\overline{\alpha}_2^{(1)}\right)+(1+2\nu_2)\overline{\beta}_2^{(2)}-(1+2\nu_1)\overline{\beta}_2^{(1)}=0 \tag{l}$$

$$\Delta\sigma_{13}^{A_nV}=\Delta\sigma_{13}^{A_n\infty}\Rightarrow$$

$$\sqrt{k_1^2+\kappa_n^2}\left(\overline{\alpha}_2^{(2)}+\overline{\alpha}_2^{(1)}\right)+\overline{\beta}_2^{(2)}-\overline{\beta}_2^{(1)}=0 \tag{m}$$

$$\sqrt{k_1^2+\kappa_n^2}\left(\overline{\alpha}_2^{(2)}-\overline{\alpha}_2^{(1)}\right)+2(\overline{\beta}_2^{(2)}+\overline{\beta}_2^{(1)})$$
$$=\frac{i\xi_n[(Q_c-2Q_b)k_1^2+(Q_c-Q_b)\kappa_n^2]}{k_1(k_1^2+\kappa_n^2)} \tag{n}$$

$$\Delta\sigma_{23}^{A_nV} = \Delta\sigma_{23}^{A_n\infty} \Rightarrow$$

$$\sqrt{k_1^2+\kappa_n^2}\left(\overline{\alpha}_2^{(2)} - \overline{\alpha}_2^{(1)}\right) + 2(\nu_2\overline{\beta}_2^{(2)} + \nu_1\overline{\beta}_2^{(1)}) = -\frac{iQ_c\xi_n k_1}{k_1^2+\kappa_n^2} \quad (o)$$

$$\sqrt{k_1^2+\kappa_n^2}\left(\overline{\alpha}_2^{(2)} + \overline{\alpha}_2^{(1)}\right) + (1+2\nu_2)\overline{\beta}_2^{(2)} - (1+2\nu_1)\overline{\beta}_2^{(1)} = 0 \quad (p) \qquad (5.2.15)$$

Next, we are concerned with satisfying boundary conditions: (5.2.14) leads to the displacement and stress fields due to an interface straight edge dislocation ($\vec{b} = (0, b, 0)$) parallel to the x_3-direction at the origin; the interface is the Ox_1x_3-plane. (5.2.15) provides the complementary terms (to first order in ξ_n) in the elastic fields of an interfacial sinusoidal edge dislocation.

5.2.3 Elastic Fields Due to an Interface Straight Edge Dislocation

Partial Elastic Fields Associated with Interface Boundary Conditions

Three distinct couples of values for $(\overline{\alpha}_2^{(m)}, \overline{\beta}_2^{(m)})$ are extracted from (5.2.14); these are:

$$\begin{aligned}
&(a)\,\overline{\alpha}_2^{(m)} = \frac{\nu_m C_m Q_a}{k_1^3} \equiv \overline{\alpha}_{2a}^{(m)};\ \overline{\beta}_2^{(m)} \\
&\quad = (-1)^{m-1}\frac{C_m Q_a}{2}\frac{\operatorname{sgn}(k_1)}{k_1^2} \equiv \overline{\beta}_{2a}^{(m)} \\
&(b)\,\overline{\alpha}_2^{(m)} = \frac{2\nu_m Q_b}{k_1^3} \equiv \overline{\alpha}_{2b}^{(m)};\ \overline{\beta}_2^{(m)} \\
&\quad = (-1)^{m-1} Q_b \frac{\operatorname{sgn}(k_1)}{k_1^2} \equiv \overline{\beta}_{2b}^{(m)} \\
&(c)\,\overline{\alpha}_2^{(m)} = \frac{\overline{V}_c}{k_1^3} \equiv \overline{\alpha}_{2c},\ \text{independent of } m;\ \overline{\beta}_2^{(m)} \\
&\quad = (-1)^{m-1}\frac{Q_c}{\nu_m}\frac{\operatorname{sgn}(k_1)}{k_1^2} \equiv \overline{\beta}_{2c}^{(m)}
\end{aligned} \qquad (5.2.16)$$

where

$$Q_a = i(\nu_1 - \nu_2)/[(1-2\nu_2)(1-\nu_1) + (1-2\nu_1)(1-\nu_2)],$$
$$\overline{V}_c = Q_b + 2Q_c[1 - (\nu_1+\nu_2)/4\nu_1\nu_2].$$

None of these couples satisfies the entire conditions (5.2.14). For each couple, we display below the associated elastic fields $\vec{u}^{(0)(m)V}$ and $(\sigma)^{(0)(m)V}$ defined in Sect. 5.2.2. A superposition of these partial fields will provide the complete form of solution. The couple $(\overline{\alpha}_{2a}^{(m)}, \overline{\beta}_{2a}^{(m)})$ is obtained from (5.2.14a–d) associated with the displacement. We have at position $\vec{x} = (x_1, x_2, x_3)$ $(\vec{u}^{(0)(m)V} \equiv \vec{u}_a^{(0)(m)V}$, $(\sigma)^{(0)(m)V} \equiv (\sigma)_a^{(0)(m)V})$

$$
\begin{aligned}
u_{1a}^{(0)(m)V} &= \frac{iC_m Q_a}{2\mu_m}\left((-1)^{m-1}(1-2\nu_m)\ln|x_1|\delta_A(x_2) + \frac{x_2^2}{r^2}\right),\\
u_{2a}^{(0)(m)V} &= \frac{ibQ_a}{2\pi}\left(-\tan^{-1}\left(\frac{x_1}{|x_2|}\right) + (-1)^m\frac{1}{2(1-\nu_m)}\frac{x_1x_2}{r^2}\right),\\
\sigma_{11a}^{(0)(m)V} &= iC_mQ_a\left((-1)^{m-1}\left[\frac{x_1}{r^2} + \left(\overline{J}_1 - \frac{x_1}{r^2}\right)\delta_A(x_2)\right] - 2\mathrm{sgn}(x_2)\frac{x_1x_2^2}{r^4}\right),\\
\sigma_{22a}^{(0)(m)V} &= iC_mQ_a\left((-1)^{m-1}\left[\frac{x_1}{r^2} + \left(\overline{J}_1 - \frac{x_1}{r^2}\right)\delta_A(x_2)\right] + 2\mathrm{sgn}(x_2)\frac{x_1x_2^2}{r^4}\right),\\
\sigma_{33a}^{(0)(m)V} &= (-1)^{m-1}2i\nu_mC_mQ_a\left[\frac{x_1}{r^2} + \left(\overline{J}_1 - \frac{x_1}{r^2}\right)\delta_A(x_2)\right],\\
\sigma_{12a}^{(0)(m)V} &= (-1)^m iC_mQ_a\frac{x_2\left(x_2^2 - x_1^2\right)}{r^4};\\
\overline{J}_1 &= \int\limits_0^\infty \sin k_1x_1 \mathrm{d}k_1.
\end{aligned}
\tag{5.2.17}
$$

Here, δ_A has the following definition: $\delta_A(x_2) = 0$ when $x_2 \neq 0$ and $\delta_A(x_2) = 1$ when $x_2 = 0$; $r^2 = x_1^2 + x_2^2$: Constant terms are omitted in the displacement. It can easily be verified that $u_{1a}^{(0)(m)} \equiv u_1^{(0)(m)\infty} - u_{1a}^{(0)(m)V}$ have the equal expression with m on the interface $x_2 = 0$; but the stresses $\sigma_{12a}^{(0)(m)} \equiv \sigma_{12}^{(0)(m)\infty} - \sigma_{12a}^{(0)(m)V}$ and $\sigma_{22a}^{(0)(m)}$ defined in a similar manner exhibit different factors (with m) that multiply the equal spatial functions $x_2(x_1^2 - x_2^2)/r^4$ and $x_1(x_1^2 + 3x_2^2)/r^4$ respectively. The pair $(\overline{\alpha}_{2b}^{(m)}, \overline{\beta}_{2b}^{(m)})$ is obtained using (5.2.14e–h) associated with the stresses. This gives (using the similar notations)

$$
\begin{aligned}
u_{1b}^{(0)(m)V} &= \frac{iQ_b}{\mu_m}\left((-1)^{m-1}(1-2\nu_m)\ln|x_1|\delta_A(x_2) + \frac{x_2^2}{r^2}\right),\\
u_{2b}^{(0)(m)V} &= \frac{iQ_b}{\mu_m}\left(-2(1-\nu_m)\tan^{-1}\left(\frac{x_1}{|x_2|}\right) + (-1)^m\frac{x_1x_2}{r^2}\right),\\
\sigma_{11b}^{(0)(m)V} &= 2iQ_b\left((-1)^{m-1}\left[\frac{x_1}{r^2} + \left(\overline{J}_1 - \frac{x_1}{r^2}\right)\delta_A(x_2)\right] - 2\mathrm{sgn}(x_2)\frac{x_1x_2^2}{r^4}\right),\\
\sigma_{22b}^{(0)(m)V} &= 2iQ_b\left((-1)^{m-1}\left[\frac{x_1}{r^2} + \left(\overline{J}_1 - \frac{x_1}{r^2}\right)\delta_A(x_2)\right] + 2\mathrm{sgn}(x_2)\frac{x_1x_2^2}{r^4}\right),
\end{aligned}
$$

$$\sigma_{33b}^{(0)(m)V} = (-1)^{m-1}4i\nu_m Q_b\Big[\frac{x_1}{r^2} + \Big(\overline{J}_1 - \frac{x_1}{r^2}\Big)\delta_A(x_2)\Big],$$
$$\sigma_{12b}^{(0)(m)V} = (-1)^m 2iQ_b\frac{x_2\big(x_2^2 - x_1^2\big)}{r^4}. \tag{5.2.18}$$

It can be verified that $u_{1b}^{(0)(m)} \equiv u_1^{(0)(m)\infty} - u_{1b}^{(0)(m)V}$ are not identical with m on the interface; but the stresses $\sigma_{12b}^{(0)(m)} \equiv \sigma_{12}^{(0)(m)\infty} - \sigma_{12b}^{(0)(m)V}$ and $\sigma_{22b}^{(0)(m)}$ (similar notations) exhibit the identical expressions with m ($= 1$ and 2) with the equal factor $(C_1 + C_2)/2$ multiplying the spatial functions $x_2(x_1^2 - x_2^2)/r^4$ and $x_1(x_1^2 + 3x_2^2)/r^4$ respectively. The couple $(\overline{\alpha}_{2c}, \overline{\beta}_{2c}^{(m)})$ is calculated using (5.2.14g–j) associated with the stresses. This leads (similar notations apply)

$$u_{1c}^{(0)(m)V} = \frac{i}{\nu_m\mu_m}\Big((-1)^m(\nu_m\overline{V}_c - Q_c)\ln|x_1|\delta_A(x_2) + Q_c\frac{x_2^2}{r^2}\Big),$$
$$u_{2c}^{(0)(m)V} = -\frac{i}{\nu_m\mu_m}\Big([2(1-2\nu_m)Q_c + \nu_m\overline{V}_c]\tan^{-1}\Big(\frac{x_1}{|x_2|}\Big) + (-1)^{m-1}Q_c\frac{x_1x_2}{r^2}\Big),$$
$$\sigma_{11c}^{(0)(m)V} = \frac{2i}{\nu_m}\Big((-1)^{m-1}[(1+2\nu_m)Q_c - \nu_m\overline{V}_c]\Big[\frac{x_1}{r^2} + \Big(\overline{J}_1 - \frac{x_1}{r^2}\Big)\delta_A(x_2)\Big]$$
$$-2Q_c\mathrm{sgn}(x_2)\frac{x_1x_2^2}{r^4}\Big),$$
$$\sigma_{22c}^{(0)(m)V} = \frac{2i}{\nu_m}\Big((-1)^{m-1}[(1-2\nu_m)Q_c + \nu_m\overline{V}_c]\Big[\frac{x_1}{r^2} + \Big(\overline{J}_1 - \frac{x_1}{r^2}\Big)\delta_A(x_2)\Big]$$
$$+2Q_c\mathrm{sgn}(x_2)\frac{x_1x_2^2}{r^4}\Big),$$
$$\sigma_{33c}^{(0)(m)V} = (-1)^{m-1}4iQ_c\Big[\frac{x_1}{r^2} + \Big(\overline{J}_1 - \frac{x_1}{r^2}\Big)\delta_A(x_2)\Big],$$
$$\sigma_{12c}^{(0)(m)V} = \frac{2ix_2}{r^2}\Big((2Q_c - \overline{V}_c)\mathrm{sgn}(x_2) + (-1)^m\frac{Q_c}{\nu_m}\frac{(x_2^2 - x_1^2)}{r^2}\Big)$$
$$+ 2\pi i(2Q_c - \overline{V}_c)\delta(x_1)\delta_A(x_2) \tag{5.2.19}$$

where $\delta(x_1)$ is the well-known Dirac delta function in $\sigma_{12c}^{(0)(m)V}$ when $x_2 = 0$.

Boundary Conditions

We define the elastic fields $\vec{u}^{(0)(m)}(\vec{x})$ and $(\sigma)^{(0)(m)}(\vec{x})$ of an interface straight edge dislocation as

$$\vec{u}^{(0\times m)} = \vec{u}^{(0\times m)\infty} - \vec{u}^{(0\times m)W}$$
$$(\sigma)^{(0)(m)} = (\sigma)^{(0)(m)\infty} - (\sigma)^{(0)(m)W} \tag{5.2.20}$$

with

$$\vec{u}^{(0)(m)W} = \eta_a^{(m)} \vec{u}_a^{(0)(m)V} + \eta_b^{(m)} \vec{u}_b^{(0)(m)V} + \eta_c^{(m)} \vec{u}_c^{(0)(m)V}.$$
$$(\sigma)^{(0)(m)W} = \eta_a^{(m)} (\sigma)_a^{(0)(m)V} + \eta_b^{(m)} (\sigma)_b^{(0)(m)V} + \eta_c^{(m)} (\sigma)_c^{(0)(m)V} \tag{5.2.21}$$

Again $\vec{u}^{(0)(m)\infty}$ and $(\sigma)^{(0)(m)\infty}$ are due to a straight edge dislocation ($\vec{b} = (0, b, 0)$) parallel to the x_3-direction at the origin in an infinitely extended homogeneous medium (m); $\vec{u}_{atoc}^{(0)(m)V}$ and $(\sigma)_{atoc}^{(0)(m)V}$ are given in (5.2.17)–(5.2.19); $\eta_{atoc}^{(m)}$ are real to be determined by the requirement that the elastic fields satisfy the following conditions:

- $\vec{u}^{(0)(m)}(\vec{x})$ is continuous across the interface (we shall write this condition for the x_1-component).
- $\oint_\Gamma du_2^{(0)(m)} = b$ for a closed contour in x_1x_2 encircling the dislocation. We may take for Γ a square of side a centred at the origin and travelled in the direction of the corkscrew advancing in the positive x_3-direction.
- The stresses $\sigma_{ij}^{(0)(m)}$ are continuous at the crossing of the interface, i.e.

$$\sigma_{ij}^{(0)(1)}(x_1, x_2 = 0, x_3) = \sigma_{ij}^{(0)(2)}(x_1, x_2 = 0, x_3).$$

- $\vec{u}^{(0)(m)W}(\vec{x})$ vanish far from the interface (i.e. when $|x_2| \to \infty$).

All the stresses involved in $(\sigma)^{(0)(m)\infty}$ and $(\sigma)_{atoc}^{(0)(m)V}$ vanish at infinity. Under such conditions above, $\vec{u}^{(0)(m)}(\vec{x})$ and $(\sigma)^{(0)(m)}(\vec{x})$ correspond to an interface straight edge dislocation. Next, we express the quantities involved in the conditions above and proceed to satisfy these.

$$u_1^{(0)(1)}(x_1, x_2 = 0, x_3) = u_1^{(0)(2)}(x_1, x_2 = 0, x_3) \Rightarrow$$
$$\frac{1 - 2\nu_m}{2\mu_m}\Big\{C_m - \eta_a^{(m)}(-1)^{m-1}C_m i Q_a - \eta_b^{(m)}(-1)^{m-1}2iQ_b$$
$$-\eta_c^{(m)}(-1)^{m-1}\frac{2i\big(Q_c - \nu_m \overline{V}_c\big)}{\nu_m(1 - 2\nu_m)}\Big\} \equiv e_1;$$

$$\int_\Gamma du_2^{(0)(m)} = b \Rightarrow \eta_a^{(m)} iQ_a$$
$$+ \eta_b^{(m)}\frac{2iQ_b}{C_m} + \eta_c^{(m)}\frac{2\pi i}{b\mu_m}\left(\frac{2(1 - 2\nu_m)Q_c}{\nu_m} + \overline{V}_c\right) \equiv e_2;$$

$$\sigma_{22}^{(0)(1)} = \sigma_{22}^{(0)(2)} \Rightarrow C_m - \eta_a^{(m)}(-1)^{m-1}C_m iQ_a - \eta_b^{(m)}(-1)^{m-1}2iQ_b$$
$$- \eta_c^{(m)}(-1)^{m-1}2i\left(\frac{(1 - 2\nu_m)Q_c + \nu_m \overline{V}_c}{\nu_m}\right) \equiv e_3;$$

$$\sigma_{12}^{(0)(1)} = \sigma_{12}^{(0)(2)} \Rightarrow \eta_c^{(m)}$$
$$= \eta_c \equiv e_4;$$

$$\sigma_{33}^{(0)(1)} = \sigma_{33}^{(0)(2)}$$
$$\Rightarrow 2\nu_m\left\{C_m - \eta_a^{(m)}(-1)^{m-1}C_m iQ_a - \eta_b^{(m)}(-1)^{m-1}2iQ_b - \eta_c^{(m)}(-1)^{m-1}2i\frac{Q_c}{\nu_m}\right\}$$
$$\equiv e_5;$$

$$\sigma_{11}^{(0)(1)} = \sigma_{11}^{(0)(2)} \Rightarrow C_m - \eta_a^{(m)}(-1)^{m-1}$$
$$C_m iQ_a - \eta_b^{(m)}(-1)^{m-1}2iQ_b$$
$$- \eta_c^{(m)}(-1)^{m-1}2i\left(\frac{(1+2\nu_m)Q_c - \nu_m\overline{V}_c}{\nu_m}\right) \equiv e_6;$$

$\vec{u}^{(0)(m)W}(\vec{x})$ vanishes when $|x_2| \to \infty \Rightarrow$

$$\eta_a^{(m)}\nu_m C_m iQ_a - \eta_b^{(m)}\nu_m 2iQ_b + \eta_c^{(m)}2iQ_c = 0 \equiv e_7; \tag{5.2.22}$$

where all e_i are constant with $m = 1$ and 2; for the various stress conditions above, we restrict ourselves to terms with the spatial function $(1/x_1)$ only.

Satisfying Boundary Conditions

We are concerned with finding the appropriate expressions for $\eta_{a\text{ to }c}^{(m)}$ that satisfy the boundary conditions (5.2.22). We first recognized that $\sigma_{12}^{(0)(m)}$ (5.2.20) takes on the interface the very simple form involving only $\eta_c^{(m)}$

$$\sigma_{12}^{(0)(m)}(x_1, x_2 = 0, x_3) = -\eta_c^{(m)}2\pi i(2Q_c - \overline{V}_c)\delta(x_1). \tag{5.2.23}$$

$\sigma_{12}^{(0)(1)} = \sigma_{12}^{(0)(2)}$ on the interface leads to $\eta_c^{(m)} = \eta_c \equiv e_4$ constant with m. Several expressions for $\eta_c^{(m)}$ can be extracted from (5.2.22), but only one value leaves $\sigma_{12}^{(0)(m)}$ (5.2.23) unchanged on inverting the elastic constants. It is obtained as

$$\eta_c = \frac{2b\mu_1\mu_2[\mu_1(1-2\nu_2) - \mu_2(1-2\nu_1)]}{\left[\mu_1^2(3-4\nu_2) - \mu_2^2(3-4\nu_1)\right]2\pi i(2Q_c - \overline{V}_c)}. \tag{5.2.24}$$

This can be arrived at from different routes involving different equations in (5.2.22). Only one route is presented as follows. $\eta_c^{(2)}$ can be isolated using (1) e_3 and e_6 in (5.2.22), (2) e_1 and e_3 and, (3) e_1 and e_2; (1) = (2) and (1) = (3) provide, respectively, the following equations:

$$C_1 - \eta_a^{(1)}C_1 iQ_a - \eta_b^{(1)}2iQ_b$$
$$+ \eta_c\frac{2i\{\nu_1(\mu_2-\mu_1)\overline{V}_c + [-\mu_2 + \mu_1(1+2\nu_1-2\nu_2)]Q_c\}}{\nu_1[\mu_2(1-2\nu_1) - \mu_1(1-2\nu_2)]} = 0 \tag{5.2.25}$$

$$2(1-\nu_1)(1-2\nu_2)C_1 - b_1^*(C_1 - \eta_a^{(1)}C_1 iQ_a - \eta_b^{(1)}2iQ_b)$$

$$+\eta_c \frac{i}{\nu_1\mu_2}\big(2Q_c b_2^* - \nu_1 \overline{V}_c[\mu_2 - \mu_1(3-4\nu_2)]\big) = 0 \tag{5.2.26}$$

where

$$b_1^* = (1-\nu_1)(1-2\nu_2) + (1-\nu_2)(1-2\nu_1),$$
$$b_2^* = \mu_2(2 - 2\nu_1 - 3\nu_2 + 4\nu_1\nu_2) - \mu_1\nu_1(3-4\nu_2).$$

Both Eqs. (5.2.25) and (5.2.26) yield η_c (5.2.24). Using (5.2.24), Eqs. (5.2.25) and e_7 (5.2.22) allow $\eta_a^{(1)}$ and $\eta_b^{(1)}$ to be calculated; then to reach $\eta_a^{(2)}$ and $\eta_b^{(2)}$, we may associate to e_7 (5.2.22), the following Eq. (5.2.27) obtained in the same way as in (5.2.25):

$$C_2 + \eta_a^{(2)} C_2 i Q_a + \eta_b^{(2)} 2iQ_b + \eta_c \frac{2i\{\nu_2(\mu_2-\mu_1)\overline{V}_c + [\mu_1 - \mu_2(1-2\nu_1+2\nu_2)]Q_c\}}{\nu_2[\mu_1(1-2\nu_2) - \mu_2(1-2\nu_1)]} = 0. \tag{5.2.27}$$

We have

$$\eta_b^{(1)} - \eta_b^{(2)} = \frac{C_1+C_2}{C_1-C_2} = \frac{\mu_1(1-\nu_2)+\mu_2(1-\nu_1)}{\mu_1(1-\nu_2)-\mu_2(1-\nu_1)}$$
$$\eta_b^{(2)} = \frac{1}{C_2-C_1}\left(C_2 + \frac{2b\mu_1\mu_2(\mu_1-\mu_2)}{\pi\left[\mu_1^2(3-4\nu_2) - \mu_2^2(3-4\nu_1)\right]}\right). \tag{5.2.28}$$

$\eta_a^{(m)}$ depends on elastic constants, this is denoted as $\eta_a^{(m)} = \eta_a^{(m)}(\nu_1, \nu_2; \mu_1, \mu_2)$; by inverting the elastic constants, this becomes $\eta_a^{(m)}(\nu_2, \nu_1; \mu_2, \mu_1)$. We have the following results:

$$\eta_a^{(1)}(\nu_1, \nu_2; \mu_1, \mu_2) + \eta_a^{(2)}(\nu_2, \nu_1; \mu_2, \mu_1) = -\frac{i}{Q_a};$$
$$\eta_a^{(2)}(\nu_1, \nu_2; \mu_1, \mu_2) = \frac{1}{2iQ_a}\left(-1 + \frac{Nu}{De}\right) \tag{5.2.29}$$

where

$$Nu = 2b\mu_1(\nu_1(1-\nu_2)d_1^*\mu_1^2 + \nu_2(1-\nu_1)e_2^*\mu_2^2 + \mu_1\mu_2[\nu_1(1-\nu_2)e_1^* + \nu_2(1-\nu_1)d_2^*])$$
$$De = \pi(1-\nu_1)(\nu_1-\nu_2)(\nu_1 C_1 + \nu_2 C_2)\left[\mu_1^2(3-4\nu_2) - \mu_2^2(3-4\nu_1)\right]$$

in which

$$d_1^* = \nu_2 - 5\nu_1 + 8\nu_1\nu_2; \quad d_2^* = \nu_2 + 3\nu_1 - 8\nu_1\nu_2,$$

$$e_1^* = 5\nu_1 - \nu_2 - 8\nu_1^2; \quad e_2^* = -3\nu_1 - \nu_2 + 8\nu_1^2.$$

In summary, $\eta_{atoc}^{(m)}$ are determined by using (5.2.24), (5.2.28 and 5.2.29).

Perfect Elastic Fields

The elastic fields due to an interface straight edge dislocation ($\vec{b} = (0, b, 0)$) parallel to the x_3-direction at the origin, are given in (5.2.20 and 5.2.21) with values of $\eta_{atoc}^{(m)}$ calculated from (5.2.24), (5.2.28 and 5.2.29), respectively. We first display special values taken on the interface by quantities $\sigma_{12}^{(0)(m)}$ and $\sigma_{22}^{(0)(m)}$ that are frequently involved in the analyses of the propagation of the interface crack loaded in tension. $\sigma_{12}^{(0)(m)}$ is obtained from (5.2.23 and 5.2.24) as

$$\sigma_{12}^{(0)(m)}(x_1, x_2 = 0, x_3) = -\frac{2b\mu_1\mu_2[\mu_1(1-2\nu_2) - \mu_2(1-2\nu_1)]}{\left[\mu_1^2(3-4\nu_2) - \mu_2^2(3-4\nu_1)\right]}\delta(x_1), \quad (5.2.30)$$

$m = 1$and2. This quantity is unchanged by inverting the elastic constants. From (5.2.22), $\sigma_{22}^{(0)(m)}(x_1, x_2 = 0, x_3)$ is first written as

$$\sigma_{22}^{(0)(m)}(x_1, x_2 = 0, x_3) = \frac{e_3}{x_1}. \quad (5.2.31)$$

The calculation can be performed with $m = 1$. Introducing in (5.2.31) the value of $(C_1 - \eta_a^{(1)} C_1 iQ_a - \eta_b^{(1)} 2iQ_b)$ taken from (5.2.25) and making use of the value of η_c (5.2.24), we obtain

$$\sigma_{22}^{(0)(m)}(x_1, x_2 = 0, x_3) = \frac{4b\mu_1\mu_2[\mu_1(1-\nu_2) - \mu_2(1-\nu_1)]}{\pi\left[\mu_1^2(3-4\nu_2) - \mu_2^2(3-4\nu_1)\right]}\frac{1}{x_1} \quad (5.2.32)$$

$\sigma_{22}^{(0)(m)}$ (5.2.32) is unchanged by inverting the elastic constants. Similarly, other elastic fields take on the interface, the following values:

$$u_1^{(0)(m)}(x_1, x_2 = 0, x_3) = -\frac{2b\mu_1\mu_2(\nu_1 - \nu_2)}{\pi\left[\mu_1^2(3-4\nu_2) - \mu_2^2(3-4\nu_1)\right]}\ln|x_1|$$

$$\sigma_{11}^{(0)(m)}(x_1, x_2 = 0, x_3) = -\frac{4b\mu_1\mu_2(\nu_1\mu_2 - \nu_2\mu_1)}{\pi\left[\mu_1^2(3-4\nu_2) - \mu_2^2(3-4\nu_1)\right]}\frac{1}{x_1}$$

$$\sigma_{33}^{(0)(m)}(x_1, x_2 = 0, x_3) = \frac{4b\nu_m\mu_1\mu_2(\mu_1 - \mu_2)}{\pi\left[\mu_1^2(3-4\nu_2) - \mu_2^2(3-4\nu_1)\right]}\frac{1}{x_1}. \quad (5.2.33)$$

5.2.4 Elastic Fields for an Interface Sinusoidal (Glide-Type) Edge Dislocation

Partial Elastic Oscillating Fields

Four values for $(\overline{\alpha}_2^{(m)}(\kappa_n), \overline{\beta}_2^{(m)}(\kappa_n))$ are extracted from (5.2.15); these are:

$$(a)\, \overline{\alpha}_2^{(m)}(\kappa_n) = \frac{\xi_n a_{1a}^{(m)}}{2} \frac{k_1}{\left(k_1^2 + \kappa_n^2\right)^{3/2}} \equiv \overline{\alpha}_{2a}^{A_2(m)}, \quad \overline{\beta}_2^{(m)}(\kappa_n) = 0 \equiv \overline{\beta}_{2a}^{A(m)};$$

$$(b)\, \overline{\alpha}_2^{(m)}(\kappa_n) = (-1)^{m-1} \frac{iQ_b \xi_n k_1}{4\sqrt{k_1^2 + \kappa_n^2}} \left(\frac{2}{k_1^2 + \kappa_n^2} - \frac{1}{k_1^2 + \rho_m \kappa_n^2} + \frac{1 - 4\nu_m}{k_1^2 + \nu_m \kappa_n^2} \right)$$

$$\equiv \overline{\alpha}_{2b}^{A_n(m)},$$

$$\overline{\beta}_2^{(m)}(\kappa_n) = -\frac{iQ_b \xi_n k_1}{2\left(k_1^2 + \nu_m \kappa_n^2\right)} \equiv \overline{\beta}_{2b}^{A(m)};$$

$$(c)\overline{\alpha}_2^{(m)}(\kappa_n) = \frac{i\xi_n}{2} \left(\frac{a_{1c}^{(m)}}{k_1\sqrt{k_1^2 + \kappa_n^2}} + \frac{a_{2c}^{(m)} k_1}{\left(k_1^2 + \kappa_n^2\right)^{3/2}} + \frac{a_{3c}^{(m)} k_1}{\left(k_1^2 + \theta\kappa_n^2\right)\sqrt{k_1^2 + \kappa_n^2}} \right)$$

$$\equiv \overline{\alpha}_{2c}^{A_n(m)},$$

$$\overline{\beta}_2^{(m)}(\kappa_n) = \frac{i\rho_m \xi_n}{2} \left(\frac{b_{1c}}{k_1} - \frac{b_{2c} k_1}{k_1^2 + \kappa_n^2} + \frac{b_{3c} k_1}{k_1^2 + \theta \kappa_n^2} \right) \equiv \overline{\beta}_{2c}^{A_n(m)};$$

$$(d)\, \overline{\alpha}_2^{(m)}(\kappa_n) = \frac{i\xi_n}{2} \left(\frac{a_{1d}^{(m)}}{k_1\sqrt{k_1^2 + \kappa_n^2}} + \frac{a_{2d}^{(m)} k_1}{\left(k_1^2 + \kappa_n^2\right)^{3/2}} \right) \equiv \overline{\alpha}_{2d}^{A_n(m)},$$

$$\overline{\beta}_2^{(m)}(\kappa_n) = \frac{i\xi_n \overline{b}^{(m)}}{2} \left(\frac{1}{k_1} + \frac{k_1}{k_1^2 + \kappa_n^2} \right) \equiv \overline{\beta}_{2d}^{A_n(m)} \tag{5.2.34}$$

where
$a_{1a}^{(m)} = (-1)^{m-1} \mu_m b C_\nu / 8\pi$, $\rho_m = \nu_1 \delta_{m2} + \nu_2 \delta_{m1}$ (δ_{ij}: Kronecker delta), $\theta = (\nu_1 + \nu_2)/2\nu_1\nu_2$;

$$a_{1c}^{(m)} = (Q_b - Q_c)\Big([\nu_2 - \nu_1 + (-1)^{m-1} 4\nu_1\nu_2] + (-1)^{m-1} 2(\nu_1 + \nu_2)\Big)/2(\nu_1 + \nu_2),$$
$$a_{2c}^{(m)} = \Big((Q_c - Q_b)[\nu_2 - \nu_1 + (-1)^{m-1} 4\nu_1\nu_2] + (-1)^{m-1} 4\nu_1\nu_2 Q_c(\theta - 1)\Big)/4\nu_1\nu_2(\theta - 1),$$
$$a_{3c}^{(m)} = [\nu_2 - \nu_1 + (-1)^{m-1} 4\nu_1\nu_2][Q_b - Q_c - 2Q_c\theta(\theta - 1)]/2(\nu_1 + \nu_2)(\theta - 1);$$

$$b_{1c} = (Q_b - Q_c)/(\nu_1 + \nu_2), \quad b_{2c} = (Q_b - Q_c)/2\nu_1\nu_2(\theta - 1),$$
$$b_{3c} = \big[Q_b - Q_c - 2Q_c\theta(\theta - 1)\big]/2\nu_1\nu_2\theta(\theta - 1);$$

$$a_{1d}^{(m)} = \frac{(Q_c - Q_b)(-\delta_{m1}[\nu_1 - \nu_2(1+4\nu_1)] + \delta_{m2}[\nu_2 - \nu_1(1+4\nu_2)])}{2[\nu_1(1-\nu_2)+\nu_2(1-\nu_1)]},$$
$$a_{2d}^{(m)} = \frac{1}{2[\nu_1(1-\nu_2)+\nu_2(1-\nu_1)]}\big(\delta_{m1}\big[Q_c(\nu_1+3\nu_2) + Q_b(\nu_1 - \nu_2(1+4\nu_1))\big]$$
$$-\delta_{m2}\big[Q_c(\nu_2+3\nu_1) + Q_b(\nu_2 - \nu_1(1+4\nu_2))\big]\big);$$

$$\overline{b}^{(m)} = \rho_m(Q_c - Q_b)/[\nu_1(1-\nu_2)+\nu_2(1-\nu_1)]. \tag{5.2.35}$$

Similarly to (5.2.16), none of these couples satisfies the entire conditions (5.2.15). For each couple, we give below the associated oscillating elastic fields $\vec{u}^{A_n(m)V}$ and $(\sigma)^{A_n(m)V}$. A superposition of these partial fields will provide the complete form of solution (to first order in ξ_n). The couple $(\overline{\alpha}_{2a}^{A_n(m)}, \overline{\beta}_{2a}^{A_n(m)})$ is obtained from (5.2.14a–d) associated with the displacement. We have at position $\vec{x} = (x_1, x_2, x_3)$ ($\vec{u}^{A_n(m)V} \equiv \vec{u}_a^{A_n(m)V}$, $(\sigma)^{A_n(m)V} \equiv (\sigma)_a^{A_n(m)V}$):

$$u_{ia}^{A_n(m)V} = \frac{a_{1a}^{(m)}}{2\mu_m}\left\|\delta_{i1} + \delta_{i2} + \delta_{i3}\frac{\partial}{\partial x_3}\right\| A_n$$
$$\times\left(\frac{2\kappa_n K_1[\kappa_n r]}{r}\big(\delta_{i1}(-1)^m|x_2| + \delta_{i2}x_1\big) + (-1)^{m-1}\left\|\kappa_n^2\delta_{i1} + \delta_{i3}\frac{\partial}{\partial x_1}\right\|\Pi_1\right),$$
$$\sigma_{iia}^{A_n(m)V} = (-1)^m a_{1a}^{(m)} A_n\left((\delta_{i1}-\delta_{i2})\frac{\partial I_0}{\partial x_1} - \kappa_n^2(\delta_{i1}-\delta_{i3})\frac{\partial \Pi_1}{\partial x_1}\right),$$
$$\sigma_{12a}^{A_n(m)V} = -a_{1a}^{(m)} A_n\left(\mathrm{sgn}(x_2)\frac{\partial I_0}{\partial x_2} + 2\kappa_n^2 K_0[\kappa_n r]\right),$$
$$\sigma_{j3a}^{A_n(m)V} = a_{1a}^{(m)}\frac{\partial A_n}{\partial x_3}\left(\delta_{j1}(-1)^m(I_0 - \kappa_n^2\Pi_1) + \delta_{j2}\frac{2\kappa_n x_1 K_1}{r}\right); \tag{5.2.36}$$

$$\Pi_z = \int_{-\infty}^{\infty}\frac{e^{(-1)^m\sqrt{k_1^2+\kappa_n^2}x_2}}{k_1^2 + z\kappa_n^2}e^{ik_1x_1}\mathrm{d}k_1 \equiv \int_{-\infty}^{\infty}\frac{e^{-\sqrt{k_1^2+\kappa_n^2}|x_2|}}{k_1^2 + z\kappa_n^2}e^{ik_1x_1}\mathrm{d}k_1,$$

$$I_0 = \int_{-\infty}^{\infty} e^{(-1)^m\sqrt{k_1^2+\kappa_n^2}x_2}e^{ik_1x_1}\mathrm{d}k_1 \equiv \int_{-\infty}^{\infty} e^{-\sqrt{k_1^2+\kappa_n^2}|x_2|}e^{ik_1x_1}\mathrm{d}k_1 = \frac{2\kappa_n|x_2|}{r}K_1.$$

Terms in brackets $\|\ \|$ are operators acting on A_n and Π_1 (i.e. Π_z with $z = 1$), separately; $r^2 = x_1^2 + x_2^2$ and subscripts $i = 1$ to 3 and $j = 1$ and 2; $K_n[x]$ is the nth-order modified Bessel function usually so denoted and δ_{ij} is the Kronecker delta. We stress that the various integrations (such as in I_0 and Π_z) performed in the present study are given for spatial positions satisfying the condition $(-1)^m x_2 < 0$ (*i.e.* $(-1)^{m-1} = \mathrm{sgn}(x_2)$) with $m = 1$ when $x_2 > \xi_n \sin\kappa_n x_3$ (half-space 1) and $m = 2$ when $x_2 < \xi_n \sin\kappa_n x_3$ (half-space 2). However, this makes no difference in the elastic fields to first order in ξ_n. The pair $(\overline{\alpha}_{2b}^{A_n(m)}, \overline{\beta}_{2b}^{A_n(m)})$ is obtained using (5.2.14e–h) associated with the stresses. We have (using the similar notations)

$$
\begin{aligned}
u_{1b}^{A_n(m)V} &= -\frac{iQ_bA_n}{4\mu_m}\left(2\left\|2(1-\nu_m)+x_2\frac{\partial}{\partial x_2}\right\|I_0 - 2\kappa_n^2\Pi_1 + \rho_m\kappa_n^2\Pi_{\rho m}\right.\\
&\quad\left.-\nu_m\kappa_n^2\left\|3-4\nu_m+2x_2\frac{\partial}{\partial x_2}\right\|\Pi_{\nu m}\right),\\
u_{2b}^{A_n(m)V} &= \frac{iQ_bA_n}{4\mu_m}\left(\frac{2}{x_2}\left\|x_1+x_2^2\frac{\partial}{\partial x_1}\right\|I_0 - \frac{\partial^2\Pi_{\rho m}}{\partial x_1\partial x_2}\right.\\
&\quad\left.+\left\|2(1-\nu_m)\kappa_n^2x_2-(3-4\nu_m)\frac{\partial}{\partial x_2}\right\|\frac{\partial\Pi_{\nu m}}{\partial x_1}\right),\\
u_{3b}^{A_n(m)V} &= \frac{iQ_b}{4\mu_m}\frac{\partial A_n}{\partial x_3}\frac{\partial}{\partial x_1}\left(2\Pi_1 - \Pi_{\rho m} + \left\|3-4\nu_m+2x_2\frac{\partial}{\partial x_2}\right\|\Pi_{\nu m}\right),\\
\sigma_{iib}^{A_n(m)V} &= -\frac{iQ_bA_n}{2}\frac{\partial}{\partial x_1}\left(2\left\|2(\delta_{i1}+\nu_m\delta_{i3})+(\delta_{i1}-\delta_{i2})x_2\frac{\partial}{\partial x_2}\right\|I_0\right.\\
&\quad+2\kappa_n^2(-\delta_{i1}+\delta_{i3})\Pi_1+\kappa_n^2[\rho_m\delta_{i1}+(1-\rho_m)\delta_{i2}-\delta_{i3}]\Pi_{\rho m}\\
&\quad\left.+\kappa_n^2\left\|\nu_m\delta_{i1}+(1-\nu_m)\delta_{i2}+(3-4\nu_m^2)\delta_{i3}-2[\nu_m\delta_{i1}+(1-\nu_m)\delta_{i2}-\delta_{i3}]x_2\frac{\partial}{\partial x_2}\right\|\Pi_{\nu m}\right)\\
\sigma_{12b}^{A_n(m)V} &= -\frac{iQ_bA_n}{2}\left(4\mathrm{sgn}(x_2)\frac{\partial^2K_0}{\partial x_1^2}+2x_2\left\|(1-\nu_m)\kappa_n^2-\frac{\partial^2}{\partial x_1^2}\right\|I_0+\rho_m\kappa_n^2\frac{\partial\Pi_{\rho m}}{\partial x_2}\right.\\
&\quad\left.-\nu_m\kappa_n^2\left\|2(1-\nu_m)\kappa_n^2x_2+\frac{\partial}{\partial x_2}\right\|\Pi_{\nu m}\right),\\
\sigma_{13b}^{A_n(m)V} &= -\frac{iQ_b}{2}\frac{\partial A_n}{\partial x_3}\left(2\left\|2(1-\nu_m)+x_2\frac{\partial}{\partial x_2}\right\|I_0-2\kappa_n^2\Pi_1\right.\\
&\quad\left.+\rho_m\kappa_n^2\Pi_{\rho m}-\nu_m\kappa_n^2\left\|3-4\nu_m+2x_2\frac{\partial}{\partial x_2}\right\|\Pi_{\nu m}\right),\\
\sigma_{23b}^{A_n(m)V} &= \frac{iQ_b}{2}\frac{\partial A_n}{\partial x_3}\frac{\partial}{\partial x_1}\left(-4\mathrm{sgn}(x_2)K_0+2x_2I_0-\frac{\partial\Pi_{\rho m}}{\partial x_2}\right.\\
&\quad\left.+\left\|2(1-\nu_m)\kappa_n^2x_2+\frac{\partial}{\partial x_2}\right\|\Pi_{\nu m}\right).
\end{aligned}
\tag{5.2.37}
$$

The couple $(\overline{\alpha}_{2c}^{A_n(m)}, \overline{\beta}_{2c}^{A_n(m)})$ is calculated from (5.2.15i–l) associated with stresses. We obtain, with the similar notations,

$$
\begin{aligned}
u_{1c}^{A_n(m)V} &= \frac{iA_n}{2\mu_m}\left(-2b_{2c}\rho_m\kappa_n^2|x_2|K_0+\sum_{s=1}^{3}\left[(-1)^m a_{sc}^{(m)}+(-1)^{s-1}b_{sc}\rho_m\left\|1+x_2\frac{\partial}{\partial x_2}\right\|\right]I_0\right.\\
&\quad\left.+\kappa_n^2\left(\rho_mb_{2c}+(-1)^{m-1}a_{2c}^{(m)}\right)\Pi_1+\theta\kappa_n^2\left[(-1)^{m-1}a_{3c}^{(m)}-\rho_mb_{3c}\left\|1+x_2\frac{\partial}{\partial x_2}\right\|\right]\Pi_\theta\right),\\
u_{2c}^{A_n(m)V} &= \frac{iA_n}{2\mu_m}\left(\left[\left((-1)^{m-1}a_{2c}^{(m)}-2\rho_m(1-2\nu_m)b_{2c}\right)\frac{x_1}{x_2}+\rho_m\sum_{s=1}^{3}\left\|(-1)^sb_{sc}x_2\frac{\partial}{\partial x_1}\right\|\right]I_0\right.
\end{aligned}
$$

$$+2\kappa_n \mathrm{sgn}(x_1)\Big[-b_{1c}\rho_m\kappa_n^2 \mathrm{sgn}(x_2)x_2^2 + \Big(a_{1c}^{(m)} + (-1)^{m-1}2\rho_m(1-2\nu_m)b_{1c}\Big)\Big\| 1 + x_2\frac{\partial}{\partial x_2}\Big\|\Big]\overline{\overline{J}}$$
$$+\Big[\rho_m(\theta-1)b_{3c}\kappa_n^2 x_2 + \Big\|\Big((-1)^{m-1}a_{3c}^{(m)} + 2\rho_m(1-2\nu_m)b_{3c}\Big)\frac{\partial}{\partial x_2}\Big\|\Big]\frac{\partial \Pi_\theta}{\partial x_1}\Big),$$

$$u_{3c}^{A_n(m)V} = \frac{i}{2\mu_m}\frac{\partial A_n}{\partial x_3}\Big(b_{2c}\rho_m x_1 I_0 - 2\kappa_n \mathrm{sgn}(x_1)|x_2|\Big[(-1)^m a_{1c}^{(m)} + \rho_m b_{1c}\Big\| 2 + x_2\frac{\partial}{\partial x_2}\Big\|\Big]\overline{\overline{J}}$$
$$+\Big((-1)^{m-1}a_{2c}^{(m)} + \rho_m b_{2c}\Big)\frac{\partial \Pi_1}{\partial x_1} - \Big[(-1)^m a_{3c}^{(m)} + \rho_m b_{3c}\Big\| 1 + x_2\frac{\partial}{\partial x_2}\Big\|\Big]\frac{\partial \Pi_\theta}{\partial x_1}\Big),$$

$$\sigma_{11c}^{A_n(m)V} = -iA_n\Bigg(-\rho_m b_{2c}\kappa_n^2 x_1 I_0 + \Bigg[\sum_{s=1}^{3}\Big((-1)^{m-1}a_{sc}^{(m)} + \rho_m(1+2\nu_m)(-1)^s b_{sc}\Big)$$
$$+\rho_m|x_2|\Big\| b_{3c} - (b_{1c}-b_{2c})\mathrm{sgn}(x_2)\frac{\partial}{\partial x_2}\Big\|\Bigg]\frac{\partial I_0}{\partial x_1} - 4\nu_m\rho_m b_{1c}\kappa_n^3|x_2|\mathrm{sgn}(x_1)\overline{\overline{J}}$$
$$+\kappa_n^2\Big((-1)^m a_{2c}^{(m)} - \rho_m b_{2c}\Big)\frac{\partial \Pi_1}{\partial x_1}$$
$$-\kappa_n^2\Big[(-1)^{m-1}a_{3c}^{(m)}\theta + \rho_m b_{3c}\Big\| 2\nu_m - \theta\Big(1 + 2\nu_m + x_2\frac{\partial}{\partial x_2}\Big)\Big\|\Big]\frac{\partial \Pi_\theta}{\partial x_1}\Bigg),$$

$$\sigma_{22c}^{A_n(m)V} = iA_n\Bigg(\sum_{s=1}^{3}\Big[(-1)^{m-1}a_{sc}^{(m)} + (-1)^{s-1}b_{sc}\rho_m\Big\| 1 - 2\nu_m - x_2\frac{\partial}{\partial x_2}\Big\|\Big]\frac{\partial I_0}{\partial x_1}$$
$$+2\kappa_n^3 \mathrm{sgn}(x_1)|x_2|\Big[(-1)^{m-1}a_{1c}^{(m)} - \rho_m b_{1c}\Big\| 2\nu_m + x_2\frac{\partial}{\partial x_2}\Big\|\Big]\overline{\overline{J}}$$
$$+\kappa_n^2(1-\theta)\Big[(-1)^{m-1}a_{3c}^{(m)} + \rho_m b_{3c}\Big\| 1 - 2\nu_m - x_2\frac{\partial}{\partial x_2}\Big\|\Big]\frac{\partial \Pi_\theta}{\partial x_1}\Bigg),$$

$$\sigma_{33c}^{A_n(m)V} = -iA_n\Bigg(\rho_m\Big\| b_{2c}\kappa_n^2 x_1 + 2\nu_m\sum_{s=1}^{3}(-1)^s b_{sc}\frac{\partial}{\partial x_1}\Big\| I_0 - \kappa_n^2\Big((-1)^m a_{2c}^{(m)} - b_{2c}\rho_m\Big)\frac{\partial \Pi_1}{\partial x_1}$$
$$-2\kappa_n^3 \mathrm{sgn}(x_1)|x_2|\Big[(-1)^m a_{1c}^{(m)} + b_{1c}\rho_m\Big\| 2(1+\nu_m) + x_2\frac{\partial}{\partial x_2}\Big\|\Big]\overline{\overline{J}}$$
$$-\kappa_n^2\Big[(-1)^m a_{3c}^{(m)} + \rho_m b_{3c}\Big\| 1 + 2\nu_m(1-\theta) + x_2\frac{\partial}{\partial x_2}\Big\|\Big]\frac{\partial \Pi_\theta}{\partial x_1}\Bigg),$$

$$\sigma_{12c}^{A_n(m)V} = iA_n\Bigg(2\kappa_n^2\Big(-a_{2c}^{(m)} + (-1)^m 2\nu_m\rho_m b_{2c}\Big)K_0 + \Big[\rho_m\kappa_n^2(b_{1c} + b_{3c}(1-\theta))x_2$$
$$+\sum_{s=1}^{3}\Big\|\Big((-1)^m a_{sc}^{(m)} + 2\rho_m\nu_m(-1)^{s-1}b_{sc}\Big)\frac{\partial}{\partial x_2} + (-1)^s b_{sc}\rho_m x_2\frac{\partial^2}{\partial x_1^2}\Big\|\Big]I_0$$
$$+\theta\kappa_n^2\Big[\rho_m(\theta-1)b_{3c}\kappa_n^2 x_2 + \Big\|\Big((-1)^{m-1}a_{3c}^{(m)} - 2\rho_m\nu_m b_{3c}\Big)\frac{\partial}{\partial x_2}\Big\|\Big]\Pi_\theta\Bigg),$$

$$\sigma_{13c}^{A_n(m)V} = i\frac{\partial A_n}{\partial x_3}\Bigg(-2b_{2c}\rho_m\kappa_n^2|x_2|K_0 + \sum_{s=1}^{3}\Big[(-1)^m a_{sc}^{(m)} + \rho_m(-1)^{s-1}b_{sc}\Big\| 1 + x_2\frac{\partial}{\partial x_2}\Big\|\Big]I_0$$
$$+\kappa_n^2\Big(\rho_m b_{2c} + (-1)^{m-1}a_{2c}^{(m)}\Big)\Pi_1 + \theta\kappa_n^2\Big[(-1)^{m-1}a_{3c}^{(m)} - \rho_m b_{3c}\Big\| 1 + x_2\frac{\partial}{\partial x_2}\Big\|\Big]\Pi_\theta\Bigg),$$

$$\sigma_{23c}^{A_n(m)V} = i\frac{\partial A_n}{\partial x_3}\Bigg(\Bigg[\Big((-1)^{m-1}a_{2c}^{(m)} + 2\rho_m\nu_m b_{2c}\Big)\frac{x_1}{x_2} + \rho_m\sum_{s=1}^{3}\Big\|(-1)^s b_{sc}x_2\frac{\partial}{\partial x_1}\Big\|\Bigg]I_0$$
$$+ 2\kappa_n\mathrm{sgn}(x_1)\Big[-b_{1c}\rho_m\kappa_n^2\mathrm{sgn}(x_2)x_2^2 + \Big(a_{1c}^{(m)} + (-1)^m 2\rho_m\nu_m b_{1c}\Big)\Big\|1 + x_2\frac{\partial}{\partial x_2}\Big\|\Big]\overline{\overline{J}}$$
$$+\Big[\rho_m(\theta - 1)b_{3c}\kappa_n^2 x_2 + \Big\|\Big((-1)^{m-1}a_{3c}^{(m)} - 2\rho_m\nu_m b_{3c}\Big)\frac{\partial}{\partial x_2}\Big\|\Big]\frac{\partial \Pi_\theta}{\partial x_1}\Bigg)$$
$$\overline{\overline{J}} = \int_{|x_1|}^{\infty}\frac{K_1[\kappa_n\sqrt{x^2 + x_2^2}]}{\sqrt{x^2 + x_2^2}}\mathrm{d}x. \tag{5.2.38}$$

The pair $(\overline{\alpha}_{2d}^{A_n(m)}, \overline{\beta}_{2d}^{A_n(m)})$ is calculated from (5.2.15m–p) associated with stresses. In a similar way, we obtain

$$u_{1d}^{A_n(m)V} = \frac{iA_n}{2\mu_m}\Bigg(2\overline{b}^{(m)}\kappa_n^2|x_2|K_0 + \Big[(-1)^m(a_{1d}^{(m)} + a_{2d}^{(m)}) + 2\overline{b}^{(m)}\Big\|1 + x_2\frac{\partial}{\partial x_2}\Big\|\Big]I_0$$
$$-\kappa_n^2\Big((-1)^m a_{2d}^{(m)} + \overline{b}^{(m)}\Big)\Pi_1\Bigg),$$

$$u_{2d}^{A_n(m)V} = \frac{iA_n}{2\mu_m}\Bigg(\Big\|\Big((-1)^{m-1}a_{2d}^{(m)} + 2(1 - 2\nu_m)\overline{b}^{(m)}\Big)\frac{x_1}{x_2} - 2\overline{b}^{(m)}x_2\frac{\partial}{\partial x_1}\Big\|I_0$$
$$- 2\kappa_n\mathrm{sgn}(x_1)\Big[\overline{b}^{(m)}\kappa_n^2\mathrm{sgn}(x_2)x_2^2 + \Big((-1)^m 2(1 - 2\nu_m)\overline{b}^{(m)} - a_{1d}^{(m)}\Big)\Big\|1 + x_2\frac{\partial}{\partial x_2}\Big\|\Big]\overline{\overline{J}}\Bigg),$$

$$u_{3d}^{A_n(m)V} = \frac{i}{2\mu_m}\frac{\partial A_n}{\partial x_3}\Bigg(-\overline{b}^{(m)}x_1 I_0 + \Big((-1)^{m-1}a_{2d}^{(m)} - \overline{b}^{(m)}\Big)\frac{\partial \Pi_1}{\partial x_1}$$
$$- 2\kappa_n\mathrm{sgn}(x_1)|x_2|\Big[(-1)^m a_{1d}^{(m)} + \overline{b}^{(m)}\Big\|2 + x_2\frac{\partial}{\partial x_2}\Big\|\Big]\overline{\overline{J}}\Bigg),$$

$$\sigma_{11d}^{A_n(m)V} = iA_n\Bigg(\kappa_n^2\Big((-1)^{m-1}a_{2d}^{(m)} - \overline{b}^{(m)}\Big)\frac{\partial \Pi_1}{\partial x_1} + 4\overline{b}^{(m)}\nu_m\kappa_n^3\mathrm{sgn}(x_1)|x_2|\overline{\overline{J}}$$
$$+ \Big\|\Big((-1)^m(a_{1d}^{(m)} + a_{2d}^{(m)}) + 2(1 + 2\nu_m)\overline{b}^{(m)}\Big)\frac{\partial}{\partial x_1} - \overline{b}^{(m)}\Big[x_1\kappa_n^2 - 2x_2\frac{\partial^2}{\partial x_1\partial x_2}\Big]\Big\|I_0\Bigg),$$

$$\sigma_{22d}^{A_n(m)V} = iA_n\Bigg(\Big[(-1)^{m-1}(a_{1d}^{(m)} + a_{2d}^{(m)}) + 2\overline{b}^{(m)}\Big\|1 - 2\nu_m - x_2\frac{\partial}{\partial x_2}\Big\|\Big]\frac{\partial I_0}{\partial x_1}$$
$$+ 2\kappa_n^3\mathrm{sgn}(x_1)|x_2|\Big[(-1)^{m-1}a_{1d}^{(m)} - \overline{b}^{(m)}\Big\|2\nu_m + x_2\frac{\partial}{\partial x_2}\Big\|\Big]\overline{\overline{J}}\Bigg),$$

$$\sigma_{33d}^{A_n(m)V} = iA_n\Bigg(\overline{b}^{(m)}\Big\|\kappa_n^2 x_1 + 4\nu_m\frac{\partial}{\partial x_1}\Big\|I_0 + \kappa_n^2\Big((-1)^m a_{2d}^{(m)} + \overline{b}^{(m)}\Big)\frac{\partial \Pi_1}{\partial x_1}$$
$$+ 2\kappa_n^3\mathrm{sgn}(x_1)|x_2|\Big[(-1)^m a_{1d}^{(m)} + \overline{b}^{(m)}\Big\|2(1 + \nu_m) + x_2\frac{\partial}{\partial x_2}\Big\|\Big]\overline{\overline{J}}\Bigg),$$

$$\sigma_{12d}^{A_n(m)V} = iA_n\Bigg(2\kappa_n^2\Big((-1)^{m-1}2\nu_m\overline{b}^{(m)} - a_{2d}^{(m)}\Big)K_0 + \Big\|\Big((-1)^m(a_{1d}^{(m)} + a_{2d}^{(m)}) + 4\nu_m\overline{b}^{(m)}\Big)\frac{\partial}{\partial x_2}$$

$$+\overline{b}^{(m)}x_2\left[\kappa_n^2 - 2\frac{\partial^2}{\partial x_1^2}\right]\Big\| I_0\Big),$$

$$\sigma_{13d}^{A_n(m)V} = i\frac{\partial A_n}{\partial x_3}\Bigg(\Bigg[(-1)^m(a_{1d}^{(m)} + a_{2d}^{(m)}) + 2\overline{b}^{(m)}\left\| 1 + x_2\frac{\partial}{\partial x_2}\right\|\Bigg] I_0$$
$$-\kappa_n^2\Big((-1)^m a_{2d}^{(m)} + \overline{b}^{(m)}\Big)\Pi_1 + 2\overline{b}^{(m)}\kappa_n^2|x_2|K_0\Big),$$

$$\sigma_{23d}^{A_n(m)V} = i\frac{\partial A_n}{\partial x_3}\Bigg(\Bigg[(-1)^{m-1}a_{2d}^{(m)}\frac{x_1}{x_2} - 2\overline{b}^{(m)}\left\|\nu_m\frac{x_1}{x_2} + x_2\frac{\partial}{\partial x_1}\right\|\Bigg] I_0$$
$$+ 2\kappa_n\mathrm{sgn}(x_1)\Big[-\overline{b}^{(m)}\kappa_n^2\mathrm{sgn}(x_2)x_2^2 + \Big((-1)^m 2\nu_m\overline{b}^{(m)} + a_{1d}^{(m)}\Big)\left\|1 + x_2\frac{\partial}{\partial x_2}\right\|\Big]\overline{\overline{J}}\Bigg). \tag{5.2.39}$$

Boundary Conditions and Corresponding Elastic Fields

To first order with respect to the perturbation A_n (i.e. ξ_n), the elastic fields $\vec{u}^{(m)}(\vec{x})$ and $(\sigma)^{(m)}(\vec{x})$ of an interface sinusoidal edge dislocation, Fig. 5.4, as deduced from (5.2.3), (5.2.6), (5.2.13), and (5.2.20), may be written as

$$\vec{u}^{(m)} = \vec{u}^{(0)(m)} + \vec{u}^{A_n(m)}$$
$$(\sigma)^{(m)} = (\sigma)^{(0)(m)} + (\sigma)^{A_n(m)}; \tag{5.2.40}$$

$\vec{u}^{(0)(m)}$ and $(\sigma)^{(0)(m)}$ correspond to the fields of a straight edge dislocation lying on a planar interface; $\vec{u}^{A_n(m)}$ and $(\sigma)^{A_n(m)}$ are oscillating expressions proportional to either the sinusoid $A_n(x_3)$ or its spatial derivative $\partial A_n/\partial x_3$ in the forms

$$\vec{u}^{A_n(m)} = \vec{u}^{A_n(m)\infty} - \vec{u}^{A_n(m)W}$$
$$(\sigma)^{A_n(m)} = (\sigma)^{A_n(m)\infty} - (\sigma)^{A_n(m)W}; \tag{5.2.41}$$

with

$$\vec{u}^{A_n(m)W} = \eta_a^{A_n(m)}\vec{u}_a^{A_n(m)V} + \eta_b^{A_n(m)}\vec{u}_b^{A_n(m)V}$$
$$+ \eta_c^{A_n(m)}\vec{u}_c^{A_n(m)V} + \eta_d^{A_n(m)}\vec{u}_d^{A_n(m)V}$$
$$(\sigma)^{A_n(m)W} = \eta_a^{A_n(m)}(\sigma)_a^{A_n(m)V} + \eta_b^{A_n(m)}(\sigma)_b^{A_n(m)V}$$
$$+ \eta_c^{A_n(m)}(\sigma)_c^{A_n(m)V} + \eta_d^{A_n(m)}(\sigma)_d^{A_n(m)V}. \tag{5.2.42}$$

$\vec{u}_{atod}^{A_n(m)V}$ and $(\sigma)_{atod}^{A_n(m)V}$ are given in (5.2.36)–(5.2.39) [for $\vec{u}_{atod}^{A_n(m)\infty}$ and $(\sigma)_{atod}^{A_n(m)\infty}$, see (5.2.6)]; $\eta_{atod}^{A_n(m)}$ are real to be determined by the requirement that the elastic fields satisfy the following conditions:

- $\vec{u}^{(m)}$(hence $\vec{u}^{A_n(m)}$) is continuous across the interface. Actually we shall write this condition for the x_1 and x_3 components (i.e. $u_1^{A_n(m)}$ and $u_3^{A_n(m)}$).

- $\oint_\Gamma d\vec{u}^{(m)} = \vec{b}$ for a closed contour Γ encircling the dislocation. With the continuity condition above, imposed on $u_1^{(m)}$ and $u_3^{(m)}$, it is sufficient to satisfy this condition for the x_2-component $u_2^{(m)}$ only; Because $\oint_\Gamma du_2^{(0)(m)} = b$, we shall just have to satisfy the condition $\oint_\Gamma du_2^{A_n(m)} = 0$. Again, we may take for Γ a square of side a in x_1x_2-planes surrounding the dislocation and travelled in the direction of the corkscrew advancing in the positive x_3-direction.
- $\sigma_{ij}^{(m)}$ (hence $\sigma_{ij}^{A_n(m)}$) are continuous at the crossing of the interface, i.e.

  $$\sigma_{ij}^{A_n(1)}(P_s) = \sigma_{ij}^{A_n(2)}(P_s).$$

- $\vec{u}^{(m)W} = \vec{u}^{(0)(m)W} + \vec{u}^{A_n(m)W}$ (hence $\vec{u}^{A_n(m)W}$) vanish far from the interface (i.e. when $|x_2| \to \infty$). This condition is automatically fulfilled because $\vec{u}^{(0)(m)W}$ tends to zero when $|x_2| \to \infty$ and $u_{ij}^{A_n(m)V}$ ($i = 1$ to 3 and $j = a$ to d) equally [see their various expressions in (5.2.36)–(5.2.39)]

All the stresses involved in $(\sigma)^{A_n(m)\infty}$ and $(\sigma)^{A_n(m)V}_{atod}$ are vanishing at infinity. Under such conditions above, $\vec{u}^{(m)}(\vec{x})$ and $(\sigma)^{(m)}(\vec{x})$ correspond to an interface sinusoidal edge dislocation (up to terms of first order (zero and one) with respect to the perturbation A_n). Next, we express the quantities involved in the various conditions above and then proceed to satisfy these. Before displaying the equations describing the above boundary conditions, we stress the followings: (1) For point $P_S(x_1, x_2 = \xi_n \sin\kappa_n x_3, x_3)$ on the interface, expanding the elastic fields to terms of first order with respect to $x_2 = \xi$, i.e.

$$\vec{u}^{A_n(m)}(x_1, x_2 = \xi, x_3) = \vec{u}^{A_n(m)}(x_1, 0, x_3) + \frac{\partial \vec{u}^{A_n(m)}}{\partial x_2}(x_1, 0, x_3)\xi$$

$$(\sigma)^{A_n(m)}(x_1, x_2 = \xi, x_3) = (\sigma)^{A_n(m)}(x_1, 0, x_3) + \frac{\partial (\sigma)^{A_n(m)}}{\partial x_2}(x_1, 0, x_3)\xi,$$

reveals that second terms (on the right of the equations above) are proportional to ξ_n^2 because the elastic fields are proportional to $A_n = \xi_n \sin\kappa_n x_3$ or $\partial A_n/\partial x_3$. Hence continuity conditions for the elastic fields on the interface will be imposed on the first terms only, i.e.

$$\vec{u}^{A_n(1)}(x_1, x_2 = 0, x_3) = \vec{u}^{A_n(2)}(x_1, 0, x_3)$$
$$(\sigma)^{A_n(1)}(x_1, 0, x_3) = (\sigma)^{A_n(2)}(x_1, 0, x_3).$$

(2) In most cases, the elastic fields on the interface are exponentially decreasing functions with x_1; these are bounded functions. Under such conditions, we have considered their linear forms with respect to x_1 and posed the coefficients proportional to x_1 constant with m. However, for $\sigma_{12}^{A_n(m)}$ and $\sigma_{23}^{A_n(m)}$, in addition to bounded functions, there exist terms proportional to $K_0[\kappa_n|x_1|]$ and $K_1[\kappa_n|x_1|]$ (modified Bessel functions, unbounded); both (more precisely $K_1[\kappa_n|x_1|]/|x_1|$ and $K_0[\kappa_n|x_1|]$) appear

in $\sigma_{12}^{A_n(m)}$ and $K_1[\kappa_n|x_1|]$ only in $\sigma_{23}^{A_n(m)}$. For both stresses, we have imposed the coefficients of the unbounded functions constant with m separately; the coefficient of $K_0[\kappa_n|x_1|]$ in $\sigma_{12}^{A_n(m)}$ correspond to that of $K_1[\kappa_n|x_1|]$ in $\sigma_{23}^{A_n(m)}$. The contributions of their associated bounded functions are not considered. It is interesting to mention that there is no singularity of the type $1/x_1$ in $\sigma_{12}^{A_n(m)}$. Hence, this stress does not contribute to the crack extension force when a crack (i.e. a continuous distribution of infinitesimal dislocations) in the interface is loaded in tension. On the contrary, because the singularity in $K_1[\kappa_n|x_1|]$ is of the $1/x_1$ type, $\sigma_{23}^{A_n(m)}$ does (5.2.3). Corresponding to the condition $\oint_\Gamma \mathrm{d}u_2^{A_n(m)} = 0$, is written below the relation that cancels the coefficient of $K_1[\kappa_n a/2]$ only, in the mathematical expression of $\oint_\Gamma \mathrm{d}u_2^{A_n(m)}$; all other terms are proportional to the side a of the square Γ that may be neglected for small values of a. We may write:

$$\begin{aligned}
&u_1^{A_n(1)}(x_1, x_2 = 0, x_3) = u_1^{A_n(2)}(x_1, x_2 = 0, x_3) \Rightarrow\\
&\frac{1}{\mu_m}\Big\{\eta_a^{A_n(m)}(-1)^{m-1}2a_{1a}^{(m)} + \eta_b^{A_n(m)} iQ_b\,[2 + \nu_m(3 - 4\nu_m) - \rho_m]\\
&+ \eta_c^{A_n(m)} 2i\Big[\rho_m b_{2c} + (-1)^{m-1}a_{2c}^{(m)} + \theta\Big((-1)^{m-1}a_{3c}^{(m)} - \rho_m b_{3c}\Big)\Big]\\
&+\eta_d^{A_n(m)} 2i\Big[(-1)^{m-1}a_{2d}^{(m)} - \overline{b}^{(m)}\Big]\Big\} \equiv e_1^{A_n};
\end{aligned}$$

$$\begin{aligned}
&u_3^{A_n(1)} = u_3^{A_n(2)} \Rightarrow\\
&\frac{1}{\mu_m}\Big\{\eta_a^{A_n(m)}(-1)^{m-1}2ia_{1a}^{(m)} + \eta_b^{A_n(m)} Q_b\big[-2 + \sqrt{\rho_m} - \sqrt{\nu_m}(3 - 4\nu_m)\big]\\
&+ \eta_c^{A_n(m)} 2\Big[-\rho_m b_{2c} + (-1)^m a_{2c}^{(m)} + \sqrt{\theta}\Big((-1)^m a_{3c}^{(m)} + \rho_m b_{3c}\Big)\Big]\\
&+\eta_d^{A_n(m)} 2\Big[(-1)^m a_{2d}^{(m)} + \overline{b}^{(m)}\Big]\Big\} \equiv e_2^{A_n};
\end{aligned}$$

$$\begin{aligned}
&\oint_\Gamma \mathrm{d}u_2^{A_n(m)} = 0 \Rightarrow\\
&(-1)^m bC_\nu + \eta_a^{A_n(m)}(-1)^{m-1}bC_\nu + \eta_b^{A_n(m)}(-1)^{m-1}\frac{8\pi iQ_b}{\mu_m}\\
&+ \eta_c^{A_n(m)}\frac{8\pi i}{\mu_m}\Big[(-1)^m 2\rho_m(1 - 2\nu_m)b_{2c} + a_{2c}^{(m)}\Big]\\
&+ \eta_d^{A_n(m)}\frac{8\pi i}{\mu_m}\Big[(-1)^{m-1}2(1 - 2\nu_m)\overline{b}^{(m)} + a_{2d}^{(m)}\Big] \equiv e_3^{A_n};
\end{aligned}$$

$$\begin{aligned}
&\sigma_{11}^{A_n(1)} = \sigma_{11}^{A_n(2)} \Rightarrow \eta_a^{A_n(m)}(-1)^{m-1}2a_{1a}^{(m)} + \eta_b^{A_n(m)} iQ_b\big[2 - \rho_m^{3/2} - \nu_m^{3/2}\big]\\
&+ \eta_c^{A_n(m)} 2i[\rho_m b_{2c} + (-1)^{m-1}a_{2c}^{(m)}\\
&+ \sqrt{\theta}((-1)^{m-1}a_{3c}^{(m)}\theta + \rho_m b_{3c}(2\nu_m(1 - \theta) - \theta))]
\end{aligned}$$

$$+\eta_d^{A_n(m)}2i\Big[(-1)^{m-1}a_{2d}^{(m)}-\overline{b}^{(m)}\Big]\equiv e_4^{A_n};$$

$$\sigma_{22}^{A_n(1)}=\sigma_{22}^{A_n(2)}\Rightarrow\eta_b^{A_n(m)}Q_b\big[(1-\nu_m)\sqrt{\nu_m}+(1-\rho_m)\sqrt{\rho_m}\big]$$
$$+\eta_c^{A_n(m)}2(1-\theta)\sqrt{\theta}\Big[(-1)^m a_{3c}^{(m)}+(-1+2\nu_m)\rho_m b_{3c}\Big]$$
$$\equiv e_5^{A_n}$$

$$\sigma_{33}^{A_n(1)}=\sigma_{33}^{A_n(2)}\Rightarrow\eta_a^{A_n(m)}(-1)^{m-1}2a_{1a}^{(m)}$$
$$+\eta_b^{A_n(m)}iQ_b\Big[2-\sqrt{\rho_m}+(3-4\nu_m^2)\sqrt{\nu_m}\Big]$$
$$+\eta_c^{A_n(m)}2i\Big[\rho_m b_{2c}+(-1)^{m-1}a_{2c}^{(m)}+\sqrt{\theta}\Big((-1)^{m-1}a_{3c}^{(m)}-\rho_m b_{3c}(1+2\nu_m(1-\theta))\Big)\Big]$$
$$+\eta_d^{A_n(m)}2i\Big[(-1)^{m-1}a_{2d}^{(m)}-\overline{b}^{(m)}\Big]\equiv e_6^{A_n};$$

$$\sigma_{12}^{A_n(1)}=\sigma_{12}^{A_n(2)}\Rightarrow$$
$$-C_m+\eta_a^{A_n(m)}2a_{1a}^{(m)}+\eta_b^{A_n(m)}(-1)^{m-1}2iQ_b$$
$$+\eta_c^{A_n(m)}2i\sum_{s=1}^{3}\big[a_{sc}^{(m)}+(-1)^m 2\nu_m\rho_m(-1)^{s-1}b_{sc}\big]$$
$$+\eta_d^{A_n(m)}2i\Big[a_{1d}^{(m)}+a_{2d}^{(m)}+(-1)^m 4\nu_m\overline{b}^{(m)}\Big]\equiv e_{71}^{A_n},$$

$$-\nu_m C_m+\eta_a^{A_n(m)}2a_{1a}^{(m)}+\eta_b^{A_n(m)}(-1)^{m-1}2iQ_b$$
$$+\eta_c^{A_n(m)}2i\Big[a_{2c}^{(m)}+(-1)^{m-1}2\nu_m\rho_m b_{2c}\Big]$$
$$+\eta_d^{A_n(m)}2i\Big[a_{2d}^{(m)}+(-1)^m 2\nu_m\overline{b}^{(m)}\Big]\equiv e_{72}^{A_n};$$

$$\sigma_{13}^{A_n(1)}=\sigma_{13}^{A_n(2)}\Rightarrow\eta_a^{A_n(m)}(-1)^{m-1}2a_{1a}^{(m)}$$
$$+\eta_b^{A_n(m)}iQ_b[2+\nu_m(3-4\nu_m)-\rho_m]$$
$$+\eta_c^{A_n(m)}2i\Big[\rho_m b_{2c}+(-1)^{m-1}a_{2c}^{(m)}+\theta\Big((-1)^{m-1}a_{3c}^{(m)}-\rho_m b_{3c}\Big)\Big]$$
$$+\eta_d^{A_n(m)}2i\Big[(-1)^{m-1}a_{2d}^{(m)}-\overline{b}^{(m)}\Big]\equiv e_8^{A_n};$$

$$\sigma_{23}^{A_n(1)}=\sigma_{23}^{A_n(2)}\Rightarrow$$

$$\text{expression }\sigma_{72}^{4}\text{ above;}\qquad(5.2.43)$$

where all $e_i^{A_n}(m)$ are constant with $m = 1$ and 2 (i.e. $e_i^{A_n}(1) = e_i^{A_n}(2)$). The Eqs. (5.2.43) have eight unknowns $\eta_i^{A_n(m)}$ ($i = a$ to d; $m = 1$ and 2). The methodology of the solution may be: (i) express the $\eta_i^{A_n(2)}$ as a function of the $\eta_i^{A_n(1)}$ giving four relations, (ii) report these relations in four independent equations in (5.2.43); we have then a linear system of four equations with unknowns $\eta_i^{A_n(1)}$ that can be resolved by the usual classical method with determinants. The result is four expressions linking the $\eta_i^{A_n(1)}$ with the elastic constants of the mediums $m = 1$ and 2. (iii) Come back to the $\eta_i^{A_n(2)}$ relations to find their respective values as a function of the elastic constants. The pair $(i\mu_2 e_1^{A_n}(1) - \mu_2 e_2^{A_n}(1); e_5^{A_n}(1))$ is a linear system of two equations with unknowns $\eta_b^{A_n(2)}$ and $\eta_c^{A_n(2)}$ that provides relations of the formers with respect to the $\eta_i^{A_n(1)}$. The result is that the relations for both $\eta_b^{A_n(2)}$ and $\eta_c^{A_n(2)}$ only depend on $\eta_b^{A_n(1)}$ and $\eta_c^{A_n(1)}$. These are introduced in the pair $(\mu_2 e_1^{A_n}(1); e_3^{A_n}(1))$ that becomes a linear system of two equations with two unknowns $\eta_a^{A_n(2)}$ and $\eta_d^{A_n(2)}$, the resolution of which, provide the relations for these parameters. It is found that both parameters depend on all the $\eta_i^{A_n(1)}$ ($i = a$ to d). Hence, we can write

$$\begin{pmatrix} \eta_a^{A_n(2)} \\ \eta_b^{A_n(2)} \\ \eta_c^{A_n(2)} \\ \eta_d^{A_n(2)} \end{pmatrix} = \begin{pmatrix} A \\ 0 \\ 0 \\ D \end{pmatrix} + \begin{pmatrix} A_a & A_b & A_c & A_d \\ 0 & B_b & B_c & 0 \\ 0 & C_b & C_c & 0 \\ D_a & D_b & D_c & D_d \end{pmatrix} \begin{pmatrix} \eta_a^{A_n(1)} \\ \eta_b^{A_n(1)} \\ \eta_c^{A_n(1)} \\ \eta_d^{A_n(1)} \end{pmatrix} \tag{5.2.44}$$

where A and D and the elements of the 4×4 matrix depend on the elastic constants of both media; these are listed in the Appendix. Next, the following notations are used:

$$e_i^{A_n}(m) = t_i^{(m)} + t_{ia}^{(m)} \eta_a^{A_n(m)} + t_{ib}^{(m)} \eta_b^{A_n(m)} + t_{ic}^{(m)} \eta_c^{A_n(m)} + t_{id}^{(m)} \eta_d^{A_n(m)}. \tag{5.2.45}$$

The comparison of (5.2.45) with (5.2.43) gives without ambiguity the values of the coefficients $t_i^{(m)}$ and $t_{ij}^{(m)}$ required by the resolution methodology. The general $e_i^{A_n}(2) = e_i^{A_n}(1)$ becomes

$$\begin{aligned} &\eta_a^{A_n(1)}\left(t_{ia}^{(2)} A_a + t_{id}^{(2)} D_a - t_{ia}^{(1)}\right) \\ &+ \eta_b^{A_n(1)}\left(t_{ia}^{(2)} A_b + t_{ib}^{(2)} B_b + t_{ic}^{(2)} C_b + t_{id}^{(2)} D_b - t_{ib}^{(1)}\right) \\ &+ \eta_c^{A_n(1)}\left(t_{ia}^{(2)} A_c + t_{ib}^{(2)} B_c + t_{ic}^{(2)} C_c + t_{id}^{(2)} D_c - t_{ic}^{(1)}\right) \\ &+ \eta_d^{A_n(1)}\left(t_{ia}^{(2)} A_d + t_{id}^{(2)} D_d - t_{id}^{(1)}\right) \\ &= t_i^{(1)} - \left(t_i^{(2)} + t_{ia}^{(2)} A + t_{id}^{(2)} D\right). \end{aligned} \tag{5.2.46}$$

For $i = 1$ to 4, we arrive at a linear nonhomogeneous system of four equations with unknowns $\eta_i^{A_n(1)}$ ($i = a$ to d), to be resolved to provide values to $\eta_i^{A_n(1)}$. The system has the form

$$\begin{cases} a_{11}\eta_a^{A_n(1)} + a_{12}\eta_b^{A_n(1)} + a_{13}\eta_c^{A_n(1)} + a_{14}\eta_d^{A_n(1)} = b_1 \\ a_{21}\eta_a^{A_n(1)} + a_{22}\eta_b^{A_n(1)} + a_{23}\eta_c^{A_n(1)} + a_{24}\eta_d^{A_n(1)} = b_2 \\ a_{31}\eta_a^{A_n(1)} + a_{32}\eta_b^{A_n(1)} + a_{33}\eta_c^{A_n(1)} + a_{34}\eta_d^{A_n(1)} = b_3 \\ a_{41}\eta_a^{A_n(1)} + a_{42}\eta_b^{A_n(1)} + a_{43}\eta_c^{A_n(1)} + a_{44}\eta_d^{A_n(1)} = b_4 \end{cases}. \tag{5.2.47}$$

In matrix notation, (5.2.47) is associated with a 4 × 4 matrix (A) with elements a_{ij} $(i,j = 1 \text{ to } 4)$ and (B) a 4 × 1 matrix with elements b_i $(i = 1 \text{ to } 4)$. Thus,

$$\begin{pmatrix} \eta_a^{A_n(1)} \\ \eta_b^{A_n(1)} \\ \eta_c^{A_n(1)} \\ \eta_d^{A_n(1)} \end{pmatrix} = (A^{-1}) \begin{pmatrix} b_1 \\ b_2 \\ b_3 \\ b_4 \end{pmatrix} \tag{5.2.48}$$

where (A^{-1}) is the inverse matrix of (A), with elements

$$(A^{-1})_{ij} = \frac{1}{\det(A)}(-1)^{i+j}\Delta_{ji}; \tag{5.2.49}$$

$\det(A)$ is a 4th-order determinant associated with matrix (A) and Δ_{ji}, a 3rd-order determinant, associated with a 3 × 3 matrix obtained from (A) by removing the row (j) and column (i) of element a_{ji}. With the help of relation $e_i^{A_n}(2) = e_i^{A_n}(1)$ (5.2.46), the following identification holds:

$$\begin{aligned} b_i &= t_i^{(1)} - \left(t_i^{(2)} + t_{ia}^{(2)}A + t_{id}^{(2)}D\right) \\ a_{i1} &= t_{ia}^{(2)}A_a + t_{id}^{(2)}D_a - t_{ia}^{(1)} \\ a_{i2} &= t_{ia}^{(2)}A_b + t_{ib}^{(2)}B_b + t_{ic}^{(2)}C_b + t_{id}^{(2)}D_b - t_{ib}^{(1)} \\ a_{i3} &= t_{ia}^{(2)}A_c + t_{ib}^{(2)}B_c + t_{ic}^{(2)}C_c + t_{id}^{(2)}D_c - t_{ic}^{(1)} \\ a_{i4} &= t_{ia}^{(2)}A_d + t_{id}^{(2)}D_d - t_{id}^{(1)}. \end{aligned} \tag{5.2.50}$$

We carefully choose the following four equations (E_i), $i = 1$ to 4:

$$\begin{aligned} &(E1): \quad e_{71}^{A}(2) - e_{72}^{A}(2) = e_{71}^{A}(1) - e_{72}^{A}(1) \\ &(E2): \quad e_{71}^{A}(2) = e_{71}^{A}(1) \\ &(E3): \quad e_5^{A_n}(2) = e_5^{A_n}(1) \\ &(E4): \quad e_1^{A_n}(2) = e_1^{A_n}(1); \end{aligned} \tag{5.2.51}$$

these lead to following identifications:

$$b_1 = C_2(1-\nu_2) - C_1(1-\nu_1) - 2i\left(a_{1d}^{(2)} + 2\nu_2\overline{b}^{(2)}\right)D,$$

$$a_{11} = 2i\left(a_{1d}^{(2)} + 2\nu_2\overline{b}^{(2)}\right)D_a,$$

$$a_{12} = 2i\left(a_{1c}^{(2)} + a_{3c}^{(2)} + 2\nu_1\nu_2(b_{1c} + b_{3c})\right)C_b + 2i\left(a_{1d}^{(2)} + 2\nu_2\overline{b}^{(2)}\right)D_b,$$

$$a_{13} = 2i\left(a_{1c}^{(2)} + a_{3c}^{(2)} + 2\nu_1\nu_2(b_{1c} + b_{3c})\right)C_c + 2i\left(a_{1d}^{(2)} + 2\nu_2\overline{b}^{(2)}\right)D_c - 2i\left(a_{1c}^{(1)} + a_{3c}^{(1)} - 2\nu_1\nu_2(b_{1c} + b_{3c})\right),$$

$$a_{14} = 2i\left(a_{1d}^{(2)} + 2\nu_2\overline{b}^{(2)}\right)D_d - 2i\left(a_{1d}^{(1)} - 2\nu_1\overline{b}^{(1)}\right);$$

$$b_2 = C_2 - C_1 - 2a_{1a}^{(2)}A - 2i\left(a_{1d}^{(2)} + a_{2d}^{(2)} + 4\nu_2\overline{b}^{(2)}\right)D,$$

$$a_{21} = 2a_{1a}^{(2)}A_a + 2i\left(a_{1d}^{(2)} + a_{2d}^{(2)} + 4\nu_2\overline{b}^{(2)}\right)D_a - 2a_{1a}^{(1)},$$

$$a_{22} = 2a_{1a}^{(2)}A_b - 2iQ_bB_b + 2i\sum_{s=1}^{3}\left[a_{sc}^{(2)} + 2\nu_1\nu_2(-1)^{s-1}b_{sc}\right]C_b + 2i\left(a_{1d}^{(2)} + a_{2d}^{(2)} + 4\nu_2\overline{b}^{(2)}\right)D_b - 2iQ_b,$$

$$a_{23} = 2a_{1a}^{(2)}A_c - 2iQ_bB_c + 2i\sum_{s=1}^{3}\left[a_{sc}^{(2)} + 2\nu_1\nu_2(-1)^{s-1}b_{sc}\right]C_c + 2i\left(a_{1d}^{(2)} + a_{2d}^{(2)} + 4\nu_2\overline{b}^{(2)}\right)D_c - 2i\sum_{s=1}^{3}\left[a_{sc}^{(1)} + 2\nu_1\nu_2(-1)^{s}b_{sc}\right],$$

$$a_{24} = 2a_{1a}^{(2)}A_d + 2i\left(a_{1d}^{(2)} + a_{2d}^{(2)} + 4\nu_2\overline{b}^{(2)}\right)D_d - 2i\left(a_{1d}^{(1)} + a_{2d}^{(1)} - 4\nu_1\overline{b}^{(1)}\right);$$

$$b_3 = 0,$$

$$a_{31} = 0,$$

$$\begin{aligned} a_{32} = {} & Q_b\big((1-\nu_2)\sqrt{\nu_2} + (1-\nu_1)\sqrt{\nu_1}\big)B_b \\ & + 2(1-\theta)\sqrt{\theta}\left(a_{3c}^{(2)} - (1-2\nu_2)\nu_1 b_{3c}\right)C_b \\ & - Q_b\big((1-\nu_2)\sqrt{\nu_2} + (1-\nu_1)\sqrt{\nu_1}\big), \end{aligned}$$

$$\begin{aligned} a_{33} = {} & Q_b\big((1-\nu_2)\sqrt{\nu_2} + (1-\nu_1)\sqrt{\nu_1}\big)B_c \\ & + 2(1-\theta)\sqrt{\theta}\left(a_{3c}^{(2)} - (1-2\nu_2)\nu_1 b_{3c}\right)C_c \\ & + 2(1-\theta)\sqrt{\theta}\left(a_{3c}^{(1)} + (1-2\nu_1)\nu_2 b_{3c}\right), \end{aligned}$$

$$a_{34} = 0;$$

$$b_4 = \frac{1}{\mu_2}\left\{2a_{1a}^{(2)}A + 2i\left(a_{2d}^{(2)} + \overline{b}^{(2)}\right)D\right\},$$

$$a_{41} = -\frac{1}{\mu_2}\left\{2a_{1a}^{(2)}A_a + 2i\left(a_{2d}^{(2)} + \overline{b}^{(2)}\right)D_a\right\} - 2\frac{a_{1a}^{(1)}}{\mu_1},$$

$$\begin{aligned} a_{42} = {} & \frac{1}{\mu_2}\Big\{-2a_{1a}^{(2)}A_b + iQ_b[2+\nu_2(3-4\nu_2)-\nu_1]B_b \\ & +2i\left[\nu_1 b_{2c} - a_{2c}^{(2)} - \theta\left(a_{3c}^{(2)} + \nu_1 b_{3c}\right)\right]C_b - 2i\left(a_{2d}^{(2)} + \overline{b}^{(2)}\right)D_b\Big\} \\ & - \frac{iQ_b}{\mu_1}(2+\nu_1(3-4\nu_1)-\nu_2), \end{aligned}$$

$$\begin{aligned} a_{43} = {} & \frac{1}{\mu_2}\Big\{-2a_{1a}^{(2)}A_c + iQ_b(2+\nu_2(3-4\nu_2)-\nu_1)B_c \\ & +2i\left(\nu_1 b_{2c} - a_{2c}^{(2)} - \theta\left(a_{3c}^{(2)} + \nu_1 b_{3c}\right)\right)C_c - 2i\left(a_{2d}^{(2)} + \overline{b}^{(2)}\right)D_c\Big\} \\ & - \frac{2i}{\mu_1}\left(\nu_2 b_{2c} + a_{2c}^{(1)} + \theta\left(a_{3c}^{(1)} - \nu_2 b_{3c}\right)\right), \end{aligned}$$

$$a_{44} = -\frac{1}{\mu_2}\left\{2a_{1a}^{(2)}A_d + 2i\left(a_{2d}^{(2)} + \overline{b}^{(2)}\right)D_d\right\} - \frac{2i}{\mu_1}\left(a_{2d}^{(1)} - \overline{b}^{(1)}\right). \tag{5.2.52}$$

(5.2.52) gives the matrices (A) and (B) and (5.2.48) the corresponding solutions $\eta_i^{A_n(1)}$ $(i = a$ to $d)$. It is then possible to reach the associated $\eta_j^{A_n(2)}$ $(j = a$ to $d)$ with the help of (5.2.44). The solution $\eta_i^{A_n(m)}$ $(i = a$ to d; $m = 1$ and $2)$ must satisfies the entire (5.2.43). Next, we shall restrict ourselves to demonstrate that we have effectively $e_5^{A_n}(1) = e_5^{A_n}(2)$ that only involves two parameters $\eta_i^{A_n(m)}(i = b$ and $c)$. We find

$$\eta_b^{A_n(1)} = \frac{a_{33}}{\det(A)} \left\{ \begin{array}{l} b_1(a_{41}a_{24} - a_{21}a_{44}) \\ +b_2(a_{11}a_{44} - a_{41}a_{14}) + b_4(a_{21}a_{14} - a_{11}a_{24}) \end{array} \right\}$$

$$\eta_c^{A_n(1)} = -\frac{a_{32}}{a_{33}}\eta_b^{A_n(1)}. \tag{5.2.53}$$

Using (5.2.44), we get

$$\eta_b^{A_n(2)} = \frac{1}{a_{33}}(a_{33}B_b - a_{32}B_c)\eta_b^{A_n(1)}$$

$$\eta_c^{A_n(2)} = \frac{1}{a_{33}}(a_{33}C_b - a_{32}C_c)\eta_b^{A_n(1)} \tag{5.2.54}$$

so that

$$\eta_c^{A_n(2)} = \frac{a_{33}C_b - a_{32}C_c}{a_{33}B_b - a_{32}B_c}\eta_b^{A_n(2)}. \tag{5.2.55}$$

It is then easy to show that

$$\begin{aligned} e_5^{A_n}(1) &= \Big(Q_b\big[(1-\nu_1)\sqrt{\nu_1} + (1-\nu_2)\sqrt{\nu_2}\big] \\ &\quad + \frac{2(1-\theta)\sqrt{\theta}a_{32}}{a_{33}}\Big[a_{3c}^{(1)} + (1-2\nu_1)\nu_2 b_{3c}\Big]\Big)\eta_b^{A_n(1)} \\ &= e_5^{A_n}(2). \end{aligned} \tag{5.2.56}$$

(5.2.56) is in favour of the rightness of our calculations. The associated fields $\vec{u}^{A_n(m)}$ and $(\sigma)^{A_n(m)}$ (5.2.41) must be continuous at the crossing of the interface, at $P_S = (x_1, x_2 = \xi, x_3)$, to first order in ξ_n. The continuity of $\sigma_{23}^{A_n(m)}$ on the interface permits to write the associated singularity term (to first order in ξ_n) as

$$\sigma_{23}^{A_n(m)}(x_1, 0, x_3) \equiv -\frac{\partial A_n}{\partial x_3}\frac{e_{72}^{A_n}}{x_1}. \tag{5.2.57}$$

(5.2.57) is useful because it contributes a non-zero value to the crack extension force when a non-planar interface crack with a wavy front is loaded in tension (see Anongba 1996, 2009; Anongba and Vitek 2003 for non-planar wavy crack propagation in a homogeneous medium, the analogous case). In summary, the elastic fields [$\vec{u}^{(m)}$ and $(\sigma)^{(m)}$: (5.2.40)] due to an interfacial sinusoidal edge dislocation (Fig. 5.4) to first order in the perturbation, is the sum of two terms: the first one corresponds to the elastic fields [$\vec{u}^{(0)(m)}$ and $(\sigma)^{(0)(m)}$: (5.2.20)] produced by an interface straight edge dislocation, lying on a planar interface, whose expressions have been provided in Sect. 3.1, under continuity requirement at the crossing of the interface; the second term is an oscillating expression [$\vec{u}^{A_n(m)}$,$(\sigma)^{A_n(m)}$: (5.2.41)], proportional to the perturbation $A_n(x_3) = \xi_n \sin\kappa_n x_3$ or its partial derivative $\partial A_n/\partial x_3$. This latter expression can be written as a linear combination (5.2.42) of partial elastic fields $\vec{u}_i^{A_n(m)V}$ ($i = a$ to d) (5.2.36)–(5.2.39). The associated proportionality coefficients $\eta_i^{A_n(m)}$ (5.2.44), (5.2.48) have been determined under the assumption that the elastic fields ($\vec{u}^{(m)}$, $(\sigma)^{(m)}$) must be continuous on crossing the sinusoidal interface.

Discussion and Confrontation with Previous Results

We would like first to mention that the values of the elastic fields [$\vec{u}^{(0)(m)}$ and $(\sigma)^{(0)(m)}$: (5.2.20)] on the interface points ($x_1, x_2 = 0, x_3$) are independent of $\overline{V}_c$ [see (5.2.30)–(5.2.33)]. This suggests that $\overline{V}_c$ could be set equal to zero with no change in the elastic fields; the associated couple [see (5.2.16)] ($\overline{\alpha}_{2c} \equiv 0, \overline{\beta}_{2c}^{(m)}$) corresponds to calculating $\overline{\beta}_2^{(m)}$ using (5.2.14i and 5.2.14j) and imposing $\overline{\alpha}_2^{(m)} \equiv 0$. Expressions for the displacement and stress fields of interface straight edge dislocations with their associated Airy stress functions have been given (Dundurs and Mura 1964; Mura 1968). In the geometry of the Fig. 5.4, the calculated shear stress $\sigma_{12}(x_1, x_2 = 0, x_3)$ is zero on the interface for an edge dislocation with Burgers vector $\vec{b} = (0, b, 0)$ perpendicular to the interface. Later, Comninou (1977a) has stressed that this shear stress contains a term proportional to the Dirac delta function; this result has been incorporated in several analyses of the propagation of the interface crack under load (Comninou 1977b; Comninou and Dundurs 1980; Anongba 2014a, b; among others). Incorporating a Dirac delta function in the value of the shear stress on the interface is a clear indication that the elasticity solutions given by Dundurs and Mura (1964) are partial. In the present study, a Dirac delta function is present in the complete elastic fields (Sect. 5.2.3) and pertains to the partial elastic fields $\vec{u}_c^{(0)(m)V}$ $(\sigma)_c^{(0)(m)V}$ only [see (5.2.19)]. This suggests that the results of the present study are potentially more general; however, discrepancies between expressions for $\sigma_{12}^{(0)(m)}(x_1, x_2 = 0, x_3)$ (5.2.30) and $\sigma_{22}^{(0)(m)}(x_1, x_2 = 0, x_3)$ (5.2.32) and those given below by Comninou and Dundurs (1980) are observed. In the geometry of the Fig. 5.4, their results are (solid $S1$):

$$\sigma_{12}^{(1)}(x_1, x_2 = 0, x_3) = -\frac{2b\mu_1\mu_2[\mu_1(1-2\nu_2) - \mu_2(1-2\nu_1)]}{[\mu_1 + \mu_2(3-4\nu_1)][\mu_2 + \mu_1(3-4\nu_2)]}\delta(x_1);$$

$$\sigma_{22}^{(1)}(x_1, x_2 = 0, x_3) = \frac{4b\mu_1\mu_2[\mu_1(1-2\nu_2) + \mu_2(1-2\nu_1)]}{\pi[\mu_1 + \mu_2(3-4\nu_1)][\mu_2 + \mu_1(3-4\nu_2)]}\frac{1}{x_1}.$$

Their shear stress $\sigma_{12}^{(1)}$ changes sign on inverting the elastic constants indicating that it is discontinuous on crossing the interface from solid $S1$ ($\sigma_{12}^{(1)}$) to solid $S2$ ($\sigma_{12}^{(2)}$). In contrast $\sigma_{12}^{(0)(m)}$ (5.2.30) is continuous at the crossing of the interface. This indicates an essential difference between the interface boundary conditions in both studies. Their $\sigma_{22}^{(1)}$ above is continuous across the interface like (5.2.32).

SUMMARY: In the present, Galerkin vectors have been used and associated interface boundary conditions have been displayed. These conditions can be decomposed into two groups: (1) the first group corresponds to a planar interface with a straight edge dislocation at the origin (5.2.13); (2) the second group corresponds to terms proportional to the sinusoid or its spatial derivative with respect to x_3 in the elastic fields expressed to first order with respect to the perturbation $A_n(x_3) = \xi_n \sin \kappa_n x_3$ (5.2.14). Satisfying both boundary conditions leads to terms of first order with respect to ξ_n in the elastic fields of a sinusoidal edge dislocation (see Fig. 5.4). The terms of zero order ($\vec{u}^{(0)(m)}$ and $(\sigma)^{(0)(m)}$) correspond to the elastic fields of an interface straight edge dislocation. These results contain the Dirac delta function in the shear stresses on the interface. We have compared these findings with those previously published on similar problems. The terms of first order are proportional to A_n or its partial derivative $\partial A_n/\partial x_3$. Among the additional oscillating stresses, $\sigma_{23}^{(m)}$ is the only one that possesses a singularity of the $1/x_1$ type on the interface. This stress contributes to the expression of the crack extension force when a non-planar interface crack with a wavy front is loaded in tension.

Appendix

The relation (5.2.44) expresses the $\eta_i^{A_n(2)}(i = a \text{ to } d)$ as linear combinations of the $\eta_i^{A_n(1)}$. The associated proportionality coefficients have been determined by a procedure expounded in the text. We obtain

$$A = \frac{\mu_2^2 t_{1d}^{(2)}\left(t_3^{(2)} - t_3^{(1)}\right)}{\Delta_{34}^{(2)}}, \quad A_a = \frac{\mu_2^2\left(t_{3d}^{(2)} t_{1a}^{(1)} - t_{1d}^{(2)} t_{3a}^{(1)}\right)}{\Delta_{34}^{(2)}},$$

$$A_d = \frac{\mu_2^2\left(t_{3d}^{(2)} t_{1d}^{(1)} - t_{1d}^{(2)} t_{3d}^{(1)}\right)}{\Delta_{34}^{(2)}},$$

$$\begin{aligned} A_b = \frac{\mu_2^2}{\mu_1 \Delta_{12}^{(2)} \Delta_{34}^{(2)}} & \left\{\mu_1 \Delta_{12}^{(2)}\left(t_{3d}^{(2)} t_{1b}^{(1)} - t_{1d}^{(2)} t_{3b}^{(1)}\right) + \left[\mu_1 R_{12}^{(2)} R_{21}^{(1)} - \mu_2 R_{22}^{(2)} R_{11}^{(1)}\right]\right. \\ & \left. \times\left(t_{3d}^{(2)} t_{1b}^{(2)} - t_{1d}^{(2)} t_{3b}^{(2)}\right) + \left[\mu_2 R_{21}^{(2)} R_{11}^{(1)} - \mu_1 R_{11}^{(2)} R_{21}^{(1)}\right]\left(t_{3d}^{(2)} t_{1c}^{(2)} - t_{1d}^{(2)} t_{3c}^{(2)}\right)\right\}, \end{aligned}$$

$$\begin{aligned} A_c = \frac{\mu_2^2}{\mu_1 \Delta_{12}^{(2)} \Delta_{34}^{(2)}} & \left\{\mu_1 \Delta_{12}^{(2)}\left(t_{3d}^{(2)} t_{1c}^{(1)} - t_{1d}^{(2)} t_{3c}^{(1)}\right) + [\mu_1 R_{12}^{(2)} R_{22}^{(1)} - \mu_2 R_{22}^{(2)} R_{12}^{(1)}]\right. \\ & \left. \times(t_{3d}^{(2)} t_{1b}^{(2)} - t_{1d}^{(2)} t_{3b}^{(2)}) + [\mu_2 R_{21}^{(2)} R_{12}^{(1)} - \mu_1 R_{11}^{(2)} R_{22}^{(1)}](t_{3d}^{(2)} t_{1c}^{(2)} - t_{1d}^{(2)} t_{3c}^{(2)})\right\}; \end{aligned}$$

$$B_b = \frac{\mu_2 R_{22}^{(2)} R_{11}^{(1)} - \mu_1 R_{12}^{(2)} R_{21}^{(1)}}{\mu_1 \Delta_{12}^{(2)}}, \quad B_c = \frac{\mu_2 R_{22}^{(2)} R_{12}^{(1)} - \mu_1 R_{12}^{(2)} R_{22}^{(1)}}{\mu_1 \Delta_{12}^{(2)}};$$

$$C_b = \frac{\mu_1 R_{11}^{(2)} R_{21}^{(1)} - \mu_2 R_{21}^{(2)} R_{11}^{(1)}}{\mu_1 \Delta_{12}^{(2)}}, \quad C_c = \frac{\mu_1 R_{11}^{(2)} R_{22}^{(1)} - \mu_2 R_{21}^{(2)} R_{12}^{(1)}}{\mu_1 \Delta_{12}^{(2)}};$$

$$D = \frac{\mu_2^2 t_{1a}^{(2)}\left(t_3^{(1)} - t_3^{(2)}\right)}{\Delta_{34}^{(2)}}, \quad D_a = \frac{\mu_2^2\left(t_{1a}^{(2)} t_{3a}^{(1)} - t_{3a}^{(2)} t_{1a}^{(1)}\right)}{\Delta_{34}^{(2)}},$$

$$D_d = \frac{\mu_2^2\left(t_{1a}^{(2)}t_{3d}^{(1)} - t_{3a}^{(2)}t_{1d}^{(1)}\right)}{\Delta_{34}^{(2)}},$$

$$D_b = -\frac{\mu_2^2}{\mu_1\Delta_{12}^{(2)}\Delta_{34}^{(2)}}\left\{\mu_1\Delta_{12}^{(2)}\left(t_{3a}^{(2)}t_{1b}^{(1)} - t_{1a}^{(2)}t_{3b}^{(1)}\right)4 + [\mu_1 R_{12}^{(2)}R_{21}^{(1)} - \mu_2 R_{22}^{(2)}R_{11}^{(1)}]\right.$$
$$\left.\times(t_{3a}^{(2)}t_{1b}^{(2)} - t_{1a}^{(2)}t_{3b}^{(2)}) + [\mu_2 R_{21}^{(2)}R_{11}^{(1)} - \mu_1 R_{11}^{(2)}R_{21}^{(1)}](t_{3a}^{(2)}t_{1c}^{(2)} - t_{1a}^{(2)}t_{3c}^{(2)})\right\},$$

$$D_c = -\frac{\mu_2^2}{\mu_1\Delta_{12}^{(2)}\Delta_{34}^{(2)}}\left\{\mu_1\Delta_{12}^{(2)}\left(t_{3a}^{(2)}t_{1c}^{(1)} - t_{1a}^{(2)}t_{3c}^{(1)}\right) + [\mu_1 R_{12}^{(2)}R_{22}^{(1)} - \mu_2 R_{22}^{(2)}R_{12}^{(1)}]\right.$$
$$\left.\times(t_{3a}^{(2)}t_{1b}^{(2)} - t_{1a}^{(2)}t_{3b}^{(2)}) + [\mu_2 R_{21}^{(2)}R_{12}^{(1)} - \mu_1 R_{11}^{(2)}R_{22}^{(1)}](t_{3a}^{(2)}t_{1c}^{(2)} - t_{1a}^{(2)}t_{3c}^{(2)})\right\};$$

where

$$\Delta_{12}^{(2)} = R_{11}^{(2)}R_{22}^{(2)} - R_{21}^{(2)}R_{12}^{(2)},$$
$$\Delta_{34}^{(2)} = \mu_2^2\left(t_{1a}^{(2)}t_{3d}^{(2)} - t_{3a}^{(2)}t_{1d}^{(2)}\right);$$

$$R_{11}^{(m)} = Q_b\left[\rho_m - \sqrt{\rho_m} - (3-4\nu_m)(\nu_m - \sqrt{\nu_m})\right],$$
$$R_{12}^{(m)} = 2(\theta - \sqrt{\theta})\left[(-1)^m a_{3c}^{(m)} + \rho_m b_{3c}\right],$$
$$R_{22}^{(m)} = 2(1-\theta)\sqrt{\theta}\left[(-1)^m a_{3c}^{(m)} - (1-2\nu_m)\rho_m b_{3c}\right],$$
$$R_{21}^{(m)} = Q_b\left[(1-\nu_m)\sqrt{\nu_m} + (1-\rho_m)\sqrt{\rho_m}\right],\ m = 1 \text{ and } 2.$$

We recall, for $t_{ij}^{(m)}$ and $t_i^{(m)}$, to see (5.2.45); all the various other parameters are defined previously in the text.

5.3 Non-planar Interface Cracks

We consider two elastic solids $R1$ and $R2$ (ν_m and μ_m, $m = 1$ and 2, being their Poisson's ratio and shear modulus, respectively) firmly welded along a non-planar interface R of arbitrary shape. The goal is to provide crack-tip stress and crack extension force expressions for a crack propagating along R. For large specimens and cracks, the following model (Fig. 5.5) in a Cartesian axis system x_i is investigated (Anongba 2018): the crack along the x_1 - axis (direction of fracture propagation) extends from $x_1 = -a$ to a; its shape in the x_2x_3 planes can be developed in the form of a series of Fourier

$$\xi = \sum_n (\xi_n \sin\kappa_n x_3 + \delta_n \cos\kappa_n x_3) \equiv \sum_n A_n \tag{5.3.1}$$

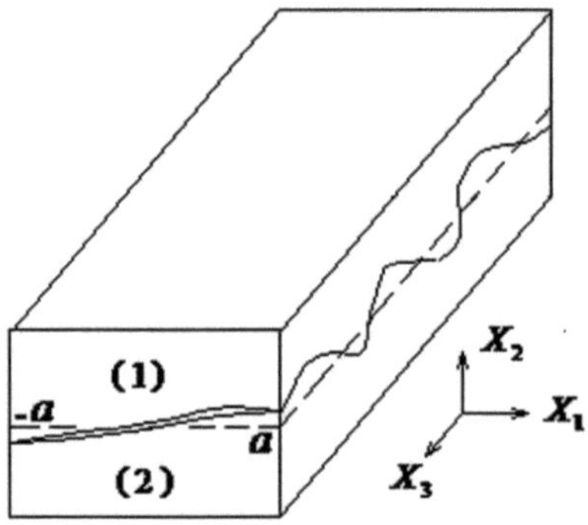

Fig. 5.5 Illustration of a crack front in two elastic solids (1) and (2) welded along a non-planar wavy surface that contains an interface crack. The crack fronts lie in x_2x_3-planes in the form ξ (5.3.1); in this geometry, the system is subjected to mixed mode $I + II + III$ loading with the applied tension in the x_2-direction. The average fracture surface (dashed) is shown perpendicular to that direction

where n is a positive integer; ξ_n, δ_n and κ_n are real numbers that depend on position x_1 along the crack length. The media ($m = 1$ and 2) are assumed to be infinitely large, and the crack front extends indefinitely in the x_3-direction.

The crack is represented by three dislocation families (I, II et III) whose common shape ξ (5.3.1) is perpendicular to x_1 with Burgers vectors $\vec{b}_I = (0, b, 0)$, $\vec{b}_{II} = (b, 0, 0)$ and $\vec{b}_{III} = (0, 0, b)$, parallel to the directions of the applied tension σ_{22}^a and shears σ_{12}^a and σ_{23}^a, respectively. In previous works (Anongba 2016 and 2017, reported in Sect. 5.4), elastic field expressions (displacement and stress) of sinusoidal interface dislocations (Fig. 5.6) perpendicular to x_1, of Burgers vectors $\vec{b}_I$, $\vec{b}_{II}$ and $\vec{b}_{III}$, were given; the fields corresponding to dislocations of form ξ (5.3.1) follow from these by superposition. It emerges from these studies that continuity of the elastic fields (displacement and stress) when crossing the interface between media (1) and (2) is not systematic. This is because there are linear relationships between some elastic field components with proportionality coefficients that depend on the elastic constants; when two components are concerned by such relations, the continuity of one at the passage of the interface can cause the discontinuity of the other. Here, we shall provide expressions for crack-tip stresses and crack extension force (per unit length of the crack front) for a non-planar crack system (Fig. 5.5), subjected to general loading, under the requirement that the stresses due to crack dislocations (involved in the crack analysis) be continuous at the crossing of the interface. In what follows we present the methodology for determining the elastic fields of dislocations and crack analysis. Then, are listed expressions of dislocation elastic fields, distribution functions of interface crack dislocations, crack-tip stresses and crack extension force. Special interface cracks are detailed. Discussion and conclusion end this report.

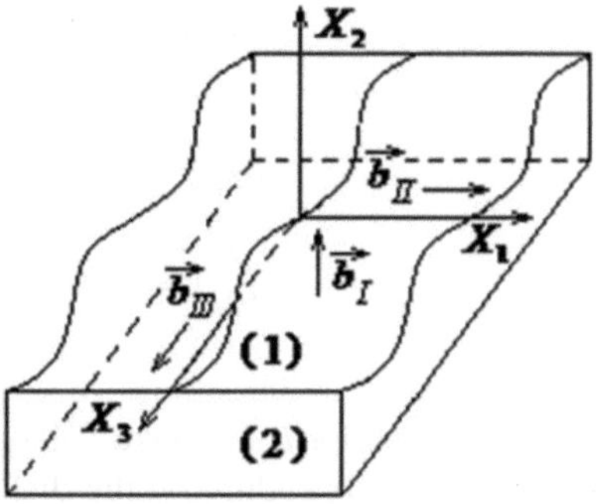

Fig. 5.6 Two elastic mediums (1) and (2) welded along a non-planar sinusoidal surface and containing an interface sinusoidal dislocation at the origin. The dislocation lies in the Ox_2x_3-plane and runs indefinitely in the x_3-direction. The Burgers vectors $\vec{b}_I$, $\vec{b}_{II}$, $\vec{b}_{III}$ are present

5.3.1 Methodology for Determining Elastic Fields of Crack Dislocations

The three types J ($J = I, II$ and III) of crack dislocation considered have equal shape ξ (5.3.1) in the Ox_2x_3-plane at the origin, Burgers vectors $\vec{b}_I = (0, b, 0), \vec{b}_{II} = (b, 0, 0)$ and $\vec{b}_{III} = (0, 0, b)$, and run indefinitely in the x_3-direction. Types I and II are of edge character on average and type III screw. The elastic fields (displacement $\vec{u}^{(J)(m)}$ and stress $(\sigma)^{(J)(m)}$) in the bi-material (media $m = 1$ and 2) may be deduced from those (sinusoidal dislocations, Fig. 5.6) with simple form $A_n = \xi_n \sin \kappa_n x_3$. For sinusoidal dislocations, the elastic fields are (to linear terms in amplitude ξ_n):

$$\vec{u}^{(J)(m)} = \vec{u}^{(J)(0)(m)} + \vec{u}^{(J)A_n(m)}$$
$$(\sigma)^{(J)(m)} = (\sigma)^{(J)(0)(m)} + (\sigma)^{(J)A_n(m)}.$$

Here $\vec{u}^{(J)(0)(m)}$ and $(\sigma)^{(J)(0)(m)}$ are of zero order; they correspond to the fields of interface straight dislocation (parallel to x_3) at the origin on Ox_1x_3 planar interface with Burgers vector $\vec{b}_J$; $\vec{u}^{(J)A_n(m)}$ and $(\sigma)^{(J)A_n(m)}$ are oscillating parts proportional to A_n or its spatial derivative $\partial A_n/\partial x_3$. When the dislocations exhibit shape ξ (5.3.1), the elastic fields take the form

$$\begin{aligned}
\vec{u}^{(J)(m)} &= \vec{u}^{(J)(0)(m)} + \vec{u}_\xi^{(J)(m)} \\
(\sigma)^{(J)(m)} &= (\sigma)^{(J)(0)(m)} + (\sigma)_\xi^{(J)(m)}; \\
\vec{u}_\xi^{(J)(m)} &= \sum_n \vec{u}^{(J)A_n(m)} \\
(\sigma)_\xi^{(J)(m)} &= \sum_n (\sigma)^{(J)A_n(m)}.
\end{aligned} \tag{5.3.2}$$

Here, A_n is that defined in (5.3.1). Because both displacement and stress have similar decompositions, we just write for the stress below; the description is equally valid for the displacement.

$$
\begin{aligned}
(\sigma)^{(J)(0)(m)} &= (\sigma)^{(J)(0)(m)\infty} - (\sigma)^{(J)(0)(m)W} \\
(\sigma)^{(J)A_n(m)} &= (\sigma)^{(J)A_n(m)\infty} - (\sigma)^{(J)A_n(m)W}; \\
(\sigma)^{(J)(0)(m)W} &= \sum_{l=a \text{ up to } e} \eta_l^{(J)(0)(m)} (\sigma)_l^{(J)(0)(m)V} \\
(\sigma)^{(J)A_n(m)W} &= \sum_{l=a \text{ up to } e} \eta_l^{(J)A_n(m)} (\sigma)_l^{(J)A_n(m)V}.
\end{aligned} \tag{5.3.3}
$$

The subscript l may take values a up to e (minimum two, a and b; maximum five, a to e) depending on the dislocation J. Expressions with ∞ ($(\sigma)^{(J)(0)(m)\infty} + (\sigma)^{(J)A_n(m)\infty} \equiv (\sigma)^{(J)(m)\infty}$) represent the fields of a sinusoidal dislocation with Burgers vector $\vec{b}_J$ in isotropic infinite medium (m) (Anongba 1995, 2009, 2010). Those with W ($(\sigma)^{(J)(0)(m)W} + (\sigma)^{(J)A_n(m)W} \equiv (\sigma)^{(J)(m)W}$) satisfy the equations of equilibrium and are constructed in such a way that the elastic fields of the interface sinusoidal dislocation are continuous at the crossing of the interface and tend to those due to the sinusoidal dislocation in an infinite medium far from the dislocation and interface. $(\sigma)^{(J)(m)W}$ is linear combination of fields with V [Anongba 2015a, b (Sect. 5.2), 2016, 2017) (i.e. $(\sigma)_l^{(J)(0)(m)V}$ and $(\sigma)_l^{(J)A_n(m)V}$] that we refer to as "partial elastic fields". They are obtained with the help of Galerkin vectors with three-dimensional biharmonic functions. The proportionality coefficients $\eta_l^{(J)(0)(m)}$ and $\eta_l^{(J)A_n(m)}$ are real numbers that are determined by the requirement that the elastic fields of the interface dislocations (that is to say $(\sigma)^{(J)(0)(m)}$ and $(\sigma)^{(J)A_n(m)}$) are continuous across the interface; we have considered the corresponding equations for spatial positions $P(x_1, x_2 = 0, x_3)$ on the average fracture plane only. When these elastic fields are bounded, we have required their linear terms proportional to x_1 to be constant with $m = 1$ and 2. When the elastic fields contain terms with singularities (of the types Dirac delta $\delta(x_1)$, $1/x_1$, $\ln(x_1)$...), the coefficient of each singularity is set constant with m. In this way, we get many coefficients denoted

$$
\begin{aligned}
e_j^{(J)(0)}(m) &= e_j^{(J)(0)}(\eta_l^{(J)(0)(m)}, \mu_m, \nu_m) \\
e_j^{(J)A_n}(m) &= e_j^{(J)A_n}(\eta_l^{(J)A_n(m)}, \mu_m, \nu_m)
\end{aligned} \tag{5.3.4}
$$

that depend on $\eta_l^{(J)(0)(m)}$, $\eta_l^{(J)A_n(m)}$ and the elastic constants μ_m shear modulus and ν_m Poisson's ratio. The subscript j may take values from minimum three ($j = 1$ to 3) up to maximum thirteen ($j = 1$ to 13). We mention that certain $e_j^{(J)(0)}(m)$ and $e_j^{(J)A_n}(m)$(in limited numbers) will not be constant with m because they are proportional to others, imposed constant with m. In the present, the components of dislocation stress fields that contribute a non-zero value to the crack extension force or are involved in the crack dislocation distribution functions will be imposed constant with m. These are stresses that contain singularities of the types $\delta(x_1)$ and $1/x_1$. A general approach is to impose on all terms with singularities in the elastic fields (displacement and stress) to be constant with m at the crossing of the interface. There are several ways to find suitable values for $\eta_l^{(J)(0)(m)}$ and $\eta_l^{(J)A_n(m)}$ that make most $e_j^{(J)(0)}(m)$ and $e_j^{(J)A_n}(m)$ constant with m. We can build a system of independent linear equations with as many equations as unknowns $\eta_l^{(J)(0)(m)}$ and $\eta_l^{(J)A_n(m)}$ that can be solved by the classical method with determinant, as performed in Sect. 5.2 (Anongba 2015b). There is another method that allows one to choose the elastic fields one wants imperatively continuous when crossing the interface. First, by direct inspection, linear combinations between the $e_j^{(J)(0)}(m)$ or $e_j^{(J)A_n}(m)$ are sought; then we impose, in progressive manner in number, to some of them to be continuous when crossing the interface. We thus succeed in constructing expressions of $\eta_l^{(J)(0)(m)}$ and $\eta_l^{(J)A_n(m)}$ which ensure an optimum continuity of the elastic fields at the interface (see Sect. 5.4, for example). Frequently, at different stages of this construction procedure, it is required that shear modules and Poisson's ratios of media (1) and (2) be different; hence, the elastic field expressions found in the present study are valid under the conditions $\mu_1 \neq \mu_2$ and $\nu_1 \neq \nu_2$.

5.3.2 *Crack Analysis*

The crack in Fig. 5.5 is assumed to be filled continuously over the interval x_1 = - a to a with three families J ($J = I, II$ and III) of dislocations with shape ξ (1) in x_2x_3-planes and Burgers vectors $\vec{b}_J$ ($\vec{b}_I = (0, b, 0)$, $\vec{b}_{II} = (b, 0, 0)$ and $\vec{b}_{III} = (0, 0, b)$). It is understood in our crack analysis that ξ_n, δ_n and κ_n [introduced in (5.3.1)] are position dependent along x_1 in the dislocation distributions. The surrounded media (m = 1 and 2) are taken infinite, isotropic, elastic and subjected to uniform remote applied loadings, tension σ_{22}^a and shears σ_{12}^a and σ_{23}^a, at infinity. In addition, we shall consider induced normal and shear stresses originating from the Poisson effect that acts perpendicularly to the tension x_2- direction. The first task is to determine the dislocation distribution functions D_J at equilibrium. D_J (x_1) gives the number of dislocations of family J in a small interval dx_1 about x_1 as D_J (x_1) dx_1. We generally arrive at a system of singular integral equations involving these functions; when the distribution functions of the dislocations have been found, we can obtain by integration the relative displacement of the faces of the crack, the crack-tip stress and the crack extension force. The surfing point on the interface is P_R (x_1, ξ, x_3). To find the equilibrium dislocation distributions, we may ask for zero total force on the crack faces; this gives

$$\begin{cases} \overline{\sigma}_{12}^{(m)} - \partial\xi/\partial x_1 \overline{\sigma}_{11}^{(m)} - \partial\xi/\partial x_3 \overline{\sigma}_{13}^{(m)} = 0 \\ \overline{\sigma}_{22}^{(m)} - \partial\xi/\partial x_1 \overline{\sigma}_{12}^{(m)} - \partial\xi/\partial x_3 \overline{\sigma}_{23}^{(m)} = 0 \\ \overline{\sigma}_{23}^{(m)} - \partial\xi/\partial x_1 \overline{\sigma}_{13}^{(m)} - \partial\xi/\partial x_3 \overline{\sigma}_{33}^{(m)} = 0 \end{cases} . \tag{5.3.5}$$

$\overline{\sigma}_{ij}^{(m)}$ stands for the total stress at any point (x_1, x_2, x_3) in the surrounding media (m = 1 and 2) and is linked to D_J. In (5.3.5), we are concerned with the points of the crack faces only. We write $\overline{\sigma}_{ij}^{(m)}$ as

$$\overline{\sigma}_{ij}^{(m)} = \sigma_{ij}^{A(m)} + \overline{\sigma}_{ij}^{(I)(m)} + \overline{\sigma}_{ij}^{(II)(m)} + \overline{\sigma}_{ij}^{(III)(m)}. \quad (5.3.6)$$

We define $\sigma_{ij}^{A(m)}$ and $\overline{\sigma}_{ij}^{(J)(m)}$ successively below. ($\sigma^{A(m)}$) includes the externally applied stresses and induced normal and shear stresses due to the Poisson effect.

$$(\sigma^{A(m)}) = \begin{pmatrix} -\nu_m\sigma_{22}^a & \sigma_{12}^a + \tau_{12}^a & 0 \\ \sigma_{12}^a + \tau_{12}^a & \sigma_{22}^a & \sigma_{23}^a + \tau_{23}^a \\ 0 & \sigma_{23}^a + \tau_{23}^a & -\nu_m\sigma_{22}^a \end{pmatrix}. \quad (5.3.7)$$

τ_{12}^a and τ_{23}^a are internal shearing stresses transmitted in the media by the interface as the result of the existence of internal uniform Poisson stresses ($-\nu_1\sigma_{22}^a$) and ($-\nu_2\sigma_{22}^a$) in the x_1 and x_3 directions in media (1) and (2), respectively. For spatial position P_R (x_1, ξ, x_3) on the interface, they read (Sect. 5.1; Anongba 2014a, b)

$$\begin{aligned} \tau_{12}^a &= -\overline{\alpha}x_1 \\ \tau_{23}^a &= -\overline{\alpha}x_3; \\ \overline{\alpha} &= \frac{\nu_s\mu_s}{a_s}\left(\frac{\nu_1}{E_1} - \frac{\nu_2}{E_2}\right)\sigma_{22}^a. \end{aligned} \quad (5.3.8)$$

The coefficient $\nu_s\mu_s/a_s$ is a characteristic of the interface, E_m ($m = 1$ and 2) is Young's modulus; we may view μ_s as a shear modulus about the interface, ν_s a constant in the crack analysis (about 0.5) and a_s a slab of material at P_R embracing the interface (measured in the x_2-direction of the applied tension) that suffers the magnitude of the shearing stress under consideration. $\overline{\sigma}_{ij}^{(J)(m)}$ has the form

$$\overline{\sigma}_{ij}^{(J)(m)}(x_1, x_2, x_3) = \int_{-a}^{a} \sigma_{ij}^{(J)(m)}(x_1 - x_1', x_2, x_3)D_J(x_1')\mathrm{d}x_1'; \quad (5.3.9)$$

$\sigma_{ij}^{(J)(m)}$ is the stress field produced by a dislocation of family J located at the origin O. In its general form (5.3.5) requires a numerical resolution. Only simple forms of ξ are tractable. For the planar interface crack, the crack dislocations are straight parallel to x_3; analytical solutions have been obtained in Sect. 5.1 under modes of applied loading I and $I + II$ based on dislocation stresses taken from Comninou and Dundurs (1980). The dislocation distribution functions D_I and D_{II} display an oscillatory singularity at the tip of the crack. When the crack is in infinitely extended homogeneous medium, we have given approximate solutions D_I and D_{III} under mixed mode $I + III$ loading when $\xi = \xi(x_3)$ depends on x_3 only. Analytical D_J values for a planar crack tilted around x_3 by angle θ_0 with respect to Ox_1x_3 in homogeneous medium under mixed mode $I + II + III$ are also reported (see Anongba 2010, 2017; for example). In the present problem of a non-planar interface crack in a bi-material, we can fortunately give approximate expressions of the crack-tip stress and crack extension force with ξ given by (5.3.1), taking for D_J, dislocation distributions

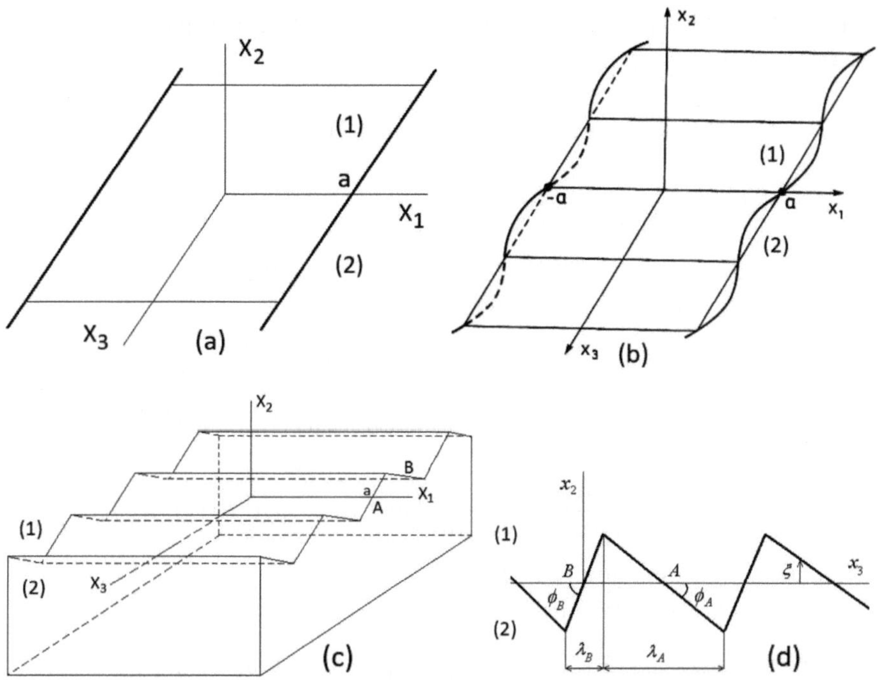

Fig. 5.7 Simple special cracks. **a** Planar interface crack with a straight front parallel to x_3 extending from $x_1 = -$ a to a. **b** Wavy crack (fluctuating about Ox_1x_3) in the form of a corrugated sheet with a sinusoidal front independent of x_1. The average plane is illustrated. **c** Non-planar crack fluctuating about Ox_1x_3 and consisting of planar facets; its fronts at $x_1 = \pm a$ lie in x_2x_3-planes. At $x_1 = a$, the crack front is characterized by inclination angles ϕ_A and ϕ_B (see (**d**)) at positions A and B located on Ox_1x_3. **d** Sketch of the crack front in (**c**) with B taken as origin. In this geometry [from (**a**)–(**c**)], the general loading of the crack systems corresponds to uniform applied σ_{22}^a, σ_{12}^a and σ_{23}^a at infinity in the x_2, x_1 and x_3 directions, respectively

of planar straight dislocation arrangements (see below). Figure 5.7 is a schematic representation of simple special cracks captured by the modelling. The cracks extend in the x_1-direction, from $x_1 = -a$ to a, and must be considered to run indefinitely in the x_3-direction. The crack shape in planes perpendicular to x_1 is described by ξ (see Fig. 5.7 c, d). The average fracture plane is Ox_1x_3. When $\xi = 0$, the crack dislocations are straight parallel to x_3 and distributed over Ox_1x_3- interface plane.

5.3.3 *Elastic Fields Due to Interface Dislocation of Edge I Average Character*

We consider an interface dislocation with Burgers vector $\vec{b}_I = (0, b, 0)$ lying indefinitely in the x_3-direction and spreading in the x_2x_3-plane at the origin in Fourier series (5.3.1). The elastic fields take the form (5.3.2). We get at spatial position (x_1, x_2, x_3) $(x_2 \neq 0, r^2 = x_1^2 + x_2^2, \mathrm{sgn}(x_2) = |x_2|/x_2)$:

$$
\begin{aligned}
u_1^{(I)(0)(m)} &= \frac{C_m(1-2\nu_m)}{2\mu_m}\ln r - \frac{C_m}{2\mu_m r^2}\left(x_1^2 + \mathrm{sgn}(x_2)x_2^2\right) \\
&\quad + \frac{x_2^2}{\mu_m r^2}\frac{b\mu_1\mu_2(\mu_1-\mu_2)}{\pi\left[\mu_1^2(3-4\nu_2)-\mu_2^2(3-4\nu_1)\right]}, \\
u_2^{(I)(0)(m)} &= -\frac{b\mu_1\mu_2}{\pi\mu_m\left[\mu_1^2(3-4\nu_2)-\mu_2^2(3-4\nu_1)\right]}\left((\mu_1-\mu_2)\frac{x_1|x_2|}{r^2}\right. \\
&\quad \left. +(\mu_1[3-2(\nu_m+\nu_2)]-\mu_2[3-2(\nu_m+\nu_1)])\tan^{-1}\frac{x_1}{|x_2|}\right) \\
\sigma_{11}^{(I)(0)(m)} &= \frac{4b\mu_1\mu_2\mathrm{sgn}(x_2)x_1}{\pi\left[\mu_1^2(3-4\nu_2)-\mu_2^2(3-4\nu_1)\right]r^2}\left(\mu_1\nu_2-\mu_2\nu_1-(\mu_1-\mu_2)\frac{x_2^2}{r^2}\right), \\
\sigma_{22}^{(I)(0)(m)} &= \frac{4b\mu_1\mu_2\mathrm{sgn}(x_2)x_1}{\pi\left[\mu_1^2(3-4\nu_2)-\mu_2^2(3-4\nu_1)\right]r^2}\left(\mu_1(1-\nu_2)-\mu_2(1-\nu_1)+(\mu_1-\mu_2)\frac{x_2^2}{r^2}\right) \\
\sigma_{33}^{(I)(0)(m)} &= \frac{4b\nu_m\mu_1\mu_2(\mu_1-\mu_2)\mathrm{sgn}(x_2)x_1}{\pi\left[\mu_1^2(3-4\nu_2)-\mu_2^2(3-4\nu_1)\right]r^2}, \\
\sigma_{12}^{(I)(0)(m)} &= \frac{4b\mu_1\mu_2|x_2|}{\pi\left[\mu_1^2(3-4\nu_2)-\mu_2^2(3-4\nu_1)\right]r^2}\left(\mu_1\nu_2-\mu_2\nu_1-(\mu_1-\mu_2)\frac{x_2^2}{r^2}\right);
\end{aligned}
$$

for $x_2 = 0$,

$$
\begin{aligned}
u_1^{(I)(0)(m)} &= -\frac{2b\mu_1\mu_2(\nu_1-\nu_2)}{\pi\left[\mu_1^2(3-4\nu_2)-\mu_2^2(3-4\nu_1)\right]}\ln|x_1| \\
u_2^{(I)(0)(m)} &= -\mathrm{sgn}(x_1)\frac{b\mu_1\mu_2(\mu_1[3-2(\nu_m+\nu_2)]-\mu_2[3-2(\nu_m+\nu_1)])}{2\mu_m\left[\mu_1^2(3-4\nu_2)-\mu_2^2(3-4\nu_1)\right]} \\
\sigma_{11}^{(I)(0)(m)} &= -\frac{4b\mu_1\mu_2(\nu_1\mu_2-\nu_2\mu_1)}{\pi\left[\mu_1^2(3-4\nu_2)-\mu_2^2(3-4\nu_1)\right]}\frac{1}{x_1} \equiv -\frac{e_6^{(I)(0)}}{x_1} \\
\sigma_{22}^{(I)(0)(m)} &= \frac{4b\mu_1\mu_2[\mu_1(1-\nu_2)-\mu_2(1-\nu_1)]}{\pi\left[\mu_1^2(3-4\nu_2)-\mu_2^2(3-4\nu_1)\right]}\frac{1}{x_1} \equiv \frac{e_3^{(I)(0)}}{x_1} \\
\sigma_{33}^{(I)(0)(m)} &= \frac{4b\nu_m\mu_1\mu_2(\mu_1-\mu_2)}{\pi\left[\mu_1^2(3-4\nu_2)-\mu_2^2(3-4\nu_1)\right]}\frac{1}{x_1} \equiv \frac{e_5^{(I)(0)}(m)}{x_1} \\
\sigma_{12}^{(I)(0)(m)} &= -\frac{2b\mu_1\mu_2[\mu_1(1-2\nu_2)-\mu_2(1-2\nu_1)]}{\left[\mu_1^2(3-4\nu_2)-\mu_2^2(3-4\nu_1)\right]}\delta(x_1) \equiv -\eta_c^{(I)(0)}4\pi i Q_c\delta(x_1);
\end{aligned}
\tag{5.3.10}
$$

$C_m = b\mu_m/2\pi(1-\nu_m)$, $Q_c = i(\nu_2C_2 - \nu_1C_1)/4$, other elastic fields are zero and constant terms are omitted. $\vec{u}^{(I)A_n(m)\infty}$ and $(\sigma)^{(I)A_n(m)\infty}$ are taken from Anongba (2009, 2010).

$$\vec{u}^{(I)A_n(m)W} = \eta_a^{(I)A_n(m)}\vec{u}_a^{(I)A_n(m)V} + \eta_d^{(I)A_n(m)}\vec{u}_d^{(I)A_n(m)V}$$
$$(\sigma)^{(I)A_n(m)W} = \eta_a^{(I)A_n(m)}(\sigma)_a^{(I)A_n(m)V} + \eta_d^{(I)A_n(m)}(\sigma)_d^{(I)A_n(m)V}; \qquad (5.3.11)$$

$\vec{u}_{a\text{and}d}^{(I)A_n(m)V}$ and $(\sigma)_{a\text{and}d}^{(I)A_n(m)V}$ are the partial elastic fields given in Sect. 5.2.

$$\eta_d^{(I)A_n(2)} = \frac{\text{num}(\eta_d^{(I)A_n(2)})}{\text{denom}(\eta_d^{(I)A_n(2)})},$$
$$\text{num}(\eta_d^{(I)A_n(2)}) = \left[a_{1d}^{(1)} - 2\nu_1\overline{b}^{(1)}\right][C_1\mu_2(1-2\nu_1) - C_2\mu_1(1-2\nu_2)]$$
$$+ \mu_2\left[a_{1d}^{(1)} + \overline{b}^{(1)}(5-8\nu_1)\right][C_1(1-\nu_1) - C_2(1-\nu_2)]$$
$$\text{denom}(\eta_d^{(I)A_n(2)}) = 2i\left\{\mu_1\left[a_{1d}^{(2)} - \overline{b}^{(2)}(5-8\nu_2)\right]\left[a_{1d}^{(1)} - 2\nu_1\overline{b}^{(1)}\right]\right.$$
$$\left.-\mu_2\left[a_{1d}^{(1)} + \overline{b}^{(1)}(5-8\nu_1)\right]\left[a_{1d}^{(2)} + 2\nu_2\overline{b}^{(2)}\right]\right\};$$
$$\eta_d^{(I)A_n(1)} = (\eta_d^{(I)A_n(2)})^{Inv};$$
$$\eta_a^{(I)A_n(2)} = -\eta_d^{(I)A_n(2)}\frac{i\left[\overline{b}^{(2)} + a_{2d}^{(2)}\right]}{a_{1a}^{(2)}},$$
$$\eta_a^{(I)A_n(1)} = (\eta_a^{(I)A_n(2)})^{Inv}. \qquad (5.3.12)$$

Here, the notation $(\ldots)^{Inv}$ means the value obtained from the expression into the bracket $(\ldots)$ by inverting the elastic constants. The various parameters $a_{1a}^{(m)}$, $a_{1d}^{(m)}$, $a_{2d}^{(m)}$ and $\overline{b}^{(m)}$ are defined in Sect. 5.2. $\sigma_{23}^{(I)A_n(m)}(x_1, 0, x_3)$ is constant with $m = 1$ and 2; it is involved in the crack analysis. Restricting ourselves to terms with $1/x_1$ only, we obtain (to linear terms in A_n or its spatial x_3-derivative)

$$\sigma_{23}^{(I)A_n(m)}(x_1, A_n, x_3) = \frac{\partial A_n}{\partial x_3}\frac{e_9^{(I)A_n}}{x_1},$$
$$e_9^{(I)A_n} = \frac{8i\overline{\overline{b}}\left[\eta_d^{(I)A_n(2)}\mu_1\nu_1(1-\nu_2) + \eta_d^{(I)A_n(1)}\mu_2\nu_2(1-\nu_1)\right]}{\mu_2 - \mu_1}; \qquad (5.3.13)$$

$$\overline{\overline{b}} = (Q_c - Q_b)/[\nu_1(1-\nu_2) + \nu_2(1-\nu_1)],\ Q_b = i(C_2 - C_1)/4.$$

5.3.4 Elastic Fields Due to Interface Dislocation of Edge II Average Character

The interface dislocation, with Burgers vector $\vec{b}_{II} = (b, 0, 0)$, lies in the x_3-direction and spreads in the Ox_2x_3-plane in the form of a Fourier series (5.3.1). We obtain:

$$
\begin{aligned}
u_1^{(II)(0)(m)} &= \frac{b}{2\pi}\tan^{-1}\frac{x_2}{x_1} - ie_1^{(II)(0)}\tan^{-1}\frac{x_1}{|x_2|} \\
&\quad + \frac{x_1x_2}{r^2}\left(\frac{C_m}{2\mu_m} + (-1)^{m-1}\frac{1}{1-2\nu_m}iV_R^{(II)(0)}\right), \\
u_2^{(II)(0)(m)} &= -\frac{(1-2\nu_m)C_m}{2\mu_m}\ln r \\
&\quad + 2\ln|x_1|\delta_A(x_2)(-1)^m\left(\frac{C_2-C_1}{4(\mu_2-\mu_1)} + \frac{1-\nu_m}{1-2\nu_m}iV_R^{(II)(0)}\right) \\
&\quad + \frac{x_2^2}{r^2}\left(\frac{C_m}{2\mu_m} + \frac{1}{1-2\nu_m}iV_R^{(II)(0)}\right), \\
\sigma_{11}^{(II)(0)(m)} &= -C_m\frac{x_2(x_2^2+3x_1^2)}{r^4} \\
&\quad - \left(\frac{x_2}{r^2} + \pi\delta(x_1)\delta_A(x_2)\right)2ie_3^{(II)(0)}(m) \\
&\quad + \frac{x_2(x_1^2-x_2^2)}{r^4}(-1)^m\frac{\mu_m}{1-2\nu_m}2iV_R^{(II)(0)}, \\
\sigma_{22}^{(II)(0)(m)} &= \frac{x_2(x_1^2-x_2^2)}{r^4}\left[C_m + \frac{(-1)^m\mu_m 2iV_R^{(II)(0)}}{1-2\nu_m}\right] \\
&\quad + \left(\frac{x_2}{r^2} + \pi\delta(x_1)\delta_A(x_2)\right)2ie_4^{(II)(0)}, \\
\sigma_{33}^{(II)(0)(m)} &= -2\nu_m C_m\frac{x_2}{r^2} - \left(\frac{x_2}{r^2} + \pi\delta(x_1)\delta_A(x_2)\right)\frac{\nu_m\mu_m}{1-2\nu_m}4iV_R^{(II)(0)}, \\
\sigma_{12}^{(II)(0)(m)} &= \frac{C_m x_1(x_1^2-x_2^2)}{r^4} \\
&\quad + \frac{x_1(-1)^m}{r^2}\left(\frac{\mu_m(C_2-C_1)}{\mu_2-\mu_1} + \frac{x_2^2}{r^2}\frac{\mu_m 4iV_R^{(II)(0)}}{1-2\nu_m}\right), x_2 \neq 0, \\
\left(\frac{\Delta V}{V}\right)^{(II)(0)(m)} &= \frac{1-2\nu_m}{2(1+\nu_m)\mu_m}\left(\sigma_{11}^{(II)(0)(m)} + \sigma_{22}^{(II)(0)(m)} + \sigma_{33}^{(II)(0)(m)}\right) \\
&= -\left(\frac{x_2}{r^2} + \pi\delta(x_1)\delta_A(x_2)\right)2iV_R^{(II)(0)} \\
&\quad + \frac{x_2(x_1^2-x_2^2)}{r^4}(-1)^m\frac{2iV_R^{(II)(0)}}{1+\nu_m} \\
&\quad - \frac{(1-2\nu_m)C_m}{\mu_m}\frac{x_2}{r^2};
\end{aligned}
\tag{5.3.14}
$$

$$
\begin{aligned}
V_R^{(II)(0)} &= \frac{ib(1-2\nu_1)(1-2\nu_2)[\mu_2(1-\nu_1)-\mu_1(1-\nu_2)]}{4\pi(1-\nu_1)(1-\nu_2)[\mu_2(1-2\nu_1)-\mu_1(1-2\nu_2)]},\\
e_1^{(II)(0)} &= \frac{i(C_2-C_1)}{2(\mu_2-\mu_1)}+V_R^{(II)(0)},\\
e_3^{(II)(0)}(m) &= \frac{i\mu_m(C_2-C_1)}{2(\mu_2-\mu_1)}+\frac{\mu_m}{1-2\nu_m}V_R^{(II)(0)},\\
e_4^{(II)(0)} &= \frac{ib\mu_1\mu_2(\nu_2-\nu_1)[\mu_2(1-\nu_1)-\mu_1(1-\nu_2)]}{2\pi(1-\nu_1)(1-\nu_2)(\mu_2-\mu_1)[\mu_2(1-2\nu_1)-\mu_1(1-2\nu_2)]}.
\end{aligned}
\tag{5.3.15}
$$

δ_A has the following definition: $\delta_A(x_2) = 0$ when $x_2 \neq 0$ and $\delta_A(x_2) = 1$ when $x_2 = 0$. The only stress component with a singularity $1/x_1$ on the Ox_1x_3-plane is $\sigma_{12}^{(II)(0)(m)}$. Dirac delta function $\delta(x_1)$ appears in $\sigma_{ii}^{(II)(0)(m)}$, $i = 1$ to 3 and the relative volume variation $(\Delta V/V)^{(II)(0)(m)}$. We have

$$
\sigma_{12}^{(II)(0)(m)}(x_1,0,x_3) = \frac{b\mu_1\mu_2(\nu_1-\nu_2)}{2\pi(1-\nu_1)(1-\nu_2)(\mu_2-\mu_1)}\frac{1}{x_1} \equiv \frac{e_6^{(II)(0)}}{x_1}. \tag{5.3.16}
$$

Expressions (5.3.14) correspond to the elastic fields of a straight interface dislocation (parallel to x_3) with a Burgers vector $\vec{b}_{II} = (b, 0, 0)$. Section 5.4 (Anongba 2017) gives similar expressions for the same dislocation. In both cases, the relative variation of volume is non-zero and continuous at the crossing of the interface. What changes is that in Sect. 5.4, $\sigma_{22}^{(II)(0)(m)}$ is discontinuous and $u_2^{(II)(0)(m)}$ continuous across the interface, while in the present work, $\sigma_{22}^{(II)(0)(m)}$ is continuous and $u_2^{(II)(0)(m)}$ discontinuous. In both, however, there is an optimum number of continuous elastic fields across the interface. For the oscillating parts of the elastic fields, we have the followings: $\vec{u}^{(II)A_n(m)\infty}$ and $(\sigma)^{(II)A_n(m)\infty}$ are taken from Anongba 2010;

$$
\begin{aligned}
\vec{u}^{(II)A_n(m)W} &= \eta_a^{(II)A_n(m)}\vec{u}_a^{(II)A_n(m)V} + \eta_d^{(II)A_n(m)}\vec{u}_d^{(II)A_n(m)V},\\
(\sigma)^{(II)A_n(m)W} &= \eta_a^{(II)A_n(m)}(\sigma)_a^{(II)A_n(m)V} + \eta_d^{(II)A_n(m)}(\sigma)_d^{(II)A_n(m)V}.
\end{aligned}
\tag{5.3.17}
$$

$\vec{u}^{(II)A_n(m)W}$ and $(\sigma)^{(II)A_n(m)W}$ are described in Anongba 2017 (Sect. 5.4.2); $\vec{u}_{a\text{and}d}^{(II)A_n(m)V}$ and $(\sigma)_{a\text{and}d}^{(II)A_n(m)V}$ can be calculated from the Appendix B of this work using relations (B.1) and (B.4), respectively; $\eta_d^{(II)A_n(m)}$ and $\eta_a^{(II)A_n(m)}$ are given by relations (42) and (47) there. We have, to linear terms in the amplitudes of oscillations,

$$
\sigma_{13}^{(II)A_n(m)}(x_1, A_n, x_3) = \frac{\partial A_n}{\partial x_3}\frac{e_8^{(II)A_n}}{x_1};
$$

$$
e_8^{(II)A_n} = \frac{Nu(e_8^{(II)A_n})}{De(e_8^{(II)A_n})},
$$

$$Nu(e_8^{(II)A_n}) = 2\mu_1\mu_2\big[(1-2\nu_1)(1-2\nu_2) - (1-\nu_1-\nu_2)^2\big]$$
$$\{2Q_c(1-\nu_1)(1-\nu_2)(C_1-C_2)$$
$$- Q_b[\nu_1(1-\nu_2)+\nu_2(1-\nu_1)][C_1(1-\nu_1)-C_2(1-\nu_2)]\},$$

$$De(e_8^{(II)A_n}) = (\nu_1-\nu_2)(\mu_1-\mu_2)$$
$$\{Q_c(1-\nu_1)(1-\nu_2)[\mu_1(1-2\nu_2)-\mu_2(1-2\nu_1)]$$
$$- Q_b[\nu_1(1-\nu_2)+\nu_2(1-\nu_1)][\mu_1(1-\nu_1)(1-2\nu_2)-\mu_2(1-\nu_2)(1-2\nu_1)]\}. \tag{5.3.18}$$

Associated $e_7^{(II)A_n}$, appearing below in the expressions of the crack-tip stress and crack extension force, is given in Sect. 5.4.2.

5.3.5 Elastic Fields Due to Interface Dislocation of Screw Average Character

The interface dislocation has Burgers vector $\vec{b}_{III} = (0, 0, b)$, lies indefinitely in the x_3-direction and spreads in the x_2x_3-plane at the origin in Fourier series (5.3.1). We obtain (Sect. 5.4.1):

$$u_3^{(III)(0)(m)}(\vec{x}) = \frac{b}{2\pi}\tan^{-1}\frac{x_2}{x_1} + \frac{D_1-D_2}{\mu_1+\mu_2}\tan^{-1}\frac{x_1}{|x_2|},$$
$$\sigma_{23}^{(III)(0)(m)}(\vec{x}) = \frac{b\mu_1\mu_2}{\pi(\mu_1+\mu_2)}\frac{x_1}{r^2} \equiv e_3^{(III)(0)}\frac{x_1}{r^2},$$
$$\sigma_{13}^{(III)(0)(m)}(\vec{x}) = -e_3^{(III)(0)}\frac{x_2}{r^2} + \delta_A(x_2)\frac{\mu_m(D_1-D_2)}{\mu_1+\mu_2}\pi\delta(x_1). \tag{5.3.19}$$

Associated other elastic field components are zero; $D_m = b\mu_m/2\pi$. $\vec{u}^{(III)A_n(m)\infty}$ and $(\sigma)^{(III)A_n(m)\infty}$ are taken from Anongba 2009, 2010.

$$\vec{u}^{(III)A_n(m)W} = \sum_{l=ctoe}\eta_l^{(III)A_n(m)}\vec{u}_l^{(III)A_n(m)V},$$
$$(\sigma)^{(III)A_n(m)W} = \sum_{l=ctoe}\eta_l^{(III)A_n(m)}(\sigma)_l^{(III)A_n(m)V}; \tag{5.3.20}$$

$\vec{u}^{(III)A_n(m)W}$ and $(\sigma)^{(III)A_n(m)W}$ are described in (Sect. 5.4.1): $\vec{u}_{ctoe}^{(III)A_n(m)V}$ and $(\sigma)_{ctoe}^{(III)A_n(m)V}$ can be calculated; $\eta_{ctod}^{(III)A_n(m)}$ correspond to $\eta_{ctod}^{A_n(m)}$ in this study where both $\eta_a^{A_n(m)}$ and $\eta_b^{A_n(m)}$ are zero. Compressive stresses $\sigma_{ii}^{(III)A_n(m)}$ $(i = 1, 2, 3)$ have singularities $1/x_1$ when $x_2 = 0$, implying that they are involved in the calculation of the crack extension force below. Associated coefficients $e_j^{(III)A_n}$, constant with $m =$ 1 and 2, correspond exactly to $e_j^{A_n}$ in Sect. 5.4.1. We write

$$\sigma_{ii}^{(III)A_n(m)}(x_1, A_n, x_3) = \frac{\partial A_n}{\partial x_3} 4\left(-e_{41}^{(III)A_n}\delta_{i1} + e_{51}^{(III)A_n}\delta_{i2} - e_6^{(III)A_n}\delta_{i3}\right)\frac{1}{x_1};$$

$$\begin{aligned} e_{52}^{(III)A_n} &= \eta_c^{(III)A_n(m)} s_{1c}^{(m)}\left[(2 - \nu_m)Q_r - \nu_m\right] \\ &= \frac{1}{3(\Gamma - 1)(\nu_2 - \nu_1)}\{(\nu_2 C_2 - \nu_1 C_1)[(1 - \nu_2)(2 - \nu_1) - (1 - \nu_1)(2 - \nu_2)\Gamma] \\ &\quad + (C_2 - C_1)[(1 - \nu_2)(1 - 2\nu_1) - \Gamma(1 - \nu_1)(1 - 2\nu_2)]\}, \end{aligned}$$

$$e_{51}^{(III)A_n} = -e_{52}^{(III)A_n}/2,$$

$$\begin{aligned} e_{41}^{(III)A_n} &= \frac{\nu_1 C_1(1 - \nu_1)(1 - 2\nu_2)\Gamma - \nu_2 C_2(1 - \nu_2)(1 - 2\nu_1)}{2[\Gamma(1 - \nu_1)(1 - 2\nu_2) - (1 - \nu_2)(1 - 2\nu_1)]} \\ &\quad + \frac{\nu_1(1 - 2\nu_2)\Gamma - \nu_2(1 - 2\nu_1)}{2[\Gamma(1 - \nu_1)(1 - 2\nu_2) - (1 - \nu_2)(1 - 2\nu_1)]} e_{52}^{(III)A_n}, \end{aligned}$$

$$\begin{aligned} e_6^{(III)A_n} &= \frac{\Gamma C_1 - C_2}{2(\Gamma - 1)} + \frac{\Gamma(\nu_1 C_1 - \nu_2 C_2)[(1 - \nu_1)(2 - \nu_2) - (1 - \nu_2)(2 - \nu_1)]}{2(\Gamma - 1)[\Gamma(1 - \nu_1)(1 - 2\nu_2) - (1 - \nu_2)(1 - 2\nu_1)]} \\ &\quad + \frac{\nu_1(1 - 2\nu_2)\Gamma - \nu_2(1 - 2\nu_1)}{2[\Gamma(1 - \nu_1)(1 - 2\nu_2) - (1 - \nu_2)(1 - 2\nu_1)]} e_{52}^{(III)A_n}. \end{aligned} \tag{5.3.21}$$

Here δ_{ij} is the Kronecker delta, $\Gamma = \mu_2/\mu_1$, $s_{1c}^{(m)} = (-1)^m iQ_b/2$, $Q_r = Q_c/Q_b$.

5.3.6 Planar Dislocation Distributions

It is assumed here that the dislocations are straight parallel to the x_3- direction (ξ − 0). We thus have a planar interface crack of finite extension, with straight fronts running indefinitely along x_3 (Fig. 5.7a). The crack is subjected to mixed mode $I + II + III$ with loadings applied at infinity. The condition (5.3.5) for the crack faces to be free from tractions becomes:

$$\begin{cases} \sigma_{22}^a + 2\pi i e_4^{(II)(0)} D_{II}(x_1) + e_3^{(I)(0)} \int\limits_{-a}^{a} \frac{D_I(x_1')\mathrm{d}x_1'}{x_1 - x_1'} = 0 \\ \sigma_{12}^a + \tau_{12}^a - \eta_c^{(I)(0)} 4\pi i Q_c D_I(x_1) + e_6^{(II)(0)} \int\limits_{-a}^{a} \frac{D_{II}(x_1')\mathrm{d}x_1'}{x_1 - x_1'} = 0\,. \\ \sigma_{23}^a + \tau_{23}^a + e_3^{(III)(0)} \int\limits_{-a}^{a} \frac{D_{III}(x_1')\mathrm{d}x_1'}{x_1 - x_1'} = 0 \end{cases} \tag{5.3.22}$$

We arrive at a system of three integral equations with Cauchy-type kernels with unknowns D_J. The Cauchy principal values of the integrals will be taken. The solution of the third equation with D_{III} is well known and is given below for zero dislocation content after unloading. We can manage the two other equations to obtain (Sect. 5.1)

$$
\begin{aligned}
&\int_{-a}^{a}\left(\sqrt{a^2-x_1^2}-\beta_A^2\sqrt{a^2-x_1'^2}\right)\frac{D_{II}(x_1')\mathrm{d}x_1'}{x_1'-x_1}=f(x_1);\\
&f(x_1)=\frac{\sigma_{12}^a+\tau_{12}^a}{e_6^{(II)(0)}}\sqrt{a^2-x_1^2}+\frac{i\beta_A^2\sigma_{22}^a}{2e_4^{(II)(0)}}x_1,\\
&\beta_A^2=-\frac{e_4^{(II)(0)}\eta_c^{(I)(0)}8Q_c}{e_3^{(I)(0)}e_6^{(II)(0)}}=1.
\end{aligned}
\tag{5.3.23}
$$

We observe that introducing the values for $\eta_c^{(I)(0)}8Q_c$ and $e_3^{(I)(0)}$ (5.3.10) on one hand, $e_4^{(II)(0)}$ (5.3.15) and $e_6^{(II)(0)}$ (5.3.16) on the other, from the dislocation stress fields, we have obtained the surprising value one to β_A^2 independent of the elastic constants. We temporary modify our notation and write (5.3.23) as

$$
\begin{aligned}
&\frac{\sqrt{1-s^2}}{\pi}\int_{-1}^{1}\frac{g(t)\mathrm{d}t}{t-s}+\frac{\lambda_A^2}{\pi}\int_{-1}^{1}\frac{\sqrt{1-t^2}g(t)\mathrm{d}t}{t-s}=\overline{f}(s);\\
&g(t)=D_{II}(at),\\
&\overline{f}(s)=\frac{1}{\pi}\left(\frac{(\sigma_{12}^a+\tau_{12}^a)\sqrt{1-s^2}}{e_6^{(II)(0)}}+\frac{i\sigma_{22}^a}{2e_4^{(II)(0)}}s\right),\\
&\lambda_A=i.
\end{aligned}
\tag{5.3.24}
$$

The integral equation in (5.3.24) takes the form of "Example 43" (page 158) in Estrada and Kanwal (2000); by a convincing operational approach, it is shown there that the solution in (5.3.24) can be written as

$$
g(s)=\frac{\overline{\psi}_1(s)-\overline{\psi}_2(s)}{2\lambda_A\sqrt{1-s^2}}
\tag{5.3.25}
$$

where $\overline{\psi}_1$ and $\overline{\psi}_2$ are solutions of the pair of Cauchy type integral equations

$$
\begin{aligned}
&\overline{\psi}_n(s)+(\delta_{n1}-\delta_{n2})\frac{\lambda_A}{\pi}\int_{-1}^{1}\frac{\overline{\psi}_n(t)}{t-s}\mathrm{d}t\\
&=\overline{f}(s)+A\delta_{n1}+B\delta_{n2}(n=1\text{ and }2)
\end{aligned}
\tag{5.3.26}
$$

$g(s)$ satisfies the additional requirement

$$
\frac{1}{\pi}\int_{-1}^{1}g(t)\ln\left(\frac{1+t}{1-t}\right)\mathrm{d}t=\int_{-1}^{1}\frac{\overline{f}(t)}{\sqrt{1-t^2}}\mathrm{d}t;
\tag{5.3.27}
$$

A and B are arbitrary constants. Solutions for (5.3.26), making use of the Plemelj formula (Estrada and Kanwal 2000; Muskhelishvili 1992), are

$$\begin{aligned} \overline{\psi}_1(s) &= \overline{f}(s)/2 \\ \overline{\psi}_2(s) &= c_0^* \overline{f}(s); \end{aligned} \tag{5.3.28}$$

c_0^* is an arbitrary constant. From (5.3.25), we have $g(s) = c_0^{**}\overline{f}(s)/2i\sqrt{1-s^2}$, and the constant c_0^{**} may be given a value by using (5.3.27). We can identify the distribution function $D_{II}(x_1)$ of the dislocation family II to the singularity term in $g(s)$ because bounded terms will be negligible sufficiently closer to the crack tips. Then, we integrate to obtain the associated D_I value in a similar way as shown in Sect. 5.1. We gather our results in the following form:

$$\begin{aligned} D_I(x_1) &= \frac{\sigma_{22}^a}{\pi e_3^{(I)(0)}} \frac{x_1}{\sqrt{a^2 - x_1^2}} \\ D_{II}(x_1) &= \frac{\sigma_{12}^a}{\pi e_6^{(II)(0)*}} \frac{x_1}{\sqrt{a^2 - x_1^2}} \\ D_{III}(x_1) &\equiv \frac{\sigma_{23}^a + \overline{\tau}_{23}^a}{\pi e_3^{(III)(0)}} \frac{x_1}{\sqrt{a^2 - x_1^2}}; \end{aligned} \tag{5.3.29}$$

$$\overline{\tau}_{23}^a \equiv \frac{1}{2a} \int_{-a}^{a} \tau_{23}^a \mathrm{d}x_3 = 0, \ e_6^{(II)(0)*} = e_6^{(II)(0)}(1 + 2a\theta_0\overline{\alpha}_0),$$

$$\theta_0 = \frac{\mu_2(1-\nu_1) - \mu_1(1-\nu_2)}{\pi[\mu_2(1-2\nu_1) - \mu_1(1-2\nu_2)]}, \ \overline{\alpha}_0 = \frac{\nu_s\mu_s}{a_s}\left(\frac{\nu_1}{E_1} - \frac{\nu_2}{E_2}\right).$$

The consequence of setting $\overline{\tau}_{23}^a = 0$ is that the shearing stress τ_{23}^a plays no role in the theory. The corresponding relative displacements ϕ_J of the faces of the crack in the x_2 ($J = I$), x_1 ($J = II$) and x_3 ($J = III$) directions are

$$\begin{aligned} \phi_I(x_1) &= \frac{\sigma_{22}^a b}{\pi e_3^{(I)(0)}} \sqrt{a^2 - x_1^2} \\ \phi_{II}(x_1) &= \frac{\sigma_{12}^a b}{\pi e_6^{(II)(0)*}} \sqrt{a^2 - x_1^2} \\ \phi_{III}(x_1) &= \frac{\sigma_{23}^a b}{\pi e_3^{(III)(0)}} \sqrt{a^2 - x_1^2}. \end{aligned} \tag{5.3.30}$$

5.3.7 Stresses About the Crack-Tip

Ahead of the crack-tip at spatial position $P_R(x_1 = a + s_1, x_2 = \xi, x_3), 0 < s_1 << a$, the total stress $\overline{\sigma}_{ij}(P_R)$ (5.3.6) is identified to the following formula

$$\overline{\sigma}_{ij}^{(m)}(s_1, x_2, x_3) = \sum_{J=I \text{ to III}} \int_{a-\delta a}^{a} \sigma_{ij}^{(J)(m)}\big(a + s_1 - x_1', x_2, x_3\big) D_J\big(x_1'\big) \mathrm{d}x_1', \; \delta a \ll a. \tag{5.3.31}$$

This stress expression means that only those dislocations located about the crack front in x_1-interval $[a - \delta a, a]$ will contribute significantly to the stress at $x_1 = a + s_1$ ahead of the crack tip as s tends to zero; any other contribution will be negligible for a sufficiently small value of s_1. Only dominant terms with singularities in the dislocation stress fields may contribute. Because P_R is outside the crack, stress terms with the Dirac delta function contribute nothing. Only dislocation stress terms with $1/x_1$ contribute and provide well-defined values to the crack extension force. Hence, we restrict ourselves to these singularities only in (5.3.31) and take the MacLaurin series expansion of $\sigma_{ij}^{(J)(m)}$ up to terms of first order with respect to x_2 (taken small). Under such conditions, the involved integrals are of the type $\int D_J(x_1')/(a + s_1 - x_1')\mathrm{d}x_1'$ which is calculated approximately by taking for D_J the straight edge and screw dislocation distributions (5.3.29) corresponding to a planar crack. We obtain

$$\begin{aligned}
\overline{\sigma}_{ii}^{(m)}(s_1, x_2, x_3) &= \frac{K_I^0}{e_3^{(I)(0)}\sqrt{2\pi s_1}}\Bigg(e_6^{(I)(0)}\delta_{i1} + e_3^{(I)(0)}\delta_{i2} + e_5^{(I)(0)}\delta_{i3} + x_2\frac{\partial^2 \xi}{\partial x_3^2}(C_m[\delta_{i3} - \delta_{i1}] \\
&\quad - \eta_d^{(I)A_n(m)} 2i\Big\{a_{1d}^{(m)}(\delta_{i3} - \delta_{i2}) + (-1)^m \overline{b}^{(m)}[3\delta_{i3} - \delta_{i1} + 2\nu_m(\delta_{i1} - \delta_{i2} + \delta_{i3})]\Big\})\Bigg) \\
&\quad + \frac{\partial \xi}{\partial x_3}\frac{K_{III}^0}{e_3^{(III)(0)}\sqrt{2\pi s_1}} 4\Big(\delta_{i2} e_{51}^{(III)A_n} - \delta_{i1} e_{41}^{(III)A_n} - \delta_{i3} e_6^{(III)A_n}\Big),
\end{aligned}$$

$$\begin{aligned}
\overline{\sigma}_{12}^{(m)}(s_1, x_2, x_3) &= \frac{K_{II}^0}{e_6^{(II)(0)*}\sqrt{2\pi s_1}} \\
&\quad \left(e_6^{(I)(0)} - x_2\frac{\partial^2 \xi}{\partial x_3^2}\Big[C_m + (-1)^m 2(e_7^{(II)A_n} - \eta_d^{(II)A_n(m)} 2t_{1d}^{(m)})\Big]\right),
\end{aligned}$$

$$\overline{\sigma}_{13}^{(m)}(s_1, x_2, x_3) = \frac{\partial \xi}{\partial x_3}\frac{e_8^{(II)A_n} K_{II}^0}{e_6^{(II)(0)*}\sqrt{2\pi s_1}},$$

$$\begin{aligned}
\overline{\sigma}_{23}^{(m)}(s_1, x_2, x_3) &= \frac{K_{III}^0}{e_3^{(III)(0)}\sqrt{2\pi s_1}}\left(e_3^{(III)(0)} + x_2\frac{\partial^2 \xi}{\partial x_3^2}\left[4e_{41}^{(III)A_n} + \frac{e_{52}^{(III)A_n} 4\nu_m(1 - Q_r)}{(2 - \nu_m)Q_r - \nu_m}\right.\right. \\
&\quad \left.\left. + \eta_d^{(III)A_n(m)}(-1)^{m-1}\frac{4iQ_b[\nu_m r_m Q_r + 1 - \nu_m]}{1 - 2\nu_m}\right]\right) + \frac{\partial \xi}{\partial x_3}\frac{e_9^{(I)A_n} K_I^0}{e_3^{(I)(0)}\sqrt{2\pi s_1}}.
\end{aligned} \tag{5.3.32}$$

Again s_1, x_2 and x_3 are arbitrary; $s_1 = x_1 - a \ll a$ ($s_1 > 0$) and x_2 is small. $K_I^0 = \sigma_{22}^a\sqrt{a\pi}$, $K_{II}^0 = \sigma_{12}^a\sqrt{a\pi}$ and $K_{III}^0 = \sigma_{23}^a\sqrt{a\pi}$.

5.3.8 Crack Extension Force

Consider a planar interface crack in the Ox_1x_3-plane, extending from $x_1 = -a$ to a, with straight fronts parallel to x_3. Under pure modes J of applied loading only ($J =$ *I, II* and *III*), the crack extension forces $G_{0s}^{J\text{pure}}$ at ($x_1 = a, x_2 = 0, x_3$) read

$$G_{0s}^{I\text{ pure}} = \frac{bK_I^{0^2}}{4\pi e_3^{(I)(0)}},\ G_{0s}^{II\text{ pure}} = \frac{bK_{II}^{0^2}}{4\pi e_6^{(II)(0)}},\ G_{0s}^{III\text{ pure}} = \frac{bK_{III}^{0^2}}{4\pi e_3^{(III)(0)}}. \tag{5.3.33}$$

To the same planar crack under general loading (mixed mode $I + II + III$) corresponds the crack extension force G_{0s} given by

$$G_{0s} = G_{0s}^a + G_{0s}^b + G_{0s}^c; \tag{5.3.34}$$

$$G_{0s}^a = \frac{bK_I^{0^2}}{4\pi e_3^{(I)(0)}},\ G_{0s}^b = \frac{bK_{II}^{0'^2}}{4\pi e_6^{(II)(0)}},$$

$$G_{0s}^c = \frac{bK_{III}^{0^2}}{4\pi e_3^{(III)(0)}},\ K_{II}^{0'} = K_{II}^0/(1 + 2a\theta_0\overline{\alpha}_0).$$

Under general loading of a non-planar crack (Fig. 5.5) with crack front ξ (5.3.1), we obtain the reduced crack extension force $\tilde{G}(P_0)$, $P_0 = (a, \xi, x_3)$, as

$$\tilde{G}(P_0) = G(P_0)/(G_{0s}^a + G_{0s}^b + G_{0s}^c) = \sum_{i,j=1}^{3} \tilde{G}_j^{(i)}(P_0);$$

$$\tilde{G}_1^{(1)} = -\frac{\partial\xi/\partial x_1 M_{12}(\Delta_{123}e_6^{(II)(0)})^{-1}}{\sqrt{1 + (\partial\xi/\partial x_1)^2 + (\partial\xi/\partial x_3)^2}}\Bigg(e_6^{(I)(0)} + \xi\frac{\partial^2\xi}{\partial x_3^2}[-C_m$$
$$+\eta_d^{(I)A_n(m)}2i(-1)^m(1 - 2\nu_m)\overline{b}^{(m)}\Big] - \frac{\partial\xi}{\partial x_3}\frac{4e_{41}^{(III)A_n}e_3^{(I)(0)}}{e_3^{(III)(0)}}M_{13}\Bigg),$$

$$\tilde{G}_2^{(1)} = -\frac{M_{12}^2 e_3^{(I)(0)}(\Delta_{123}e_6^{(II)(0)})^{-1}}{\sqrt{1 + (\partial\xi/\partial x_1)^2 + (\partial\xi/\partial x_3)^2}}\Bigg(1 + \frac{\xi}{e_6^{(II)(0)}}\frac{\partial^2\xi}{\partial x_3^2}[-C_m$$
$$+(-1)^m 2(\eta_d^{(II)A_n(m)}2t_{1d}^{(m)}\ e_7^{(II)A_n})\Big]\Bigg),$$

$$\tilde{G}_3^{(1)} = -\frac{(\partial\xi/\partial x_3)^2 M_{12}^2}{\sqrt{1+(\partial\xi/\partial x_1)^2+(\partial\xi/\partial x_3)^2}}\frac{e_8^{(II)A_n}e_3^{(I)(0)}}{\Delta_{123}e_6^{(II)(0)^2}},$$

$$\tilde{G}_1^{(2)} = -\frac{\partial\xi}{\partial x_1}\frac{e_6^{(II)(0)}}{e_3^{(I)(0)}}\frac{1}{M_{12}}\tilde{G}_2^{(1)},$$

$$\tilde{G}_2^{(2)} = \frac{(\Delta_{123}e_3^{(I)(0)})^{-1}}{\sqrt{1+(\partial\xi/\partial x_1)^2+(\partial\xi/\partial x_3)^2}}\left(e_3^{(I)(0)} + \xi\frac{\partial^2\xi}{\partial x_3^2}2i\eta_d^{(I)A_n(m)}\right.$$
$$\left.\times\left[a_{1d}^{(m)} + (-1)^m 2\nu_m\overline{b}^{(m)}\right] + \frac{\partial\xi}{\partial x_3}\frac{4e_{51}^{(III)A_n}e_3^{(I)(0)}}{e_3^{(III)(0)}}M_{13}\right),$$

$$\tilde{G}_3^{(2)} = -\frac{\partial\xi/\partial x_3 M_{13}(\Delta_{123}e_3^{(III)(0)})^{-1}}{\sqrt{1+(\partial\xi/\partial x_1)^2+(\partial\xi/\partial x_3)^2}}\left(e_3^{(III)(0)} + \xi\frac{\partial^2\xi}{\partial x_3^2}\left[4e_{41}^{(III)A_n} + \frac{e_{52}^{(III)A_n}4\nu_m(1-Q_r)}{(2-\nu_m)Q_r-\nu_m}\right.\right.$$
$$\left.\left.+\eta_d^{(III)A_n(m)}(-1)^{m-1}\frac{4iQ_b[\nu_m r_m Q_r + 1 - \nu_m]}{1-2\nu_m}\right] + \frac{\partial\xi}{\partial x_3}\frac{e_3^{(III)(0)}e_9^{(I)A_n}}{e_3^{(I)(0)}M_{13}}\right),$$

$$\tilde{G}_1^{(3)} = \frac{\partial\xi/\partial x_1}{\partial\xi/\partial x_3}\frac{e_6^{(II)(0)}}{e_3^{(III)(0)}}\frac{M_{13}}{M_{12}}\tilde{G}_3^{(1)},$$

$$\tilde{G}_2^{(3)} = -\frac{M_{13}}{\partial\xi/\partial x_3}\frac{e_3^{(I)(0)}}{e_3^{(III)(0)}}\tilde{G}_3^{(2)},$$

$$\tilde{G}_3^{(3)} = -\frac{\partial\xi/\partial x_3 M_{13}(\Delta_{123}e_3^{(III)(0)})^{-1}}{\sqrt{1+(\partial\xi/\partial x_1)^2+(\partial\xi/\partial x_3)^2}}\left(e_5^{(I)(0)} + \xi\frac{\partial^2\xi}{\partial x_3^2}\left[C_m - \eta_d^{(I)A_n(m)}2i\right.\right.$$
$$\left.\left.\times\left\{a_{1d}^{(m)} + (-1)^m\overline{b}^{(m)}(3+2\nu_m)\right\}\right] - \frac{\partial\xi}{\partial x_3}\frac{4e_6^{(III)A_n}e_3^{(I)(0)}}{e_3^{(III)(0)}}M_{13}\right),$$

$$\Delta_{123} = 1 + (e_3^{(I)(0)}/e_6^{(II)(0)})M_{12}^2 + (e_3^{(I)(0)}/e_3^{(III)(0)})M_{13}^2. \quad (5.3.35)$$

In (5.3.35), $M_{12} = K_{II}^{0'}/K_I^0$, $M_{13} = K_{III}^0/K_I^0$ and defined is $M_{23} = \sigma_{23}^a/\sigma_{12}^a$. Next, we describe special cracks with numerical applications using epoxy(1)/glass(2) bi-material.

5.3.9 Special Cracks

Planar Crack with a Straight Front

The simple case corresponds to a planar crack of length $2a$ in the Ox_1x_3-plane, extending from $x_1 = -a$ to a, with a straight front parallel to x_3. When the crack is in a totally homogeneous isotropic medium (m), the crack extension forces $G_0^{J(m)}$ under pure modes J of applied loading only ($J = I, II$ and III) read

$$G_0^{J(m)} = \frac{b}{4\pi C_m}\left(\delta_{JI}K_I^{0^2} + \delta_{JII}K_{II}^{0^2} + \delta_{JIII}K_{III}^{0^2}/(1-\nu_m)\right). \tag{5.3.36}$$

Hence, with $G_{0s}^{J\text{pure}}$ (5.3.33), we have

$$G_{0s}^{J\text{pure}}/G_0^{J(m)} = C_m\left(\delta_{JI}\frac{1}{e_3^{(I)(0)}} + \delta_{JII}\frac{1}{e_6^{(II)(0)}} + \delta_{JIII}\frac{1-\nu_m}{e_3^{(III)(0)}}\right). \tag{5.3.37}$$

For numerical application throughout, we take media epoxy/glass as system (1)/(2). The parameters taken from Liechti and Chai (1992) are: $E_1 = 2.03$ GPa, $\nu_1 = 0.37$; $E_2 = 68.95$ GPa, $\nu_2 = 0.20$; $E_i = 2\mu_i(1+\nu_i)$, $a/a_s = 10$, $\mu_s = (\mu_1+\mu_2)/2$, $\nu_s = 0.5$ (Anongba 2014a). We have

$$G_{0s}^{J\text{pure}}/G_0^{J(1)} = 0.5\delta_{JI} + 4.6\delta_{JII} + 0.5\delta_{JIII} \tag{5.3.38}$$

$$G_{0s}^{J\text{pure}}/G_0^{J(2)} = 15.1\delta_{JI} + 140\delta_{JII} + 20\delta_{JIII}. \tag{5.3.39}$$

Under such conditions, further extension of the crack initially located on the interface would be in the epoxy (outside the interface), except for pure mode II. Again, with the same interface planar crack (as defined above) but now under general loading (mixed mode $I+II+III$), the crack extension force G_{0_s} is given by (5.3.34). Defining the normalized crack extension force $\tilde{G}_{0s}^{(m)}$ as

$$\tilde{G}_{0s}^{(m)} = G_{0s}/\left(G_0^{I(m)} + G_0^{II(m)} + G_0^{III(m)}\right),$$

we have

$$\tilde{G}_{0s}^{(m)} = \frac{C_m\left[1 + (e_3^{(I)(0)}/e_6^{(II)(0)})M_{12}^2 + (e_3^{(I)(0)}/e_3^{(III)(0)})M_{13}^2\right]}{e_3^{(I)(0)}\left[1 + m_{12}^2 + M_{13}^2/(1-\nu_m)\right]}; \tag{5.3.40}$$

$$m_{12} = \sigma_{12}^a/\sigma_{22}^a = K_{II}^0/K_I^0,\ M_{12} = K_{II}^{0'}/K_I^0 = m_{12}/(1+2a\theta_0\overline{\alpha}_0).$$

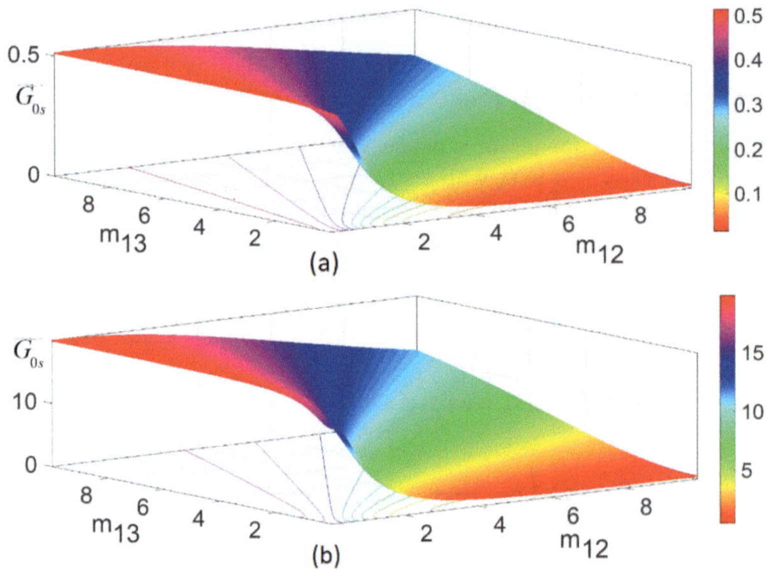

Fig. 5.8 Normalized crack extension force $\tilde{G}_{0s}^{(m)}$ (5.3.40) (planar crack under mixed mode $I + II + III$ loading including Poisson effects) as a function of applied shear/tension ratios $m_{12} = \sigma_{12}^a/\sigma_{22}^a$ and $m_{13} = \sigma_{23}^a/\sigma_{22}^a = M_{13}$ for **a** $m = 1$ epoxy and **b** $m = 2$ glass (see text)

$\tilde{G}_{0s}^{(m)}$ is displayed in Fig. 5.8 for epoxy(1)/glass(2) bi-material. Pure mode I loading is achieved when both m_{12} and M_{13} are equal to zero. Pure modes II and III are obtained when m_{12} and M_{13} tend to infinity, respectively. It is concluded that the value of the crack extension force is smaller along the interface as compared to its values in the homogeneous medium (1).

Non-planar Crack with a Segmented Front

The considered crack has a front that consists of straight segments. It fluctuates around the Ox_1x_3-plane and $\xi = \xi(x_3)$ is independent of x_1 (Fig. 5.7c, d). We describe ξ below taking locally B as origin as in Fig. 5.7d. ξ is then odd and $(2\lambda = \lambda_A + \lambda_B)-$periodical with respect to x_3 where λ_A and λ_B(Fig. 5.7d) are the projected length along x_3 of planar facets at A and B respectively. ξ is given by:

$$\begin{aligned} \xi &= \tan\phi_B x_3, \quad |x_3| \le \lambda_B/2 \\ &= \tan\phi_A(-x_3 + \lambda), \quad x_3 \in [\lambda_B/2, \lambda_B/2 + \lambda_A]. \end{aligned} \tag{5.3.41}$$

We assume general loading (mixed mode $I + II + III$), and next consider successively the reduced crack extension force $\tilde{G}(P_0)$ (5.3.35), now denoted $\tilde{G}_{sv}$, express the spatial average $\langle\tilde{G}_{sv}\rangle$ defined as $\langle\tilde{G}_{sv}\rangle = (1/2\lambda)\int_0^{2\lambda}\tilde{G}_{sv}dx_3$, and ultimately the condition for an extremum of $\langle\tilde{G}_{sv}\rangle$. $\tilde{G}_{sv}$ on a period reads: for $|x_3| < \lambda_B/2$,

$$\tilde{G}_{sv} = \frac{\cos\phi_B}{\Delta_{123}}\left(1 + \frac{e_3^{(I)(0)}}{e_6^{(II)(0)}}M_{12}^2 + \frac{e_3^{(I)(0)}}{e_3^{(III)(0)}}M_{13}^2 + \left[-e_3^{(III)(0)} + 4e_{51}^{(III)A_n} + e_9^{(I)A_n}\right.\right.$$
$$\left.\left. -e_5^{(I)(0)}\right]\frac{M_{13}}{e_3^{(III)(0)}}\tan\phi_B - \left[\frac{e_9^{(I)A_n}}{e_3^{(I)(0)2}} + \frac{e_8^{(II)A_n}}{e_6^{(II)(0)2}}M_{12}^2 - \frac{4e_6^{(III)A_n}}{e_3^{(III)(0)2}}M_{13}^2\right]e_3^{(I)(0)}\tan^2\phi_B\right);$$

for $x_3 \in \,]\lambda_B/2, \lambda_B/2 + \lambda_A]$,

$$\tilde{G}_{sv} = \frac{\cos\phi_A}{\Delta_{123}}\left(1 + \frac{e_3^{(I)(0)}}{e_6^{(II)(0)}}M_{12}^2 + \frac{e_3^{(I)(0)}}{e_3^{(III)(0)}}M_{13}^2 - \left[-e_3^{(III)(0)} + 4e_{51}^{(III)A_n} + e_9^{(I)A_n}\right.\right.$$
$$\left.\left. -e_5^{(I)(0)}\right]\frac{M_{13}}{e_3^{(III)(0)}}\tan\phi_A - \left[\frac{e_9^{(I)A_n}}{e_3^{(I)(0)2}} + \frac{e_8^{(II)A_n}}{e_6^{(II)(0)2}}M_{12}^2 - \frac{4e_6^{(III)A_n}}{e_3^{(III)(0)2}}M_{13}^2\right]e_3^{(I)(0)}\tan^2\phi_A\right). \tag{5.3.42}$$

For given M_{12}, M_{13} and crack profile (ϕ_A, ϕ_B), (5.3.42) provides the reduced crack extension force at an arbitrary spatial position $P_0(a, \xi, x_3)$ on the segmented crack front. $\tilde{G}_{sv}$ takes constant values on facets at A and B and does not depend on media numeration m, except for a value v_m in $e_5^{(I)(0)}$ (5.3.10); hence, it is almost unchanged by inverting the elastic constants. The average $\langle\tilde{G}_{sv}\rangle$ may be written as

$$\langle\tilde{G}_{sv}\rangle = \frac{1}{\Delta_{123}}\left(\Delta_{123}v_0\left[e_3^{(III)(0)} - 4e_{51}^{(III)A_n} - e_9^{(I)A_n}\right.\right.$$
$$\left.\left. +e_5^{(I)(0)}\right]\frac{M_{13}}{e_3^{(III)(0)}}v_1 - \left[\frac{e_9^{(I)A_n}}{e_3^{(I)(0)2}} + \frac{e_8^{(II)A_n}}{e_6^{(II)(0)2}}M_{12}^2 - \frac{4e_6^{(III)A_n}}{e_3^{(III)(0)2}}M_{13}^2\right]e_3^{(I)(0)}v_2\right); \tag{5.3.43}$$

$$v_0 = (1/(p_A + p_B))\left(p_A/\sqrt{1 + p_B^2} + p_B/\sqrt{1 + p_A^2}\right),$$
$$v_1 = (p_A p_B/(p_A + p_B))\left(1/\sqrt{1 + p_B^2} - 1/\sqrt{1 + p_A^2}\right),$$
$$v_2 = (p_A p_B/(p_A + p_B))\left(p_A/\sqrt{1 + p_A^2} + p_B/\sqrt{1 + p_B^2}\right).$$

Here $p_A = \tan\varphi_A, p_B = \tan\varphi_B$. We restrict ourselves to the condition for an extremum for $\langle\tilde{G}_{sv}\rangle$ with respect to ϕ_A by cancelling $\partial\langle\tilde{G}_{sv}\rangle/\partial\phi_A$. We have

$$\partial\langle\tilde{G}_{sv}\rangle/\partial\phi_A = \frac{\partial p_A/\partial\phi_A}{\Delta_{123}}\left(\Delta_{123}v_0'\left[e_3^{(III)(0)} - 4e_{51}^{(III)A_n} - e_9^{(I)A_n}\right.\right.$$
$$\left.\left. +e_5^{(I)(0)}\right]\frac{M_{13}}{e_3^{(III)(0)}}v_1' - \left[\frac{e_9^{(I)A_n}}{e_3^{(I)(0)2}} + \frac{e_8^{(II)A_n}}{e_6^{(II)(0)2}}M_{12}^2 - \frac{4e_6^{(III)A_n}}{e_3^{(III)(0)2}}M_{13}^2\right]e_3^{(I)(0)}v_2'\right); \tag{5.3.44}$$
$$v_0' = (p_B/(p_A + p_B))\left((p_A + p_B)^{-1}\left[1/\sqrt{1 + p_B^2} - 1/\sqrt{1 + p_A^2}\right] - p_A/(1 + p_A^2)^{3/2}\right),$$

$$\nu_1' = (p_B/(p_A+p_B))\left(p_B(p_A+p_B)^{-1}\left[1/\sqrt{1+p_B^2} - 1/\sqrt{1+p_A^2}\right] + p_A^2/(1+p_A^2)^{3/2}\right),$$
$$\nu_2' = (p_B/(p_A+p_B))\left(p_B(p_A+p_B)^{-1}\left[p_A/\sqrt{1+p_A^2} + p_B/\sqrt{1+p_B^2}\right] + p_A/(1+p_A^2)^{3/2}\right).$$

$\partial\langle\tilde{G}_{sv}\rangle/\partial\phi_A = 0$ leads to finding the roots of a polynomial of order 2 in M_{13}; this gives

$$M_{13} = (-b^* + \sqrt{\Delta^*})/2a^*; \tag{5.3.45}$$

$$\begin{aligned}
\Delta^* &= b^{*^2} - 4a^*c^*,\\
a^* &= (e_3^{(I)(0)} e_6^{(II)(0)})^2 \left(e_3^{(III)(0)} \nu_0' + 4e_6^{(III)A_n} \nu_2'\right),\\
b^* &= e_3^{(I)(0)} e_3^{(III)(0)} e_6^{(II)(0)^2} \left(-e_3^{(III)(0)} - e_5^{(I)(0)} + 4e_{51}^{(III)A_n} + e_9^{(I)A_n}\right) \nu_1',\\
c^* &= e_3^{(I)(0)} e_6^{(II)(0)} e_3^{(III)(0)^2} \left(e_6^{(II)(0)} + e_3^{(I)(0)} M_{12}^2\right) \nu_0'\\
&\quad - e_3^{(III)(0)^2} \left(e_6^{(II)(0)^2} e_9^{(I)A_n} + e_3^{(I)(0)^2} e_8^{(II)A_n} M_{12}^2\right) \nu_2'.
\end{aligned}$$

The numerical application is for the bi-material, epoxy (1)/glass (2). We recall that $a/a_s = 10$, $M_{13} = \sigma_{23}^a/\sigma_{22}^a$, $M_{12} = m_{12}/22$ and $m_{12} = \sigma_{12}^a/\sigma_{22}^a$. For M_{12} lower than certain value (not determined but, say 0.5 approximately), $\langle\tilde{G}_{sv}\rangle$ (5.3.43) increases with ϕ_A and ϕ_B from values close to one to values as large as 60, Fig. 5.9. No extremum is present. For larger M_{12}, maximums are present at relatively small M_{13}, Fig. 5.10a, b. Negative values correspond to forbidden motion of the crack. The behaviour at larger M_{13} is presented on Fig. 5.10c; extremum as given by (5.3.45) could be minimum there. Values (M_{13}, ϕ_A) (5.3.45) at fixed ϕ_B and M_{12}, corresponding to extremums on the curves $\langle\tilde{G}_{sv}\rangle$ (ϕ_A) are displayed on Fig. 5.11.

Non-planar Crack with a Sinusoidal Front

Here, the crack has a sinusoidal front $\xi = \xi(x_3)$ in x_2x_3-plane that is independent of x_1 (Fig. 5.7b), in the form

$$\xi/\xi_c = \sin\kappa_c x_3 = \sin 2\pi\bar{x}_3$$

where $\bar{x}_3 = x_3/\lambda_c$ and $\lambda_c = 2\pi/\kappa_c$ is the wavelength. In the following, we express the reduced crack extension force (5.3.35), denoted $\tilde{G}_{sc}$ for the sinusoidal crack, the spatial average $\langle\tilde{G}_{sc}\rangle = (1/\lambda_c)\int_0^{\lambda_c} \tilde{G}_{sc}\mathrm{d}x_3$ and finally conditions under which $\langle\tilde{G}_{sc}\rangle$ has extremum. $\tilde{G}_{sc}$ depends on the product $\xi_c\kappa_c = \tan\phi_c$; ϕ_c, the crack-front inclination angle, is the acute angle measured in the plane perpendicular to the crack propagation direction between the crack front and the average fracture plane. We obtain

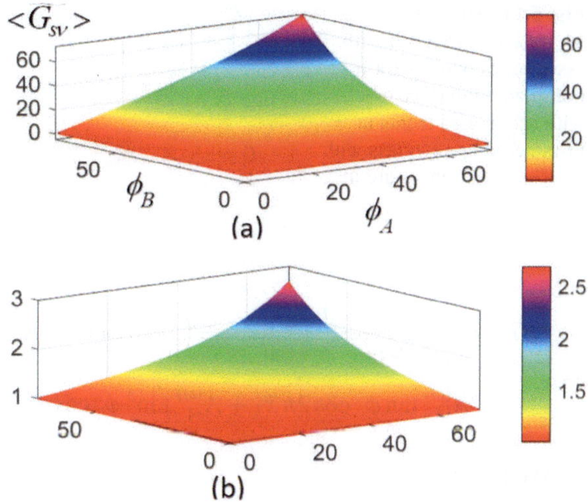

Fig. 5.9 Surfaces $\langle \tilde{G}_{sv} \rangle (\phi_A, \phi_B)$ (5.3.43) with associated contours for a non-planar interface crack with segmented fronts in bi-material epoxy/glass: **a** $M_{12} = 0.001$ and $M_{13} = 1$; **b** $M_{12} = 0.5$ and $M_{13} = 0.001$. ϕ_A and ϕ_B are in degrees

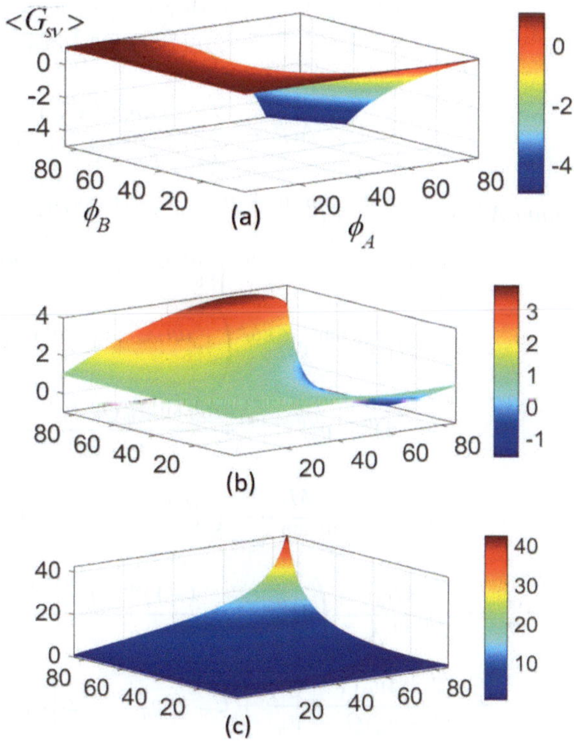

Fig. 5.10 Surfaces $\langle \tilde{G}_{sv} \rangle (\phi_A, \phi_B)$ (5.3.43) for a non-planar interface crack with segmented fronts in bi-material epoxy/glass with $M_{12} = 1$: **a** $M_{13} = 0.3$; **b** $M_{13} = 0.6$ and **c** $M_{13} = 1$. ϕ_A and ϕ_B are in degrees

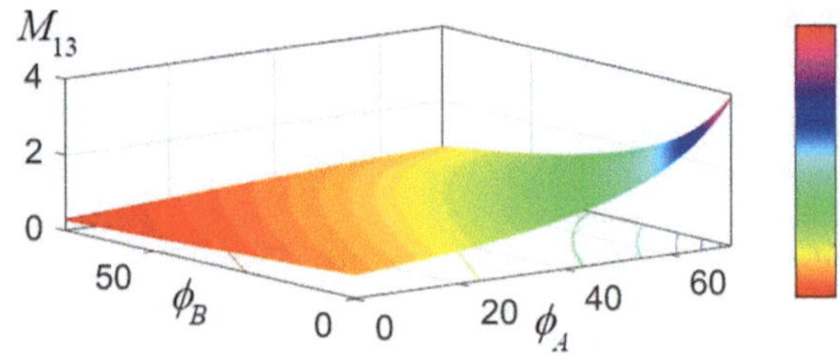

Fig. 5.11 Surface $M_{13}(\phi_A, \phi_B)$ (5.3.45) with associated contours for a non-planar interface crack with segmented fronts in bi-material epoxy/glass with $M_{12} = 1$. ϕ_A and ϕ_B are in degrees

$$\tilde{G}_{sc} = \frac{1}{\sqrt{1+\tan^2\phi_c \cos^2 2\pi\overline{x}_3}}\Big(1 + A_{s1}\tan^2\phi_c + 2A_{s2}^{(m)}\tan^2\phi_c \sin^2 2\pi\overline{x}_3$$
$$+A_{s3}^{(m)}\tan\phi_c\cos 2\pi\overline{x}_3 + A_{s4}^{(m)}\tan^3\phi_c\cos 2\pi\overline{x}_3\sin^2 2\pi\overline{x}_3\Big). \tag{5.3.46}$$

We have:

$$A_{s1} = -\frac{e_9^{(I)A_n} e_3^{(III)(0)^2} e_6^{(II)(0)^2} + e_8^{(II)A_n} e_3^{(I)(0)^2} e_3^{(III)(0)^2} M_{12}^2 - 4e_6^{(III)A_n} e_3^{(I)(0)^2} e_6^{(II)(0)^2} M_{13}^2}{e_3^{(I)(0)} e_3^{(III)(0)^2} e_6^{(II)(0)^2}\Delta_{123}};$$

$$A_{s2}^{(m)} = \frac{num(A_{s2}^{(m)})}{2e_3^{(I)(0)} e_3^{(III)(0)^2} e_6^{(II)(0)^2}\Delta_{123}},$$

$$num(A_{s2}^{(m)}) = e_6^{(II)(0)^2} e_3^{(III)(0)^2}\Big\{e_9^{(I)A_n} - \eta_d^{(I)A_n(m)} 2i\Big[a_{1d}^{(m)} + (-1)^m 2\nu_m \overline{b}^{(m)}\Big]\Big\}$$
$$+ M_{12}^2 e_3^{(I)(0)^2} e_3^{(III)(0)^2}\Big\{e_8^{(II)A_n} + (-1)^m 2e_7^{(II)A_n} + 2\mu_m V_{R1}^{(II)A_n}/(1-2\nu_m)\Big\}$$
$$- M_{13}^2 e_3^{(I)(0)^2} e_6^{(II)(0)^2}\Bigg(4e_6^{(III)A_n} + 4e_{41}^{(III)A_n} + \frac{e_{52}^{(III)A_n} 4\nu_m(1-Q_r)}{(2-\nu_m)Q_r - \nu_m}$$
$$+(-1)^{m-1}\eta_d^{(III)A_n(m)}\frac{4iQ_b(\nu_m r_m Q_r + 1 - \nu_m)}{1-2\nu_m}\Bigg);$$

$$A_{s3}^{(m)} = \frac{M_{13}}{e_3^{(III)(0)}\Delta_{123}}\Big(e_9^{(I)A_n} - e_5^{(I)(0)} + 4e_{51}^{(III)A_n} - e_3^{(III)(0)}\Big);$$

$$A_{s4}^{(m)} = \frac{M_{13}}{e_3^{(III)(0)}\Delta_{123}}\Bigg(4e_{41}^{(III)A_n} + \frac{(-1)^{m-1}\eta_d^{(III)A_n(m)} 4iQ_b(\nu_m r_m Q_r + 1 - \nu_m)}{1-2\nu_m}$$
$$+\frac{e_{52}^{(III)A_n} 4\nu_m(1-Q_r)}{(2-\nu_m)Q_r - \nu_m} + C_m - \eta_d^{(I)A_n(m)} 2i\Big[a_{1d}^{(m)} + (-1)^m \overline{b}^{(m)}(3+2\nu_m)\Big]\Bigg). \tag{5.3.47}$$

For given M_{12}, M_{13} and crack profile ϕ_c, (5.3.46) gives the reduced crack extension force at arbitrary spatial position $P_0(a, \xi_c \sin\kappa_c x_3, x_3)$ on the crack front. We observe that A_{s1} is unchanged by inverting the elastic constants, as also is $A_{s3}^{(m)}$ except in the latter case for a factor ν_m in $e_5^{(I)(0)}$ (5.3.10). $A_{s2}^{(m)}$ and $A_{s4}^{(m)}$, in the forms we arrived at, seem not unchanged by inverting the elastic constants. We observe that by inverting the elastic constants in $A_{s2}^{(1)}$ and $A_{s4}^{(1)}$, we obtain $A_{s2}^{(2)}$ and $A_{s4}^{(2)}$, respectively. Hence, one can work with $m = 1$ for half-space (1) in the crack extension force expression (5.3.46) and invert the elastic constants to obtain the results applicable for half-space (2). One expects both expressions to have similar magnitudes. In the calculation of $\langle\tilde{G}_{sc}\rangle$, terms with $\tan\phi_c$ and $\tan^3\phi_c$ (odd powers) in $\tilde{G}_{sc}$ contribute nothing. We get

$$\begin{aligned}\langle\tilde{G}_{sc}\rangle = \cos\phi_c\big(&[1 + A_{s1}\tan^2\phi_c]F(1/2, 1/2; 1; \sin^2\phi_c)\\ &+A_{s2}^{(m)}\tan^2\phi_c F(3/2, 1/2; 2; \sin^2\phi_c)\big)\end{aligned} \tag{5.3.48}$$

where F is Gauss's hypergeometric function. The condition for extremum of $\langle\tilde{G}_{sc}\rangle$ (ϕ_c) is $\partial\langle\tilde{G}_{sc}\rangle/\partial\phi_c = 0$. We have

$$\partial\langle\tilde{G}_{sc}\rangle/\partial\phi_c = \cos\phi_c\left(u_{s0} + u_{s1}A_{s1} + u_{s2}A_{s2}^{(m)}\right); \tag{5.3.49}$$

$$\begin{aligned}u_{s0} = &-\tan\phi_c F(1/2, 1/2; 1; \sin^2\phi_c)\\ &+\sin 2\phi_c F(3/2, 3/2; 2; \sin^2\phi_c)/4,\end{aligned}$$

$$\begin{aligned}u_{s1} = &\tan\phi_c\big[(2 + \tan^2\phi_c)F(1/2, 1/2; 1; \sin^2\phi_c)\\ &+\sin 2\phi_c\tan\phi_c F(3/2, 3/2; 2; \sin^2\phi_c)/4\big],\end{aligned}$$

$$\begin{aligned}u_{s2} = &\tan\phi_c\big[(2 + \tan^2\phi_c)F(3/2, 1/2; 2; \sin^2\phi_c)\\ &+3\sin 2\phi_c\tan\phi_c F(5/2, 3/2; 3; \sin^2\phi_c)/8\big].\end{aligned}$$

Given M_{12}, $\partial\langle\tilde{G}_{sc}\rangle/\partial\phi_c = 0$ corresponds to finding the roots of a polynomial of order 2 in M_{13}; this gives

$$M_{13}^2 = -\frac{1}{num_{13}}\left(num_0 + M_{12}^2 num_{12}\right); \tag{5.3.50}$$

$$\begin{aligned}num_{13} = e_3^{(I)(0)^2} e_6^{(II)(0)^2}\Bigg(&u_{s0}2e_3^{(III)(0)} + u_{s1}8e_6^{(III)A_n} - u_{s2}\Bigg[\frac{e_{52}^{(III)A_n}4\nu_m(1-Q_r)}{(2-\nu_m)Q_r - \nu_m}\\ &+4e_6^{(III)A_n} + 4e_{41}^{(III)A_n} + \frac{(-1)^{m-1}\eta_d^{(III)A_n(m)}4iQ_b(\nu_m r_m Q_r + 1 - \nu_m)}{1-2\nu_m}\Bigg]\Bigg),\end{aligned}$$

$$\begin{aligned}num_0 = e_3^{(III)(0)^2} e_6^{(II)(0)^2}\Big(&u_{s0}2e_3^{(I)(0)} - u_{s1}2e_9^{(I)A_n}\\ &+u_{s2}\Big\{e_9^{(I)A_n} - \eta_d^{(I)A_n(m)}2i\Big[a_{1d}^{(m)} + (-1)^m 2\nu_m\overline{b}^{(m)}\Big]\Big\}\Big),\end{aligned}$$

$$num_{12} = e_3^{(I)(0)^2} e_3^{(III)(0)^2} \Big(u_{s0} 2 e_6^{(II)(0)} - u_{s1} 2 e_8^{(II)A_n}$$
$$+ u_{s2} \Big\{ e_8^{(II)A_n} + (-1)^m 2 e_7^{(II)A_n} + 2\mu_m V_{R1}^{(II)A_n} / (1 - 2\nu_m) \Big\} \Big).$$

Given M_{12} and M_{13}, $\langle \tilde{G}_{sc} \rangle$ (5.3.48) is a function of ϕ_c that may contain extremums, the associated ϕ_c values of which are given by (5.3.50). Figures 5.12 and 5.13 are plots of $\langle \tilde{G}_{sc} \rangle$ and pairs (ϕ_c, M_{13}) (given M_{12}) respectively. Gauss's hypergeometric functions are restricted to the nine first terms of their respective series. On Fig. 5.12, for each value of M_{12}, two behaviours of $\langle \tilde{G}_{sc} \rangle$ are clearly distinguished: at the lowest values of M_{13}, $\langle \tilde{G}_{sc} \rangle$ increases with ϕ_c, from the value 1 ($\phi_c = 0$) to values that can exceed 100 ($\phi_c \approx 80°$); at the highest values of M_{13}, $\langle \tilde{G}_{sc} \rangle$ decreases with ϕ_c, from 1 to negative values. Cracks that produce negative values cannot propagate. At intermediate values of M_{13}, extremums of $\langle \tilde{G}_{sc} \rangle$ are expected, presumably minimums. These extremums are given in Fig. 5.13.

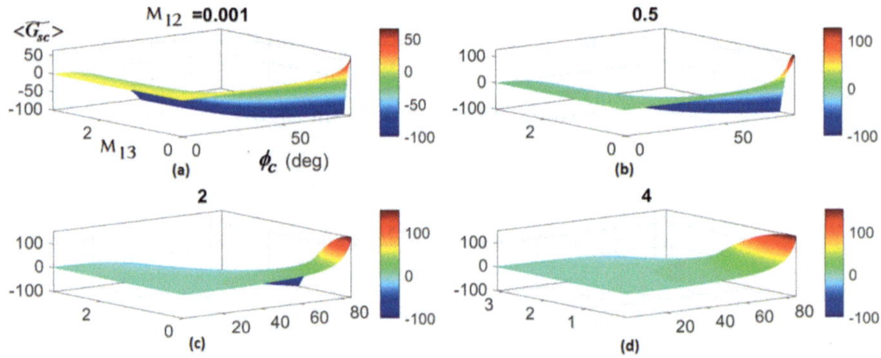

Fig. 5.12 $\langle \tilde{G}_{sc} \rangle$ (5.3.48) ($m = 1$), for a non-planar crack with sinusoidal fronts, as a function of (ϕ_c, M_{13}) in bi-material epoxy(1)/glass(2) with four different values of M_{12}: **a** $M_{12} = 0.001$, **b** 0.5, **c** 2 and **d** 4

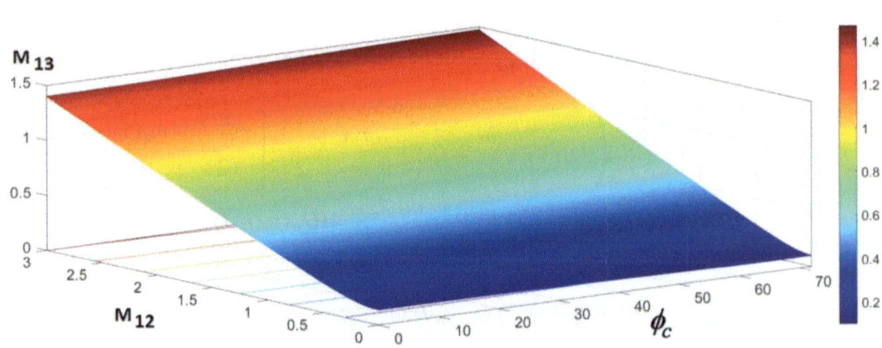

Fig. 5.13 M_{13} (5.3.50) ($m = 1$) as a function of (ϕ_c, M_{12}) for non-planar crack with sinusoidal fronts in bi- material epoxy/glass

5.3.10 Discussion

The expressions for the crack dislocation distributions D_J (5.3.29), crack tip stresses (5.3.32) and crack extension force (5.3.35) are similar in forms to those for a crack in a totally homogeneous medium. We stress that when $\beta_A^2 = 1$ (5.3.23), no oscillatory solution of the elastic fields at the tip of the interface crack exists; this can be checked directly by introducing a function of the form $\overline{\psi}_n(s) = c_0^*(1 + s/1 - s)^{i\delta}\overline{f}(s)$ in (5.3.26). We can write β_A^2 as

$$\beta_A^2 = -\frac{1}{(\pi x_1 \delta(x_1))^2} \frac{\sigma_{12}^{(I)(0)(m)} \sigma_{22}^{(II)(0)(m)}}{\sigma_{12}^{(II)(0)(m)} \sigma_{22}^{(I)(0)(m)}} = 1, \tag{5.3.51}$$

independent of media (1) and (2). In the relation (5.3.51) above, the dislocation stresses are measured on the interface plane at $P(x_1, x_2 = 0, x_3)$ and identified to their singular terms proportional to $1/x_1$ and $\delta(x_1)$. In Sect. 5.1, we have analysed a planar interface crack under mode I and mixed mode $I + II$ loadings (considering shearing stresses on the interface originating from Poisson effect) and represented the crack by continuous distributions of straight edge dislocations. The dislocation stress fields on the interface (positions $P(x_1, x_2 = 0, x_3)$) were taken from Comninou and Dundurs (1980):

- Edges with $\vec{b}_{II} = (b, 0, 0)$ (Family II)

$$\sigma_{12}^{(II)}(x_1, 0, x_3) = \frac{bC}{\pi}\frac{1}{x_1},$$
$$\sigma_{22}^{(II)}(x_1, 0, x_3) = -b\beta C\delta(x_1);$$

- Edges with $\vec{b}_I = (0, b, 0)$ (Family I)

$$\sigma_{12}^{(I)}(x_1, 0, x_3) = b\beta C\delta(x_1),$$
$$\sigma_{22}^{(I)}(x_1, 0, x_3) = \frac{bC}{\pi}\frac{1}{x_1}. \tag{5.3.52}$$

Moreover,

$$C = \frac{2\mu_1(1-\alpha)}{(\kappa_1+1)(1-\beta^2)} = \frac{2\mu_2(1+\alpha)}{(\kappa_2+1)(1-\beta^2)}$$

with

$$\alpha = \frac{\mu_1(\kappa_2+1) - \mu_2(\kappa_1+1)}{\mu_1(\kappa_2+1) + \mu_2(\kappa_1+1)}, \quad -1 \le \alpha \le 1$$

$$\beta = \frac{\mu_1(\kappa_2 - 1) - \mu_2(\kappa_1 - 1)}{\mu_1(\kappa_2 + 1) + \mu_2(\kappa_1 + 1)}, \; -\frac{1}{2} \le \beta \le \frac{1}{2} \tag{5.3.53}$$

where $\kappa_i = 3 - 4\nu_i$. We arrived at a singular integral equation identical to that in (5.3.23), but with β^2 in place of β_A^2. The corresponding dislocation distributions, D_I and D_{II}, and the stresses carry oscillatory singularities at the tip of the cracks. In addition, individual crack extension forces, associated with modes I and II, are ill defined, oscillating indefinitely in their limiting values. However, the total energy release rate G is well defined and reads in mixed mode $I + II$:

$$G = \frac{a\pi(1 + 4\delta^2)}{4C}\left((\sigma_{22}^a - a\delta\overline{\alpha})^2 + \left(\sigma_{12}^a - \frac{a\overline{\alpha}}{2}\right)^2\right); \tag{5.3.54}$$

$$\delta = \frac{1}{2\pi}\ln\left(\frac{1+\beta}{1-\beta}\right).$$

The oscillatory crack-tip elastic fields and crack extension force agree with previous studies in absence of Poisson effect (Williams 1959; Erdogan 1963; Rice and Sih 1965; England 1965; Hutchinson 1990). For certain combinations of materials, $\beta = 0$: under such conditions, the oscillatory character of the elastic fields at the tip of the interface cracks disappears and we recover a behaviour like that described in the present study. We observe (5.3.52) that the same spatial dependence is given to $\sigma_{22}^{(I)}$ and $\sigma_{12}^{(II)}$ on the interface. We are facing two different interface straight edges I and II. The ways these independent dislocations are introduced into the bi-material are very different: for the interface climb edge I, one can displace one part of material (1) only (say $x_1 > 0, x_2 > 0$) with respect to $(x_1 < 0, x_2 > 0)$ by b whilst for the interface glide edge II, the displacement may be performed along the interface (say, $x_1 < 0, x_2 > 0$) with respect to $(x_1 < 0, x_2 < 0)$ by b. Now, taking each dislocation separately in the bi-material, why their stresses must have the identical spatial dependencies on the interface, i.e., σ_{12} (interface glide edge II) $= \sigma_{22}$ (interface climb edge I)? This behavior is observed when both dislocations are in an infinitely extended isotropic medium. For the bi-material, this result is not obvious. In a similar way (5.3.52), except for a minus sign, why σ_{22} (interface glide edge II) $= -\,\sigma_{12}$ (interface climb edge I)? These are so regardless of the types of material involved in the bimaterial. This contrasts with the results of the present study where $\sigma_{22}^{(I)(0)}(x_1, 0, x_3) = e_3^{(I)(0)}/x_1$ (5.3.10) and $\sigma_{12}^{(II)(0)}(x_1, 0, x_3) = e_6^{(II)(0)}/x_1$ (5.3.16) are different; a similar remark applies with $\sigma_{12}^{(I)(0)}$ (5.3.10) and $\sigma_{22}^{(II)(0)}$ (5.3.16). Both studies, (Comninou and Dundurs 1980; Anongba 2014a, b) and the present one, use similar spatial dependences (i.e. $1/x_1$ and $\delta(x_1)$) for the stresses but with different coefficients. Here the coefficients $e_3^{(I)(0)}$, $\eta_c^{(I)(0)} 4\pi i Q_c$ (5.3.10), $e_6^{(II)(0)}$ (5.3.16) and $e_4^{(II)(0)}$ (5.3.15) are unchanged by inverting the elastic constants; this ensures continuity of the stresses at the crossing of the interface. In contrast, β (5.3.53) changes sign on inverting the elastic constants in stress coefficients (5.3.52) suggesting that associated stresses are discontinuous on the interface. Hence, the presence of Dirac delta function $\delta(x_1)$ on $x_2 = 0$ in the interface dislocation stress fields and imposed continuity conditions when crossing the interface provide the result $\beta_A^2 = 1$. We assume the planar crack to be submitted to mixed mode $I + II$ loading only and defined $\hat{G}_{0s} = G_{0_s}/G$ where G_{0s} and G are given by (5.3.34) and (5.3.54), respectively. This gives

$$\hat{G}_{0s} = \frac{bC}{\pi e_3^{(I)(0)}\left(1+4\delta^2\right)}\left(\frac{1+M_{12}^2\left(e_3^{(I)(0)}/e_6^{(II)(0)}\right)}{(1-a\delta\overline{\alpha}_0)^2+(m_{12}-a\overline{\alpha}_0/2)^2}\right); \tag{5.3.55}$$

$m_{12} = \sigma_{12}^a/\sigma_{22}^a, M_{12} = m_{12}/(1+2a\theta_0\overline{\alpha}_0)$. Figure 5.14 is a plot of $\hat{G}_{0s}(m_{12})$ (5.3.55). $\hat{G}_{0s}$ is smaller than 1.

Figure 5.15 displays $\hat{G}_{0s}$ (5.3.55), neglecting Poisson shearing stresses in x_1 direction. Remarkably, $\hat{G}_{0s} \cong 0.98$ under mode I loading only ($m_{12} = 0$) suggesting that both, G_{0s} and G, predict the same behaviour of the interface planar crack in epoxy/glass under mode I loading, without Poisson effect. In fact, from (5.3.55) and $\hat{G}_{0s} \cong 0.98$, the coefficient $bC/(\pi e_3^{(I)(0)}(1+4\delta^2)) \cong 0.98$ is close to one, implying that

$$e_3^{(I)(0)} \simeq \frac{bC}{\pi\left(1+4\delta^2\right)}. \tag{5.3.56}$$

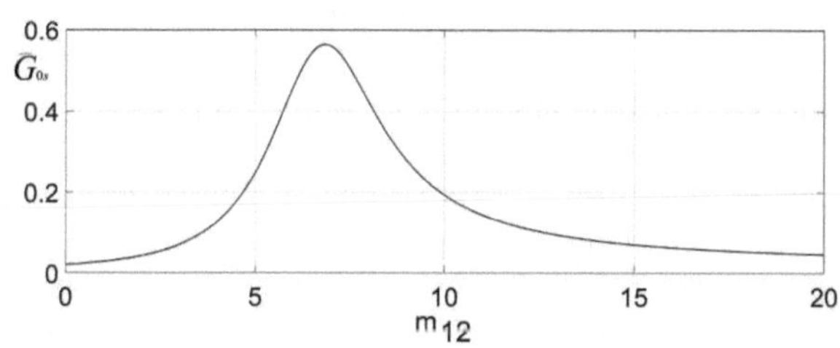

Fig. 5.14 $\hat{G}_{0s}$ (5.3.55) as a function of $m_{12} = \sigma_{12}^a/\sigma_{22}^a$; mixed mode I + II loading of planar interface crack in bi-material epoxy (1)/glass (2)

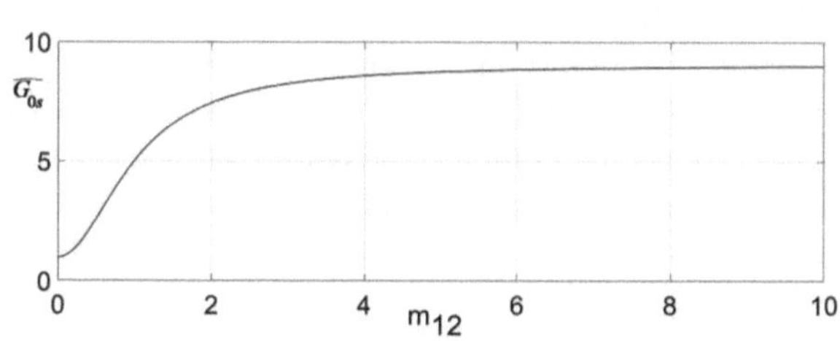

Fig. 5.15 $\hat{G}_{0s}$ (5.3.55) as a function of $m_{12} = \sigma_{12}^a/\sigma_{22}^a$, in absence of Poisson shearing stresses along x_1 on the interface, $M_{12} = m_{12}$; mixed mode $I + II$ loading of planar interface crack in bi-material epoxy (1)/glass (2)

Other material combination alumina (1)/nickel (2) confirms (5.3.56). This indicates that (5.3.34) (i.e.G^{a}_{0s}) and (5.3.54) are nearly identical under pure mode *I* loading, in absence of Poisson effect. Because for the planar crack, the mode that opens the crack faces is what matters for crack extension, the present study and those that involve oscillatory singularities of elastic fields at the tip of the interface crack, predict the same behaviour of planar interface cracks in real materials. The theory (5.3.38) predicts that further extension of a crack initially located on the planar interface of epoxy/glass would be in epoxy; this is confirmed by the observation (Comninou 1990).

Now, under pure mode *II* loading of planar interface crack in absence of Poisson effect, (5.3.55) gives

$$\hat{G}_{0s} = \frac{bC}{\pi e_6^{(II)(0)}\left(1 + 4\delta^2\right)} \; (Pumore\ mode\ II\ loading). \tag{5.3.57}$$

This cannot be close to one because, as seen from Fig. 5.15 for m_{12} large, $e_3^{(I)(0)}$ and $e_6^{(II)(0)}$ have different magnitudes.

SUMMARY: We have studied a model of cracking of a non-plane interface of a bi-elastic solid material. The crack is macroscopic in size with an oscillatory form front ξ (small in magnitude) located in a plane perpendicular to the x_1 direction of fracture propagation. We considered general loading (mixed mode $I + II + III$) and induced normal and shearing stresses in the bi-material as the results of the contractions due to the Poisson effect in the mediums (1) and (2) acting perpendicularly to the x_2 direction of the applied tension. The shearing stress τ^{a}_{23} plays no role in the theory, on average. We represented the crack by a continuous distribution made of 3 families J of dislocation having the form ξ. Expressions of the elastic fields (displacement and stress) have been given in the linear approximation with respect to the oscillation amplitudes associated with ξ with the concern that the dislocation stresses involved in the dislocation distribution functions D_J at the equilibrium and the crack extension force are continuous at the crossing of the interface. The analytical expressions of the D_J obtained are those of planar arrangements of straight dislocations corresponding to a plane interface crack. They do not contain oscillatory singularities at the end of the crack; it is rather the type of singularity (5.3.29) observed in cracking in a homogeneous medium. Adopting these expressions, we could give approximate expressions of crack-tip stresses and crack extension force for non-planar interface crack with front $\xi\,(x_1 = a, x_3)$. We compared this extension force of the crack with that obtained with elastic fields containing oscillatory singularities at the tip of the crack; it is the planar crack which is best documented in mode *I* loading. A remarkable agreement has been found.

5.4 Appendix to Sect. 5.3

5.4.1 Glide-Type Edge and Screw Sinusoidal Dislocations

The content of this section has been published as: P.N.B. ANONGBA, Non-planar interface crack under general loading-*I*. Glide type edge and screw sinusoidal dislocation, *Rev. Ivoir. Sci. Technol.*, 28 (2016) 24–59. The main objective of this study is to analyse the conditions for the propagation of a crack, along a non-planar interface *R* of arbitrary shape, in a pair of two firmly welded different solids *R*1 and *R*2. This work is fundamental by the fact that most of the materials, used in practice, are composite materials that can deteriorate in service by the propagation of interface cracks. To a large crack under general applied loading, fluctuating about average fracture plane (e.g. Ox_1x_3 of a Cartesian coordinate system x_i) normal to the applied tension direction, the following description applies locally: x_1, average crack propagation direction in that plane; x_2x_3, local plane of the crack front and x_3, average crack-front direction. Hence, we can define a simple model by specifying that the crack extends from $x_1 = -a$ to a, with a front lying in the x_2x_3-plane in a general form $x_2 = f(x_1, x_3)$ and an average direction that runs indefinitely along x_3. This is the model of interface crack (Fig. 5.5, Sect. 5.3) that we shall adopt in the present study. The method of analysis consists of representing the crack by a continuous array of infinitesimal dislocations having the same shape as the crack front. f can be expanded in the form of a Fourier series as

$$f = \sum_n (\xi_n \sin \kappa_n x_3 + \delta_n \cos \kappa_n x_3) + h \equiv \xi + h \tag{1}$$

where n is a positive integer; h, ξ_n, δ_n and κ_n are real numbers that depend on position x_1 along the crack length. From the stress fields of the dislocations, the crack-tip stresses and crack extension force (per unit length of the crack front) can be evaluated. In the case of an interface crack, mode *I* loading causes shear stresses corresponding to mode *II* and vice versa (for example, see Anongba 2014). It is mandatory to have the stress fields of three types of dislocation before undertaking an analysis of the conditions of non-planar interface crack motion. In the present part *I* of this study, the elastic fields of sinusoidal dislocations, edges and screws with Burgers vectors $\vec{b}_I = (0, b, 0)$ and $\vec{b}_{III} = (0, 0, b)$ are described, with special attention to stress terms with singularities $1/x_1$ and Dirac delta function $\delta(x_1)$ that come into play in crack analyses (calculation of the crack extension force, for example). In what follows, the methodology for obtaining the dislocation elastic fields and associated calculation results is first presented. Discussion and concluding remarks then follow where the passage from the elastic fields of dislocations with a sinusoidal shape to those having the form f is indicated.

Methodology

We consider a dislocation lying on a non-planar interface S having the form of a corrugated sheet that separates two firmly welded elastic solids $S1$ and $S2$ of infinite sizes (Fig. 5.6, Sect. 5.3). S is defined by the point $P_S(x_1, x_2 = \xi_n \sin \kappa_n x_3, x_3)$ and $S1$ and $S2$ occupy the regions $x_2 > \xi_n \sin \kappa_n x_3$ and $x_2 < \xi_n \sin \kappa_n x_3$, respectively. The dislocation is located at the origin, runs indefinitely in the x_3-direction and spreads in the x_2x_3-plane in the form

$$A_n = \xi_n \sin \kappa_n x_3. \tag{2}$$

When its Burgers vector $\vec{b}_I = (0, b, 0)$ is in the x_2-direction, the dislocation is edge on average. Because $\vec{b}_I$ is in the plane of location of the dislocation, this is a glide-type edge dislocation. With the Burgers vector $\vec{b}_{III} = (0, 0, b)$ in the x_3-direction, the dislocation is screw on average. This is the aim of the present study to provide displacement $\vec{u}^{(m)}$ and stress $(\sigma)^{(m)}$ fields of these types of interface dislocation. The solution methodology is that of Anongba 2015. The elastic fields ($\vec{u}^{(m)}$, $(\sigma)^{(m)}$) are assumed to be the difference between two quantities ($\vec{u}^{(m)\infty}$, $(\sigma)^{(m)\infty}$) and ($\vec{u}^{(m)W}$, $(\sigma)^{(m)W}$):

$$\begin{aligned} \vec{u}^{(m)} &= \vec{u}^{(m)\infty} - \vec{u}^{(m)W} \\ (\sigma)^{(m)} &= (\sigma)^{(m)\infty} - (\sigma)^{(m)W}. \end{aligned} \tag{3}$$

The former with ∞ corresponds to the fields of a sinusoidal dislocation (edge or screw) in an infinitely extended homogeneous solid (m); the latter with W satisfies the equations of equilibrium and is constructed in such a way that:

(a) ($\vec{u}^{(m)}$, $(\sigma)^{(m)}$) are continuous at the crossing of the interface, implying that

$$\begin{aligned} \Delta\vec{u}^{\infty}(P_S) &\equiv \vec{u}^{(2)\infty} - \vec{u}^{(1)\infty} = \vec{u}^{(2)W} - \vec{u}^{(1)W} \equiv \Delta\vec{u}^{W}(P_S) \\ (\Delta\sigma)^{\infty}(P_S) &\equiv (\sigma)^{(2)\infty} - (\sigma)^{(1)\infty} = (\sigma)^{(2)W} - (\sigma)^{(1)W} \equiv (\Delta\sigma)^{W}(P_S); \end{aligned} \tag{4}$$

(b) ($\vec{u}^{(m)}$, $(\sigma)^{(m)}$) tends to ($\vec{u}^{(m)\infty}$, $(\sigma)^{(m)\infty}$) when one moves far away from the interface in the x_2-direction. This means that

$$\begin{aligned} \vec{u}^{(m)W} &\to 0 \\ (\sigma)^{(m)W} &\to 0 \end{aligned} \tag{5}$$

when $|x_2| \to \infty$. ($\vec{u}^{(m)\infty}$, $(\sigma)^{(m)\infty}$) may be taken from Anongba 2009 and 2010; they are given to linear expressions with respect to ξ_n. The associated terms ($\vec{u}^{(0)(m)\infty}$, $(\sigma)^{(0)(m)\infty}$) of zero order correspond to the fields of a straight dislocation and second terms ($\vec{u}^{A_n(m)\infty}$, $(\sigma)^{A_n(m)\infty}$) are proportional to either A_n or its spatial derivative $\partial A_n/x_3$. Hence, we have

$$\Delta\vec{u}^{\infty}(P_S) = \Delta\vec{u}^{(0)\infty} + \Delta\vec{u}^{A_n\infty}$$
$$(\Delta\sigma)^{\infty}(P_S) = (\Delta\sigma)^{(0)\infty} + (\Delta\sigma)^{A_n\infty}, \tag{6}$$

on the interface point $P_S(x_1, x_2 = \xi_n \sin\kappa_n x_3, x_3)$; Appendix A below gives the complete list, component by component, for the sinusoidal screw dislocation; the corresponding values for the glide-type sinusoidal edge dislocation have been given in Anongba (2015). $(\vec{u}^{(m)W}, (\sigma)^{(m)W})$ are obtained with the help of Galerkin vectors; these are available for the edge dislocation (Anongba 2015). For the screw dislocation, we arrive at Galerkin vectors $\vec{V}$ with only one non-zero x_3-component, arranged in the form

$$V_3(\vec{x}) = \overline{\alpha}_3(\vec{k})e^{i\vec{k}.\vec{x}} + \overline{\beta}_3(\vec{k})x_2 e^{i\vec{k}.\vec{x}} \tag{7}$$

under the condition $\vec{k}^2 = k_1^2 + k_2^2 + k_3^2 = 0$ that ensures the biharmonicity of V_3. For V_3 to cancel far from the interface, we write

$$k_2 = k_2^{(m)} \equiv (-1)^{m-1} i\sqrt{k_1^2 + k_3^2} \tag{8}$$

with $m = 1$ when $x_2 > \xi_n \sin\kappa_n x_3$(half-space 1) and $m = 2$ when $x_2 < \xi_n \sin\kappa_n x_3$ (half-space 2). We use the notations

$$\vec{k}^{(m)} \equiv \left(k_1, k_2^{(m)}, k_3\right), \overline{\alpha}_3^{(m)} \equiv \overline{\alpha}_3\left(\vec{k}^{(m)}\right), \overline{\beta}_3^{(m)} \equiv \overline{\beta}_3\left(\vec{k}^{(m)}\right);$$

hence for half-space 1 ($x_2 > \xi_n \sin\kappa_n x_3$), solid (1)

$$V_3(\vec{x}) \equiv V_3^{(1)}(\vec{x}) = \overline{\alpha}_3^{(1)} e^{i\vec{k}^{(1)}.\vec{x}} + \overline{\beta}_3^{(1)} x_2 e^{i\vec{k}^{(1)}.\vec{x}}$$

and for half-space 2 ($x_2 < \xi_n \sin\kappa_n x_3$), solid (2)

$$V_3(\vec{x}) \equiv V_3^{(2)}(\vec{x}) = \overline{\alpha}_3^{(2)} e^{i\vec{k}^{(2)}.\vec{x}} + \overline{\beta}_3^{(2)} x_2 e^{i\vec{k}^{(2)}.\vec{x}}.$$

The elastic fields corresponding to V_3 (7) may be first calculated; then, more general forms $\vec{u}^{(m)V}$ and $(\sigma)^{(m)V}$ are constructed from the previous ones by superposition over k_1 and k_3; we may write

$$\vec{u}^{(m)V} = \vec{u}^{(m)A} + \vec{u}^{(m)B} = \vec{u}^{(m)A+B}$$
$$(\sigma)^{(m)V} = (\sigma)^{(m)A} + (\sigma)^{(m)B} = (\sigma)^{(m)A+B} \tag{9}$$

where terms with A and B refer to $\overline{\alpha}_3$ and $\overline{\beta}_3$ in (7) respectively. For $\vec{u}^{(m)V}$ and $(\sigma)^{(m)V}$ to conform with $\vec{u}^{(m)\infty}$ and $(\sigma)^{(m)\infty}$, the summation over k_1 is continuous and that over k_3 is discrete. k_3 takes three values: $-\kappa_n$, 0, κ_n. The fields corresponding to $k_3 = 0$ are denoted $\vec{u}^{(0)(m)V}$ and $(\sigma)^{(0)(m)V}$ and terms associated with $k_3 = -\kappa_n$ and

κ_n are merged to form expressions denoted by $\vec{u}^{A_n(m)V}$ and $(\sigma)^{A_n(m)V}$; this is made possible by requiring that

$$\overline{\alpha}_3^{(m)}(\kappa_n) \equiv -\overline{\alpha}_3^{(m)}(-\kappa_n),\ \overline{\beta}_3^{(m)}(\kappa_n) \equiv -\overline{\beta}_3^{(m)}(-\kappa_n). \tag{10}$$

In (10), $\overline{\alpha}_3^{(m)}(\kappa_n)$ stands for $\overline{\alpha}_3(k_1, k_2^{(m)}, \kappa_n)$. We write

$$\begin{aligned} \vec{u}^{(m)V} &= \vec{u}^{(0)(m)V} + \vec{u}^{A_n(m)V} \\ (\sigma)^{(m)V} &= (\sigma)^{(0)(m)V} + (\sigma)^{A_n(m)V} \end{aligned} \tag{11}$$

introducing subsequently the notation $\vec{u}^{(0)(m)A}$, $\vec{u}^{A_n(m)A}$, $\vec{u}^{(0)(m)B}$, $\vec{u}^{A_n(m)B}$ and even for the stress. $\vec{u}^{(0)(m)V}$ and $(\sigma)^{(0)(m)V}$ are x_3-independent; $\vec{u}^{A_n(m)V}$ and $(\sigma)^{A_n(m)V}$ are proportional to the sinusoid $A_n(x_3)$ or to its spatial derivative $\partial A_n/\partial x_3$. Here also, for points P_S on the interface, $\Delta\vec{u}^V$ and $(\Delta\sigma)^V$ are expanded up to terms of first order with respect to $x_2 = \xi$ in a similar manner as in (A.2) (see Appendix A) for $\Delta\vec{u}^\infty$ and $(\Delta\sigma)^\infty$. Requiring $\Delta\vec{u}^V = \Delta\vec{u}^\infty$ and $(\Delta\sigma)^V = (\Delta\sigma)^\infty$ lead to the following equations, writing first the conditions corresponding to $k_3 = 0$ (i.e. $\Delta u_i^{(0)V} = \Delta u_i^{(0)\infty}$ and $\Delta\sigma_{ij}^{(0)V} = \Delta\sigma_{ij}^{(0)\infty}$).

$$\Delta u_3^{(0)V} = \Delta u_3^{(0)\infty} \Rightarrow$$

$$\frac{\overline{\beta}_3^{(2)}}{C_2} + \frac{\overline{\beta}_3^{(1)}}{C_1} = 0 \tag{a}$$

$$\Delta\sigma_{13}^{(0)V} = \Delta\sigma_{13}^{(0)\infty} \text{ and } \Delta\sigma_{23}^{(0)V} = \Delta\sigma_{23}^{(0)\infty} \Rightarrow$$

$$(1-\nu_2)\overline{\beta}_3^{(2)} + (1-\nu_1)\overline{\beta}_3^{(1)} = 0 \tag{b}$$

$$(1-\nu_2)\overline{\beta}_3^{(2)} - (1-\nu_1)\overline{\beta}_3^{(1)} = (Q_c - Q_b)\frac{\text{sgn}(k_1)}{k_1^2} \tag{12c}$$

where $C_m = b\mu_m/2\pi(1-\nu_m)$, $Q_b = i(C_2 - C_1)/4$, $Q_c = i(\nu_2 C_2 - \nu_1 C_1)/4$;

$\text{sgn}(k_1) = k_1/|k_1|$; μ_m and ν_m are shear modulus and Poisson's ratio. In Eqs. (12a–c) above, $\overline{\beta}_3^{(m)}$ stands for $\overline{\beta}_3(k_1, k_2^{(m)}, k_3 = 0)$. Other elastic field components are zero.

The conditions corresponding to $\Delta u_i^{A_n V} = \Delta u_i^{A_n\infty}$ and $\Delta\sigma_{ij}^{A_n V} = \Delta\sigma_{ij}^{A_n\infty}$ are now listed as:

$$\Delta u_1^{A_n V} = \Delta u_1^{A_n\infty} \Rightarrow$$

$$\frac{\overline{\alpha}_3^{(2)}}{\mu_2} - \frac{\overline{\alpha}_3^{(1)}}{\mu_1} = \frac{bC_\nu \xi_n}{8\pi} \frac{k_1}{\left(k_1^2 + \kappa_n^2\right)^{3/2}} \tag{a}$$

$$\sqrt{k_1^2 + \kappa_n^2}\left(\frac{\overline{\alpha}_3^{(2)}}{\mu_2} + \frac{\overline{\alpha}_3^{(1)}}{\mu_1}\right) + \frac{\overline{\beta}_3^{(2)}}{\mu_2} - \frac{\overline{\beta}_3^{(1)}}{\mu_1} = 0 \tag{b}$$

$\Delta u_2^{A_n V} = \Delta u_2^{A_n \infty} \Rightarrow$ (13b) above and

$$\sqrt{k_1^2 + \kappa_n^2}\left(\frac{\overline{\alpha}_3^{(2)}}{\mu_2} - \frac{\overline{\alpha}_3^{(1)}}{\mu_1}\right) + 2\left(\frac{\overline{\beta}_3^{(2)}}{\mu_2} + \frac{\overline{\beta}_3^{(1)}}{\mu_1}\right) = -\frac{bC_\nu \xi_n}{8\pi} \frac{k_1}{k_1^2 + \kappa_n^2} \tag{c}$$

$$\Delta u_3^{A_n V} = \Delta u_3^{A_n \infty} \Rightarrow$$

$$\kappa_n^2\left(\frac{\overline{\alpha}_3^{(2)}}{\mu_2} - \frac{\overline{\alpha}_3^{(1)}}{\mu_1}\right) + 4\sqrt{k_1^2 + \kappa_n^2}\left(\frac{(1-\nu_2)\overline{\beta}_3^{(2)}}{\mu_2} + \frac{(1-\nu_1)\overline{\beta}_3^{(1)}}{\mu_1}\right)$$
$$= \frac{bC_\nu \xi_n \kappa_n^2}{8\pi} \frac{k_1}{\left(k_1^2 + \kappa_n^2\right)^{3/2}} \tag{d}$$

$$\kappa_n^2\sqrt{k_1^2 + \kappa_n^2}\left(\frac{\overline{\alpha}_3^{(2)}}{\mu_2} + \frac{\overline{\alpha}_3^{(1)}}{\mu_1}\right) + \frac{\overline{\beta}_3^{(2)}}{\mu_2}\left[\kappa_n^2 + 4(1-\nu_2)(k_1^2 + \kappa_n^2)\right]$$
$$- \frac{\overline{\beta}_3^{(1)}}{\mu_1}\left[\kappa_n^2 + 4(1-\nu_1)(k_1^2 + \kappa_n^2)\right] = 0 \tag{e}$$

$$\Delta \sigma_{11}^{A_n V} = \Delta \sigma_{11}^{A_n \infty} \Rightarrow$$

$$k_1^2\left(\overline{\alpha}_3^{(2)} - \overline{\alpha}_3^{(1)}\right) + 2\sqrt{k_1^2 + \kappa_n^2}\left(\nu_2\overline{\beta}_3^{(2)} + \nu_1\overline{\beta}_3^{(1)}\right) = -\frac{iQ_b \xi_n k_1 (k_1^2 + 2\kappa_n^2)}{\left(k_1^2 + \kappa_n^2\right)^{3/2}} \tag{f}$$

$$k_1^2\sqrt{k_1^2 + \kappa_n^2}\left(\overline{\alpha}_3^{(2)} + \overline{\alpha}_3^{(1)}\right) + [(1+2\nu_2)k_1^2 + 2\nu_2\kappa_n^2]\overline{\beta}_3^{(2)}$$
$$- [(1+2\nu_1)k_1^2 + 2\nu_1\kappa_n^2]\overline{\beta}_3^{(1)} = 0 \tag{g}$$

$$\Delta \sigma_{22}^{A_n V} = \Delta \sigma_{22}^{A_n \infty} \Rightarrow$$

$$\sqrt{k_1^2 + \kappa_n^2}\left(\overline{\alpha}_3^{(2)} - \overline{\alpha}_3^{(1)}\right) + 2(1-\nu_2)\overline{\beta}_3^{(2)}$$
$$+ 2(2-\nu_1)\overline{\beta}_3^{(1)} = \frac{i(2Q_c - Q_b)\xi_n k_1}{k_1^2 + \kappa_n^2} \tag{h}$$

$$\sqrt{k_1^2+\kappa_n^2}\left(\overline{\alpha}_3^{(2)}+\overline{\alpha}_3^{(1)}\right)+(3-2\nu_2)\overline{\beta}_3^{(2)}-(3-2\nu_1)\overline{\beta}_3^{(1)}=0 \tag{i}$$

$$\Delta\sigma_{33}^{A_nV}=\Delta\sigma_{33}^{A_n\infty}\Rightarrow$$

$$\kappa_n^2\left(\overline{\alpha}_3^{(2)}-\overline{\alpha}_3^{(1)}\right)+2\sqrt{k_1^2+\kappa_n^2}\left((2-\nu_2)\overline{\beta}_3^{(2)}+(2-\nu_1)\overline{\beta}_3^{(1)}\right)=-\frac{iQ_b\xi_nk_1(2k_1^2+\kappa_n^2)}{\left(k_1^2+\kappa_n^2\right)^{3/2}} \tag{j}$$

$$\kappa_n^2\sqrt{k_1^2+\kappa_n^2}\left(\overline{\alpha}_3^{(2)}+\overline{\alpha}_3^{(1)}\right)+[(5-2\nu_2)\kappa_n^2+2(2-\nu_2)k_1^2]\overline{\beta}_3^{(2)}-[(5-2\nu_1)\kappa_n^2+2(2-\nu_1)k_1^2]\overline{\beta}_3^{(1)}=0 \tag{k}$$

$$\Delta\sigma_{12}^{A_nV}=\Delta\sigma_{12}^{A_n\infty}\Rightarrow$$

$$\sqrt{k_1^2+\kappa_n^2}\left(\overline{\alpha}_3^{(2)}-\overline{\alpha}_3^{(1)}\right)+2(\overline{\beta}_3^{(2)}+\overline{\beta}_3^{(1)})=\frac{i\xi_n[(Q_c-2Q_b)k_1^2+(Q_c-Q_b)\kappa_n^2]}{k_1(k_1^2+\kappa_n^2)} \tag{l}$$

$$\sqrt{k_1^2+\kappa_n^2}\left(\overline{\alpha}_3^{(2)}+\overline{\alpha}_3^{(1)}\right)+\overline{\beta}_3^{(2)}-\overline{\beta}_3^{(1)}=0 \tag{m}$$

$$\Delta\sigma_{13}^{A_nV}=\Delta\sigma_{13}^{A_n\infty}\Rightarrow$$

$$\kappa_n^2\left(\overline{\alpha}_3^{(2)}-\overline{\alpha}_3^{(1)}\right)+2\sqrt{k_1^2+\kappa_n^2}\left((1-\nu_2)\overline{\beta}_3^{(2)}+(1-\nu_1)\overline{\beta}_3^{(1)}\right)=\frac{i\xi_n\left[Q_b\kappa_n^2k_1^2+(Q_c-Q_b)(k_1^2+\kappa_n^2)^2\right]}{k_1\left(k_1^2+\kappa_n^2\right)^{3/2}} \tag{n}$$

$$\kappa_n^2\sqrt{k_1^2+\kappa_n^2}\left(\overline{\alpha}_3^{(2)}+\overline{\alpha}_3^{(1)}\right)+[(3-2\nu_2)\kappa_n^2+2(1-\nu_2)k_1^2]\overline{\beta}_3^{(2)}-[(3-2\nu_1)\kappa_n^2+2(1-\nu_1)k_1^2]\overline{\beta}_3^{(1)}=0 \tag{o}$$

$\Delta\sigma_{23}^{A_nV}=\Delta\sigma_{23}^{A_n\infty}\Rightarrow$(o) above and

$$\kappa_n^2\sqrt{k_1^2+\kappa_n^2}\left(\overline{\alpha}_3^{(2)}-\overline{\alpha}_3^{(1)}\right)+2[(2-\nu_2)\kappa_n^2+(1-\nu_2)k_1^2]\overline{\beta}_3^{(2)}$$

$$+\,2[(2-\nu_1)\kappa_n^2+(1-\nu_1)k_1^2]\overline{\beta}_3^{(1)}=\frac{i\xi_n k_1[(Q_c-Q_b)k_1^2+(Q_c-2Q_b)\kappa_n^2]}{k_1^2+\kappa_n^2}\quad(p)$$

(13)

Next, we are concerned with satisfying boundary conditions: (12) leads to the displacement and stress fields due to an interface straight screw dislocation ($\vec{b}_{III}=(0,0,b)$) parallel to the x_3-direction at the origin; the interface is the Ox_1x_3-plane. (13) provides the complementary terms (to first order in ξ_n) in the elastic fields of an interfacial sinusoidal screw dislocation.

Calculation Results

Elastic Fields of an Interface Glide-Type Sinusoidal Edge Dislocation

When ξ_n is small, the elastic fields (displacement $\vec{u}^{(m)}$ and stress $(\sigma)^{(m)}$) at an arbitrary position $\vec{x}=(x_1,x_2,x_3)$ may be expressed (to linear terms in ξ_n) as

$$\begin{aligned}\vec{u}^{(m)}&=\vec{u}^{(0)(m)}+\vec{u}^{A_n(m)}\\(\sigma)^{(m)}&=(\sigma)^{(0)(m)}+(\sigma)^{A_n(m)},\end{aligned}\qquad(14)$$

with $m=1$ and 2 for medium $S1$ and $S2$, respectively. $\vec{u}^{(0)(m)}$ and $(\sigma)^{(0)(m)}$ are of zero order, independent of x_3, and correspond to the elastic fields of a straight edge dislocation lying on the planar Ox_1x_3-interface. These are given under continuity requirement of the fields on crossing the interface (Anongba 2015). $\vec{u}^{A_n(m)}$ and $(\sigma)^{A_n(m)}$ are oscillating fields proportional to either A_n or its spatial derivative $\partial A_n/\partial x_3$ with respect to x_3, written in the forms

$$\begin{aligned}\vec{u}^{A_n(m)}&=\vec{u}^{A_n(m)\infty}-\vec{u}^{A_n(m)W}\\(\sigma)^{A_n(m)}&=(\sigma)^{A_n(m)\infty}-(\sigma)^{A_n(m)W}.\end{aligned}\qquad(15)$$

Here, expressions with ∞ are associated with a sinusoidal edge dislocation in an infinitely extended homogeneous medium (m) with the equal elastic constants; second terms with W read

$$\begin{aligned}\vec{u}^{A_n(m)W}&=\eta_a^{A_n(m)}\vec{u}_a^{A_n(m)V}+\eta_b^{A_n(m)}\vec{u}_b^{A_n(m)V}\\&\quad+\eta_c^{A_n(m)}\vec{u}_c^{A_n(m)V}+\eta_d^{A_n(m)}\vec{u}_d^{A_n(m)V}\\(\sigma)^{A_n(m)W}&=\eta_a^{A_n(m)}(\sigma)_a^{A_n(m)V}+\eta_b^{A_n(m)}(\sigma)_b^{A_n(m)V}\\&\quad+\eta_c^{A_n(m)}(\sigma)_c^{A_n(m)V}+\eta_d^{A_n(m)}(\sigma)_d^{A_n(m)V}.\end{aligned}\qquad(16)$$

$\vec{u}_{atod}^{A_n(m)V}$ and $(\sigma)_{atod}^{A_n(m)V}$ are given in Anongba 2015; $\eta_{atod}^{A_n(m)}$ are real determined essentially by continuity conditions of the fields $\vec{u}^{A_n(m)}$ and $(\sigma)^{A_n(m)}$ on the interface. These lead to several equations denoted by $e_i^{A_n}$ (see relation (43) in Anongba 2015). We follow the treatment in Anongba 2015 (same definitions and notations) but introduce some changes that follow; in this section, we use the notation *Eqn* (*N*), *N* integer, to designate an equation (*N*) in Anongba 2015:

(i) Retaining only terms proportional to $1/x_1$, $u_2^{A_n(m)}$ is written as

$$u_2^{A_n(m)}(x_1, 0, x_3) = -\frac{A_n}{2} e_3^{A_n*} \frac{1}{x_1} \tag{17}$$

were

$$e_3^{A_n*} \equiv \frac{1}{\mu_m}\Big\{C_m(1-2\nu_m) + \eta_a^{A_n(m)} 2a_{1a}^{(m)} + \eta_b^{A_n(m)}(-1)^m(1-2\nu_m)2iQ_b$$
$$+ \eta_c^{A_n(m)} 2i\left[\sum_{i=1}^{3} a_{ic}^{(m)} + (-1)^{m-1}2\rho_m(1-2\nu_m)(b_{1c}-b_{2c}+b_{3c})\right]$$
$$+\eta_d^{A_n(m)} 2i\left[a_{1d}^{(m)} + a_{2d}^{(m)} + (-1)^{m-1}4(1-2\nu_m)\overline{b}^{(m)}\right]\Big\}.$$

$e_3^{A_n*}$ is constant with $m = 1$ and 2. $e_3^{A_n*}$ is used in place of $e_3^{A_n}$ in Eq. (43). We stress that $e_3^{A_n}$ is proportional to the modified Bessel function $K_1[\kappa_n|x_1|]$ only that has a singularity of the type $1/x_1$ but additional terms with $1/x_1$ do exist in $u_2^{A_n(m)}$. Considering all the singularity terms with $1/x_1$ leads to $e_3^{A_n*}$ above.

(ii) $e_{72}^{A_n}$ in Eq. (43) is proportional to the modified Bessel function $K_0[\kappa_n|x_1|]$ only that has a singularity of the type $\ln|x_1|$; however, there are other terms with $\ln|x_1|$ in $\sigma_{12}^{A_n(m)}$. Collecting all the terms with $\ln|x_1|$ and setting the coefficient of $\ln|x_1|$ constant with m corresponds to $e_{72}^{A_n*}$ constant with m with

$$e_{72}^{A_n*} \equiv -\nu_m C_m + \eta_a^{A_n(m)} 2a_{1a}^{(m)} + \eta_b^{A_n(m)}(-1)^{m-1} iQ_b(2+\nu_m-\rho_m)$$
$$+ \eta_c^{A_n(m)} 2i\left[a_{2c}^{(m)} + \theta a_{3c}^{(m)} + (-1)^{m-1}2\nu_m\rho_m(b_{2c}-\theta b_{3c})\right]$$
$$+ \eta_d^{A_n(m)} 2i\left[a_{2d}^{(m)} + (-1)^m 2\nu_m\overline{b}^{(m)}\right]. \tag{18}$$

$e_{72}^{A_n*}$ is at present considered in place of $e_{72}^{A_n}$.

(iii) We add a new equation $e_9^{A_n*}$ corresponding to the condition that $\sigma_{23}^{A_n(m)}(x_1, 0, x_3)$ is constant with $m = 1$ and 2. Restricting ourselves to all the terms with $1/x_1$ only, we obtain

$$\sigma_{23}^{A_n(m)}(x_1, 0, x_3) \equiv \frac{\partial A_n}{\partial x_3}\frac{e_9^{A_n*}}{x_1} \tag{19}$$

with

$$e_9^{A_n*} \equiv \nu_m C_m - \eta_a^{A_n(m)} 2a_{1a}^{(m)} + \eta_b^{A_n(m)}(-1)^m 2iQ_b$$

$$- \eta_c^{A_n(m)} 2i \left[\sum_{i=1}^{3} a_{ic}^{(m)} + (-1)^m 2\rho_m \nu_m (b_{1c} - b_{2c} + b_{3c}) \right]$$
$$- \eta_d^{A_n(m)} 2i \left[a_{1d}^{(m)} + a_{2d}^{(m)} + (-1)^m 4\nu_m \overline{b}^{(m)} \right].$$

We stress that all the others $e_i^{A_n}$ in Eq. (43) are unchanged.

(iv) In the resolution procedure, it is required to choose four equations (Ei), $i = 1$ to 4, corresponding to four independent equations $e_j^{A_n}(1) = e_j^{A_n}(2)$. A choice corresponds to Eq. (51); ($E1$) and ($E2$), there, are at present replaced by ($E1$)* and ($E2$)* below.

$$(E1)^* : e_9^{A_n^*}(2) + e_{72}^{A_n^*}(2) = e_9^{A_n^*}(1) + e_{72}^{A_n^*}(1)$$
$$(E2)^* : e_9^{A_n*}(2) = e_9^{A_n*}(1). \qquad (20)$$

($E3$) and ($E4$) are unchanged. The identifications corresponding to ($E1$)* and ($E2$)*, to be used in place of ($E1$) and ($E2$) in Eq. (52), are:

$$b_1 = 2i\left(a_{1d}^{(2)} + 2\nu_2 \overline{b}^{(2)}\right) D,$$

$$a_{11} = -2i\left(a_{1d}^{(2)} + 2\nu_2 \overline{b}^{(2)}\right) D_a,$$

$$\begin{aligned} a_{12} = & -iQ_b(\nu_2 - \nu_1) B_b \\ & - 2i\left(a_{1c}^{(2)} + (1-\theta) a_{3c}^{(2)} + 2\nu_1\nu_2 [b_{1c} - 2b_{2c} + (1+\theta) b_{3c}]\right) C_b \\ & - 2i\left(a_{1d}^{(2)} + 2\nu_2 \overline{b}^{(2)}\right) D_d + iQ_b(\nu_2 - \nu_1), \end{aligned}$$

$$\begin{aligned} a_{13} = & -iQ_b(\nu_2 - \nu_1) B_c \\ & - 2i\left(a_{1c}^{(2)} + (1-\theta) a_{3c}^{(2)} + 2\nu_1\nu_2 [b_{1c} - 2b_{2c} + (1+\theta) b_{3c}]\right) C_c \\ & - 2i\left(a_{1d}^{(2)} + 2\nu_2 \overline{b}^{(2)}\right) D_c \\ & - 2i\left(a_{1c}^{(1)} + (1-\theta) a_{3c}^{(1)} - 2\nu_1\nu_2 [b_{1c} - 2b_{2c} + (1+\theta) b_{3c}]\right), \end{aligned}$$

$$a_{14} = -2i\left(a_{1d}^{(2)} + 2\nu_2 \overline{b}^{(2)}\right) D_d - 2i\left(a_{1d}^{(1)} - 2\nu_1 \overline{b}^{(1)}\right);$$

$$b_2 = \nu_1 C_1 - \nu_2 C_2 + 2a_{1a}^{(2)} A + 2i\left(a_{1d}^{(2)} + a_{2d}^{(2)} + 4\nu_2 \overline{b}^{(2)}\right) D,$$

$$a_{21} = -2a_{1a}^{(2)} A_a - 2i\left(a_{1d}^{(2)} + a_{2d}^{(2)} + 4\nu_2 \overline{b}^{(2)}\right) D_a - 2a_{1a}^{(1)},$$

$$a_{22} = -2a_{1a}^{(2)}A_b + 2iQ_bB_b - 2i\sum_{s=1}^{3}\left[a_{sc}^{(2)} + 2\nu_1\nu_2(-1)^{s-1}b_{sc}\right]C_b$$
$$- 2i\left(a_{1d}^{(2)} + a_{2d}^{(2)} + 4\nu_2\overline{b}^{(2)}\right)D_b - 2iQ_b,$$

$$a_{23} = -2a_{1a}^{(2)}A_c + 2iQ_bB_c - 2i\sum_{s=1}^{3}\left[a_{sc}^{(2)} + 2\nu_1\nu_2(-1)^{s-1}b_{sc}\right]C_c$$
$$- 2i\left(a_{1d}^{(2)} + a_{2d}^{(2)} + 4\nu_2\overline{b}^{(2)}\right)D_c + 2i\sum_{s=1}^{3}\left[a_{sc}^{(1)} + 2\nu_1\nu_2(-1)^{s}b_{sc}\right],$$

$$a_{24} = -2a_{1a}^{(2)}A_d - 2i\left(a_{1d}^{(2)} + a_{2d}^{(2)} + 4\nu_2\overline{b}^{(2)}\right)D_d + 2i\left(a_{1d}^{(1)} + a_{2d}^{(1)} - 4\nu_1\overline{b}^{(1)}\right). \quad (21)$$

With changes (i) to ($i4$) above, the resolution methodology (Anongba 2015) can be followed, step by step, with no more modifications in the various listed relations there (i.e. Eqs. (43)–(56) included, and in addition, all the expressions in the associated Appendix). Again, we stress that the singularity term in $\sigma_{23}^{A_n(m)}(x_1, 0, x_3)$ is given by (19) above with $e_9^{A_n*}$ and not Eq. (57) with $e_{72}^{A_n}$.

The solutions $\eta_b^{A_n(m)}$ and $\eta_c^{A_n(m)}$ ($m = 1$ and 2) have been listed in Eqs. (5.53)–(5.55). Using the two first equations in Eq. (5.47), we can write

$$\eta_a^{A_n(1)} = \frac{a_{33}(a_{24}b_1 - a_{14}b_2) + [a_{33}(a_{14}a_{22} - a_{24}a_{12}) - a_{32}(a_{14}a_{23} - a_{24}a_{13})]\eta_b^{A_n(1)}}{a_{33}(a_{24}a_{11} - a_{14}a_{21})},$$
$$\eta_d^{A_n(1)} = \frac{a_{33}(a_{21}b_1 - a_{11}b_2) + [a_{33}(a_{11}a_{22} - a_{21}a_{12}) - a_{32}(a_{11}a_{23} - a_{21}a_{13})]\eta_b^{A_n(1)}}{a_{33}(a_{14}a_{21} - a_{24}a_{11})}. \quad (22)$$

$\eta_a^{A_n(2)}$ and $\eta_d^{A_n(2)}$ are obtained using Eq. (44):

$$\eta_a^{A_n(2)} = A + A_a\eta_a^{A_n(1)} + A_b\eta_b^{A_n(1)} + A_c\eta_c^{A_n(1)} + A_d\eta_d^{A_n(1)},$$
$$\eta_d^{A_n(2)} = D + D_a\eta_a^{A_n(1)} + D_b\eta_b^{A_n(1)} + D_c\eta_c^{A_n(1)} + D_d\eta_d^{A_n(1)}. \quad (23)$$

In summary, all the $\eta_{atod}^{A_n(m)}$ ($m = 1$ and 2) have been given, permitting one to express the oscillating parts (15) of the elastic fields $\vec{u}^{(m)}$ and $(\sigma)^{(m)}$ (14). We stress that in our previous work (Anongba 2015), we have imposed coefficients of modified Bessel functions $K_0[\kappa_n|x_1|]$ and $K_1[\kappa_n|x_1|]$ constant with m in the elastic fields $u_2^{A_n(m)}(x_1, 0, x_3)$, $\sigma_{12}^{A_n(m)}(x_1, 0, x_3)$ and $\sigma_{23}^{A_n(m)}(x_1, 0, x_3)$ whilst the present study focuses on functions $\ln|x_1|$ and $1/x_1$, respectively. Both procedures are self-consistent and should yield the equal $\eta_{atod}^{A_n(m)}$.

Elastic Fields of an Interface Sinusoidal Screw Dislocation

Interface Straight Screw Dislocation

Two distinct values for $\overline{\beta}_3^{(m)}$ are extracted from (12); these are:

$$
\begin{aligned}
(a)\,\overline{\beta}_3^{(m)} &= \left(\delta_{1m} - \frac{C_2}{C_1}\delta_{2m}\right)\frac{\mathrm{sgn}(k_1)}{k_1^2} \equiv \overline{\beta}_{3a}^{(m)} \\
(b)\,\overline{\beta}_3^{(m)} &= (-1)^{m-1}\frac{(Q_b - Q_c)}{2(1-\nu_m)}\frac{\mathrm{sgn}(k_1)}{k_1^2} \equiv \overline{\beta}_{3b}^{(m)}
\end{aligned} \tag{24}
$$

where δ_{ij} is the Kronecker delta. $\overline{\beta}_{3a}^{(m)}$ is obtained using (12a) and continuity requirement for $\vec{u}^{(0)(m)}$ at the crossing of the interface at $x_2 = 0$. $\overline{\beta}_{3b}^{(m)}$ is obtained from (12b and c and 12c) associated with the stresses. None of these values satisfies the entire (2). The associated elastic fields denoted $\vec{u}_{a\,\mathrm{and}\,b}^{(0)(m)V}$ and $(\sigma)_{a\,\mathrm{and}\,b}^{(0)(m)V}$ are displayed below. A superposition of these partial fields will provide the complete form of the solution. We have at position $\vec{x} = (x_1, x_2, x_3)$:

$$
\begin{aligned}
\delta_{ja}u_{3a}^{(0)(m)V} + \delta_{jb}u_{3b}^{(0)(m)V} &= \frac{1}{\mu_m}\left(\delta_{ja}(-1)^m V_a^{(m)} + \delta_{jb}2i(Q_c - Q_b)\right)\tan^{-1}\frac{x_1}{|x_2|}, \\
\delta_{ja}\sigma_{23a}^{(0)(m)V} + \delta_{jb}\sigma_{23b}^{(0)(m)V} &= \left(\delta_{ja}V_a^{(m)} + \delta_{jb}(-1)^m 2i(Q_c - Q_b)\right)\frac{x_1}{r^2}, \\
\delta_{ja}\sigma_{13a}^{(0)(m)V} + \delta_{jb}\sigma_{13b}^{(0)(m)V} &= \left(\delta_{ja}(-1)^m V_a^{(m)} + \delta_{jb}2i(Q_c - Q_b)\right)\left(\frac{|x_2|}{r^2} + \pi\delta_A(x_2)\delta(x_1)\right)
\end{aligned} \tag{25}
$$

where $V_a^{(m)} = 4i(1-\nu_m)(\delta_{1m} - \delta_{2m}C_2/C_1)$, $j = a$ and b and $\delta(x_1)$ is the Dirac delta function; here δ_A has the following definition: $\delta_A(x_2) = 0$ when $x_2 \neq 0$ and $\delta_A(x_2) = 1$ when $x_2 = 0$; $r^2 = x_1^2 + x_2^2$. The other elastic fields are zero.

We define the elastic fields $\vec{u}^{(0)(m)}(\vec{x})$ and $(\sigma)^{(0)(m)}(\vec{x})$ of an interface straight screw dislocation as

$$
\begin{aligned}
\vec{u}^{(0)(m)} &= \vec{u}^{(0)(m)\infty} - \vec{u}^{(0)(m)W} \\
(\sigma)^{(0)(m)} &= (\sigma)^{(0)(m)\infty} - (\sigma)^{(0)(m)W}
\end{aligned} \tag{26}
$$

with

$$
\begin{aligned}
\vec{u}^{(0)(m)W} &= \eta_a^{(m)}\vec{u}_a^{(0)(m)V} + \eta_b^{(m)}\vec{u}_b^{(0)(m)V} \\
(\sigma)^{(0)(m)W} &= \eta_a^{(m)}(\sigma)_a^{(0)(m)V} + \eta_b^{(m)}(\sigma)_b^{(0)(m)V}.
\end{aligned} \tag{27}
$$

Again $\vec{u}^{(0)(m)\infty}$ and $(\sigma)^{(0)(m)\infty}$ are due to a straight screw $\vec{b}_{III} = (0, 0, b)$ parallel to the x_3-direction at the origin in an infinite medium; $\vec{u}_{a\mathrm{and}b}^{(0)(m)V}$ and $(\sigma)_{a\mathrm{and}b}^{(0)(m)V}$ are given in (25). $\eta^{(m)}_{a\text{and}b}$ are real, to be determined by the condition that the elastic fields satisfy the following relations:

- $\vec{u}^{(0)(m)}(\vec{x})$ and $(\sigma)^{(0)(m)}(\vec{x})$ are continuous when crossing the Ox_1x_3-plane.
- $\oint_\Gamma du_3^{(0)(m)} = b$ for a closed contour Γ in x_1x_2 encircling the dislocation.
- $\vec{u}^{(0)(m)W}$ vanish far from the interface (i.e. when $|x_2| \to \infty$).

The last condition is fulfilled because all the $\vec{u}^{(0)(m)V}_{a\text{and}b}$ (25) vanish when $|x_2| \to \infty$. Also $(\sigma)^{(0)(m)\infty}$ and $(\sigma)^{(0)(m)V}_{a\text{and}b}$ vanish at infinity. Under such conditions, $\vec{u}^{(0)(m)}(\vec{x})$ and $(\sigma)^{(0)(m)}(\vec{x})$ do correspond to an interface straight screw dislocation. Next, we express the quantities involved in the above-mentioned requirements and proceed to satisfy these.

$$
\begin{aligned}
&u_3^{(0)(1)} = u_3^{(0)(2)} \text{ and } \oint_\Gamma du_3^{(0)(m)} = b \Rightarrow \\
&\left(\eta_a^{(m)}(-1)^m V_a^{(m)} + \eta_b^{(m)} 2i(Q_c - Q_b)\right)/\mu_m \equiv e_1' \\
&\sigma_{13}^{(0)(1)} = \sigma_{13}^{(0)(2)} \Rightarrow \\
&\eta_a^{(m)}(-1)^m V_a^{(m)} + \eta_b^{(m)} 2i(Q_c - Q_b) \equiv e_2' \\
&\sigma_{23}^{(0)(1)} = \sigma_{23}^{(0)(2)} \Rightarrow \\
&D_m - \eta_a^{(m)} V_a^{(m)} + \eta_b^{(m)}(-1)^{m-1} 2i(Q_c - Q_b) \equiv e_3'
\end{aligned} \tag{28}
$$

where $D_m = b\mu_m/2\pi = C_m(1 - \nu_m)$; all e_i' are constant with $m = 1$ and 2. Only expressions interconnecting the $\eta^{(m)}_{a\text{and}b}$ may be derived from (28). But only one expression leaves unchanged elastic fields $\vec{u}^{(0)(m)}$ and $(\sigma)^{(0)(m)}$ by inverting the elastic constants. This is written as

$$
\begin{aligned}
&\eta_a^{(m)}(-1)^m V_a^{(m)} + \eta_b^{(m)} 2i(Q_c - Q_b) \\
&\quad \equiv \frac{(\mu_1\delta_{1m} + \mu_2\delta_{2m})(D_2 - D_1)}{\mu_1 + \mu_2}, \; m = 1 \text{ and } 2.
\end{aligned} \tag{29}
$$

It may be obtained using $e'_{1\text{and}3}$ (28) in the form

$$
D_m + (-1)^{m-1}\mu_m e_1'(m) = e_3'(m);
$$

then, taking $m = 2$ and introducing $e'_{1\text{and}3}(1)$ [using (28)] in place of $e'_{1\text{and}3}(2)$, we obtain (29) for $m = 1$. The case $m = 2$ in (29) is obtained in a similar manner. With (29), the elastic fields are:

$$
\begin{aligned}
u_3^{(0)(m)}(\vec{x}) &= \frac{b}{2\pi}\tan^{-1}\frac{x_2}{x_1} + \frac{D_1 - D_2}{\mu_1 + \mu_2}\tan^{-1}\frac{x_1}{|x_2|}, \\
\sigma_{23}^{(0)(m)}(\vec{x}) &= \frac{b\mu_1\mu_2}{\pi(\mu_1 + \mu_2)}\frac{x_1}{r^2},
\end{aligned}
$$

$$\sigma_{13}^{(0)(m)}(\vec{x}) = -\frac{b\mu_1\mu_2}{\pi(\mu_1+\mu_2)}\frac{x_2}{r^2} + \delta_A(x_2)\frac{\mu_m(D_1-D_2)}{\mu_1+\mu_2}\pi\delta(x_1). \tag{30}$$

These are in complete agreement with previous works (see Mura 1968 and references therein). Other elastic field components are zero.

Interface Sinusoidal Screw Dislocation

Five values for $(\overline{\alpha}_3^{(m)}(\kappa_n), \overline{\beta}_3^{(m)}(\kappa_n))$ are extracted from (13); these are

(a) $\overline{\alpha}_3^{(m)} = \frac{\xi_n s_{1a}^{(m)}}{2}\frac{k_1}{(k_1^2+\kappa_n^2)^{3/2}} \equiv \overline{\alpha}_{3a}^{A_n(m)}$, $\overline{\beta}_3^{(m)} = 0 \equiv \overline{\beta}_{3a}^{A_n(m)}$;

(b)
$$\overline{\alpha}_3^{(m)} = \frac{\xi_n s_{1b}^{(m)} k_1}{\sqrt{k_1^2+\kappa_n^2}}\left(\frac{4-\nu_m-3\rho_m}{k_1^2+\Omega_{1b}\kappa_n^2} - \frac{2(1-\rho_m)}{k_1^2+\kappa_n^2}\right) \equiv \overline{\alpha}_{3b}^{A_n(m)},$$
$$\overline{\beta}_3^{(m)} = \xi_n r_{1b}^{(m)} k_1\left(\frac{1}{k_1^2+\Omega_{1b}\kappa_n^2} - \frac{1}{k_1^2+\kappa_n^2}\right) \equiv \overline{\beta}_{3b}^{A_n(m)};$$

(c)
$$\overline{\alpha}_3^{(m)} = \frac{\xi_n s_{1c}^{(m)}}{\kappa_n^2}\left((2-\nu_m)(1+Q_r)\frac{\sqrt{k_1^2+\kappa_n^2}}{k_1} - \frac{2(2-\nu_m)\kappa_n^2}{k_1\sqrt{k_1^2+\kappa_n^2}}\right.$$
$$\left. - \frac{2k_1}{\sqrt{k_1^2+\kappa_n^2}} + \frac{\kappa_n^2 k_1}{(k_1^2+\kappa_n^2)^{3/2}}\right) \equiv \overline{\alpha}_{3c}^{A_n(m)}$$

$\overline{\beta}_3^{(m)} = \frac{\xi_n r_{1c}}{k_1}\left(1+Q_r - \frac{2\kappa_n^2}{k_1^2+\kappa_n^2}\right) \equiv \overline{\beta}_{3c}^{A_n}$ independent of m;

(d)
$$\overline{\alpha}_3^{(m)} = \frac{\xi_n s_{1d}^{(m)} k_1}{(k_1^2+\kappa_n^2)^{3/2}}\left(s_{2d}^{(m)} + \frac{\kappa_n^2 s_{3d}^{(m)}}{k_1^2 - s_m\kappa_n^2} - \frac{\kappa_n^2 s_{4d}^{(m)}}{k_1^2 - r_m\kappa_n^2}\right) \equiv \overline{\alpha}_{3d}^{A_n(m)},$$
$$\overline{\beta}_3^{(m)} = \xi_n r_{1d}^{(m)} k_1\left(\frac{r_m+\tilde{Q}_r}{k_1^2 - r_m\kappa_n^2} + \frac{1-\tilde{Q}_r}{k_1^2+\kappa_n^2}\right) \equiv \overline{\beta}_{3d}^{A_n(m)};$$

(e) $\overline{\alpha}_3^{(m)} = \frac{\xi_n s_{1e}^{(m)} k_1}{\sqrt{k_1^2+\kappa_n^2}}\left(\frac{s_{3e}^{(m)}}{k_1^2} + \frac{s_{2e}^{(m)} - s_{3e}^{(m)}}{k_1^2+\kappa_n^2}\right) \equiv \overline{\alpha}_{3e}^{A_n(m)}$,

$$\overline{\beta}_3^{(m)} = \xi_n r_{1e}^{(m)} k_1\left(-\frac{1}{k_1^2} + \frac{2}{k_1^2+\kappa_n^2}\right) \equiv \overline{\beta}_{3e}^{A_n(m)} \tag{31}$$

where

$$\begin{aligned}
s_{1a}^{(m)} &= (-1)^m \mu_m b C_\nu / 8\pi,\ C_\nu \\
&= [1/(1-\nu_1) - 1/(1-\nu_2)],\ Q_r = Q_c/Q_b = 1/\tilde{Q}_r, \\
r_m &= \nu_m/(1-2\nu_m),\ \rho_m = \nu_1\delta_{m2} \\
&\quad + \nu_2\delta_{m1},\ s_m = \rho_m/(1-2\rho_m), \\
\Omega_{1b} &= [(1-\nu_1)(1-2\nu_2) + (1-\nu_2)(1-2\nu_1)]/4(1-\nu_1)(1-\nu_2);
\end{aligned}$$

$$s_{1b}^{(m)} = (-1)^{m-1} C_v D_m / 8(2 - \nu_1 - \nu_2),\ r_{1b}^{(m)} = C_v C_m / 16(1 - \Omega_{1b});$$

$$s_{1c}^{(m)} = (-1)^m i Q_b / 2,\ r_{1c} = -i Q_b / 4;$$

$$s_{1d}^{(m)} = (-1)^m i Q_c / 2,\ s_{2d}^{(m)} = -2\tilde{Q}_r \\ + 2[\rho_m - \nu_m - 2\nu_m(1 - 2\rho_m)]/(1 - 2\nu_m)(1 - 2\rho_m),$$

$$s_{3d}^{(m)} = (s_m + \tilde{Q}_r)/(1 - 2\rho_m),\ s_{4d}^{(m)} = (r_m + \tilde{Q}_r)(5 - 4\nu_m)/(1 - 2\nu_m),$$

$$r_{1d}^{(m)} = i Q_c / 2(1 - 2\nu_m)(1 + r_m);$$

$$s_{1e}^{(m)} = (-1)^m i Q_b / 8(1 - \nu_m)(1 - \rho_m),\ r_{1e}^{(m)} = i(Q_c - Q_b)/4(1 - \nu_m),$$

$$s_{2e}^{(m)} = Q_r(-\rho_m - 3\nu_m + 4\nu_m\rho_m) - 4 + 5\rho_m + 7\nu_m - 8\nu_m\rho_m,$$

$$s_{3e}^{(m)} = (Q_r - 1)(8 - 7\rho_m - 5\nu_m + 4\nu_m\rho_m). \tag{32}$$

None of these couples satisfies the entire conditions (13). For each couple, we give in Appendix B the associated oscillating elastic fields $\vec{u}^{A_n(m)V} = \vec{u}^{A_n(m)A} + \vec{u}^{A_n(m)B}$ and $(\sigma)^{A_n(m)V} = (\sigma)^{A_n(m)A} + (\sigma)^{A_n(m)B}$ defined in (9). A superposition of these partial fields will provide the complete form of solution (to first order in ξ_n).

The elastic fields $\vec{u}^{(m)}(\vec{x})$ and $(\sigma)^{(m)}(\vec{x})$ of an interface sinusoidal screw dislocation may be written as

$$\vec{u}^{(m)} = \vec{u}^{(0)(m)} + \vec{u}^{A_n(m)} \\ (\sigma)^{(m)} = (\sigma)^{(0)(m)} + (\sigma)^{A_n(m)}; \tag{33}$$

$\vec{u}^{(0)(m)}$ and $(\sigma)^{(0)(m)}$ (30) correspond to the fields of a straight screw dislocation lying on a planar interface; $\vec{u}^{A_n(m)}$ and $(\sigma)^{A_n(m)}$ are oscillating expressions proportional to either the sinusoid $A_n(x_3)$ or its spatial derivative $\partial A_n / \partial x_3$ in the forms

$$\vec{u}^{A_n(m)} = \vec{u}^{A_n(m)\infty} - \vec{u}^{A_n(m)W} \\ (\sigma)^{A_n(m)} = (\sigma)^{A_n(m)\infty} - (\sigma)^{A_n(m)W}; \tag{34}$$

with

$$\vec{u}^{A_n(m)W} = \sum_{j=a\,to\,e} \eta_j^{A_n(m)} \vec{u}_j^{A_n(m)V}$$

$$(\sigma)^{A_n(m)W} = \sum_{j=atoe} \eta_j^{A_n(m)} (\sigma)_j^{A_n(m)V}. \tag{35}$$

$\vec{u}_{atoe}^{A_n(m)V}$ and $(\sigma)_{atoe}^{A_n(m)V}$ are given in Appendix B (for $\vec{u}^{A_n(m)\infty}$ and $(\sigma)^{A_n(m)\infty}$, see Anongba 2009, 2010); $\eta_{atoe}^{A_n(m)}$ are real to be determined by the requirement that the elastic fields be continuous when crossing the interface. It is sufficient to write this condition for points on the average interface plane. Before displaying the corresponding equations, we stress what follows.

- Both $u_1^{A_n(m)}(x_1, 0, x_3)$ and $\sigma_{13}^{A_n(m)}(x_1, 0, x_3)$ contain two terms with singularities $\ln|x_1|$ and $1/x_1^2$; the associated coefficients ($e_{11}^{A_n}$ and $e_{81}^{A_n}$) and ($e_{12}^{A_n}$ and $e_{82}^{A_n}$), respectively, have been taken constant.
- $u_2^{A_n(m)}(x_1, 0, x_3)$, $\sigma_{12}^{A_n(m)}(x_1, 0, x_3)$ and $\sigma_{23}^{A_n(m)}(x_1, 0, x_3)$ are bounded functions. Under such conditions, we have considered their linear forms with respect to x_1 and posed the terms proportional to x_1 constant with $m = 1$ and 2.
- $\sigma_{11}^{A_n(m)}(x_1, 0, x_3)$ and $\sigma_{22}^{A_n(m)}(x_1, 0, x_3)$ have terms with singularities $1/x_1$ and those of $\partial^2 I_0/\partial|x_2|\partial x_1 \equiv \left(2/x_1^2 + \kappa_n^2 \ln \kappa_n|x_1|\right)/x_1$; the associated coefficients ($e_{41}^{A_n}$ and $e_{51}^{A_n}$) and ($e_{42}^{A_n}$ and $e_{52}^{A_n}$), respectively, have been set constant.
- $u_3^{A_n(m)}(x_1, 0, x_3)$ and $\sigma_{33}^{A_n(m)}(x_1, 0, x_3)$ contain terms with the singularity $1/x_1$ only. The associated coefficients $e_3^{A_n}$ and $e_6^{A_n}$ are taken constant.

We may write:

$$u_1^{A_n(1)}(x_1, 0, x_3) = u_1^{A_n(2)}(x_1, 0, x_3)$$
$$\Rightarrow \frac{1}{\mu_m}\left\{\frac{C_m(1-2\nu_m)}{4} - \eta_a^{A_n(m)}\frac{s_{1a}^{(m)}}{2} + \eta_b^{A_n(m)}\frac{s_{1a}^{(m)}}{2} + \eta_c^{A_n(m)} s_{1c}^{(m)}(1-2\nu_m)\right.$$
$$\left.-\eta_d^{A_n(m)} s_{1d}^{(m)} s_{2d}^{(m)} - \eta_e^{A_n(m)} s_{1e}^{(m)} s_{2e}^{(m)}\right\} \equiv e_{11}^{A_n},$$
$$\eta_c^{A_n(m)}\frac{1}{\mu_m} s_{1c}^{(m)}\left[(2-\nu_m)Q_r - \nu_m\right] \equiv e_{12}^{A_n};$$
$$u_2^{A_n(1)} = u_2^{A_n(2)}$$
$$\Rightarrow \frac{1}{\mu_m}\left\{\eta_a^{A_n(m)}(-1)^m s_{1a}^{(m)}/2 - \eta_b^{A_n(m)}\left((-1)^m s_{1b}^{(m)}\left[2(1-\rho_m) - \sqrt{\Omega_{1b}}(4-\nu_m-3\rho_m)\right]\right.\right.$$
$$\left.+r_{1b}^{(m)}\left(1-\sqrt{\Omega_{1b}}\right)\right) + \eta_d^{A_n(m)}\left((-1)^m s_{1d}^{(m)}\left[s_{2d}^{(m)} - \frac{s_{3d}^{(m)}}{1+s_m} + \frac{s_{4d}^{(m)}}{1+r_m}\right] + r_{1d}^{(m)}(1-\tilde{Q}_r)\right)$$
$$\left.+\eta_e^{A_n(m)}\left(2r_{1e}^{(m)} + (-1)^m s_{1e}^{(m)}\left|s_{2e}^{(m)} - s_{3e}^{(m)}\right|\right)\right\} \equiv e_2^{A_n};$$
$$u_3^{A_n(1)} = u_3^{A_n(2)}$$
$$\Rightarrow \frac{1}{\mu_m}\left\{D_m/2 - \eta_c^{A_n(m)} s_{1c}^{(m)}\left[-2+\nu_m(1+Q_r)\right] + \eta_d^{A_n(m)}(-1)^{m-1}4(1-\nu_m)(1+r_m)\, r_{1d}^{(m)}\right.$$
$$\left.+\eta_e^{A_n(m)}(-1)^{m-1}4(1-\nu_m) r_{1e}^{(m)}\right\} \equiv e_3^{A_n};$$

$$
\begin{aligned}
&\sigma_{11}^{A_n(1)} = \sigma_{11}^{A_n(2)} \\
&\Rightarrow \frac{C_m}{4} - \eta_a^{A_n(m)} \frac{s_{1a}^{(m)}}{2} + \eta_b^{A_n(m)} \frac{s_{1a}^{(m)}}{2} + \eta_c^{A_n(m)} s_{1c}^{(m)} \left[1 - \nu_m (1 - Q_r)\right] \\
&- \eta_d^{A_n(m)} \left[s_{1d}^{(m)} s_{2d}^{(m)} + (-1)^m 2\nu_m (1 + r_m) r_{1d}^{(m)} \right] \\
&- \eta_e^{A_n(m)} \left[s_{1e}^{(m)} s_{2e}^{(m)} + (-1)^m 2\nu_m r_{1e}^{(m)} \right] \equiv e_{41}^{A_n}, \\
&\eta_c^{A_n(m)} s_{1c}^{(m)} \left[(\nu_m - 2) Q_r + \nu_m \right] \equiv e_{42}^{A_n};
\end{aligned}
$$

$$
\begin{aligned}
&\sigma_{22}^{A_n(1)} = \sigma_{22}^{A_n(2)} \\
&\Rightarrow \frac{C_m (1 - 2\nu_m)}{4} - \eta_a^{A_n(m)} \frac{s_{1a}^{(m)}}{2} + \eta_b^{A_n(m)} \frac{s_{1a}^{(m)}}{2} \\
&- \eta_c^{A_n(m)} s_{1c}^{(m)} \left[-2(1 - \nu_m) + Q_r \right] \\
&- \eta_d^{A_n(m)} \left[s_{1d}^{(m)} s_{2d}^{(m)} + (-1)^m 2(1 - \nu_m)(1 + r_m) r_{1d}^{(m)} \right] \\
&- \eta_e^{A_n(m)} \left[s_{1e}^{(m)} s_{2e}^{(m)} + (-1)^m 2(1 - \nu_m) r_{1e}^{(m)} \right] \equiv e_{51}^{A_n}, \\
&\eta_c^{A_n(m)} s_{1c}^{(m)} \left[(2 - \nu_m) Q_r - \nu_m \right] \equiv e_{52}^{A_n};
\end{aligned}
$$

$$
\begin{aligned}
&\sigma_{33}^{A_n(1)} = \sigma_{33}^{A_n(2)} \\
&\Rightarrow \frac{C_m}{2} + \eta_c^{A_n(m)} 2 s_{1c}^{(m)} \\
&+ \eta_d^{A_n(m)} (-1)^{m-1} 2(2 - \nu_m)(1 + r_m) r_{1d}^{(m)} \\
&+ \eta_e^{A_n(m)} (-1)^{m-1} 2(2 - \nu_m) r_{1e}^{(m)} \equiv e_6^{A_n};
\end{aligned}
$$

$$
\begin{aligned}
&\sigma_{12}^{A_n(1)} = \sigma_{12}^{A_n(2)} \\
&\Rightarrow \eta_a^{A_n(m)} (-1)^{m-1} \frac{s_{1a}^{(m)}}{2} \\
&+ \eta_b^{A_n(m)} \Big((-1)^m s_{1b}^{(m)} \left[2(1 - \rho_m) - \Omega_{1b} (4 - \nu_m - 3\rho_m) \right] \\
&+ r_{1b}^{(m)} (1 - \Omega_{1b}) \Big) + \eta_d^{A_n(m)} \Big((-1)^{m-1} s_{1d}^{(m)} \left[s_{2d}^{(m)} - s_{3d}^{(m)} + s_{4d}^{(m)} \right] + r_{1d}^{(m)} (1 + r_m)(r_m - 1 + \tilde{Q}_r) \Big) \\
&+ \eta_e^{A_n(m)} \Big((-1)^m s_{1e}^{(m)} \left[s_{3e}^{(m)} - s_{2e}^{(m)} \right] - 2 r_{1e}^{(m)} \Big) \equiv e_7^{A_n};
\end{aligned}
$$

$$
\begin{aligned}
&\sigma_{13}^{A_n(1)} = \sigma_{13}^{A_n(2)} \\
&\Rightarrow -\frac{C_m}{4} - \eta_a^{A_n(m)} \frac{s_{1a}^{(m)}}{2} - \eta_c^{A_n(m)} s_{1c}^{(m)} \\
&- \eta_e^{A_n(m)} \left[s_{1e}^{(m)} s_{2e}^{(m)} + (-1)^{m-1} 4(1 - \nu_m) r_{1e}^{(m)} \right] \\
&- \eta_d^{A_n(m)} \left[s_{1d}^{(m)} s_{2d}^{(m)} + (-1)^{m-1} 2(1 - \nu_m)(1 + r_m)(1 - r_m - \tilde{Q}_r) r_{1d}^{(m)} \right] \equiv e_{81}^{A_n},
\end{aligned}
$$

$$-\frac{D_m}{4}+\eta_c^{A_n(m)}s_{1c}^{(m)}(Q_r-1)+\eta_d^{A_n(m)}(-1)^m 2(1-\nu_m)(1+r_m)r_{1d}^{(m)}$$
$$+\eta_e^{A_n(m)}(-1)^m 2(1-\nu_m)r_{1e}^{(m)}\equiv e_{82}^{A_n};$$

$$\sigma_{23}^{A_n(1)}=\sigma_{23}^{A_n(2)}$$
$$\Rightarrow \eta_a^{A_n(m)}(-1)^m\frac{s_{1a}^{(m)}}{2}$$
$$-\eta_b^{A_n(m)}\Big((-1)^m s_{1b}^{(m)}\Big[2(1-\rho_m)-\sqrt{\Omega_{1b}}(4-\nu_m-3\rho_m)\Big]$$
$$+r_{1b}^{(m)}\Big[1-\sqrt{\Omega_{1b}}-2(1-\nu_m)(1-\Omega_{1b})\sqrt{\Omega_{1b}}\Big]\Big)$$
$$+\eta_d^{A_n(m)}\left((-1)^m s_{1d}^{(m)}\left[s_{2d}^{(m)}-\frac{s_{3d}^{(m)}}{1+s_m}+\frac{s_{4d}^{(m)}}{1+r_m}\right]\mid r_{1d}^{(m)}(1-\tilde{Q}_r)\right)$$
$$+\eta_e^{A_n(m)}\Big((-1)^m s_{1e}^{(m)}\Big[s_{2e}^{(m)}-s_{3e}^{(m)}\Big]+2r_{1e}^{(m)}\Big)\equiv e_9^{A_n}. \quad (36)$$

All $e_i^{A_n}(m)$ are constant with $m=1$ and 2 (i.e. $e_i^{A_n}(1)=e_i^{A_n}(2)$). We can see that $e_{42}^{A_n}=-e_{52}^{A_n}=-\mu_m e_{12}^{A_n}$. There are twelve equations in (36) with ten unknowns $\eta_{atoe}^{A_n(m)}$. A solution can be found with ten independent equations. A methodology of the solution may be (Anongba 2015): (i) express the $\eta_i^{A_n(2)}$ as a function of the $\eta_i^{A_n(1)}$ giving five relations, (ii) report these relations in five independent equations in (36); we have then a linear system of five equations with unknowns $\eta_i^{A_n(1)}$ that can be resolved by the usual classical method with determinants. The result is five expressions linking the $\eta_i^{A_n(1)}$ with the elastic constants of the mediums $m=1$ and 2. (iii) Come back to the $\eta_i^{A_n(2)}$ relations to find their respective values as a function of the elastic constants. Here however, we shall proceed differently by first calculating values of several $e_i^{A_n}$, including those associated with the singularity $1/x_1$ in the stress fields. When a crack is represented by a continuous distribution of infinitesimal dislocations, stress terms that have a singularity $1/x_1$ contribute a non-zero value to the crack extension force. We have, using (36),

$$\nu_m C_m/2+(-1)^m 2(1-2\nu_m)\big[\eta_c^{A_n(m)}(1+Q_r)r_{1c}$$
$$+\eta_d^{A_n(m)}(1+r_m)r_{1d}^{(m)}+\eta_e^{A_n(m)}r_{1e}^{(m)}\big]=e_{42}^{A_n}+e_{41}^{A_n}-e_{51}^{A_n}\equiv E_1^{A_n}. \quad (37)$$

Equations $e_3^{A_n}(1)=e_3^{A_n}(2)$, $e_6^{A_n}(1)=e_6^{A_n}(2)$, $E_1^{A_n}(1)=E_1^{A_n}(2)$ and $e_{42}^{A_n}(1)=e_{42}^{A_n}(2)$ are four relations that provide the values of unknowns $\eta_c^{A_n(m)}$ and $\eta_d^{A_n(m)}(1+r_m)r_{1d}^{(m)}+\eta_e^{A_n(m)}r_{1e}^{(m)}$, $m=1$ and 2, in the forms:

$$e_{52}^{A_n}=\eta_c^{A_n(m)}s_{1c}^{(m)}\big[(2-\nu_m)Q_r-\nu_m\big]$$
$$=\frac{1}{3(\Gamma-1)(\nu_2-\nu_1)}\{(\nu_2C_2-\nu_1C_1)[(1-\nu_2)(2-\nu_1)-(1-\nu_1)(2-\nu_2)\Gamma]$$

$$+ (C_2 - C_1)[(1-\nu_2)(1-2\nu_1) - \Gamma(1-\nu_1)(1-2\nu_2)]\},$$

$$(-1)^m 4\left[\eta_d^{A_n(m)}(1+r_m)r_{1d}^{(m)} + \eta_e^{A_n(m)} r_{1e}^{(m)}\right] = \frac{(2-\rho_m)(\nu_2 C_2 - \nu_1 C_1) + (1-2\rho_m)(C_2 - C_1)}{3(\nu_2 - \nu_1)} + 2(1+Q_r)\eta_c^{A_n(m)} s_{1c}^{(m)}, \tag{38}$$

where $\Gamma = \mu_2/\mu_1$. We also have [see (36)]

$$\eta_c^{A_n(m)} s_{1c}^{(m)}(Q_r - 1) + (-1)^m 2(1-\nu_m)\left[\eta_d^{A_n(m)}(1+r_m)r_{1d}^{(m)} + \eta_e^{A_n(m)} r_{1e}^{(m)}\right] = \mu_m e_{11}^{A_n} - e_{51}^{A_n}. \tag{39}$$

(39) with $m = 1$ and 2 is a system of two equations with unknowns $e_{11}^{A_n}$ and $e_{51}^{A_n}$ that can be solved with the help of the second of (38). A direct inspection of various $e_i^{A_n}$ in (36) shows that these can be given values with the help of (38). Hence, we display following expressions [$e_{52}^{A_n}$ is given in (38)]:

$$e_{11}^{A_n} = \frac{1}{6(\mu_2-\mu_1)(\nu_2-\nu_1)}\{(\nu_1 C_1 - \nu_2 C_2)[(1-\nu_1)(2-\nu_2) - (1-\nu_2)(2-\nu_1)] + (C_1 - C_2)[(1-\nu_1)(1-2\nu_2) - (1-\nu_2)(1-2\nu_1)]\},$$

$$e_{51}^{A_n} = -e_{52}^{A_n}/2,$$

$$e_3^{A_n} = \frac{b}{4\pi} + \frac{(1-\nu_1)(1-\nu_2)(\nu_1 C_1 - \nu_2 C_2)}{\mu_1(1-\nu_2)(1-2\nu_1) - \mu_2(1-\nu_1)(1-2\nu_2)} + \frac{(1-\nu_1)(1-2\nu_2) - (1-\nu_2)(1-2\nu_1)}{\mu_1(1-\nu_2)(1-2\nu_1) - \mu_2(1-\nu_1)(1-2\nu_2)} e_{52}^{A_n},$$

$$e_{41}^{A_n} = \frac{\nu_1 C_1(1-\nu_1)(1-2\nu_2)\Gamma - \nu_2 C_2(1-\nu_2)(1-2\nu_1)}{2[\Gamma(1-\nu_1)(1-2\nu_2) - (1-\nu_2)(1-2\nu_1)]} + \frac{\nu_1(1-2\nu_2)\Gamma - \nu_2(1-2\nu_1)}{2[\Gamma(1-\nu_1)(1-2\nu_2) - (1-\nu_2)(1-2\nu_1)]} e_{52}^{A_n},$$

$$e_6^{A_n} = \frac{\Gamma C_1 - C_2}{2(\Gamma - 1)} + \frac{\Gamma(\nu_1 C_1 - \nu_2 C_2)[(1-\nu_1)(2-\nu_2) - (1-\nu_2)(2-\nu_1)]}{2(\Gamma-1)[\Gamma(1-\nu_1)(1-2\nu_2) - (1-\nu_2)(1-2\nu_1)]} + \frac{\nu_1(1-2\nu_2)\Gamma - \nu_2(1-2\nu_1)}{2[\Gamma(1-\nu_1)(1-2\nu_2) - (1-\nu_2)(1-2\nu_1)]} e_{52}^{A_n}. \tag{40}$$

These expressions are unchanged by inverting the elastic constants. At this stage, only $\eta_c^{A_n(m)}$ is given a value (38). To proceed further, we introduce two parameters

$$\eta_{b+a}^{A_n(m)} = \eta_b^{A_n(m)} + \eta_a^{A_n(m)},$$
$$\eta_{b-a}^{A_n(m)} = \eta_b^{A_n(m)} - \eta_a^{A_n(m)} \tag{41}$$

and search for these two parameters and $\eta_d^{A_n(m)}$ and $\eta_e^{A_n(m)}$, $m = 1$ and 2; we have eight unknowns and look for eight independent equations. Because the value of $e_{11}^{A_n}$ is known (40), its expressions with the associated $\eta_i^{A_n(m)}$ (36) provide two equations; two additional relations are given by the second of (38) with $\eta_d^{A_n(m)}(1+r_m)r_{1d}^{(m)} + \eta_e^{A_n(m)} r_{1e}^{(m)}$. These first four equations can be written in the form

$$c_{11}^{(m)}\eta_d^{A_n(m)} + c_{12}^{(m)}\eta_e^{A_n(m)} = b_1^{(m)},$$
$$c_{21}^{(m)}\eta_d^{A_n(m)} + c_{22}^{(m)}\eta_e^{A_n(m)} = b_2^{(m)} \tag{42}$$

where

$$c_{11}^{(m)} = -s_{1d}^{(m)}s_{2d}^{(m)},\ c_{12}^{(m)} = -s_{1e}^{(m)}s_{2e}^{(m)},$$
$$b_1^{(m)} = -\eta_{b-a}^{A_n(m)}\frac{s_{1a}^{(m)}}{2} + \mu_m e_{11}^{A_n} - \frac{C_m(1-2\nu_m)}{4} - \eta_c^{A_n(m)}s_{1c}^{(m)}(1-2\nu_m),$$
$$c_{21}^{(m)} = (1+r_m)r_{1d}^{(m)},\ c_{22}^{(m)} = r_{1e}^{(m)},$$
$$b_2^{(m)} = \frac{(-1)^m}{4}\left(\frac{(2-\rho_m)(\nu_2C_2-\nu_1C_1)+(1-2\rho_m)(C_2-C_1)}{3(\nu_2-\nu_1)}\right.$$
$$\left. +2(1+Q_r)\eta_c^{A_n(m)}s_{1c}^{(m)}\right).$$

We obtain

$$\eta_e^{A_n(m)} = \frac{c_{21}^{(m)}b_1^{(m)} - c_{11}^{(m)}b_2^{(m)}}{c_{21}^{(m)}c_{12}^{(m)} - c_{11}^{(m)}c_{22}^{(m)}},$$
$$\eta_d^{A_n(m)} = \frac{c_{22}^{(m)}b_1^{(m)} - c_{12}^{(m)}b_2^{(m)}}{c_{22}^{(m)}c_{11}^{(m)} - c_{12}^{(m)}c_{21}^{(m)}}. \tag{43}$$

Relations (43) give $\eta_d^{A_n(m)}$ and $\eta_e^{A_n(m)}$, $m = 1$ and 2, as a function of $\eta_{b-a}^{A_n(m)}$ only through $b_1^{(m)}$. It remains to evaluate $\eta_{b+a}^{A_n(m)}$ and $\eta_{b-a}^{A_n(m)}$ with four independent equations. We may use: $e_2^{A_n}(1) = e_2^{A_n}(2)$, $e_7^{A_n}(1) = e_7^{A_n}(2)$, $e_9^{A_n}(1) + e_7^{A_n}(1) = e_9^{A_n}(2) + e_7^{A_n}(2)$ and $e_{81}^{A_n}(1) = e_{81}^{A_n}(2)$. We then have four independent equations with unknown $\eta_{b\pm a}^{A_n(m)}$ that can be resolved by the usual classical method with determinants (see Anongba 2015 for analogous analysis).

In summary, the elastic fields [$\vec{u}^{(m)}$ and $(\sigma)^{(m)}$: (33)] of an interfacial sinusoidal screw dislocation (Fig. 5.6, Sect. 5.3) to linear order in ξ_n, are the sum of two terms: the first ones correspond to the elastic fields [$\vec{u}^{(0)(m)}$ and $(\sigma)^{(0)(m)}$: (30)] produced by an interface straight screw dislocation; second terms are oscillating expressions [$\vec{u}^{A_n(m)}$,$(\sigma)^{A_n(m)}$: (34)], proportional to the perturbation $A_n(x_3) = \xi_n \sin\kappa_n x_3$ or its partial derivative $\partial A_n/\partial x_3$. These latter expressions can be written as linear combinations (35) of partial elastic fields $\vec{u}_i^{A_n(m)V}$ ($i = a$ to e) (see Appendix B). The associated proportionality coefficients $\eta_i^{A_n(m)}$ fulfilled continuity requirements (36) of the elastic ($\vec{u}^{(m)}$, $(\sigma)^{(m)}$) on crossing the sinusoidal interface. These can be solved to provide values for the $\eta_i^{A_n(m)}$ [see (38) and (43) with the associated text]. Several $e_i^{A_n}$ values have been calculated (40) that include the coefficients of the various stress terms with the singularity $1/x_1$. We shall make use of certain terms in the crack analysis (Sect. 5.3).

Discussion and Concluding Remarks

The displacement and stress fields of sinusoidal dislocations (glide-type edge and screw) lying at the origin, on a non-planar interface S with the form of a corrugated sheet (Fig. 5.6, Sect. 5.3), have been determined. Stress terms with the singularity $1/x_1$ have been emphasized in view of crack analyses. As an illustration, consider the stress $\sigma_{33}^{(m)}$ (case of sinusoidal screw dislocation only): on the interface $x_2 = A_n$ is assumed small. We can take the linear form of the stress up to term with x_2, i.e.

$$\sigma_{33}^{(m)}(x_1, x_2, x_3) = \sigma_{33}^{(m)}(x_1, 0, x_3) + \frac{\partial \sigma_{33}^{(m)}(x_1, 0, x_3)}{\partial x_2} x_2.$$

We have

$$\sigma_{33}^{(m)}(x_1, 0, x_3) = \sigma_{33}^{(0)(m)}(x_1, 0, x_3) + \sigma_{33}^{A_n(m)}(x_1, 0, x_3).$$

The first term $\sigma_{33}^{(0)(m)}$ (30) is zero. Restricting ourselves to stress terms with $1/x_1$ only, it is easy to see that

$$\sigma_{33}^{A_n(m)}(x_1, 0, x_3) = -\frac{\partial A_n}{\partial x_3} \frac{4e_6^{A_n}}{x_1}. \tag{44}$$

The value of $e_6^{A_n}$ is known (40) with no reference to the $\eta_i^{A_n(m)}$. This result is sufficient in the analysis of several special crack fronts when $\sigma_{33}^{(m)}$ is involved (see Anongba 2009, 2010 for the case of homogeneous solids). But in general, it will be necessary to write down explicitly the value of all the $\eta_i^{A_n(m)}$ to have the stress singularities. This study also reveals the presence of a term with Dirac delta function $\delta(x_1)$ in $\sigma_{13}^{(0)(m)}$ (30) on the interface at $x_2 = 0$. The associated coefficient is proportional to shear modulus μ_m; hence it changes with $m = 1$ and 2. No alternative has been found to this value.

Assume now that the interface with the associated dislocation have a more general form $f(1)$ in the x_2x_3-plane. Writing $B_n = \xi_n \sin \kappa_n x_3 + \delta_n \cos \kappa_n x_3$ and assuming ξ small, the elastic fields $\vec{u}^{(m)}$ and $(\sigma)^{(m)}$ in the bi-material are simply (to linear terms in the amplitudes)

$$\vec{u}^{(m)} = \vec{u}_h^{(0)(m)} + \vec{u}_\xi^{(m)},$$

$$(\sigma)^{(m)} = (\sigma)_h^{(0)(m)} + (\sigma)_\xi^{(m)}; \tag{45}$$

where

$$\vec{u}_\xi^{(m)} = \sum_n \vec{u}^{B_n(m)},$$
$$(\sigma)_\xi^{(m)} = \sum_n (\sigma)^{B_n(m)}. \tag{46}$$

Here, $\vec{u}_h^{(0)(m)}$ and $(\sigma)_h^{(0)(m)}$ are the fields of an interface straight dislocation, displaced by $x_2 = h$ from the origin. In the various elastic field expressions obtained with sinusoidal dislocations with shape A_n, we just have to replace A_n with B_n and add the symbol $\sum$ in front of the oscillating elastic fields; in addition, x_2 has to be replaced by $x_2 - h$.

We shall proceed further by providing expressions for the elastic fields of interface climb-type edge sinusoidal dislocation ($\vec{b}_{II} = (b, 0, 0)$), crack tip stresses and crack extension force. These will be the subject of part *II* of the work (Sect. 5.4.2).

Appendix A: Difference of the Values of $\vec{u}^{(m)\infty}$ and $(\sigma)^{(m)\infty}$ (Screw) When Crossing the Interface

Our purpose here is to write down the differences $\Delta\vec{u}^\infty$ and $(\Delta\sigma)^\infty$ (4) on crossing the interface at arbitrary point $P_S(x_1, x_2 = \xi_n \sin \kappa_n x_3, x_3)$. We use the notation $x_2 = \xi$ ($\xi = \xi_n \sin \kappa_n x_3$ small) and take the MacLaurin series expansions of the elastic fields up to terms of first order with respect to ξ; this means that

$$\Delta\vec{u}^\infty(x_1, x_2 = \xi, x_3) = \Delta\vec{u}^\infty(x_1, 0, x_3) + \frac{\partial \Delta\vec{u}^\infty}{\partial x_2}(x_1, 0, x_3)\xi,$$
$$(\Delta\sigma)^\infty(x_1, x_2 = \xi, x_3) = (\Delta\sigma)^\infty(x_1, 0, x_3) + \frac{\partial(\Delta\sigma)^\infty}{\partial x_2}(x_1, 0, x_3)\xi. \tag{A1}$$

$\vec{u}^{(m)\infty}$ and $(\sigma)^{(m)\infty}$ are taken from our previous works (Anongba 2009, 2010). We obtain (u_i is the *i*-component of vector $\vec{u}$ and σ_{ij} the *ij*-element of the stress matrix (σ); $i, j = 1$ to 3)

$$\Delta u_i^\infty(x_1, x_2 = \xi, x_3) = \Delta u_i^{(0)\infty} + \Delta u_i^{A_n\infty}$$
$$\Delta\sigma_{ij}^\infty(x_1, x_2 = \xi, x_3) = \Delta\sigma_{ij}^{(0)\infty} + \Delta\sigma_{ij}^{A_n\infty}$$

as

$$\Delta u_1^{(0)\infty} = 0,$$

$$\Delta u_1^{A_n\infty} = \frac{\partial A_n}{\partial x_3}\frac{bC_\nu}{8\pi}\int_{-\infty}^{\infty}\frac{k_1^2}{\left(k_1^2+\kappa_n^2\right)^{3/2}}e^{ik_1x_1}\mathrm{d}k_1;$$

$$\Delta u_2^{(0)\infty} = 0,$$

$$\Delta u_2^{A_n\infty} = \frac{\partial A_n}{\partial x_3}\frac{bC_v}{8\pi}\int\limits_{-\infty}^{\infty}\frac{ik_1}{\sqrt{k_1^2+\kappa_n^2}}e^{ik_1x_1}\mathrm{d}k_1\xi;$$

$$\Delta u_3^{(0)\infty} = 0,$$

$$\Delta u_3^{A_n\infty} = \kappa_n^2 A_n\frac{bC_v}{8\pi}\int\limits_{-\infty}^{\infty}\frac{ik_1}{\left(k_1^2+\kappa_n^2\right)^{3/2}}e^{ik_1x_1}\mathrm{d}k_1;$$

$$\Delta\sigma_{11}^{(0)\infty} = 0,$$

$$\Delta\sigma_{11}^{A_n\infty} = 2Q_b\frac{\partial A_n}{\partial x_3}\int\limits_{-\infty}^{\infty}\frac{k_1(k_1^2+2\kappa_n^2)}{\left(k_1^2+\kappa_n^2\right)^{3/2}}e^{ik_1x_1}\mathrm{d}k_1;$$

$$\Delta\sigma_{22}^{(0)\infty} = 0,$$

$$\Delta\sigma_{22}^{A_n\infty} = 2(2Q_c-Q_b)\frac{\partial A_n}{\partial x_3}\int\limits_{-\infty}^{\infty}\frac{k_1}{\sqrt{k_1^2+\kappa_n^2}}e^{ik_1x_1}\mathrm{d}k_1;$$

$$\Delta\sigma_{33}^{(0)\infty} = 0,$$

$$\Delta\sigma_{33}^{A_n\infty} = 2Q_b\frac{\partial A_n}{\partial x_3}\int\limits_{-\infty}^{\infty}\frac{k_1(2k_1^2+\kappa_n^2)}{\left(k_1^2+\kappa_n^2\right)^{3/2}}e^{ik_1x_1}\mathrm{d}k_1;$$

$$\Delta\sigma_{12}^{(0)\infty} = 0,$$

$$\Delta\sigma_{12}^{A_n\infty} = 2i\frac{\partial A_n}{\partial x_3}\int\limits_{-\infty}^{\infty}\frac{Q_b\kappa_n^2+(Q_c-2Q_b)(k_1^2+\kappa_n^2)}{\sqrt{k_1^2+\kappa_n^2}}e^{ik_1x_1}\mathrm{d}k_1\xi;$$

$$\Delta\sigma_{13}^{(0)\infty} = 2i(Q_c-Q_b)\int\limits_{-\infty}^{\infty}|k_1|e^{ik_1x_1}dk_1\xi,$$

$$\Delta\sigma_{13}^{A_n\infty} = -2iA_n\int\limits_{-\infty}^{\infty}\frac{Q_b\kappa_n^2k_1^2+(Q_c-Q_b)(k_1^2+\kappa_n^2)^2}{\left(k_1^2+\kappa_n^2\right)^{3/2}}e^{ik_1x_1}\mathrm{d}k_1;$$

$$\Delta\sigma_{23}^{(0)\infty} = 2(Q_c-Q_b)\int\limits_{-\infty}^{\infty}\mathrm{sgn}(k_1)e^{ik_1x_1}\mathrm{d}k_1,$$

$$\Delta\sigma_{23}^{A_n\infty} = 2A_n \int_{-\infty}^{\infty} \frac{k_1\left[(Q_b - Q_c)k_1^2 + (2Q_b - Q_c)\kappa_n^2\right]}{\left(k_1^2 + \kappa_n^2\right)^{1/2}} e^{ik_1x_1}\mathrm{d}k_1\xi; \tag{A.2}$$

where

$$C_\nu = [1/(1-\nu_1) - 1/(1-\nu_2)],\ Q_b = i(C_2 - C_1)/4,$$
$$Q_c = i(\nu_2C_2 - \nu_1C_1)/4,\ C_m = b\mu_m/2\pi(1-\nu_m);$$

$\mathrm{sgn}(k_1) = k_1/|k_1|$; μ_m and ν_m are shear modulus and Poisson's ratio.

Appendix B: Partial Oscillating Elastic Fields (Screw)

The couple $(\overline{\alpha}_{3a}^{A_n(m)}, \overline{\beta}_{3a}^{A_n(m)})$ is obtained from (13a–e) associated with the displacement. We have at position $\vec{x} = (x_1, x_2, x_3)$ $(\vec{u}^{A_n(m)V} \equiv \vec{u}_a^{A_n(m)V}, (\sigma)^{A_n(m)V} \equiv (\sigma)_a^{A_n(m)V})$:

$$\begin{aligned}
u_{ia}^{A_n(m)V} &= \frac{s_{1a}^{(m)}}{2\mu_m}\left\|(\delta_{i1}+\delta_{i2})\frac{\partial}{\partial x_3} + \delta_{i3}\right\|A_n\left(2K_0[\kappa_n r]\delta_{i1} + (-1)^{m-1}\frac{\partial\Pi_1}{\partial x_1}\right.\\
&\left.+\kappa_n^2\left\|-\delta_{i1} + \delta_{i3}\frac{\partial}{\partial x_1}\right\|I_{1a}\right),\\
\sigma_{iia}^{A_n(m)V} &= \frac{\partial A_n}{\partial x_3}\kappa_n s_{1a}^{(m)}\left((-\delta_{i1}+\delta_{i2})\frac{2x_1K_1[\kappa_n r]}{r} + \kappa_n(-\delta_{i1}+\delta_{i3})\frac{\partial I_{1a}}{\partial x_1}\right),\\
\sigma_{12a}^{A_n(m)V} &= \frac{\partial A_n}{\partial x_3}\kappa_n s_{1a}^{(m)}(-1)^m\left(\frac{2|x_2|K_1[\kappa_n r]}{r} - \kappa_n\Pi_1\right),\\
\sigma_{j3a}^{A_n(m)V} &= A_n\kappa_n^2 s_{1a}^{(m)}\left(\delta_{j1}\left(\kappa_n^2 I_{1a} - 2K_0[\kappa_n r]\right) + \delta_{j2}(-1)^m\frac{\partial\Pi_1}{\partial x_1}\right);
\end{aligned} \tag{B.1}$$

$$\Pi_z = \int_{-\infty}^{\infty} \frac{e^{(-1)^m\sqrt{k_1^2+\kappa_n^2}x_2}}{k_1^2 + z\kappa_n^2} e^{ik_1x_1}\mathrm{d}k_1 \equiv \int_{-\infty}^{\infty} \frac{e^{-\sqrt{k_1^2+\kappa_n^2}|x_2|}}{k_1^2 + z\kappa_n^2} e^{ik_1x_1}\mathrm{d}k_1,$$

$$I_{1a} = \int_{-\infty}^{\infty} \frac{e^{(-1)^m\sqrt{k_1^2+\kappa_n^2}x_2}}{\left(k_1^2+\kappa_n^2\right)^{3/2}} e^{ik_1x_1}\mathrm{d}k_1 \equiv \int_{-\infty}^{\infty} \frac{e^{-\sqrt{k_1^2+\kappa_n^2}|x_2|}}{\left(k_1^2+\kappa_n^2\right)^{3/2}} e^{ik_1x_1}\mathrm{d}k_1.$$

Terms in brackets $\|\ \|$ are operators acting on A_n and I_{1a}, separately; Π_1 is the value of Π_z for $z = 1$, $r^2 = x_1^2 + x_2^2$ and subscripts $i = 1$ to 3 and $j = 1$ and 2; $K_n[x]$ is the nth-order modified Bessel function usually so denoted and δ_{ij} is the Kronecker delta. We stress that the various integrations (such as in I_{1a} and Π_z) performed in the present study are given for spatial positions satisfying the condition $(-1)^m x_2 < 0$ (*i.e.* $(-1)^{m-1} = \mathrm{sgn}(x_2)$) with $m = 1$ when $x_2 > \xi_n \sin\kappa_n x_3$ (half-space 1) and $m = 2$

when $x_2 < \xi_n \sin\kappa_n x_3$(half-space 2). However, this makes no difference in the elastic fields to first order in ξ_n.

The pair $(\overline{\alpha}_{3b}^{A_n(m)}, \overline{\beta}_{3b}^{A_n(m)})$ is obtained using (13b–e) associated with the displacement. We have (using the similar notations)

$$u_{1b}^{A_n(m)A} = \frac{\partial A_n}{\partial x_3}\frac{s_{1b}^{(m)}}{\mu_m}\left(\frac{\nu_1+\nu_2-2-2\Omega_{1b}(1-\rho_m)}{1-\Omega_{1b}}\frac{\partial \Pi_1}{\partial|x_2|} + \frac{\Omega_{1b}(4-\nu_m-3\rho_m)}{1-\Omega_{1b}}\frac{\partial \Pi_{\Omega_{1b}}}{\partial|x_2|} + 2(1-\rho_m)\kappa_n^2 I_{1a}\right),$$

$$u_{1b}^{A_n(m)B} = \frac{\partial A_n}{\partial x_3}\frac{r_{1b}^{(m)}}{\mu_m}\kappa_n^2 x_2\left(\Pi_1 - \Omega_{1b}\Pi_{\Omega_{1b}}\right);$$

$$u_{2b}^{A_n(m)A} = \frac{\partial A_n}{\partial x_3}\frac{s_{1b}^{(m)}}{\mu_m}(-1)^{m-1}\frac{\partial}{\partial x_1}\left([4-\nu_m-3\rho_m]\Pi_{\Omega_{1b}} - 2[1-\rho_m]\Pi_1\right),$$

$$u_{2b}^{A_n(m)B} = \frac{\partial A_n}{\partial x_3}\frac{r_{1b}^{(m)}}{\mu_m}\left\|\frac{\partial}{\partial x_1} + (-1)^{m-1}x_2\frac{\partial^2}{\partial x_1\partial|x_2|}\right\|(\Pi_1 - \Pi_{\Omega_{1b}});$$

$$u_{3b}^{A_n(m)A} = A_n\frac{s_{1b}^{(m)}}{\mu_m}\left(2\kappa_n^2(\rho_m-1)\frac{\partial I_{1a}}{\partial x_1} + \frac{4-\nu_m-3\rho_m}{1-\Omega_{1b}}\frac{\partial^2}{\partial x_1\partial|x_2|}(\Pi_1 - \Pi_{\Omega_{1b}})\right),$$

$$u_{3b}^{A_n(m)B} = A_n\frac{r_{1b}^{(m)}}{\mu_m}\left\|-\kappa_n^2 x_2\frac{\partial}{\partial x_1} + (-1)^m 4(1-\nu_m)\frac{\partial^2}{\partial x_1\partial|x_2|}\right\|(\Pi_1 - \Pi_{\Omega_{1b}});$$

$$\sigma_{11b}^{A_n(m)A} = \frac{\partial A_n}{\partial x_3}2s_{1b}^{(m)}\frac{\partial}{\partial x_1}\left(\frac{\partial \Pi_1}{\partial|x_2|}\frac{\nu_1+\nu_2-2-2\Omega_{1b}(1-\rho_m)}{1-\Omega_{1b}} + \frac{\partial \Pi_{\Omega_{1b}}}{\partial|x_2|}\frac{\Omega_{1b}(4-\nu_m-3\rho_m)}{1-\Omega_{1b}} + 2(1-\rho_m)\kappa_n^2 I_{1a}\right),$$

$$\sigma_{11b}^{A_n(m)B} = \frac{\partial A_n}{\partial x_3}2r_{1b}^{(m)}\frac{\partial}{\partial x_1}\left(\kappa_n^2 x_2(\Pi_1 - \Omega_{1b}\Pi_{\Omega_{1b}}) + (-1)^m 2\nu_m\frac{\partial}{\partial|x_2|}(\Pi_1 - \Pi_{\Omega_{1b}})\right);$$

$$\sigma_{22b}^{A_n(m)A} = \frac{\partial A_n}{\partial x_3}2s_{1b}^{(m)}\frac{\partial^2}{\partial x_1\partial|x_2|}\left([4-\nu_m-3\rho_m]\Pi_{\Omega_{1b}} - 2[1-\rho_m]\Pi_1\right),$$

$$\sigma_{22b}^{A_n(m)B} = -\frac{\partial A_n}{\partial x_3}2r_{1b}^{(m)}\frac{\partial}{\partial x_1}\left((1-\Omega_{1b})\kappa_n^2 x_2\Pi_{\Omega_{1b}} + (-1)^m 2(1-\nu_m)\frac{\partial}{\partial|x_2|}(\Pi_1 - \Pi_{\Omega_{1b}})\right);$$

$$\sigma_{33b}^{A_n(m)A} = \frac{\partial A_n}{\partial x_3}2s_{1b}^{(m)}\left(2\kappa_n^2(\rho_m-1)\frac{\partial I_{1a}}{\partial x_1} + \frac{4-\nu_m-3\rho_m}{1-\Omega_{1b}}\frac{\partial^2}{\partial x_1\partial|x_2|}(\Pi_1 - \Pi_{\Omega_{1b}})\right),$$

$$\sigma_{33b}^{A_n(m)B} = \frac{\partial A_n}{\partial x_3}2r_{1b}^{(m)}\left\|-\kappa_n^2 x_2\frac{\partial}{\partial x_1} + (-1)^m 2(2-\nu_m)\frac{\partial^2}{\partial x_1\partial|x_2|}\right\|(\Pi_1 - \Pi_{\Omega_{1b}});$$

$$\sigma_{12b}^{A_n(m)A} = \frac{\partial A_n}{\partial x_3} 2 s_{1b}^{(m)} (-1)^m \big(I_0(2-\nu_1-\nu_2) + 2(1-\rho_m)\kappa_n^2 \Pi_1 - [4-\nu_m-3\rho_m]\Omega_{1b}\kappa_n^2 \Pi_{\Omega_{1b}}\big),$$

$$\sigma_{12b}^{A_n(m)B} = \frac{\partial A_n}{\partial x_3} 2 r_{1b}^{(m)} \kappa_n^2 \left\| 1 + (-1)^{m-1} x_2 \frac{\partial}{\partial |x_2|} \right\| (\Pi_1 - \Omega_{1b}\Pi_{\Omega_{1b}});$$

$$\sigma_{13b}^{A_n(m)A} = -A_n 2 s_{1b}^{(m)} \kappa_n^2 \left(\frac{\partial \Pi_1}{\partial |x_2|} \frac{\nu_1+\nu_2-2-2\Omega_{1b}(1-\rho_m)}{1-\Omega_{1b}} + \frac{\partial \Pi_{\Omega_{1b}}}{\partial |x_2|} \frac{\Omega_{1b}(4-\nu_m-3\rho_m)}{1-\Omega_{1b}} + 2(1-\rho_m)\kappa_n^2 I_{1a} \right),$$

$$\sigma_{13b}^{A_n(m)B} = -A_n r_{1b}^{(m)} \kappa_n^2 \left\| 2\kappa_n^2 x_2 + (-1)^{m-1} 4(1-\nu_m) \frac{\partial}{\partial |x_2|} \right\| (\Pi_1 - \Omega_{1b}\Pi_{\Omega_{1b}});$$

$$\sigma_{23b}^{A_n(m)A} = A_n 2 s_{1b}^{(m)} (-1)^m \kappa_n^2 \frac{\partial}{\partial x_1} \big([4-\nu_m-3\rho_m]\Pi_{\Omega_{1b}} - 2[1-\rho_m]\Pi_1\big),$$

$$\sigma_{23b}^{A_n(m)B} = A_n 2 r_{1b}^{(m)} \kappa_n^2 \frac{\partial}{\partial x_1} \Big\{ [1+2(1-\nu_m)(1-\Omega_{1b})]\Pi_{\Omega_{1b}} - \Pi_1 + (-1)^m x_2 \frac{\partial}{\partial |x_2|} (\Pi_1 - \Pi_{\Omega_{1b}}) \Big\}; \tag{B.2}$$

$$I_0 = \int_{-\infty}^{\infty} e^{(-1)^m \sqrt{k_1^2+\kappa_n^2} x_2} e^{ik_1x_1} dk_1 \equiv \int_{-\infty}^{\infty} e^{-\sqrt{k_1^2+\kappa_n^2}|x_2|} e^{ik_1x_1} dk_1 = \frac{2\kappa_n |x_2|}{r} K_1.$$

The couple $(\overline{\alpha}_{3c}^{A_n(m)}, \overline{\beta}_{3c}^{A_n(m)})$ is obtained using (13j, k, n and o) associated with stresses. We obtain

$$u_{1c}^{A_n(m)A} = \frac{\partial A_n}{\partial x_3} \frac{s_{1c}^{(m)}}{\mu_m} \frac{1}{\kappa_n^2} \left((\nu_m - (2-\nu_m)Q_r) \frac{\partial I_0}{\partial |x_2|} - 2(1-2\nu_m)\kappa_n^2 K_0[\kappa_n r] - \kappa_n^4 I_{1a} \right),$$

$$u_{1c}^{A_n(m)B} = \frac{\partial A_n}{\partial x_3} \frac{r_{1c}}{\mu_m} x_2 \big((1+Q_r)I_0 - 2\kappa_n^2 \Pi_1\big);$$

$$u_{2c}^{A_n(m)A} = \frac{\partial A_n}{\partial x_3} \frac{s_{1c}^{(m)}}{\mu_m} \frac{1}{\kappa_n^2} (-1)^{m-1} \left((-\nu_m + (2-\nu_m)Q_r) \frac{\partial I_0}{\partial x_1} + \kappa_n^2 \frac{\partial \Pi_1}{\partial x_1} + (2-\nu_m)(Q_r-1)\kappa_n^2 i I_{1c} \right),$$

$$u_{2c}^{A_n(m)B} = -\frac{\partial A_n}{\partial x_3} \frac{r_{1c}}{\mu_m} \left\| 1 + (-1)^{m-1} x_2 \frac{\partial}{\partial |x_2|} \right\| \left(2\frac{\partial \Pi_1}{\partial x_1} + i(Q_r - 1) I_{1c} \right);$$

$$u_{3c}^{A_n(m)A} = A_n \frac{s_{1c}^{(m)}}{\mu_m}\left(-2\nu_m \frac{\partial K_0}{\partial x_1} + i(2-\nu_m)(1-Q_r)\frac{\partial I_{1c}}{\partial |x_2|} + \kappa_n^2 \frac{\partial I_{1a}}{\partial x_1}\right),$$

$$u_{3c}^{A_n(m)B} = A_n \frac{r_{1c}}{\mu_m}\left\| \kappa_n^2 x_2 + (-1)^{m-1} 4(1-\nu_m)\frac{\partial}{\partial |x_2|}\right\|\left(2\frac{\partial \Pi_1}{\partial x_1} + i(Q_r - 1)I_{1c}\right);$$

$$\sigma_{11c}^{A_n(m)A} = \frac{\partial A_n}{\partial x_3} 2 s_{1c}^{(m)} \frac{1}{\kappa_n^2}\left(\frac{\partial^2 I_0}{\partial x_1 \partial |x_2|}(\nu_m + Q_r(\nu_m - 2)) - \kappa_n^4 \frac{\partial I_{1a}}{\partial x_1}\right.$$
$$\left. - 2\kappa_n^2(1-2\nu_m)\frac{\partial K_0}{\partial x_1}\right),$$

$$\sigma_{11c}^{A_n(m)B} = \frac{\partial A_n}{\partial x_3} 2 r_{1c}\left(x_2 \frac{\partial}{\partial x_1}\left[(1+Q_r)I_0 - 2\kappa_n^2 \Pi_1\right]\right.$$
$$\left. + (-1)^{m-1} 2\nu_m \frac{\partial}{\partial |x_2|}\left[2\frac{\partial \Pi_1}{\partial x_1} + i(Q_r - 1)I_{1c}\right]\right);$$

$$\sigma_{22c}^{A_n(m)A} = \frac{\partial A_n}{\partial x_3} 2 s_{1c}^{(m)} \frac{1}{\kappa_n^2}\left(\frac{\partial^2 I_0}{\partial x_1 \partial |x_2|}(Q_r(2-\nu_m) - \nu_m) - 2\kappa_n^2 \frac{\partial K_0}{\partial x_1}\right.$$
$$\left. + i\kappa_n^2 (2-\nu_m)(Q_r - 1)\frac{\partial I_{1c}}{\partial |x_2|}\right),$$

$$\sigma_{22c}^{A_n(m)B} = -\frac{\partial A_n}{\partial x_3} 2 r_{1c}\left(x_2\left[(1+Q_r)\frac{\partial I_0}{\partial x_1} + i\kappa_n^2 (Q_r - 1) I_{1c}\right]\right.$$
$$\left. + (-1)^m 2(1-\nu_m)\left[4\frac{\partial K_0}{\partial x_1} + i(1-Q_r)\frac{\partial I_{1c}}{\partial |x_2|}\right]\right);$$

$$\sigma_{33c}^{A_n(m)A} = \frac{\partial A_n}{\partial x_3} 2 s_{1c}^{(m)}\left(4(1-\nu_m)\frac{\partial K_0}{\partial x_1} + i(2-\nu_m)(1-Q_r)\frac{\partial I_{1c}}{\partial |x_2|} + \kappa_n^2 \frac{\partial I_{1a}}{\partial x_1}\right),$$

$$\sigma_{33c}^{A_n(m)B} = \frac{\partial A_n}{\partial x_3} 2 r_{1c}\left(x_2 \kappa_n^2\left[2\frac{\partial \Pi_1}{\partial x_1} + i(Q_r - 1)I_{1c}\right]\right.$$
$$\left. + (-1)^m 2(2-\nu_m)\left[4\frac{\partial K_0}{\partial x_1} + i(1-Q_r)\frac{\partial I_{1c}}{\partial |x_2|}\right]\right);$$

$$\sigma_{12c}^{A_n(m)A} = \frac{\partial A_n}{\partial x_3} 2 s_{1c}^{(m)} \frac{1}{\kappa_n^2}(-1)^m\left(\frac{\partial^2 I_0}{\partial x_1^2}(\nu_m + Q_r(\nu_m - 2))\right.$$
$$\left. - \kappa_n^4 \Pi_1 + \kappa_n^2\left[(2-\nu_m)Q_r + \nu_m - 1\right]I_0\right),$$

$$\sigma_{12c}^{A_n(m)B} = \frac{\partial A_n}{\partial x_3} 2 r_{1c}\left((1+Q_r)I_0 - 2\kappa_n^2 \Pi_1\right.$$
$$\left. + (-1)^{m-1} x_2 \frac{\partial}{\partial |x_2|}\left[(1+Q_r)I_0 - 2\kappa_n^2 \Pi_1\right]\right);$$

$$\sigma_{13c}^{A_n(m)A} = -A_n 2 s_{1c}^{(m)}\left((\nu_m - (2-\nu_m)Q_r)\frac{\partial I_0}{\partial |x_2|}\right.$$

$$-2(1-2\nu_m)\kappa_n^2 K_0[\kappa_n r] - \kappa_n^4 I_{1a}\big),$$

$$\sigma_{13c}^{A_n(m)B} = -A_n r_{1c}\bigg((-1)^{m-1}4(1-\nu_m)\frac{\partial}{\partial|x_2|}\big[(1+Q_r)I_0 - 2\kappa_n^2\Pi_1\big] + x_2 2\kappa_n^2\big[(1+Q_r)I_0 - 2\kappa_n^2\Pi_1\big]\bigg);$$

$$\sigma_{23c}^{A_n(m)A} = A_n 2s_{1c}^{(m)}(-1)^m\bigg(\big[(2-\nu_m)Q_r - \nu_m\big]\frac{\partial I_0}{\partial x_1} + i\kappa_n^2(2-\nu_m)(Q_r-1)I_{1c} + \kappa_n^2\frac{\partial \Pi_1}{\partial x_1}\bigg),$$

$$\sigma_{23c}^{A_n(m)B} = A_n r_{1c}\bigg(i2\kappa_n^2(3-2\nu_m)(Q_r-1)I_{1c} + 4\kappa_n^2\frac{\partial \Pi_1}{\partial x_1} + 4(1-\nu_m)(1+Q_r)\frac{\partial I_0}{\partial x_1} + x_2 2\kappa_n^2(-1)^m\bigg[4\frac{\partial K_0}{\partial x_1} + i(1-Q_r)\frac{\partial I_{1c}}{\partial|x_2|}\bigg]\bigg); \tag{B.3}$$

$$I_{1c} \equiv \int_{-\infty}^{\infty}\frac{e^{-\sqrt{k_1^2+\kappa_n^2}|x_2|}}{k_1}e^{ik_1x_1}dk_1.$$

The pair $(\overline{\alpha}_{3d}^{A_n(m)}, \overline{\beta}_{3d}^{A_n(m)})$ is calculated from (13f–i) associated with stresses. We have

$$u_{1d}^{A_n(m)A} = \frac{\partial A_n}{\partial x_3}\frac{s_{1d}^{(m)}}{\mu_m}\Bigg(2K_0\bigg[s_{2d}^{(m)} - \frac{s_m s_{3d}^{(m)}}{(1+s_m)^2} + \frac{r_m s_{4d}^{(m)}}{(1+r_m)^2}\bigg] \kappa_n^2\bigg[s_{2d}^{(m)} - \frac{s_{3d}^{(m)}}{1+s_m} + \frac{s_{4d}^{(m)}}{1+r_m}\bigg]I_{1a} - \frac{s_m s_{3d}^{(m)}}{(1+s_m)^2}\frac{\partial \Pi_{(-s_m)}}{\partial|x_2|} + \frac{r_m s_{4d}^{(m)}}{(1+r_m)^2}\frac{\partial \Pi_{(-r_m)}}{\partial|x_2|}\Bigg),$$

$$u_{1d}^{A_n(m)B} = \frac{\partial A_n}{\partial x_3}\frac{r_{1d}^{(m)}}{\mu_m}x_2\big((1+r_m)I_0 + r_m\kappa_n^2(r_m+\tilde{Q}_r)\Pi_{(-r_m)} - \kappa_n^2(1-\tilde{Q}_r)\Pi_1\big);$$

$$u_{2d}^{A_n(m)A} = \frac{\partial A_n}{\partial x_3}\frac{s_{1d}^{(m)}}{\mu_m}(-1)^{m-1}\frac{\partial}{\partial x_1}\Bigg(\bigg[s_{2d}^{(m)} - \frac{s_{3d}^{(m)}}{1+s_m} + \frac{s_{4d}^{(m)}}{1+r_m}\bigg]\Pi_1 + \frac{s_{3d}^{(m)}}{1+s_m}\Pi_{(-s_m)} - \frac{s_{4d}^{(m)}}{1+r_m}\Pi_{(-r_m)}\Bigg),$$

$$u_{2d}^{A_n(m)B} = -\frac{\partial A_n}{\partial x_3}\frac{r_{1d}^{(m)}}{\mu_m}\bigg\|\frac{\partial}{\partial x_1} + (-1)^{m-1}x_2\frac{\partial^2}{\partial x_1\partial|x_2|}\bigg\|\big((r_m+\tilde{Q}_r)\Pi_{(-r_m)} + (1-\tilde{Q}_r)\Pi_1\big);$$

$$u_{3d}^{A_n(m)A} = A_n \frac{s_{1d}^{(m)}}{\mu_m} \frac{\partial}{\partial x_1} \left(2K_0 \left[\frac{s_{4d}^{(m)}}{(1+r_m)^2} - \frac{s_{3d}^{(m)}}{(1+s_m)^2} \right] \right.$$

$$\left. + \kappa_n^2 \left[s_{2d}^{(m)} - \frac{s_{3d}^{(m)}}{1+s_m} + \frac{s_{4d}^{(m)}}{1+r_m} \right] I_{1a} - \frac{s_{3d}^{(m)}}{(1+s_m)^2} \frac{\partial \Pi_{(-s_m)}}{\partial |x_2|} + \frac{s_{4d}^{(m)}}{(1+r_m)^2} \frac{\partial \Pi_{(-r_m)}}{\partial |x_2|} \right),$$

$$u_{3d}^{A_n(m)B} = A_n \frac{r_{1d}^{(m)}}{\mu_m} \left\| x_2 \kappa_n^2 \frac{\partial}{\partial x_1} + (-1)^{m-1} 4(1-\nu_m) \frac{\partial^2}{\partial x_1 \partial |x_2|} \right\|$$
$$\times \left((r_m + \tilde{Q}_r)\Pi_{(-r_m)} + (1-\tilde{Q}_r)\Pi_1 \right);$$

$$\sigma_{11d}^{A_n(m)A} = \frac{\partial A_n}{\partial x_3} 2s_{1d}^{(m)} \frac{\partial}{\partial x_1} \left(2K_0 \left[s_{2d}^{(m)} - \frac{s_m s_{3d}^{(m)}}{(1+s_m)^2} + \frac{r_m s_{4d}^{(m)}}{(1+r_m)^2} \right] \right.$$

$$\left. - \kappa_n^2 \left[s_{2d}^{(m)} - \frac{s_{3d}^{(m)}}{1+s_m} + \frac{s_{4d}^{(m)}}{1+r_m} \right] I_{1a} - \frac{s_m s_{3d}^{(m)}}{(1+s_m)^2} \frac{\partial \Pi_{(-s_m)}}{\partial |x_2|} + \frac{r_m s_{4d}^{(m)}}{(1+r_m)^2} \frac{\partial \Pi_{(-r_m)}}{\partial |x_2|} \right),$$

$$\sigma_{11d}^{A_n(m)B} = \frac{\partial A_n}{\partial x_3} 2r_{1d}^{(m)} \frac{\partial}{\partial x_1} \left\{ x_2 \left[(1+r_m) I_0 + r_m \kappa_n^2 (r_m + \tilde{Q}_r)\Pi_{(-r_m)} - \kappa_n^2 (1-\tilde{Q}_r)\Pi_1 \right] \right.$$
$$\left. + (-1)^{m-1} 2\nu_m \frac{\partial}{\partial |x_2|} \left((r_m + \tilde{Q}_r)\Pi_{(-r_m)} + (1-\tilde{Q}_r)\Pi_1 \right) \right\};$$

$$\sigma_{22d}^{A_n(m)A} = -\frac{\partial A_n}{\partial x_3} 2s_{1d}^{(m)} \frac{\partial}{\partial x_1} \left(2 \left[s_{2d}^{(m)} - \frac{s_{3d}^{(m)}}{1+s_m} + \frac{s_{4d}^{(m)}}{1+r_m} \right] K_0 \right.$$
$$\left. - \frac{s_{3d}^{(m)}}{1+s_m} \frac{\partial \Pi_{(-s_m)}}{\partial |x_2|} + \frac{s_{4d}^{(m)}}{1+r_m} \frac{\partial \Pi_{(-r_m)}}{\partial |x_2|} \right),$$

$$\sigma_{22d}^{A_n(m)B} = -\frac{\partial A_n}{\partial x_3} 2r_{1d}^{(m)} \left\{ x_2 (1+r_m) \frac{\partial}{\partial x_1} \left[I_0 + \kappa_n^2 (r_m + \tilde{Q}_r)\Pi_{(-r_m)} \right] \right.$$
$$\left. + (-1)^{m-1} 2(1-\nu_m) \frac{\partial^2}{\partial x_1 \partial |x_2|} \left((r_m + \tilde{Q}_r)\Pi_{(-r_m)} + (1-\tilde{Q}_r)\Pi_1 \right) \right\};$$

$$\sigma_{33d}^{A_n(m)A} = \frac{\partial A_n}{\partial x_3} 2s_{1d}^{(m)} \frac{\partial}{\partial x_1} \left(2K_0 \left[\frac{s_{4d}^{(m)}}{(1+r_m)^2} - \frac{s_{3d}^{(m)}}{(1+s_m)^2} \right] \right.$$

$$\left. + \kappa_n^2 \left[s_{2d}^{(m)} - \frac{s_{3d}^{(m)}}{1+s_m} + \frac{s_{4d}^{(m)}}{1+r_m} \right] I_{1a} - \frac{s_{3d}^{(m)}}{(1+s_m)^2} \frac{\partial \Pi_{(-s_m)}}{\partial |x_2|} + \frac{s_{4d}^{(m)}}{(1+r_m)^2} \frac{\partial \Pi_{(-r_m)}}{\partial |x_2|} \right),$$

$$\sigma_{33d}^{A_n(m)B} = \frac{\partial A_n}{\partial x_3} 2r_{1d}^{(m)} \left\| x_2 \kappa_n^2 \frac{\partial}{\partial x_1} + (-1)^{m-1} 2(2-\nu_m) \frac{\partial^2}{\partial x_1 \partial |x_2|} \right\|$$
$$\times \left((r_m + \tilde{Q}_r)\Pi_{(-r_m)} + (1-\tilde{Q}_r)\Pi_1 \right);$$

$$\sigma_{12d}^{A_n(m)A} = \frac{\partial A_n}{\partial x_3} 2s_{1d}^{(m)} (-1)^m \left(s_{2d}^{(m)} I_0 - \kappa_n^2 \left[s_{2d}^{(m)} - \frac{s_{3d}^{(m)}}{1+s_m} + \frac{s_{4d}^{(m)}}{1+r_m} \right] \Pi_1 \right.$$
$$\left. + \frac{s_m \kappa_n^2 s_{3d}^{(m)}}{1+s_m} \Pi_{(-s_m)} - \frac{r_m \kappa_n^2 s_{4d}^{(m)}}{1+r_m} \Pi_{(-r_m)} \right),$$
$$\sigma_{12d}^{A_n(m)B} = \frac{\partial A_n}{\partial x_3} 2r_{1d}^{(m)} \left\| 1 + x_2 (-1)^{m-1} \frac{\partial}{\partial |x_2|} \right\|$$
$$\times \left((r_m+1) I_0 + \kappa_n^2 r_m (r_m + \tilde{Q}_r) \Pi_{(-r_m)} - \kappa_n^2 (1-\tilde{Q}_r) \Pi_1 \right);$$

$$\sigma_{13d}^{A_n(m)A} = -A_n 2s_{1d}^{(m)} \kappa_n^2 \left(2K_0 \left[s_{2d}^{(m)} - \frac{s_m s_{3d}^{(m)}}{(1+s_m)^2} + \frac{r_m s_{4d}^{(m)}}{(1+r_m)^2} \right] \right.$$
$$\left. -\kappa_n^2 \left[s_{2d}^{(m)} - \frac{s_{3d}^{(m)}}{1+s_m} + \frac{s_{4d}^{(m)}}{1+r_m} \right] I_{1a} - \frac{s_m s_{3d}^{(m)}}{(1+s_m)^2} \frac{\partial \Pi_{(-s_m)}}{\partial |x_2|} + \frac{r_m s_{4d}^{(m)}}{(1+r_m)^2} \frac{\partial \Pi_{(-r_m)}}{\partial |x_2|} \right),$$
$$\sigma_{13d}^{A_n(m)B} = -A_n 2r_{1d}^{(m)} \left\| x_2 \kappa_n^2 + 2(1-\nu_m)(-1)^{m-1} \frac{\partial}{\partial |x_2|} \right\|$$
$$\times \left((r_m+1) I_0 + \kappa_n^2 r_m (r_m + \tilde{Q}_r) \Pi_{(-r_m)} - \kappa_n^2 (1-\tilde{Q}_r) \Pi_1 \right);$$

$$\sigma_{23d}^{A_n(m)A} = A_n 2s_{1d}^{(m)} \kappa_n^2 (-1)^m \frac{\partial}{\partial x_1} \left(\left[s_{2d}^{(m)} - \frac{s_{3d}^{(m)}}{1+s_m} + \frac{s_{4d}^{(m)}}{1+r_m} \right] \Pi_1 \right.$$
$$\left. + \frac{s_{3d}^{(m)}}{1+s_m} \Pi_{(-s_m)} - \frac{s_{4d}^{(m)}}{1+r_m} \Pi_{(-r_m)} \right),$$
$$\sigma_{23d}^{A_n(m)B} = A_n 2r_{1d}^{(m)} \frac{\partial}{\partial x_1} \left\{ 2(1-\nu_m)(1+r_m) \left[I_0 + \kappa_n^2 (r_m + \tilde{Q}_r) \Pi_{(-r_m)} \right] \right.$$
$$\left. + \kappa_n^2 \left\| 1 + x_2 (-1)^{m-1} \frac{\partial}{\partial |x_2|} \right\| \left((r_m + \tilde{Q}_r) \Pi_{(-r_m)} + (1-\tilde{Q}_r) \Pi_1 \right) \right\}; \quad \text{(B.4)}$$

$$\Pi_{(-z)} \equiv \int_{-\infty}^{\infty} \frac{e^{-\sqrt{k_1^2+\kappa_n^2}|x_2|}}{k_1^2 - z\kappa_n^2} e^{ik_1 x_1} \mathrm{d}k_1.$$

The couple $(\overline{\alpha}_{3e}^{A_n(m)}, \overline{\beta}_{3e}^{A_n(m)})$ and is obtained from (13l, m, o and p) associated with stresses. We have

$$u_{1e}^{A_n(m)A} = \frac{\partial A_n}{\partial x_3} \frac{s_{1e}^{(m)}}{\mu_m} \left(2s_{2e}^{(m)} K_0 + \kappa_n^2 \left[s_{3e}^{(m)} - s_{2e}^{(m)} \right] I_{1a} \right),$$
$$u_{1e}^{A_n(m)B} = \frac{\partial A_n}{\partial x_3} \frac{r_{1e}^{(m)}}{\mu_m} x_2 \left(I_0 - 2\kappa_n^2 \Pi_1 \right);$$

$$u_{2e}^{A_n(m)A} = \frac{\partial A_n}{\partial x_3}\frac{s_{1e}^{(m)}}{\mu_m}(-1)^{m-1}\left(is_{3e}^{(m)}I_{1c} - \left[s_{3e}^{(m)} - s_{2e}^{(m)}\right]\frac{\partial \Pi_1}{\partial x_1}\right),$$
$$u_{2e}^{A_n(m)B} = \frac{\partial A_n}{\partial x_3}\frac{r_{1e}^{(m)}}{\mu_m}\left\| 1 + x_2(-1)^{m-1}\frac{\partial}{\partial |x_2|}\right\|\left(iI_{1c} - 2\frac{\partial \Pi_1}{\partial x_1}\right);$$

$$u_{3e}^{A_n(m)A} = A_n\frac{s_{1e}^{(m)}}{\mu_m}\left(s_{3e}^{(m)}\frac{\partial}{\partial |x_2|}\left[\frac{\partial \Pi_1}{\partial x_1} - iI_{1c}\right] + \kappa_n^2\left[s_{2e}^{(m)} - s_{3e}^{(m)}\right]\frac{\partial I_{1a}}{\partial x_1}\right),$$
$$u_{3e}^{A_n(m)B} = A_n\frac{r_{1e}^{(m)}}{\mu_m}\left\| x_2\kappa_n^2 + 4(1-\nu_m)(-1)^{m-1}\frac{\partial}{\partial |x_2|}\right\|\left(2\frac{\partial \Pi_1}{\partial x_1} - iI_{1c}\right);$$

$$\sigma_{11e}^{A_n(m)A} = \frac{\partial A_n}{\partial x_3}2s_{1e}^{(m)}\frac{\partial}{\partial x_1}\left(2s_{2e}^{(m)}K_0 + \kappa_n^2\left[s_{3e}^{(m)} - s_{2e}^{(m)}\right]I_{1a}\right),$$

$$\sigma_{11e}^{A_n(m)B} = \frac{\partial A_n}{\partial x_3}2r_{1e}^{(m)}\left(x_2\frac{\partial}{\partial x_1}\left[I_0 - 2\kappa_n^2\Pi_1\right] + (-1)^{m-1}2\nu_m\frac{\partial}{\partial |x_2|}\left[2\frac{\partial \Pi_1}{\partial x_1} - iI_{1c}\right]\right);$$

$$\sigma_{22e}^{A_n(m)A} = \frac{\partial A_n}{\partial x_3}2s_{1e}^{(m)}\left(is_{3e}^{(m)}\frac{\partial I_{1c}}{\partial |x_2|} + 2\left[s_{3e}^{(m)} - s_{2e}^{(m)}\right]\frac{\partial K_0}{\partial x_1}\right),$$
$$\sigma_{22e}^{A_n(m)B} = \frac{\partial A_n}{\partial x_3}2r_{1e}^{(m)}\left(x_2\left[\frac{\partial I_0}{\partial x_1} + i\kappa_n I_{1c}\right] + (-1)^m 2(1-\nu_m)\frac{\partial}{\partial |x_2|}\left[2\frac{\partial \Pi_1}{\partial x_1} - iI_{1c}\right]\right);$$
$$\sigma_{33e}^{A_n(m)A} = \frac{\partial A_n}{\partial x_3}2s_{1e}^{(m)}\left(s_{3e}^{(m)}\frac{\partial}{\partial |x_2|}\left[\frac{\partial \Pi_1}{\partial x_1} - iI_{1c}\right] + \kappa_n^2\left[s_{2e}^{(m)} - s_{3e}^{(m)}\right]\frac{\partial I_{1a}}{\partial x_1}\right),$$
$$\sigma_{33e}^{A_n(m)B} = \frac{\partial A_n}{\partial x_3}2r_{1e}^{(m)}\left\| x_2\kappa_n^2 + 2(2-\nu_m)(-1)^{m-1}\frac{\partial}{\partial |x_2|}\right\|\left(2\frac{\partial \Pi_1}{\partial x_1} - iI_{1c}\right);$$
$$\sigma_{12e}^{A_n(m)A} = \frac{\partial A_n}{\partial x_3}2s_{1e}^{(m)}(-1)^m\left(s_{2e}^{(m)}I_0 + \kappa_n^2\left[s_{3e}^{(m)} - s_{2e}^{(m)}\right]\Pi_1\right),$$
$$\sigma_{12e}^{A_n(m)B} = \frac{\partial A_n}{\partial x_3}2r_{1e}^{(m)}\left\| 1 + x_2(-1)^{m-1}\frac{\partial}{\partial |x_2|}\right\|\left(I_0 - 2\kappa_n^2\Pi_1\right);$$

$$\sigma_{13e}^{A_n(m)A} = -A_n 2s_{1e}^{(m)}\kappa_n^2\left(2s_{2e}^{(m)}K_0 + \kappa_n^2\left[s_{3e}^{(m)} - s_{2e}^{(m)}\right]I_{1a}\right),$$
$$\sigma_{13e}^{A_n(m)B} = -A_n 2r_{1e}^{(m)}\left\| x_2\kappa_n^2 + 2(1-\nu_m)(-1)^{m-1}\frac{\partial}{\partial |x_2|}\right\|\left(I_0 - 2\kappa_n^2\Pi_1\right);$$

$$\sigma_{23e}^{A_n(m)A} = A_n 2s_{1e}^{(m)}\kappa_n^2(-1)^m\left(is_{3e}^{(m)}I_{1c} + \left[s_{2e}^{(m)} - s_{3e}^{(m)}\right]\frac{\partial \Pi_1}{\partial x_1}\right),$$
$$\sigma_{23e}^{A_n(m)B} = A_n 2r_{1e}^{(m)}\left(\frac{\partial}{\partial x_1}\left[2(1-\nu_m)I_0 + 2\kappa_n^2\Pi_1\right] - i\kappa_n^2(3-2\nu_m)I_{1c}\right.$$
$$\left. + x_2\kappa_n^2(-1)^{m-1}\frac{\partial}{\partial |x_2|}\left[2\frac{\partial \Pi_1}{\partial x_1} - iI_{1c}\right]\right). \qquad \text{(B.5)}$$

5.4.2 Climb-Type Sinusoidal Edge Dislocation

The content of this section has been published as: P.N.B. ANONGBA, Non-planar interface crack under general loading-*II*. Climb-type sinusoidal edge dislocation, *Rev. Ivoir. Sci. Technol.*, 30 (2017) 1–36.

The study is involved with the determination of the elastic fields of a non-planar interface crack model in equilibrium under an externally applied general loading. It consists of two infinitely extended elastic mediums $R1$ and $R2$ firmly welded along a non-planar interface R (Fig. 5.5, Sect. 5.3). With respect to a Cartesian coordinate system x_i, the crack extends from $x_1 = -a$ to a, fluctuates on average about Ox_1x_3-plane; its front runs indefinitely in the x_3- direction and spreads in x_2x_3-planes in the Fourier series form f, written as

$$f = \sum_n (\xi_n \sin \kappa_n x_3 + \delta_n \cos \kappa_n x_3) + h \equiv \xi + h \tag{1}$$

where n is a positive integer; h, ξ_n, δ_n and κ_n are real numbers that depend on position x_1 along the crack length. The system is subjected to remote applied general loading, corresponding to a tension along x_2, and two shears (parallel to x_1x_3) directed along x_1 and x_3, respectively. The method of analysis consists in representing the crack by a continuous distribution of three families of non-straight infinitesimal dislocations having the form f. Families 1 and 2 are edges (on average) and family 3 is screw. The associated Burgers vectors $\vec{b}_j (j = I, II, III)$ are directed along the applied tension and shears x_2, x_1 and x_3 directions, respectively. It is sufficient to investigate a simpler model where the interface is in the form of a corrugated surface S and the dislocations have a sinusoidal form A_n given by

$$A_n = \xi_n \sin \kappa_n x_3. \tag{2}$$

The applied loadings and dislocation Burgers vectors are unchanged. The case of a general shape, with crack front f (1), is achieved by superposition. From the elastic fields of the dislocations, the crack-tip stress and crack extension force can be evaluated. We plan to extend the analysis to non-planar interface cracks under general loading (mixed mode $I + II + III$). The first part I of this study (Sect. 5.4.1) has provided expressions for the displacement and stress fields of sinusoidal dislocations with $\vec{b}_I$(glide-type edges) and $\vec{b}_{III}$(screws). Those of dislocation family 2 with $\vec{b}_{II}$ are the subject of this part II. For the various types of dislocation, the general geometry is that of Fig. 5.6; because $\vec{b}_{II} = (b, 0, 0)$ is out of the Ox_2x_3-plane of location of the dislocation, this is a climb-type edge. In what follows, the methodology and elastic field expressions are given, respectively. Discussion and conclusion then follow.

Methodology

We seek the elastic fields of a dislocation ($\vec{b}_{II} = (b, 0, 0)$) lying on a non-planar interface S having the form of a corrugated sheet that separates two firmly welded elastic solids $S1$ and $S2$ of infinite sizes. S is defined by the point $P_S(x_1, x_2 = \xi_n \sin \kappa_n x_3, x_3)$ and $S1$ and $S2$ occupy the regions $x_2 > \xi_n \sin \kappa_n x_3$ and $x_2 < \xi_n \sin \kappa_n x_3$, respectively. The situation is shown in Fig. 5.6 where $S1$ and $S2$ are confined for illustration purpose in a parallelepiped of finite sizes. The dislocation is located at the origin, runs indefinitely in the x_3-direction and spreads in the x_2x_3-plane in the form A_n (2). The methodology has been detailed in different places (Anongba 2015, 2016). The elastic fields $(\vec{u}^{(m)},(\sigma)^{(m)})$ are assumed to be the difference between two quantities $(\vec{u}^{(m)\infty}, (\sigma)^{(m)\infty})$ and $(\vec{u}^{(m)W},(\sigma)^{(m)W})$:

$$\begin{aligned}\vec{u}^{(m)} &= \vec{u}^{(m)\infty} - \vec{u}^{(m)W}\\ (\sigma)^{(m)} &= (\sigma)^{(m)\infty} - (\sigma)^{(m)W}.\end{aligned} \tag{3}$$

The former with ∞ corresponds to the fields of a sinusoidal dislocation (edge or screw) in an infinitely extended homogeneous solid (m); the latter with W satisfies the equations of equilibrium and is constructed in such a way that:

(a) $(\vec{u}^{(m)}, (\sigma)^{(m)})$ are continuous at the crossing of the interface, implying that

$$\Delta\vec{u}^{\infty}(P_S) \equiv \vec{u}^{(2)\infty} - \vec{u}^{(1)\infty} = \vec{u}^{(2)W} - \vec{u}^{(1)W} \equiv \Delta\vec{u}^{W}(P_S)$$

$$(\Delta\sigma)^{\infty}(P_S) \equiv (\sigma)^{(2)\infty} - (\sigma)^{(1)\infty} = (\sigma)^{(2)W} - (\sigma)^{(1)W} \equiv (\Delta\sigma)^{W}(P_S); \tag{4}$$

(b) $(\vec{u}^{(m)},(\sigma)^{(m)})$ tends to $(\vec{u}^{(m)\infty},(\sigma)^{(m)\infty})$ when one moves far away from the interface in the x_2- direction. This means that

$$\begin{aligned}\vec{u}^{(m)W} &\to 0\\ (\sigma)^{(m)W} &\to 0\end{aligned} \tag{5}$$

when $|x_2| \to \infty$. $(\vec{u}^{(m)\infty}, (\sigma)^{(m)\infty})$ are given to linear expressions with respect to ξ_n. The associated terms $(\vec{u}^{(0)(m)\infty}, (\sigma)^{(0)(m)\infty})$ of zero order correspond to the fields of a straight dislocation and second terms $(\vec{u}^{A_n(m)\infty}, (\sigma)^{A_n(m)\infty})$ are proportional to either A_n or its spatial derivative $\partial A_n/x_3$. Hence, we have

$$\begin{aligned}\Delta\vec{u}^{\infty}(P_S) &= \Delta\vec{u}^{(0)\infty} + \Delta\vec{u}^{A_n\infty}\\ (\Delta\sigma)^{\infty}(P_S) &= (\Delta\sigma)^{(0)\infty} + (\Delta\sigma)^{A_n\infty}\end{aligned} \tag{6}$$

on the interface point $P_S(x_1, x_2 = \xi_n \sin \kappa_n x_3, x_3)$; Appendix A below gives the complete list, component by component, for the climb-type sinusoidal edge dislocation. $(\vec{u}^{(m)W}, (\sigma)^{(m)W})$ are obtained with the help of Galerkin vectors. For the climb-type sinusoidal edge, we arrive at Galerkin vectors $\vec{V}$ with only one non-zero x_1-component, arranged in the form

$$V_1(\vec{x}) = \overline{\alpha}_1(\vec{k})e^{i\vec{k}.\vec{x}} + \overline{\beta}_1(\vec{k})x_2e^{i\vec{k}.\vec{x}} \tag{7}$$

under the condition $\vec{k}^2 = k_1^2 + k_2^2 + k_3^2 = 0$ that ensures the biharmonicity of V_1. For V_1 to cancel far from the interface, we write

$$k_2 = k_2^{(m)} \equiv (-1)^{m-1} i \sqrt{k_1^2 + k_3^2} \tag{8}$$

with $m = 1$ when $x_2 > \xi_n \sin \kappa_n x_3$ (half-space 1) and $m = 2$ when $x_2 < \xi_n \sin \kappa_n x_3$ (half-space 2). We use the notations

$$\vec{k}^{(m)} \equiv \left(k_1, k_2^{(m)}, k_3\right), \overline{\alpha}_1^{(m)} \equiv \overline{\alpha}_1\left(\vec{k}^{(m)}\right), \overline{\beta}_1^{(m)} \equiv \overline{\beta}_1\left(\vec{k}^{(m)}\right);$$

hence for half-space 1 ($x_2 > \xi_n \sin \kappa_n x_3$), solid (1)

$$V_1(\vec{x}) \equiv V_1^{(1)}(\vec{x}) = \overline{\alpha}_1^{(1)} e^{i\vec{k}^{(1)}.\vec{x}} + \overline{\beta}_1^{(1)} x_2 e^{i\vec{k}^{(1)}.\vec{x}}$$

and for half-space 2 ($x_2 < \xi_n \sin \kappa_n x_3$), solid (2)

$$V_1(\vec{x}) \equiv V_1^{(2)}(\vec{x}) = \overline{\alpha}_1^{(2)} e^{i\vec{k}^{(2)}.\vec{x}} + \overline{\beta}_1^{(2)} x_2 e^{i\vec{k}^{(2)}.\vec{x}}.$$

The elastic fields corresponding to V_1 (7) may be first calculated; then, more general forms $\vec{u}^{(m)V}$ and $(\sigma)^{(m)V}$ are constructed from the previous ones by superposition over k_1 and k_3; we may write

$$\begin{aligned} \vec{u}^{(m)V} &= \vec{u}^{(m)A} + \vec{u}^{(m)B} = \vec{u}^{(m)A+B} \\ (\sigma)^{(m)V} &= (\sigma)^{(m)A} + (\sigma)^{(m)B} = (\sigma)^{(m)A+B} \end{aligned} \tag{9}$$

where terms with A and B refer to $\overline{\alpha}_1$ and $\overline{\beta}_1$ in (7) respectively. For $\vec{u}^{(m)V}$ and $(\sigma)^{(m)V}$ to conform with $\vec{u}^{(m)\infty}$ and $(\sigma)^{(m)\infty}$, the summation over k_1 is continuous and that over k_3 is discrete. k_3 takes three values: $-\kappa_n$, 0, κ_n. The fields corresponding to $k_3 = 0$ are denoted $\vec{u}^{(0)(m)V}$ and $(\sigma)^{(0)(m)V}$, and terms associated with $k_3 = -\kappa_n$ and κ_n are merged to form expressions denoted by $\vec{u}^{A_n(m)V}$ and $(\sigma)^{A_n(m)V}$; this is made possible by requiring that

$$\overline{\alpha}_1^{(m)}(\kappa_n) \equiv -\overline{\alpha}_1^{(m)}(-\kappa_n), \overline{\beta}_1^{(m)}(\kappa_n) \equiv -\overline{\beta}_1^{(m)}(-\kappa_n). \tag{10}$$

In (10), $\overline{\alpha}_1^{(m)}(\kappa_n)$ stands for $\overline{\alpha}_1(k_1, k_2^{(m)}, \kappa_n)$. We write

$$\begin{aligned} \vec{u}^{(m)V} &= \vec{u}^{(0)(m)V} + \vec{u}^{A_n(m)V} \\ (\sigma)^{(m)V} &= (\sigma)^{(0)(m)V} + (\sigma)^{A_n(m)V} \end{aligned} \tag{11}$$

introducing subsequently the notation $\vec{u}^{(0)(m)A}$, $\vec{u}^{A_n(m)A}$, $\vec{u}^{(0)(m)B}$, $\vec{u}^{A_n(m)B}$ and even for the stress. $\vec{u}^{(0)(m)V}$ and $(\sigma)^{(0)(m)V}$ are x_3-independent; $\vec{u}^{A_n(m)V}$ and $(\sigma)^{A_n(m)V}$ are proportional to the sinusoid $A_n(x_3)$ or to its spatial derivative $\partial A_n/\partial x_3$. Here also, for points P_S on the interface, $\Delta\vec{u}^V$ and $(\Delta\sigma)^V$ are expanded up to terms of first order with respect to $x_2 = \xi$ in a similar manner as in (A.2) (see Appendix A below) for $\Delta\vec{u}^\infty$ and $(\Delta\sigma)^\infty$. Requiring $\Delta\vec{u}^V = \Delta\vec{u}^\infty$ and $(\Delta\sigma)^V = (\Delta\sigma)^\infty$ lead to the following equations, writing first the conditions corresponding to $k_3 = 0$(i.e. $\Delta u_i^{(0)V} = \Delta u_i^{(0)\infty}$ and $\Delta\sigma_{ij}^{(0)V} = \Delta\sigma_{ij}^{(0)\infty}$).

$$\Delta u_1^{(0)V} = \Delta u_1^{(0)\infty} \Rightarrow$$

$$|k_1|\left(\frac{\overline{\alpha}_1^{(2)}}{\mu_2} - \frac{\overline{\alpha}_1^{(1)}}{\mu_1}\right) + 4\left(\frac{(1-\nu_2)\overline{\beta}_1^{(2)}}{\mu_2} + \frac{(1-\nu_1)\overline{\beta}_1^{(1)}}{\mu_1}\right) = 0 \tag{a}$$

$$|k_1|\left(\frac{\overline{\alpha}_1^{(2)}}{\mu_2} + \frac{\overline{\alpha}_1^{(1)}}{\mu_1}\right) + \frac{(5-4\nu_2)\overline{\beta}_1^{(2)}}{\mu_2} - \frac{(5-4\nu_1)\overline{\beta}_1^{(1)}}{\mu_1} = \frac{ibC_\nu}{4\pi}\frac{\operatorname{sgn}(k_1)}{k_1^2} \tag{b}$$

$$\Delta u_2^{(0)V} = \Delta u_2^{(0)\infty} \Rightarrow$$

$$|k_1|\left(\frac{\overline{\alpha}_1^{(2)}}{\mu_2} + \frac{\overline{\alpha}_1^{(1)}}{\mu_1}\right) + \left(\frac{\overline{\beta}_1^{(2)}}{\mu_2} - \frac{\overline{\beta}_1^{(1)}}{\mu_1}\right) = \frac{ibC_\nu}{4\pi}\frac{\operatorname{sgn}(k_1)}{k_1^2} \tag{c}$$

$$|k_1|\left(\frac{\overline{\alpha}_1^{(2)}}{\mu_2} - \frac{\overline{\alpha}_1^{(1)}}{\mu_1}\right) + 2\left(\frac{\overline{\beta}_1^{(2)}}{\mu_2} + \frac{\overline{\beta}_1^{(1)}}{\mu_1}\right) = 0 \tag{d}$$

$$\Delta\sigma_{11}^{(0)V} = \Delta\sigma_{11}^{(0)\infty} \Rightarrow$$

$$|k_1|\left(\overline{\alpha}_1^{(2)} - \overline{\alpha}_1^{(1)}\right) + 2(2-\nu_2)\overline{\beta}_1^{(2)} + 2(2-\nu_1)\overline{\beta}_1^{(1)} = 0 \tag{e}$$

$$|k_1|\left(\overline{\alpha}_1^{(2)} + \overline{\alpha}_1^{(1)}\right) + (5-2\nu_2)\overline{\beta}_1^{(2)} - (5-2\nu_1)\overline{\beta}_1^{(1)} = -6Q_b\frac{\operatorname{sgn}(k_1)}{k_1^2} \tag{f}$$

$$\Delta\sigma_{22}^{(0)V} = \Delta\sigma_{22}^{(0)\infty} \Rightarrow$$

$$|k_1|\left(\overline{\alpha}_1^{(2)} - \overline{\alpha}_1^{(1)}\right) + 2(1-\nu_2)\overline{\beta}_1^{(2)} + 2(1-\nu_1)\overline{\beta}_1^{(1)} = 0 \tag{g}$$

$$|k_1|\left(\overline{\alpha}_1^{(2)} + \overline{\alpha}_1^{(1)}\right) + (3-2\nu_2)\overline{\beta}_1^{(2)} - (3-2\nu_1)\overline{\beta}_1^{(1)} = -2Q_b\frac{\operatorname{sgn}(k_1)}{k_1^2} \tag{h}$$

$$\Delta\sigma_{33}^{(0)V} = \Delta\sigma_{33}^{(0)\infty} \Rightarrow$$

$$\nu_2\overline{\beta}_1^{(2)} + \nu_1\overline{\beta}_1^{(1)} = 0 \tag{i}$$

$$\nu_2\overline{\beta}_1^{(2)} - \nu_1\overline{\beta}_1^{(1)} = -2Q_c\frac{\text{sgn}(k_1)}{k_1^2} \tag{j}$$

$$\Delta\sigma_{12}^{(0)V} = \Delta\sigma_{12}^{(0)\infty} \Rightarrow (h) \text{ and } (e) \text{ above} \tag{12}$$

where $C_m = b\mu_m/2\pi(1-\nu_m)$, $Q_b = i(C_2 - C_1)/4$, $Q_c = i(\nu_2 C_2 - \nu_1 C_1)/4$;

$\text{sgn}(k_1) = k_1/|k_1|$; μ_m and ν_m are shear modulus and Poisson's ratio. In Eqs. (12a–j) above, $\overline{\alpha}_1^{(m)}$ and $\overline{\beta}_1^{(m)}$ stand for $\overline{\alpha}_1(k_1, k_2^{(m)}, k_3 = 0)$ and $\overline{\beta}_1(k_1, k_2^{(m)}, k_3 = 0)$, respectively. Other elastic field components are zero.

The conditions corresponding to $\Delta u_i^{A_nV} = \Delta u_i^{A_n\infty}$ and $\Delta\sigma_{ij}^{A_nV} = \Delta\sigma_{ij}^{A_n\infty}$ are now listed as:

$$\Delta u_1^{A_nV} = \Delta u_1^{A_n\infty} \Rightarrow$$

$$k_1^2\left(\frac{\overline{\alpha}_1^{(2)}}{\mu_2} - \frac{\overline{\alpha}_1^{(1)}}{\mu_1}\right) + 4\sqrt{k_1^2+\kappa_n^2}\left(\frac{(1-\nu_2)\overline{\beta}_1^{(2)}}{\mu_2} + \frac{(1-\nu_1)\overline{\beta}_1^{(1)}}{\mu_1}\right) = -\frac{bC_\nu\xi_n}{8\pi}\frac{k_1\left(k_1^2+2\kappa_n^2\right)}{\left(k_1^2+\kappa_n^2\right)^{3/2}} \tag{a}$$

$$k_1^2\sqrt{k_1^2+\kappa_n^2}\left(\frac{\overline{\alpha}_1^{(2)}}{\mu_2} + \frac{\overline{\alpha}_1^{(1)}}{\mu_1}\right) + \frac{\overline{\beta}_1^{(2)}}{\mu_2}\left[(5-4\nu_2)k_1^2 + 4(1-\nu_2)\kappa_n^2\right] - \frac{\overline{\beta}_1^{(1)}}{\mu_1}\left[(5-4\nu_1)k_1^2 + 4(1-\nu_1)\kappa_n^2\right] = 0 \tag{b}$$

$$\Delta u_2^{A_nV} = \Delta u_2^{A_n\infty} \Rightarrow$$

$$\sqrt{k_1^2+\kappa_n^2}\left(\frac{\overline{\alpha}_1^{(2)}}{\mu_2} + \frac{\overline{\alpha}_1^{(1)}}{\mu_1}\right) + \frac{\overline{\beta}_1^{(2)}}{\mu_2} - \frac{\overline{\beta}_1^{(1)}}{\mu_1} = 0 \tag{c}$$

$$\sqrt{k_1^2+\kappa_n^2}\left(\frac{\overline{\alpha}_1^{(2)}}{\mu_2} - \frac{\overline{\alpha}_1^{(1)}}{\mu_1}\right) + 2\left(\frac{\overline{\beta}_1^{(2)}}{\mu_2} + \frac{\overline{\beta}_1^{(1)}}{\mu_1}\right) = -\frac{bC_\nu\xi_n}{8\pi}\frac{\left(3k_1^2+2\kappa_n^2\right)}{k_1\left(k_1^2+\kappa_n^2\right)} \tag{d}$$

$\Delta u_3^{A_nV} = \Delta u_3^{A_n\infty} \Rightarrow$ (13c) above and

$$\frac{\overline{\alpha}_1^{(2)}}{\mu_2} - \frac{\overline{\alpha}_1^{(1)}}{\mu_1} = -\frac{bC_\nu\xi_n}{8\pi}\frac{\left(k_1^2+2\kappa_n^2\right)}{k_1\left(k_1^2+\kappa_n^2\right)^{3/2}} \tag{e}$$

$$\Delta\sigma_{11}^{A_nV} = \Delta\sigma_{11}^{A_n\infty} \Rightarrow$$

$$k_1^2\left(\overline{\alpha}_1^{(2)} - \overline{\alpha}_1^{(1)}\right) + 2\sqrt{k_1^2 + \kappa_n^2}\left[(2 - \nu_2)\overline{\beta}_1^{(2)} + (2 - \nu_1)\overline{\beta}_1^{(1)}\right] = -\frac{iQ_b\xi_n k_1(3k_1^2 + 4\kappa_n^2)}{\left(k_1^2 + \kappa_n^2\right)^{3/2}} \tag{f}$$

$$k_1^2\sqrt{k_1^2 + \kappa_n^2}\left(\overline{\alpha}_1^{(2)} + \overline{\alpha}_1^{(1)}\right) + \left[(5 - 2\nu_2)k_1^2 + 2(2 - \nu_2)\kappa_n^2\right]\overline{\beta}_1^{(2)} - \left[(5 - 2\nu_1)k_1^2 + 2(2 - \nu_1)\kappa_n^2\right]\overline{\beta}_1^{(1)} = 0 \tag{g}$$

$$\Delta\sigma_{22}^{A_nV} = \Delta\sigma_{22}^{A_n\infty} \Rightarrow$$

$$\sqrt{k_1^2 + \kappa_n^2}\left(\overline{\alpha}_1^{(2)} - \overline{\alpha}_1^{(1)}\right) + 2\left[(1 - \nu_2)\overline{\beta}_1^{(2)} + (1 - \nu_1)\overline{\beta}_1^{(1)}\right] = -\frac{i\xi_n(Q_b k_1^2 + 2Q_c\kappa_n^2)}{k_1\left(k_1^2 + \kappa_n^2\right)} \tag{h}$$

$$\sqrt{k_1^2 + \kappa_n^2}\left(\overline{\alpha}_1^{(2)} + \overline{\alpha}_1^{(1)}\right) + (3 - 2\nu_2)\overline{\beta}_1^{(2)} - (3 - 2\nu_1)\overline{\beta}_1^{(1)} = 0 \tag{i}$$

$$\Delta\sigma_{33}^{A_nV} = \Delta\sigma_{33}^{A_n\infty} \Rightarrow$$

$$\kappa_n^2\left(\overline{\alpha}_1^{(2)} - \overline{\alpha}_1^{(1)}\right) + 2\sqrt{k_1^2 + \kappa_n^2}\left(\nu_2\overline{\beta}_1^{(2)} + \nu_1\overline{\beta}_1^{(1)}\right) = -\frac{i\xi_n\left\{k_1^2\left[2Q_c k_1^2 + (4Q_c - Q_b)\kappa_n^2\right] + 2Q_c\kappa_n^4\right\}}{k_1\left(k_1^2 + \kappa_n^2\right)^{3/2}} \tag{j}$$

$$\kappa_n^2\sqrt{k_1^2 + \kappa_n^2}\left(\overline{\alpha}_1^{(2)} + \overline{\alpha}_1^{(1)}\right) + \left[2\nu_2 k_1^2 + (1 + 2\nu_2)\kappa_n^2\right]\overline{\beta}_1^{(2)} - \left[2\nu_1 k_1^2 + (1 + 2\nu_1)\kappa_n^2\right]\overline{\beta}_1^{(1)} = 0 \tag{k}$$

$$\Delta\sigma_{12}^{A_nV} = \Delta\sigma_{12}^{A_n\infty} \Rightarrow$$

$$k_1^2\sqrt{k_1^2 + \kappa_n^2}\left(\overline{\alpha}_1^{(2)} + \overline{\alpha}_1^{(1)}\right) + \left[(3 - 2\nu_2)k_1^2 + 2(1 - \nu_2)\kappa_n^2\right]\overline{\beta}_1^{(2)} - \left[(3 - 2\nu_1)k_1^2 + 2(1 - \nu_1)\kappa_n^2\right]\overline{\beta}_1^{(1)} = 0 \tag{l}$$

$$k_1^2\sqrt{k_1^2 + \kappa_n^2}\left(\overline{\alpha}_1^{(2)} - \overline{\alpha}_1^{(1)}\right) + 2\left[(2 - \nu_2)k_1^2 + (1 - \nu_2)\kappa_n^2\right]\overline{\beta}_1^{(2)} + 2\left[(2 - \nu_1)k_1^2 + (1 - \nu_1)\kappa_n^2\right]\overline{\beta}_1^{(1)} = -\frac{iQ_b\xi_n k_1(3k_1^2 + 2\kappa_n^2)}{k_1^2 + \kappa_n^2} \tag{m}$$

$\Delta\sigma_{13}^{A_nV} = \Delta\sigma_{13}^{A_n\infty} \Rightarrow$ (131) above and

$$k_1^2\left(\overline{\alpha}_1^{(2)} - \overline{\alpha}_1^{(1)}\right) + 2\sqrt{k_1^2 + \kappa_n^2}\left[(1-\nu_2)\overline{\beta}_1^{(2)} + (1-\nu_1)\overline{\beta}_1^{(1)}\right]$$
$$= -\frac{iQ_b\xi_n k_1(k_1^2 + 2\kappa_n^2)}{\left(k_1^2+\kappa_n^2\right)^{3/2}} \qquad \text{(n)}$$

$$\Delta\sigma_{23}^{A_nV} = \Delta\sigma_{23}^{A_n\infty} \Rightarrow$$

$$\sqrt{k_1^2+\kappa_n^2}\left(\overline{\alpha}_1^{(2)} + \overline{\alpha}_1^{(1)}\right) + \overline{\beta}_1^{(2)} - \overline{\beta}_1^{(1)} = 0 \qquad \text{(o)}$$

$$\sqrt{k_1^2+\kappa_n^2}\left(\overline{\alpha}_1^{(2)} - \overline{\alpha}_1^{(1)}\right) + 2\left[\overline{\beta}_1^{(2)} + \overline{\beta}_1^{(1)}\right]$$
$$= -\frac{i\xi_n\left[(2Q_c+Q_b)k_1^2 + 2Q_c\kappa_n^2\right]}{k_1\left(k_1^2+\kappa_n^2\right)}\ (p) \qquad (13)$$

Next, we are concerned with satisfying boundary conditions: (12) leads to the displacement and stress fields due to an interface straight edge dislocation ($\vec{b}_{II} = (b,0,0)$) parallel to the x_3-direction at the origin; the interface is the Ox_1x_3-plane. (13) provides the complementary terms (to first order in ξ_n) in the elastic fields of an interfacial sinusoidal climb-type edge dislocation.

Calculation Results

Displacement and Stress Fields of an Interface Straight Edge Dislocation

Four distinct couples of values for $(\overline{\alpha}_1^{(m)}, \overline{\beta}_1^{(m)})$ are extracted from (12). These are (j = a to d):

$$\overline{\alpha}_1^{(m)} = \frac{\overline{\overline{\alpha}}_j^{(m)}}{k_1^3} \equiv \overline{\alpha}_{1j}^{(m)}$$

$$\overline{\beta}_1^{(m)} = \overline{\overline{\beta}}_j^{(m)}\frac{\operatorname{sgn}(k_1)}{k_1^2} \equiv \overline{\beta}_{1j}^{(m)} \qquad (14)$$

where

$$\overline{\overline{\alpha}}_a^{(m)} = \frac{iD_mC_\nu}{4},\ \overline{\overline{\beta}}_a^{(m)} = 0;$$

$$\overline{\overline{\alpha}}_b^{(m)} = \frac{(5-6\nu_m)Q_b}{4},\ \overline{\overline{\beta}}_b^{(m)} = (-1)^{m-1}\frac{3Q_b}{4};$$

$$\overline{\overline{\alpha}}_c^{(m)} = \frac{1}{2}\left(\frac{iC_\nu D_m}{2} + \frac{\mu_2 Q_c}{\mu_1\nu_1}\delta_{m2} + \frac{\mu_1 Q_c}{\mu_2\nu_2}\delta_{m1} + \frac{(9-8\nu_m)Q_c}{\nu_m}\right),$$
$$\overline{\overline{\beta}}_c^{(m)} = (-1)^{m-1}\frac{Q_c}{\nu_m};$$

$$\overline{\overline{\alpha}}_d^{(m)} = \frac{1}{2}\left(\frac{iC_\nu D_m}{2} - \frac{\mu_2 Q_c}{\mu_1 \nu_1}\delta_{m2} - \frac{\mu_1 Q_c}{\mu_2 \nu_2}\delta_{m1} + \frac{3Q_c}{\nu_m}\right),\ \overline{\overline{\beta}}_d^{(m)} = \overline{\overline{\beta}}_c^{(m)} \tag{15}$$

where δ_{ij} is the Kronecker delta and $D_m = b\mu_m/2\pi = C_m(1-\nu_m)$. The couples $(\overline{\alpha}_{1j}^{(m)}, \overline{\beta}_{1j}^{(m)}), j = a$ to d, are obtained from (12a–d) associated with the displacement, (12e–h) associated with stresses, (12a, b, i, j) and (12c, d, i, j), respectively. None of these values satisfies the entire (12). The associated elastic fields denoted $\vec{u}_{atod}^{(0)(m)V}$ and $(\sigma)_{atod}^{(0)(m)V}$ are displayed below. A superposition of these partial fields will provide the complete form of the solution. We have at position $\vec{x} = (x_1, x_2, x_3)$ ($\vec{u}^{(0)(m)V} \equiv \vec{u}_j^{(0)(m)V}$, $(\sigma)^{(0)(m)V} \equiv (\sigma)_j^{(0)(m)V}$; $j = a$ to d):

$$u_{1j}^{(0)(m)V} = \frac{i\left(\overline{\overline{\alpha}}_j^{(m)} + (-1)^m 4(1-\nu_m)\overline{\overline{\beta}}_j^{(m)}\right)}{\mu_m}\tan^{-1}\frac{x_1}{|x_2|} + \frac{i\overline{\overline{\beta}}_j^{(m)}}{\mu_m}\frac{x_1 x_2}{r^2},$$

$$u_{2j}^{(0)(m)V} = \frac{i\left((-1)^m\overline{\overline{\alpha}}_j^{(m)} + \overline{\overline{\beta}}_j^{(m)}\right)}{\mu_m}\ln|x_1|\delta_A(x_2) - \frac{i(-1)^m\overline{\overline{\beta}}_j^{(m)}}{\mu_m}\frac{x_2^2}{r^2},$$

$$\sigma_{11j}^{(0)(m)V} = 2i\left(\overline{\overline{\alpha}}_j^{(m)} + (-1)^m 2(2-\nu_m)\overline{\overline{\beta}}_j^{(m)}\right)\left(\frac{x_2}{r^2} + \pi\delta(x_1)\delta_A(x_2)\right) - \frac{2i\overline{\overline{\beta}}_j^{(m)} x_2(x_1^2 - x_2^2)}{r^4},$$

$$\sigma_{22j}^{(0)(m)V} = -2i\left(\overline{\overline{\alpha}}_j^{(m)} + (-1)^m 2(1-\nu_m)\overline{\overline{\beta}}_j^{(m)}\right)\left(\frac{x_2}{r^2} + \pi\delta(x_1)\delta_A(x_2)\right) + \frac{2i\overline{\overline{\beta}}_j^{(m)} x_2(x_1^2 - x_2^2)}{r^4},$$

$$\sigma_{33j}^{(0)(m)V} = (-1)^m 4\nu_m i\overline{\overline{\beta}}_j^{(m)}\left(\frac{x_2}{r^2} + \pi\delta(x_1)\delta_A(x_2)\right),$$

$$\sigma_{12j}^{(0)(m)V} = 2i\left((-1)^m\overline{\overline{\alpha}}_j^{(m)} + (3-2\nu_m)\overline{\overline{\beta}}_j^{(m)}\right)\left(\frac{x_1}{r^2} + (\overline{J}_1 - \frac{x_1}{r^2})\delta_A(x_2)\right) + \frac{(-1)^m 4i\overline{\overline{\beta}}_j^{(m)}\operatorname{sgn}(x_2)x_1x_2^2}{r^4}; \tag{16}$$

$$\overline{J}_1 = \int_0^\infty \sin k_1 x_1 dk_1.$$

Here, $\text{sgn}(x_i) = x_i/|x_i|$, $\delta(x_1)$ is the Dirac delta function and δ_A has the following definition: $\delta_A(x_2) = 0$ when $x_2 \neq 0$ and $\delta_A(x_2) = 1$ when $x_2 = 0$; $r^2 = x_1^2 + x_2^2$. Constant terms are omitted in the displacement. The other elastic fields are zero. We define the elastic fields $\vec{u}^{(0)(m)}(\vec{x})$ and $(\sigma)^{(0)(m)}(\vec{x})$ of an interface straight edge dislocation as

$$\begin{aligned}\vec{u}^{(0)(m)} &= \vec{u}^{(0)(m)\infty} - \vec{u}^{(0)(m)W}\\ (\sigma)^{(0)(m)} &= (\sigma)^{(0)(m)\infty} - (\sigma)^{(0)(m)W}\end{aligned} \tag{17}$$

with

$$\begin{aligned}\vec{u}^{(0)(m)W} &= \sum_{j=a\text{to}d} \eta_j^{(m)} \vec{u}_j^{(0)(m)V}\\ (\sigma)^{(0)(m)W} &= \sum_{j=a\text{to}d} \eta_j^{(m)} (\sigma)_j^{(0)(m)V}.\end{aligned} \tag{18}$$

Again $\vec{u}^{(0)(m)\infty}$ and $(\sigma)^{(0)(m)\infty}$ are due to a straight edge $\vec{b}_{II} = (b, 0, 0)$ parallel to the x_3- direction at the origin in an infinite medium; $\vec{u}_{atod}^{(0)(m)V}$ and $(\sigma)_{atod}^{(0)(m)V}$ are given in (16). $\eta_{atod}^{(m)}$ are real, to be determined by the condition that the elastic fields satisfy the following relations:

- $\vec{u}^{(0)(m)}(\vec{x})$ and $(\sigma)^{(0)(m)}(\vec{x})$ are continuous when crossing the Ox_1x_3-plane.
- $\oint_\Gamma \mathrm{d}u_1^{(0)(m)} = b$ for a closed contour Γ in x_1x_2 encircling the dislocation.
- $\vec{u}^{(0)(m)W}$ vanishes far from the interface (i.e. when $|x_2| \to \infty$).

It is easy to show that the second condition above correspond to $u_1^{(0)(m)}$ constant with m on the interface. All the stresses involved in $(\sigma)^{(0)(m)\infty}$ and $(\sigma)_{atod}^{(0)(m)V}$ vanish at infinity. Under such conditions, $\vec{u}^{(0)(m)}(\vec{x})$ and $(\sigma)^{(0)(m)}(\vec{x})$ correspond to an interface straight edge dislocation. Next, we express the quantities involved in the above-mentioned requirements and proceed to satisfy these.

$$u_1^{(0)(1)}(x_1, x_2 = 0, x_3) = u_1^{(0)(2)}(x_1, x_2 = 0, x_3) \Rightarrow$$

$$\frac{1}{\mu_m}\left(\sum_j \eta_j^{(m)} \overline{\overline{\alpha}}_j^{(0)(m)} + 4(1 - \nu_m)e_7(m)\right) \equiv e_1;$$

$$u_2^{(0)(1)} = u_2^{(0)(2)} \Rightarrow$$

$$\frac{1}{\mu_m}\left(C_m(2\nu_m - 1) - 2i(-1)^m\left[\sum_j \eta_j^{(m)} \overline{\overline{\alpha}}_j^{(0)(m)} + e_7(m)\right]\right) \equiv e_2;$$

$$\sigma_{11}^{(0)(1)} = \sigma_{11}^{(0)(2)} \Rightarrow$$

$$\sum_j \eta_j^{(m)} \overline{\overline{\alpha}}_j^{(0)(m)} + 2(2 - \nu_m) e_7(m) \equiv e_3;$$

$$\sigma_{22}^{(0)(1)} = \sigma_{22}^{(0)(2)} \Rightarrow$$

$$\sum_j \eta_j^{(m)} \overline{\overline{\alpha}}_j^{(0)(m)} + 2(1 - \nu_m) e_7(m) \equiv e_4;$$

$$\sigma_{33}^{(0)(1)} = \sigma_{33}^{(0)(2)} \Rightarrow$$

$$\nu_m e_7(m) \equiv e_5;$$

$$\sigma_{12}^{(0)(1)} = \sigma_{12}^{(0)(2)} \Rightarrow$$

$$C_m - 2i(-1)^m \left(\sum_j \eta_j^{(m)} \overline{\overline{\alpha}}_j^{(0)(m)} + (3 - 2\nu_m) e_7(m) \right) \equiv e_6 \tag{19}$$

where

$$(-1)^m \sum_j \eta_j^{(m)} \overline{\overline{\beta}}_j^{(0)(m)} \equiv e_7(m).$$

Here, the subscript j takes the values a to d. The continuity of an elastic field requires associated e_i constant with m at the crossing of the interface. The additional condition, $\vec{u}^{(0)(m)}$ tends to $\vec{u}^{(0)(m)\infty}$ when $|x_2| \to \infty \Rightarrow$, corresponds to vanishing $\vec{u}^{(0)(m)W}(\vec{x})$ far from the interface. This gives

$\vec{u}^{(0)(m)W}(\vec{x})$ vanishes when $|x_2| \to \infty \Rightarrow$

$$u_2^{(0)(m)W} = -ie_7(m)/\mu_m = 0. \tag{20}$$

This condition is due to the x_2-component of the displacement $\vec{u}^{(0)(m)W}(\vec{x})$ only; the other x_1-component tends to zero. Because $e_7(m)$ is a constant in medium (m), imposing its value to be identically zero is not mandatory; elastic displacements, differing only by a constant, are admissible in elasticity. Instead, we shall choose a value to $e_7(m)$ that ensures strongest continuity of the elastic fields at the crossing of the interface. We have checked that $e_7(m) \equiv 0$ leads to poor continuity (i.e. only a restricted number of elastic field components, $u_i^{(0)(m)}$ and $\sigma_{ij}^{(0)(m)}$, are continuous when crossing the interface). A best continuity behaviour of the elastic fields at the interface is presented as follows. The relative volume variation $(\Delta V/V)^{(0)(m)}$ at arbitrary position $(x_1, 0, x_3)$ is obtained to be

$$\left(\frac{\Delta V}{V}\right)^{(0)(m)}(x_1, 0, x_3) = -2\pi\delta(x_1)iV_R^{(0)} \tag{21}$$

with

$$V_R^{(0)} = \frac{1-2\nu_m}{\mu_m}e_7(m) = (1-2\nu_m)iu_2^{(0)(m)W}(x_1, \infty, x_3).$$

It has a singularity given by the Dirac delta function $\delta(x_1)$. We demand $V_R^{(0)}$ to be constant with $m = 1$ and 2; this gives a value to $e_7(m)$ that is m dependent. We have

$$\frac{e_7(2)}{e_7(1)} = \Gamma\frac{1-2\nu_1}{1-2\nu_2} \tag{22}$$

where $\Gamma = \mu_2/\mu_1$. The expressions for $e_1(m)$ and $e_6(m)$ in (19) are merged to read

$$C_m - 2i(-1)^m\mu_m\left[e_1(m) - V_R^{(0)}\right] = e_6(m). \tag{23}$$

We seek continuity of the elastic fields on the interface, hence we impose $e_1(m)$ and $e_6(m)$ to be constant with m; this leads, with the help of (23), to

$$\begin{aligned} e_6 &= \frac{\mu_1C_2 - \mu_2C_1}{\mu_1 - \mu_2}, \\ e_1 &= \frac{i(C_2 - C_1)}{2(\mu_1 - \mu_2)} + V_R^{(0)}. \end{aligned} \tag{24}$$

These expressions are unchanged by inverting the elastic constants. We isolate $\sum_j \eta_j^{(m)}\overline{\overline{\alpha}}_j^{(0)(m)}$ with the expression for e_1 in (19) and obtain

$$\sum_j \eta_j^{(m)}\overline{\overline{\alpha}}_j^{(0)(m)} = \frac{i\mu_m(C_2 - C_1)}{2(\mu_2 - \mu_1)} - \frac{(3-2\nu_m)\mu_m}{1-2\nu_m}V_R^{(0)}. \tag{25}$$

Introducing (25) in the expression for $e_2(m)$ in (19) and imposing $e_2(m)$ to be constant with m, provide a value to $V_R^{(0)}$ in the form

$$V_R^{(0)} = -\frac{ib(1-2\nu_1)(1-2\nu_2)}{8\pi(1-\nu_1)(1-\nu_2)}. \tag{26}$$

This relation is unchanged by inverting the elastic constants. The other elastic parameters are obtained as

$$e_2 = -\frac{b}{\pi} + \frac{4i[\mu_1(1-\nu_1)(1-2\nu_2) - \mu_2(1-\nu_2)(1-2\nu_1)]}{(1-2\nu_1)(1-2\nu_2)(\mu_2 - \mu_1)}V_R^{(0)},$$

$$
\begin{aligned}
e_3(m) &= \frac{i\mu_m(C_2 - C_1)}{2(\mu_2 - \mu_1)} + \frac{\mu_m}{1 - 2\nu_m} V_R^{(0)}, \\
e_4(m) &= \frac{i\mu_m(C_2 - C_1)}{2(\mu_2 - \mu_1)} - \frac{\mu_m}{1 - 2\nu_m} V_R^{(0)}, \\
e_5(m) &= \frac{\nu_m \mu_m}{1 - 2\nu_m} V_R^{(0)}.
\end{aligned} \tag{27}
$$

The continuities at the interface of $u_1^{(0)(m)}$, $u_2^{(0)(m)}$, $\sigma_{12}^{(0)(m)}$ and $(\Delta V/V)^{(0)(m)}$ are achieved. The elastic fields, at position $\vec{x} = (x_1, x_2, x_3)$, are:

$$
\begin{aligned}
u_1^{(0)(m)} &= \frac{b}{2\pi} \tan^{-1} \frac{x_2}{x_1} - ie_1 \tan^{-1} \frac{x_1}{|x_2|} + \frac{x_1 x_2}{r^2} \left(\frac{C_m}{2\mu_m} + (-1)^{m-1} \frac{1}{1 - 2\nu_m} iV_R^{(0)} \right), \\
u_2^{(0)(1)} &= -\frac{(1 - 2\nu_1)C_1}{2\mu_1} \ln r - 2 \ln|x_1| \delta_A(x_2) \left(\frac{C_2 - C_1}{4(\mu_2 - \mu_1)} + \frac{1 - \nu_1}{1 - 2\nu_1} iV_R^{(0)} \right) \\
&\quad + \frac{x_2^2}{r^2} \left(\frac{C_1}{2\mu_1} + \frac{1}{1 - 2\nu_1} iV_R^{(0)} \right), \\
\sigma_{11}^{(0)(m)} &= -C_m \frac{x_2(x_2^2 + 3x_1^2)}{r^4} - \left(\frac{x_2}{r^2} + \pi\delta(x_1)\delta_A(x_2) \right) 2ie_3(m) \\
&\quad + \frac{x_2(x_1^2 - x_2^2)}{r^4} (-1)^m \frac{\mu_m}{1 - 2\nu_m} 2iV_R^{(0)}, \\
\sigma_{22}^{(0)(m)} &= \frac{x_2(x_1^2 - x_2^2)}{r^4} \left[C_m + (-1)^m \frac{\mu_m 2iV_R^{(0)}}{1 - 2\nu_m} \right] + \left(\frac{x_2}{r^2} + \pi\delta(x_1)\delta_A(x_2) \right) 2ie_4(m), \\
\sigma_{33}^{(0)(m)} &= -2\nu_m C_m \frac{x_2}{r^2} - \left(\frac{x_2}{r^2} + \pi\delta(x_1)\delta_A(x_2) \right) \frac{\nu_m \mu_m}{1 - 2\nu_m} 4iV_R^{(0)}, \\
\sigma_{12}^{(0)(1)} &= C_1 \frac{x_1(x_1^2 - x_2^2)}{r^4} - \left(\frac{x_1}{r^2} + (\overline{J}_1 - \frac{x_1}{r^2})\delta_A(x_2) \right) \frac{\mu_1(C_2 - C_1)}{\mu_2 - \mu_1} \\
&\quad - \operatorname{sgn}(x_2) \frac{x_1 x_2^2}{r^4} \frac{\mu_1}{1 - 2\nu_1} 4iV_R^{(0)}, \\
\left(\frac{\Delta V}{V} \right)^{(0)(m)} &= \frac{1 - 2\nu_m}{2(1 + \nu_m)\mu_m} \left(\sigma_{11}^{(0)(m)} + \sigma_{22}^{(0)(m)} + \sigma_{33}^{(0)(m)} \right) \\
&= -\left(\frac{x_2}{r^2} + \pi\delta(x_1)\delta_A(x_2) \right) 2iV_R^{(0)} + \frac{x_2(x_1^2 - x_2^2)}{r^4} (-1)^m \frac{2iV_R^{(0)}}{1 + \nu_m} \\
&\quad - \frac{(1 - 2\nu_m)C_m}{\mu_m} \frac{x_2}{r^2}.
\end{aligned} \tag{28}
$$

Corresponding expressions for half-space $m = 2$ are obtained by inverting the elastic constants. The relations between $e_i(m)$ and $V_R^{(0)}$ are taken from relations (21) and (27); the value of $V_R^{(0)}$ is given by (26). Other elastic fields are zero. Dirac delta function $\delta(x_1)$ is present in $\sigma_{ii}^{(0)(m)}(x_1, 0, x_3)(i = 1 \text{ to } 3)$ and $(\Delta V/V)^{(0)(m)}$, on the interface. The only stress component with a singularity $1/x_1$ on the interface is $\sigma_{12}^{(0)(m)}$.

Elastic Fields of an Interface Climb-Type Sinusoidal Edge Dislocation

Five values for $(\overline{\alpha}_1^{(m)}(\kappa_n),\overline{\beta}_1^{(m)}(\kappa_n))$ are extracted from (13); these are.

$$\text{(f)}\quad \overline{\alpha}_1^{(m)} = \frac{\xi_n l_{1a}^{(m)}}{k_1\sqrt{k_1^2+\kappa_n^2}}\left(1+\frac{\kappa_n^2}{k_1^2+\kappa_n^2}\right) \equiv \overline{\alpha}_{1a}^{A_n(m)},\ \overline{\beta}_1^{(m)} = 0 \equiv \overline{\beta}_{1a}^{A_n(m)};$$

$$\text{(g)}\quad \begin{aligned} \overline{\alpha}_1^{(m)} &= \frac{\xi_n l_{1b}^{(m)}}{k_1\sqrt{k_1^2+\kappa_n^2}}\left(l_{2b}^{(m)} + \frac{\kappa_n^2 l_{3b}^{(m)}}{k_1^2+\kappa_n^2} - \frac{\kappa_n^2 l_{4b}^{(m)}}{k_1^2+\tilde{\Omega}_{1b}\kappa_n^2}\right) \equiv \overline{\alpha}_{1b}^{A_n(m)}, \\ \overline{\beta}_1^{(m)} &= \xi_n t_{1b}^{(m)} k_1\left(\frac{1}{k_1^2+\kappa_n^2} - \frac{\tilde{\Omega}_{1b}}{k_1^2+\tilde{\Omega}_{1b}\kappa_n^2}\right) \equiv \overline{\beta}_{1b}^{A_n(m)}; \end{aligned}$$

$$\text{(h)}\quad \overline{\alpha}_1^{(m)} = \frac{\xi_n l_{1c}^{(m)}}{k_1\sqrt{k_1^2+\kappa_n^2}}\left(1-2\nu_m-\frac{\kappa_n^2}{k_1^2+\kappa_n^2}\right) \equiv \overline{\alpha}_{1c}^{A_n(m)}$$

$$\overline{\beta}_1^{(m)} = -\frac{\xi_n i Q_b}{2}\frac{k_1}{k_1^2+\kappa_n^2} \equiv \overline{\beta}_{1c}^{A_n} \text{ independent of } m;$$

$$\text{(i)}\quad \begin{aligned} \overline{\alpha}_1^{(m)} &= \frac{\xi_n l_{1d}^{(m)}}{k_1\sqrt{k_1^2+\kappa_n^2}}\left(l_{2d}^{(m)} - \frac{\kappa_n^2 l_{3d}^{(m)}}{k_1^2+\kappa_n^2}\right) \equiv \overline{\alpha}_{1d}^{A_n(m)}, \\ \overline{\beta}_1^{(m)} &= \xi_n t_{1d}^{(m)}\frac{k_1}{k_1^2+\kappa_n^2} \equiv \overline{\beta}_{1d}^{A_n(m)}; \end{aligned}$$

$$\text{(j)}\quad \overline{\alpha}_1^{(m)} = \frac{\xi_n l_{1e}^{(m)} k_1}{\sqrt{k_1^2+\kappa_n^2}}\left(\frac{l_{2e}^{(m)}}{k_1^2-r_2^*\kappa_n^2} + \frac{l_{3e}^{(m)}}{k_1^2-r_1^*\kappa_n^2} - \frac{l_{4e}^{(m)}}{k_1^2-r_m^*\kappa_n^2}\right.$$

$$\left. - \frac{2Q_b}{k_1^2+\kappa_n^2} + \frac{4Q_c}{k_1^2}\right) \equiv \overline{\alpha}_{1e}^{A_n(m)}$$

$$\overline{\beta}_1^{(m)} = -\frac{i\xi_n k_1}{2}\left(\frac{t_{1e}^{(m)}}{k_1^2-r_m^*\kappa_n^2} + \frac{Q_b}{k_1^2+\kappa_n^2}\right) \equiv \overline{\beta}_{1e}^{A_n(m)} \tag{29}$$

where

$$\begin{aligned} l_{1a}^{(m)} &= (-1)^{m-1}\mu_m b C_v/16\pi,\ C_v \\ &= [1/(1-\nu_1) - 1/(1-\nu_2)],\quad r_m^* = (1-\nu_m)/\nu_m, \\ \rho_m &= \nu_1\delta_{m2} + \nu_2\delta_{m1},\ \Omega_{1b} \\ &= 4(1-\nu_1)(1-\nu_2)/[(1-\nu_1)(1-2\nu_2) + (1-\nu_2)(1-2\nu_1)]; \end{aligned}$$

$$\begin{aligned} l_{1a}^{(m)} &= (-1)^{m-1}\mu_m b C_v/16\pi,\ C_v \\ &= [1/(1-\nu_1) - 1/(1-\nu_2)],\quad r_m^* = (1-\nu_m)/\nu_m, \end{aligned}$$

$$\rho_m = \nu_1\delta_{m2} + \nu_2\delta_{m1},\ \Omega_{1b} \\ = 4(1-\nu_1)(1-\nu_2)/[(1-\nu_1)(1-2\nu_2) + (1-\nu_2)(1-2\nu_1)];$$

$$\begin{aligned} l_{1b}^{(m)} &= (-1)^{m-1} l_{1a}^{(m)}/4(1-\nu_1)(1-\nu_2)(1-\tilde{\Omega}_{1b}), \\ l_{2b}^{(m)} &= (1-\tilde{\Omega}_{1b})\left[(\nu_1-\nu_2)\tilde{\Omega}_{1b} + (-1)^{m-1}4(1-\nu_1)(1-\nu_2)(1+2\tilde{\Omega}_{1b})\right], \\ l_{3b}^{(m)} &= (-1)^{m-1}4(1-\nu_1)(1-\nu_2)(1-\tilde{\Omega}_{1b}) - (\nu_1-\nu_2)\tilde{\Omega}_{1b}, \\ l_{4b}^{(m)} &= \tilde{\Omega}_{1b}^2\left[(-1)^{m-1}8(1-\nu_1)(1-\nu_2)(1-\tilde{\Omega}_{1b}) - (\nu_1-\nu_2)\tilde{\Omega}_{1b}\right], \\ t_{1b}^{(m)} &= 2\tilde{\Omega}_{1b}(1-\rho_m) l_{1b}^{(m)}; \end{aligned}$$

$$l_{1c}^{(m)} = (-1)^m i Q_b/2;$$

$$\begin{aligned} l_{1d}^{(m)} &= (-1)^m i/2[\nu_2(1-\nu_1) + \nu_1(1-\nu_2)], \\ l_{2d}^{(m)} &= Q_c(4 - 3\rho_m - \nu_m - 2[\nu_2(1-\nu_1) + \nu_1(1-\nu_2)]) \\ &\quad - Q_b[\nu_2(1-\nu_1) + \nu_1(1-\nu_2)], \\ l_{3d}^{(m)} &= Q_c(4 - 3\rho_m - \nu_m - 2) - Q_b[\nu_2(1-\nu_1) + \nu_1(1-\nu_2)], \\ t_{1d}^{(m)} &= (-1)^{m-1}2Q_c(1-\rho_m) l_{1d}^{(m)}; \end{aligned}$$

$$\begin{aligned} l_{1e}^{(m)} &= (-1)^{m-1} i/4,\ l_{2e}^{(m)} = (\delta_{m1}/\nu_2)\left[Q_c - Q_b/(1+r_2^*)\right], \\ l_{3e}^{(m)} &= (\delta_{m2}/\nu_1)\left[Q_c - Q_b/(1+r_1^*)\right], \\ l_{4e}^{(m)} &= (5/\nu_m)\left[Q_c - Q_b/(1+r_m^*)\right], \\ t_{1e}^{(m)} &= (1/\nu_m)\left[Q_c - Q_b/(1+r_m^*)\right]. \end{aligned} \tag{30}$$

None of these couples satisfies the entire conditions (13). For each couple, we give in Appendix B the associated oscillating elastic fields $\vec{u}^{A_n(m)V} = \vec{u}^{A_n(m)A} + \vec{u}^{A_n(m)B}$ and $(\sigma)^{A_n(m)V} = (\sigma)^{A_n(m)A} + (\sigma)^{A_n(m)B}$ defined in (11). A superposition of these partial fields will provide the complete form of solution (to first order in ξ_n). The elastic fields $\vec{u}^{(m)}(\vec{x})$ and $(\sigma)^{(m)}(\vec{x})$ of an interface sinusoidal edge dislocation may be written as

$$\vec{u}^{(m)} = \vec{u}^{(0)(m)} + \vec{u}^{A_n(m)}$$

$$(\sigma)^{(m)} = (\sigma)^{(0)(m)} + (\sigma)^{A_n(m)} \tag{31}$$

$\vec{u}^{(0)(m)}$ and $(\sigma)^{(0)(m)}$ (28) correspond to the fields of a straight edge dislocation lying on a planar interface; $\vec{u}^{A_n(m)}$ and $(\sigma)^{A_n(m)}$ are oscillating expressions proportional to either the sinusoid $A_n(x_3)$ or its spatial derivative $\partial A_n/\partial x_3$ in the forms

$$\vec{u}^{A_n(m)} = \vec{u}^{A_n(m)\infty} - \vec{u}^{A_n(m)W}$$
$$(\sigma)^{A_n(m)} = (\sigma)^{A_n(m)\infty} - (\sigma)^{A_n(m)W} \tag{32}$$

with

$$\vec{u}^{A_n(m)W} = \sum_{j=atoe} \eta_j^{A_n(m)} \vec{u}_j^{A_n(m)V}$$
$$(\sigma)^{A_n(m)W} = \sum_{j=atoe} \eta_j^{A_n(m)} (\sigma)_j^{A_n(m)V} \tag{33}$$

$\vec{u}_{atoe}^{A_n(m)V}$ and $(\sigma)_{atoe}^{A_n(m)V}$ are given in Appendix B below (for $\vec{u}^{A_n(m)\infty}$ and $(\sigma)^{A_n(m)\infty}$, see Anongba 2010);$\eta_{atoe}^{A_n(m)}$ are real to be determined by the requirement that the elastic fields be continuous when crossing the interface. It is sufficient to write this condition for points on the average interface plane. Before displaying the corresponding equations, we stress what follows.

- Both $u_1^{A_n(m)}(x_1, 0, x_3)$ and $\sigma_{13}^{A_n(m)}(x_1, 0, x_3)$ contain terms with singularity $1/x_1$ only; the associated coefficients $e_1^{A_n}$ and $e_8^{A_n}$, respectively, have been taken constant.
- $u_2^{A_n(m)}(x_1, 0, x_3)$, $\sigma_{12}^{A_n(m)}(x_1, 0, x_3)$ and $\sigma_{23}^{A_n(m)}(x_1, 0, x_3)$ are bounded functions. Under such conditions, we have considered their linear forms with respect to x_1 and posed the terms proportional to x_1 constant with $m = 1$ and 2.
- $\sigma_{11}^{A_n(m)}(x_1, 0, x_3)$, $\sigma_{22}^{A_n(m)}(x_1, 0, x_3)$ and $\sigma_{33}^{A_n(m)}(x_1, 0, x_3)$ have terms with singularities $1/x_1^2$ and $\ln|x_1|$; the associated coefficients ($e_{41}^{A_n}$,$e_{51}^{A_n}$ and $e_{61}^{A_n}$) and ($e_{42}^{A_n}$, $e_{52}^{A_n}$ and $e_{62}^{A_n}$), respectively, have been set constant.
- $u_3^{A_n(m)}(x_1, 0, x_3)$ contains a term with the singularity $\ln|x_1|$ only. The associated coefficients $e_3^{A_n}$ is taken constant.

We may write:

$$u_1^{A_n(1)}(x_1, 0, x_3) = u_1^{A_n(2)}(x_1, 0, x_3) \Rightarrow$$

$$\frac{1}{\mu_m}\Big\{C_m(3-2\nu_m) - \eta_a^{A_n(m)}4l_{1a}^{(m)} - \eta_b^{A_n(m)}4\Big[l_{1b}^{(m)}l_{2b}^{(m)} + (-1)^m 4(1-\nu_m)t_{1b}^{(m)}(1-\tilde{\Omega}_{1b})\Big]$$
$$+\eta_c^{A_n(m)}4l_{1c}^{(m)}(3-2\nu_m) - \eta_d^{A_n(m)}4\Big[l_{1d}^{(m)}l_{2d}^{(m)} + t_{1d}^{(m)}(-1)^m 4(1-\nu_m)\Big]$$
$$-\eta_e^{A_n(m)}4\Big(l_{1e}^{(m)}[l_{2e}^{(m)} + l_{3e}^{(m)} - l_{4e}^{(m)} - 2Q_b + 4Q_c] + (-1)^{m-1}i2(1-\nu_m)(t_{1e}^{(m)} + Q_b)\Big)\Big\} \equiv e_1^{A_n};$$

$$u_2^{A_n(1)} = u_2^{A_n(2)} \Rightarrow$$

$$\frac{1}{\mu_m}\Big\{\eta_a^{A_n(m)}(-1)^m 2l_{1a}^{(m)} + \eta_b^{A_n(m)}2\Big[(-1)^m l_{1b}^{(m)}(l_{3b}^{(m)} - l_{4b}^{(m)}) + t_{1b}^{(m)}(\tilde{\Omega}_{1b}^2 - 1)\Big]$$

$$+\eta_d^{A_n(m)}2\Big[(-1)^{m-1}l_{1d}^{(m)}l_{3d}^{(m)}-t_{1d}^{(m)}\Big]$$
$$+\eta_e^{A_n(m)}\Big((-1)^m 2l_{1e}^{(m)}[l_{2e}^{(m)}r_2^*+l_{3e}^{(m)}r_1^*-l_{4e}^{(m)}r_m^*+2Q_b]+i(Q_b-t_{1e}^{(m)}r_m^*)\Big)\Big\}\equiv e_2^{A_n};$$

$$u_3^{A_n(1)}=u_3^{A_n(2)}\Rightarrow$$

$$\frac{1}{\mu_m}\Big\{C_m(1-2\nu_m)+\eta_a^{A_n(m)}4l_{1a}^{(m)}+\eta_b^{A_n(m)}4l_{1b}^{(m)}l_{2b}^{(m)}+\eta_c^{A_n(m)}4l_{1c}^{(m)}(1-2\nu_m)$$
$$+\eta_d^{A_n(m)}4l_{1d}^{(m)}l_{2d}^{(m)}+\eta_e^{A_n(m)}4l_{1e}^{(m)}[l_{2e}^{(m)}+l_{3e}^{(m)}-l_{4e}^{(m)}-2Q_b+4Q_c]\Big\}\equiv e_3^{A_n};$$

$$\sigma_{11}^{A_n(1)}=\sigma_{11}^{A_n(2)}\Rightarrow$$

$$3C_m-\eta_a^{A_n(m)}4l_{1a}^{(m)}-\eta_b^{A_n(m)}4\Big[l_{1b}^{(m)}l_{2b}^{(m)}+t_{1b}^{(m)}(-1)^m 2(2-\nu_m)(1-\tilde{\Omega}_{1b})\Big]$$
$$+\eta_c^{A_n(m)}12l_{1c}^{(m)}-\eta_d^{A_n(m)}4\Big[l_{1d}^{(m)}l_{2d}^{(m)}+t_{1d}^{(m)}(-1)^m 2(2-\nu_m)\Big]$$
$$-\eta_e^{A_n(m)}4\Big(l_{1e}^{(m)}[l_{2e}^{(m)}+l_{3e}^{(m)}-l_{4e}^{(m)}-2Q_b+4Q_c]+(-1)^{m-1}i(2-\nu_m)(t_{1e}^{(m)}+Q_b)\Big)\equiv e_{41}^{A_n},$$
$$-C_m+\eta_b^{A_n(m)}2\Big[l_{1b}^{(m)}(l_{2b}^{(m)}-l_{3b}^{(m)}+l_{4b}^{(m)})+(-1)^m t_{1b}^{(m)}2(2-\nu_m)(1-\tilde{\Omega}_{1b}^2)\Big]$$
$$-\eta_c^{A_n(m)}4l_{1c}^{(m)}+\eta_d^{A_n(m)}2\Big[l_{1d}^{(m)}(l_{2d}^{(m)}+l_{3d}^{(m)})+t_{1d}^{(m)}(-1)^m 2(2-\nu_m)\Big]$$
$$+\eta_e^{A_n(m)}2\Big(l_{1e}^{(m)}[l_{2e}^{(m)}(1-r_2^*)+l_{3e}^{(m)}(1-r_1^*)$$
$$-l_{4e}^{(m)}(1-r_m^*)-4Q_b+4Q_c]+i(-1)^{m-1}(2-\nu_m)(Q_b-t_{1e}^{(m)}r_m^*)\Big)\equiv e_{42}^{A_n};$$

$$\sigma_{22}^{A_n(1)}=\sigma_{22}^{A_n(2)}\Rightarrow$$

$$-C_m+\eta_a^{A_n(m)}4l_{1a}^{(m)}+\eta_b^{A_n(m)}4\Big[l_{1b}^{(m)}l_{2b}^{(m)}+t_{1b}^{(m)}(-1)^m 2(1-\nu_m)(1-\tilde{\Omega}_{1b})\Big]$$
$$-\eta_c^{A_n(m)}4l_{1c}^{(m)}+\eta_d^{A_n(m)}4\Big[l_{1d}^{(m)}l_{2d}^{(m)}+t_{1d}^{(m)}(-1)^m 2(1-\nu_m)\Big]$$
$$+\eta_e^{A_n(m)}4\Big(l_{1e}^{(m)}[l_{2e}^{(m)}+l_{3e}^{(m)}-l_{4e}^{(m)}-2Q_b+4Q_c]+(-1)^{m-1}i(1-\nu_m)(t_{1e}^{(m)}+Q_b)\Big)\equiv e_{51}^{A_n},$$
$$C_m(1-2\nu_m)+\eta_a^{A_n(m)}4l_{1a}^{(m)}+\eta_b^{A_n(m)}4\Big[l_{1b}^{(m)}(l_{3b}^{(m)}-l_{4b}^{(m)})+(-1)^m t_{1b}^{(m)}2(1-\nu_m)(\tilde{\Omega}_{1b}^2-1)\Big]$$
$$+\eta_c^{A_n(m)}4l_{1c}^{(m)}(1-2\nu_m)-\eta_d^{A_n(m)}4\Big[l_{1d}^{(m)}l_{3d}^{(m)}+t_{1d}^{(m)}(-1)^m 2(1-\nu_m)\Big]$$
$$+\eta_e^{A_n(m)}4\Big(l_{1e}^{(m)}[l_{2e}^{(m)}r_2^*+l_{3e}^{(m)}r_1^*-l_{4e}^{(m)}r_m^*+2Q_b]+i(-1)^m(1-\nu_m)(Q_b-t_{1e}^{(m)}r_m^*)\Big)\equiv e_{52}^{A_n};$$

$$\sigma_{33}^{A_n(1)}=\sigma_{33}^{A_n(2)}\Rightarrow$$

$$\nu_m C_m+\eta_b^{A_n(m)}4t_{1b}^{(m)}(-1)^{m-1}\nu_m(1-\tilde{\Omega}_{1b})+\eta_c^{A_n(m)}4\nu_m l_{1c}^{(m)}$$

$$+\eta_d^{A_n(m)}4t_{1d}^{(m)}(-1)^{m-1}\nu_m+\eta_e^{A_n(m)}2i(-1)^m\nu_m(t_{1e}^{(m)}+Q_b)\equiv e_{61}^{A_n},$$

$$C_m+\eta_a^{A_n(m)}4l_{1a}^{(m)}+\eta_b^{A_n(m)}4\Big[l_{1b}^{(m)}l_{2b}^{(m)}+(-1)^m t_{1b}^{(m)}2\nu_m(\tilde{\Omega}_{1b}^2-1)\Big]$$
$$+\eta_c^{A_n(m)}4l_{1c}^{(m)}+\eta_d^{A_n(m)}4\Big[l_{1d}^{(m)}l_{2d}^{(m)}+t_{1d}^{(m)}(-1)^{m-1}2\nu_m\Big]$$
$$+\eta_e^{A_n(m)}4\Big(l_{1e}^{(m)}[l_{2e}^{(m)}+l_{3e}^{(m)}-l_{4e}^{(m)}-2Q_b+4Q_c]+i(-1)^m\nu_m(Q_b-t_{1e}^{(m)}r_m^*)\Big)\equiv e_{62}^{A_n};$$

$$\sigma_{12}^{A_n(1)}=\sigma_{12}^{A_n(2)}\Rightarrow$$

$$\eta_a^{A_n(m)}(-1)^m 2l_{1a}^{(m)}+\eta_b^{A_n(m)}2\Big((-1)^m l_{1b}^{(m)}(l_{3b}^{(m)}-l_{4b}^{(m)}\tilde{\Omega}_{1b}^{1/2})$$
$$+t_{1b}^{(m)}[-1-2(1-\nu_m)\tilde{\Omega}_{1b}^{3/2}+(3-2\nu_m)\tilde{\Omega}_{1b}^{5/2}]\Big)$$
$$+\eta_d^{A_n(m)}2\Big[(-1)^{m-1}l_{1d}^{(m)}l_{3d}^{(m)}-t_{1d}^{(m)}\Big]+\eta_e^{A_n(m)}iQ_b\Big[1+(-1)^{m-1}4il_{1e}^{(m)}\Big]\equiv e_7^{A_n};$$

$$\sigma_{13}^{A_n(1)}=\sigma_{13}^{A_n(2)}\Rightarrow$$

$$-C_m+\eta_a^{A_n(m)}4l_{1a}^{(m)}+\eta_b^{A_n(m)}4\Big[l_{1b}^{(m)}l_{2b}^{(m)}+t_{1b}^{(m)}(-1)^m 2(1-\nu_m)(1-\tilde{\Omega}_{1b})\Big]$$
$$-\eta_c^{A_n(m)}4l_{1c}^{(m)}+\eta_d^{A_n(m)}4\Big[l_{1d}^{(m)}l_{2d}^{(m)}+t_{1d}^{(m)}(-1)^m 2(1-\nu_m)\Big]$$
$$+\eta_e^{A_n(m)}4\Big(l_{1e}^{(m)}[l_{2e}^{(m)}+l_{3e}^{(m)}-l_{4e}^{(m)}+2Q_b-4Q_c]+(-1)^{m-1}i(1-\nu_m)(t_{1e}^{(m)}+Q_b)\Big)\equiv e_8^{A_n};$$

$$\sigma_{23}^{A_n(1)}=\sigma_{23}^{A_n(2)}\Rightarrow$$

$$\eta_a^{A_n(m)}(-1)^m 2l_{1a}^{(m)}+\eta_b^{A_n(m)}2\Big[(-1)^m l_{1b}^{(m)}(l_{3b}^{(m)}-l_{4b}^{(m)})+t_{1b}^{(m)}(\tilde{\Omega}_{1b}^2-1)\Big]$$
$$+\eta_d^{A_n(m)}2\Big[(-1)^{m-1}l_{1d}^{(m)}l_{3d}^{(m)}-t_{1d}^{(m)}\Big]$$
$$+\eta_e^{A_n(m)}\Big((-1)^m 2l_{1e}^{(m)}[l_{2e}^{(m)}r_2^*+l_{3e}^{(m)}r_1^*-l_{4e}^{(m)}r_m^*+2Q_b]+i(Q_b-t_{1e}^{(m)}r_m^*)\Big)\equiv e_9^{A_n}. \quad (34)$$

All $e_i^{A_n}(m)$ are constant with $m=1$ and 2 (i.e. $e_i^{A_n}(1)=e_i^{A_n}(2)$).We seek values for $\eta_{atoe}^{A_n(m)}$ that provide continuity at the interface for the largest number of the elastic field components $u_i^{A_n(m)}$ and $\sigma_{ij}^{A_n(m)}$. We shall proceed in a manner like that for $u_i^{(0)(m)}$ and $\sigma_{ij}^{(0)(m)}$. Restricting ourselves to terms that are singular, the trace $T_r(\sigma)^{A_n(m)}$ of the stress matrix $(\sigma)^{A_n(m)}$ takes the form

$$T_r(\sigma)^{A_n(m)}(x_1, 0, x_3) = A_n\left(t_{R1}^{A_n}(m)\frac{\partial I_0}{\partial|x_2|} + \kappa_n^2 t_{R2}^{A_n}(m)\frac{\partial \Pi_{(\pm z)}}{\partial|x_2|}\right) \tag{35}$$

With

$$t_{R1}^{A_n}(m) = \frac{1}{2}\left(e_{41}^{A_n}(m) + e_{51}^{A_n}(m) + 2e_{61}^{A_n}(m)\right),$$
$$t_{R2}^{A_n}(m) = \frac{1}{2}\left(2e_{42}^{A_n}(m) + e_{52}^{A_n}(m) - e_{62}^{A_n}(m)\right).$$

The associated oscillating part of the relative volume variation $(\Delta V/V)^{A_n(m)}$ at position $(x_1, 0, x_3)$ has the same form (35) with similar coefficients

$$V_{R1}^{A_n}(m) = \frac{(1-2\nu_m)}{2(1+\nu_m)\mu_m} t_{R1}^{A_n}(m),$$
$$V_{R2}^{A_n}(m) = \frac{(1-2\nu_m)}{2(1+\nu_m)\mu_m} t_{R2}^{A_n}(m). \tag{36}$$

Several relations, interconnecting the $e_i^{A_n}(m)$ and $\eta_{atoe}^{A_n(m)}$, are involved in the procedure under way; these are [by using (34)]

$$V_{R1}^{A_n}(m) = \frac{(1-2\nu_m)}{4\nu_m\mu_m}\left\{e_{41}^{A_n}(m) - \mu_m e_1^{A_n}(m)\right\} = \frac{(1-2\nu_m)}{8(1-\nu_m)}\left\{e_1^{A_n}(m) + e_3^{A_n}(m)\right\}; \tag{a}$$

$$e_{61}^{A_n}(m) = \frac{2\nu_m\mu_m}{1-2\nu_m} V_{R1}^{A_n}(m); \tag{b}$$

$$e_{51}^{A_n}(m) = -e_{41}^{A_n}(m) + \frac{4\mu_m}{1-2\nu_m} V_{R1}^{A_n}(m); \tag{c}$$

$$e_{51}^{A_n}(m) - e_8^{A_n}(m) = \eta_e^{A_n(m)}(-1)^{m-1}4i(2Q_c - Q_b); \tag{d}$$

$$e_9^{A_n}(m) - e_7^{A_n}(m) = \mu_m e_2^{A_n}(m) - e_7^{A_n}(m) = \eta_e^{A_n(m)}\varpi_e(m) + \eta_b^{A_n(m)}\varpi_b(m),$$

$$\varpi_e(m) = \frac{i}{2}\left(Q_b\left[r_2^*\delta_{m1} + r_1^*\delta_{m2} - 3r_m^*\right] - Q_c\left[\delta_{m1}r_2^*/\nu_2 + \delta_{m2}r_1^*/\nu_1 - 3r_m^*/\nu_m\right]\right),$$

$$\varpi_b(m) = l_{1b}^{(m)}2(1-\tilde{\Omega}_{1b}^{1/2})\tilde{\Omega}_{1b}^2\Big(4(1-\nu_1)(1-\nu_2)(2+\tilde{\Omega}_{1b}^{1/2})$$
$$- [2+3(\nu_1+\nu_2) - 4\nu_1\nu_2]\tilde{\Omega}_{1b}\Big); \tag{e}$$

$$V_{R1}^{A_n}(m) + V_{R2}^{A_n}(m) = -\frac{1-2\nu_m}{2\mu_m}\Big(\eta_b^{A_n(m)}4\tilde{\Omega}_{1b}(1-\tilde{\Omega}_{1b})(-1)^{m-1}t_{1b}^{(m)}$$

$$+\eta_e^{A_n(m)} 2i(-1)^{m-1} t_{1e}^{(m)}/\nu_m\Big);$$

$$+\eta_e^{A_n(m)} 2i(-1)^{m-1} t_{1e}^{(m)}/\nu_m\Big); \qquad \text{(f)}$$

$$V_{R1}^{A_n}(m) = \frac{1-2\nu_m}{2\mu_m}\Big(C_m - \eta_b^{A_n(m)}(-1)^m t_{1b}^{(m)} 4(1-\tilde{\Omega}_{1b}) + \eta_c^{A_n(m)} 4 l_{1c}^{(m)}$$
$$- \eta_d^{A_n(m)} 4(-1)^m t_{1d}^{(m)} + \eta_e^{A_n(m)} 2i(-1)^m (t_{1e}^{(m)} + Q_b)\Big). \, (g) \qquad (37)$$

We impose the relative volume variation to be continuous at the crossing of the interface implying that $V_{R1\text{and}2}^{A_n}$ (36) are constant with $m = 1$ and 2. This provides values to $t_{R1\text{and}2}^{A_n}(m)$ that are m dependent. Using (36) and (37a–c) under $e_1^{A_n}$ also constant with m, we obtain successively

$$e_1^{A_n} = \frac{\mu_1\nu_1(1-2\nu_2)e_{41}^{A_n}(2) - \mu_2\nu_2(1-2\nu_1)e_{41}^{A_n}(1)}{\mu_1\mu_2(\nu_1-\nu_2)},$$
$$V_{R1}^{A_n} = \frac{(1-2\nu_1)(1-2\nu_2)\Big[e_{41}^{A_n}(1)\mu_2 - e_{41}^{A_n}(2)\mu_1\Big]}{4\mu_1\mu_2(\nu_1-\nu_2)},$$
$$t_{R1}^{A_n}(m) = 2\frac{\delta_{m2}\mu_2(1+\nu_2) + \delta_{m1}\mu_1(1+\nu_1)}{\delta_{m2}(1-2\nu_2) + \delta_{m1}(1-2\nu_1)} V_{R1}^{A_n},$$
$$e_{61}^{A_n}(m) = \frac{2\nu_m[\delta_{m2}\mu_2(1+\nu_2) + \delta_{m1}\mu_1(1+\nu_1)]}{(1+\nu_m)[\delta_{m2}(1-2\nu_2) + \delta_{m1}(1-2\nu_1)]} V_{R1}^{A_n},$$
$$e_{51}^{A_n}(m) = -e_{41}^{A_n}(m) + \frac{4[\delta_{m2}\mu_2(1+\nu_2) + \delta_{m1}\mu_1(1+\nu_1)]}{(1+\nu_m)[\delta_{m2}(1-2\nu_2) + \delta_{m1}(1-2\nu_1)]} V_{R1}^{A_n}. \qquad (38)$$

These expressions depend on $e_{41}^{A_n}(1)$ and $e_{41}^{A_n}(2)$; $e_1^{A_n}$ and $V_{R1}^{A_n}$ are unchanged by inverting the elastic constants, as required. The only oscillating stress that has a singularity $1/x_1$ at $(x_1, 0, x_2)$ is $\sigma_{13}^{A_n}$; hence, this stress contributes a non-zero value to the crack extension force when a non-planar interface crack is subjected to general loading (Sect. 5.3). Consequently, we demand $\sigma_{13}^{A_n}$ to be continuous at the interface; this requires the associated coefficient $e_8^{A_n}$ (34) to be constant with $m = 1$ and 2. Inspection of (37d) tells us that taking $\eta_e^{A_n(m)} \equiv 0$ leads to $e_8^{A_n}(m) = e_{51}^{A_n}(m)$, an expression of which is displayed in (38). Imposing $e_8^{A_n}(m) = e_{51}^{A_n}(m)$ constant opens the possibility to find a relation between $e_{41}^{A_n}(1)$ and $e_{41}^{A_n}(2)$. In this way, we find

$$e_{41}^{A_n}(2) = -e_{41}^{A_n}(1)\frac{\mu_2[\mu_1(1-\nu_1-\nu_2) - \mu_2(1-2\nu_1)]}{\mu_1[\mu_2(1-\nu_1-\nu_2) - \mu_1(1-2\nu_2)]}. \qquad (39)$$

Further progression comes from (37e). It tells us that first $e_9^{A_n}(m) = \mu_m e_2^{A_n}(m)$ and second $e_9^{A_n}(m) = e_7^{A_n}(m)$; to see this, it is sufficient to write down three equations corresponding successively to $m = 1$ and 2, and (third) the one that results from the inversion of the elastic constants in equation for $m = 1$; summing, member by member, the two last equations yields $e_9^{A_n}(2) = e_7^{A_n}(2)$ and in like manner $e_9^{A_n}(1) = e_7^{A_n}(1)$. Furthermore, we work with the value zero for $\eta_e^{A_n}$, this implies that $\eta_b^{A_n}$ must be zero. A link between $V_{R1}^{A_n}$ and $V_{R2}^{A_n}$ emerges from (37f). Hence, we write

$$\begin{aligned} e_9^{A_n}(m) &= e_7^{A_n}(m) = \mu_m e_2^{A_n}(m), \\ \eta_e^{A_n(m)} &\equiv 0 \Rightarrow \eta_b^{A_n(m)} = 0, \\ V_{R2}^{A_n} &= -V_{R1}^{A_n}. \end{aligned} \tag{40}$$

Under $\eta_e^{A_n}$ and $\eta_b^{A_n}$ equal to zero, we have the following relation:

$$\begin{aligned} &-2\mu_m V_{R1}^{A_n}(m)/(1-2\nu_m) + \eta_d^{A_n(m)} 4\left[l_{1d}^{(m)}(l_{2d}^{(m)} + l_{3d}^{(m)}) + t_{1d}^{(m)}(-1)^m 2(1-\nu_m)\right] \\ &= e_8^{A_n}(m) + 2(-1)^{m-1} e_7^{A_n}(m). \end{aligned} \tag{41}$$

It remains to find values to $e_{41}^{A_n}(1)$ and $e_7^{A_n}(m)$. This can be done by providing first a value to $\eta_d^{A_n(m)}$; for this purpose, we use (37g) in which $\eta_c^{A_n(m)} \equiv 0$ [$\eta_e^{A_n}$ and $\eta_b^{A_n}$ are equally zero from (40)] and obtain

$$\eta_d^{A_n(m)} = \frac{\nu_1(1-\nu_2) + \nu_2(1-\nu_1)}{(\nu_1 C_1 - \nu_2 C_2)(1-\rho_m)}(-1)^{m-1}\left(C_m - \frac{2\mu_m}{1-2\nu_m} V_{R1}^{A_n}\right). \tag{42}$$

Introducing this value in (41) and imposing $e_7^{A_n}(m)$ (as also $e_8^{A_n}$) to be constant with $m = 1$ and 2 yields a value to $e_{41}^{A_n}(1)$ in the form

$$e_{41}^{A_n}(1) = \frac{Nu(e_{41}^{A_n})}{De(e_{41}^{A_n})} \tag{43}$$

with

$$\begin{aligned} Nu(e_{41}^{A_n}) = {} & 2\mu_1[\mu_1(1-2\nu_2) - \mu_2(1-\nu_1-\nu_2)]\{2Q_c(1-\nu_1)(1-\nu_2)(C_1 - C_2) \\ & - Q_b[\nu_1(1-\nu_2) + \nu_2(1-\nu_1)][C_1(1-\nu_1) - C_2(1-\nu_2)]\}, \end{aligned}$$

$$\begin{aligned} De(e_{41}^{A_n}) = {} & (\mu_1 - \mu_2)\{Q_c(1-\nu_1)(1-\nu_2)[\mu_1(1-2\nu_2) - \mu_2(1-2\nu_1)] \\ & - Q_b[\nu_1(1-\nu_2) + \nu_2(1-\nu_1)][\mu_1(1-\nu_1)(1-2\nu_2) - \mu_2(1-\nu_2)(1-2\nu_1)]\}. \end{aligned}$$

Using (43) and (39), (38) provides values

$$V_{R1}^{A_n} = \frac{Nu(V_{R1}^{A_n})}{De(V_{R1}^{A_n})} = -V_{R2}^{A_n} \tag{44}$$

with

$$Nu(V_{R1}^{A_n}) = (1-2\nu_1)(1-2\nu_2)\{2Q_c(1-\nu_1)(1-\nu_2)(C_1-C_2) - Q_b[\nu_1(1-\nu_2)+\nu_2(1-\nu_1)][C_1(1-\nu_1)-C_2(1-\nu_2)]\},$$

$$De(V_{R1}^{A_n}) = 2\{Q_c(1-\nu_1)(1-\nu_2)[\mu_1(1-2\nu_2)-\mu_2(1-2\nu_1)]$$

$$-Q_b[\nu_1(1-\nu_2)+\nu_2(1-\nu_1)][\mu_1(1-\nu_1)(1-2\nu_2)-\mu_2(1-\nu_2)(1-2\nu_1)]\};$$

$$e_8^{A_n} = e_{51}^{A_n} = \frac{Nu(e_8^{A_n})}{De(e_8^{A_n})} \qquad (45)$$

With

$$Nu(e_8^{A_n}) = 2\mu_1\mu_2\Big[(1-2\nu_1)(1-2\nu_2)-(1-\nu_1-\nu_2)^2\Big]\{2Q_c(1-\nu_1)(1-\nu_2)(C_1-C_2) - Q_b[\nu_1(1-\nu_2)+\nu_2(1-\nu_1)][C_1(1-\nu_1)-C_2(1-\nu_2)]\},$$

$$De(e_8^{A_n}) = (\nu_1-\nu_2)(\mu_1-\mu_2)\{Q_c(1-\nu_1)(1-\nu_2)[\mu_1(1-2\nu_2)-\mu_2(1-2\nu_1)] - Q_b[\nu_1(1-\nu_2)+\nu_2(1-\nu_1)][\mu_1(1-\nu_1)(1-2\nu_2)-\mu_2(1-\nu_2)(1-2\nu_1)]\}.$$

These values are unchanged by inverting the elastic constants, as expected. The associated $e_7^{A_n}$ and $e_2^{A_n}(m)$ are

$$e_7^{A_n} = e_9^{A_n} = \mu_m e_2^{A_n}(m) = -\frac{e_8^{A_n}}{2} + \frac{V_{R1}^{A_n}}{2}\left(\frac{\mu_1\Big[C_2(-\nu_1+2\nu_1\nu_2-\nu_2^2)+\nu_2(1-\nu_1)C_1\Big]}{(1-\nu_2)(1-2\nu_1)(\nu_2C_2-\nu_1C_1)}\right.$$
$$\left.+\frac{\mu_2\Big[C_1(-\nu_2+2\nu_1\nu_2-\nu_1^2)+\nu_1(1-\nu_2)C_2\Big]}{(1-\nu_1)(1-2\nu_2)(\nu_1C_1-\nu_2C_2)}\right) - \frac{\nu_2-\nu_1}{\nu_2C_2-\nu_1C_1}\left(\frac{C_1[C_2(1-2\nu_2)+C_1]}{4(1-\nu_2)}\right.$$
$$\left.+\frac{C_2[C_1(1-2\nu_1)+C_2]}{4(1-\nu_1)}\right). \qquad (46)$$

Beginning with five parameters $\eta_{atoe}^{A_n(m)}$, only survive $\eta_a^{A_n(m)}$ and $\eta_d^{A_n(m)}$; the others are zero. Using (34), we have

$$\eta_a^{A_n(m)} = (-1)^m\frac{1}{2l_{1a}^{(m)}}\left(e_7^{A_n} - \eta_d^{A_n(m)}2\Big[(-1)^{m-1}l_{1d}^{(m)}l_{3d}^{(m)} - t_{1d}^{(m)}\Big]\right). \qquad (47)$$

Here, $\eta_d^{A_n(m)}$ and $e_7^{A_n}$ are given by (42) and (46). With $\eta_a^{A_n(m)}$ (47) and $\eta_d^{A_n(m)}$ (42), the elastic fields $\vec{u}^{A_n(m)}$ and $(\sigma)^{A_n(m)}$ (32) are completely determined, thus providing the elastic fields $\vec{u}^{(m)}(\vec{x})$ and $(\sigma)^{(m)}(\vec{x})$ (31) of an interface sinusoidal climb-type edge dislocation.

Discussion

In the present study, we considered two welded infinitely extended elastic media $S1$ and $S2$ (Fig. 5.6), with an interface S having the form of a corrugated sheet. We have considered sinusoidal dislocation running indefinitely on this surface and have studied the elastic displacement and stress fields ($\vec{u}^{(m)}(\vec{x})$ and $(\sigma)^{(m)}(\vec{x})$) produced in the surrounding media. The elastic fields are decomposed individually, in linear approximation with respect to the amplitude of the sinusoid, into two terms. The first terms correspond to the elastic fields ($\vec{u}^{(0)(m)}(\vec{x})$ and $(\sigma)^{(0)(m)}(\vec{x})$) of a straight dislocation on a flat interface and the second ones ($\vec{u}^{A_n(m)}$ and $(\sigma)^{A_n(m)}$) constitute an oscillating part, proportional to the sinusoid or to its spatial derivative with respect to the direction where the dislocation runs. We left with an initial vision that the elastic fields are continuous at the crossing of the interface and join those of a sinusoidal dislocation in a homogeneous infinite medium, far from the interface. Expressions (28) for $\vec{u}^{(0)(m)}(\vec{x})$ and $(\sigma)^{(0)(m)}(\vec{x})$ have been obtained. The result is that in order to ensure the continuity of the elastic fields at the crossing of the interface, the relative variation of volume at the interface $(\Delta V/V)^{(0)(m)}$ exhibits a spatial dependence proportional to the delta function of Dirac $\delta(x_1)$, with a coefficient of proportionality $(-2\pi i V_R^{(0)})$ (26), well-defined, dependent on the Poisson ratios and which remains unchanged at the crossing of the interface. In addition, the following result holds

$$u_2^{(0)(m)}(x_1,\infty,x_3) = u_2^{(0)(m)\infty}(x_1,\infty,x_3) + \frac{b(1-2\nu_1)(1-2\nu_2)}{8\pi(1-2\nu_m)(1-\nu_1)(1-\nu_2)}. \tag{48}$$

In other words, under conditions of continuity at the interface, the displacement component $u_2^{(0)(m)}$ (far from the interface) differs from that of a straight dislocation $u_2^{(0)(m)\infty}$ in an infinitely extended homogeneous medium. The difference observed is constant, well defined, depending on the Poisson ratios of the two media and changes from medium $S1$ to medium $S2$. Moreover, this difference is related to the relative variation of volume at the interface, in the vicinity of dislocation. The properties which we have just described apply only when the conditions of continuity are observed. In the contrary case, for example if $V_R^{(0)} = 0$, the expression of $u_2^{(0)(m)}$ (28) is not constant with $m = 1$ and 2 on the interface; as a result, $u_2^{(0)(m)}$ and $u_2^{(0)(m)\infty}$ meet far from the interface. The elastic fields $\vec{u}^{A_n(m)}$ and $(\sigma)^{A_n(m)}$ are completely determined, thus providing the elastic fields $\vec{u}^{(m)}(\vec{x})$ and $(\sigma)^{(m)}(\vec{x})$ of an interface sinusoidal climb-type edge dislocation. One may wonder if the proposed solution is the best under condition of continuity of the elastic fields at the interface. The answer is yes because it is not possible to increase the number of elastic field components constant with $m = 1$ and 2 at the interface. This is because relations interconnecting the parameters $e_i^{A_n}(m)$ exist. The relation (37a) shows that if $V_{R1}^{A_n}$ and $e_1^{A_n}$ are constant with m, then $e_{41}^{A_n}(m)$ and $e_3^{A_n}(m)$ will depend on m; similarly, $e_{61}^{A_n}(m)$ depends on m if $V_{R1}^{A_n}$ is constant with m, by the relation (37b). It will also be noted that the relation $e_9^{A_n}(m) = \mu_m e_2^{A_n}(m)$ (40) indicates that if $e_9^{A_n}$ is constant with m then $e_2^{A_n}(m)$ depends on m. In our study, the constant oscillating elastic fields with m at the interface are:

$u_1^{A_n(m)}$, $(\Delta V/V)^{A_n(m)}$, $\sigma_{22}^{A_n(m)}$, $\sigma_{12}^{A_n(m)}$, $\sigma_{13}^{A_n(m)}$ and $\sigma_{23}^{A_n(m)}$; the others $u_2^{A_n(m)}$, $u_3^{A_n(m)}$, $\sigma_{11}^{A_n(m)}$ and $\sigma_{33}^{A_n(m)}$ are not.

Conclusion

This study considers two elastic solids $S1$ and $S2$, of infinite sizes, firmly welded along a non-planar interface S having the form of a corrugated sheet and provides expressions for the elastic fields (displacement and stress) of a dislocation lying indefinitely on that interface in the sinusoidal direction with a Burgers vector $\vec{b}_{II}$ (Fig. 5.6, Sect. 5.3) perpendicular to its shape (climb-type sinusoidal edge dislocation). The proposed solutions ($\vec{u}^{(m)}$, $(\sigma)^{(m)}$) are sums of terms ($\vec{u}^{(0)(m)}$, $(\sigma)^{(0)(m)}$) corresponding to a straight interfacial edge dislocation and ($\vec{u}^{A_n(m)}$, $(\sigma)^{A_n(m)}$) proportional to the sinusoid A_n (2) or its spatial derivative $\partial A_n/\partial x_3$. It is shown that optimum continuity of the elastic fields at the crossing of the interface is achieved when the relative variation of volume is continuous at the interface and expressions of the elastic fields are written under such conditions.

Appendix A: Difference of the Values of $\vec{u}^{(m)\infty}$ and $(\sigma)^{(m)\infty}$ (Climb-Type Edge) When Crossing the Interface

Our purpose here is to write down the differences $\Delta\vec{u}^{\infty}$ and $(\Delta\sigma)^{\infty}$ (4) on crossing the interface at arbitrary point $P_S(x_1, x_2 = \xi_n \sin\kappa_n x_3, x_3)$. We use the notation $x_2 = \xi$ ($\xi = \xi_n \sin\kappa_n x_3$ small) and take the MacLaurin series expansions of the elastic fields up to terms of first order with respect to ξ; this means that

$$\Delta\vec{u}^{\infty}(x_1, x_2 = \xi, x_3) = \Delta\vec{u}^{\infty}(x_1, 0, x_3) + \frac{\partial\Delta\vec{u}^{\infty}}{\partial x_2}(x_1, 0, x_3)\xi,$$
$$(\Delta\sigma)^{\infty}(x_1, x_2 = \xi, x_3) = (\Delta\sigma)^{\infty}(x_1, 0, x_3) + \frac{\partial(\Delta\sigma)^{\infty}}{\partial x_2}(x_1, 0, x_3)\xi. \quad \text{(A.1)}$$

$\vec{u}^{(m)\infty}$ and $(\sigma)^{(m)\infty}$ are taken from our previous works (Anongba 2010). We obtain (u_i is the i-component of vector $\vec{u}$ and σ_{ij} the ij-element of the stress matrix (σ); i, j = 1 to 3)

$$\Delta u_i^{\infty}(x_1, x_2 = \xi, x_3) = \Delta u_i^{(0)\infty} + \Delta u_i^{A_n\infty}$$

$$\Delta\sigma_{ij}^{\infty}(x_1, x_2 = \xi, x_3) = \Delta\sigma_{ij}^{(0)\infty} + \Delta\sigma_{ij}^{A_n\infty}$$

$$\Delta u_1^{(0)\infty} = \frac{ibC_\nu}{8\pi}\int_{-\infty}^{\infty} \operatorname{sgn}(k_1)e^{ik_1x_1}dk_1\xi,$$

$$\Delta u_1^{A_n\infty} = -A_n \frac{ibC_\nu}{8\pi} \int_{-\infty}^{\infty} \frac{k_1\left(k_1^2 + 2\kappa_n^2\right)}{\left(k_1^2 + \kappa_n^2\right)^{3/2}} e^{ik_1x_1} \mathrm{d}k_1;$$

$$\Delta u_2^{(0)\infty} = \frac{bC_\nu}{8\pi} \int_{-\infty}^{\infty} \frac{1}{|k_1|} e^{ik_1x_1} \mathrm{d}k_1,$$

$$\Delta u_2^{A_n\infty} = -A_n \frac{bC_\nu}{8\pi} \int_{-\infty}^{\infty} \frac{3k_1^2 + 2\kappa_n^2}{\sqrt{k_1^2 + \kappa_n^2}} e^{ik_1x_1} \mathrm{d}k_1 \xi;$$

$$\Delta u_3^{(0)\infty} = 0,$$

$$\Delta u_3^{A_n\infty} = -\frac{\partial A_n}{\partial x_3} \frac{bC_\nu}{8\pi} \int_{-\infty}^{\infty} \frac{k_1^2 + 2\kappa_n^2}{\left(k_1^2 + \kappa_n^2\right)^{3/2}} e^{ik_1x_1} \mathrm{d}k_1;$$

$$\Delta \sigma_{11}^{(0)\infty} = -6iQ_b \int_{-\infty}^{\infty} |k_1| e^{ik_1x_1} \mathrm{d}k_1 \xi,$$

$$\Delta \sigma_{11}^{A_n\infty} = 2iQ_b A_n \int_{-\infty}^{\infty} \frac{k_1^2(3k_1^2 + 4\kappa_n^2)}{\left(k_1^2 + \kappa_n^2\right)^{3/2}} e^{ik_1x_1} \mathrm{d}k_1;$$

$$\Delta \sigma_{22}^{(0)\infty} = 2iQ_b \int_{-\infty}^{\infty} |k_1| e^{ik_1x_1} \mathrm{d}k_1 \xi,$$

$$\Delta \sigma_{22}^{A_n\infty} = -2iA_n \int_{-\infty}^{\infty} \frac{Q_b k_1^2 + 2Q_c \kappa_n^2}{\sqrt{k_1^2 + \kappa_n^2}} e^{ik_1x_1} \mathrm{d}k_1;$$

$$\Delta \sigma_{33}^{(0)\infty} = -4iQ_c \int_{-\infty}^{\infty} |k_1| e^{ik_1x_1} \mathrm{d}k_1 \xi,$$

$$\Delta \sigma_{33}^{A_n\infty} = 4iA_n \int_{-\infty}^{\infty} \frac{k_1^2\left[2Q_c k_1^2 + \kappa_n^2(-Q_b + 4Q_c)\right] + 2Q_c \kappa_n^4}{2\left(k_1^2 + \kappa_n^2\right)^{3/2}} e^{ik_1x_1} \mathrm{d}k_1;$$

$$\Delta \sigma_{12}^{(0)\infty} = -2Q_b \int_{-\infty}^{\infty} \operatorname{sgn}(k_1) e^{ik_1x_1} \mathrm{d}k_1,$$

$$\Delta\sigma_{12}^{A_n\infty} = A_n 2Q_b \int_{-\infty}^{\infty} k_1 \frac{3k_1^2 + 2\kappa_n^2}{\sqrt{k_1^2 + \kappa_n^2}} e^{ik_1x_1} \mathrm{d}k_1 \xi;$$

$$\Delta\sigma_{13}^{(0)\infty} = 0,$$

$$\Delta\sigma_{13}^{A_n\infty} = \frac{\partial A_n}{\partial x_3} 2Q_b \int_{-\infty}^{\infty} k_1 \frac{k_1^2 + 2\kappa_n^2}{\left(k_1^2 + \kappa_n^2\right)^{3/2}} e^{ik_1x_1} \mathrm{d}k_1;$$

$$\Delta\sigma_{23}^{(0)\infty} = 0,$$

$$\Delta\sigma_{23}^{A_n\infty} = -\frac{\partial A_n}{\partial x_3} 2i \int_{-\infty}^{\infty} \frac{(2Q_c + Q_b)k_1^2 + 2Q_c\kappa_n^2}{\left(k_1^2 + \kappa_n^2\right)^{1/2}} e^{ik_1x_1} \mathrm{d}k_1 \xi; \tag{A.2}$$

$$C_\nu = [1/(1 - \nu_1) - 1/(1 - \nu_2)],\ Q_b = i(C_2 - C_1)/4,$$

$$Q_c = i(\nu_2 C_2 - \nu_1 C_1)/4,\ C_m = b\mu_m/2\pi(1 - \nu_m);$$

$\mathrm{sgn}(k_1) = k_1/|k_1|$; μ_m and ν_m are shear modulus and Poisson's ratio.

Appendix B: Partial Oscillating Elastic Fields (Climb-Type Edge)

The couple $(\overline{\alpha}_{3a}^{A_n(m)}, \overline{\beta}_{3a}^{A_n(m)})$ is obtained from (13a–e) associated with the displacement. We have at position $\vec{x} = (x_1, x_2, x_3)$ $(\vec{u}^{A_n(m)V} \equiv \vec{u}_a^{A_n(m)V}, (\sigma)^{A_n(m)V} \equiv (\sigma)_a^{A_n(m)V})$:

$$u_{ia}^{A_n(m)V} = \frac{l_{1a}^{(m)}}{\mu_m} \left\| \delta_{i1}\frac{\partial}{\partial x_1} + \delta_{i2} + \delta_{i3}\frac{\partial}{\partial x_3} \right\| A_n\Big((\delta_{i1} + \delta_{i3})\left[-\frac{\partial \Pi_1}{\partial |x_2|} + \kappa_n^2 I_{1a} \right] - \delta_{i2}(-1)^{m-1}\left[I_0 + \kappa_n^2 \Pi_1\right] \Big),$$

$$\sigma_{iia}^{A_n(m)V} = 2A_n l_{1a}^{(m)} \left((\delta_{i1} - \delta_{i2})\frac{\partial I_0}{\partial |x_2|} + (\delta_{i1} - \delta_{i3})\kappa_n^4 I_{1a} + (\delta_{i3} - \delta_{i2})\kappa_n^2 \frac{\partial \Pi_1}{\partial |x_2|} \right),$$

$$\sigma_{12a}^{A_n(m)V} = 2A_n l_{1a}^{(m)} (-1)^m \frac{\partial}{\partial x_1}\left(I_0 + \kappa_n^2 \Pi_1\right),$$

$$\sigma_{j3a}^{A_n(m)V} = \frac{\partial A_n}{\partial x_3} 2 l_{1a}^{(m)} \left(\delta_{j1}\frac{\partial}{\partial x_1}\left[-\frac{\partial \Pi_1}{\partial |x_2|} + \kappa_n^2 I_{1a} \right] + \delta_{j2}(-1)^m\left[I_0 + \kappa_n^2 \Pi_1\right] \right); \tag{B.1}$$

$$\Pi_z = \int_{-\infty}^{\infty} \frac{e^{(-1)^m\sqrt{k_1^2+\kappa_n^2}x_2}}{k_1^2+z\kappa_n^2} e^{ik_1x_1}dk_1 \equiv \int_{-\infty}^{\infty} \frac{e^{-\sqrt{k_1^2+\kappa_n^2}|x_2|}}{k_1^2+z\kappa_n^2} e^{ik_1x_1}dk_1,$$

$$I_0 = \int_{-\infty}^{\infty} e^{(-1)^m\sqrt{k_1^2+\kappa_n^2}x_2} e^{ik_1x_1}dk_1 \equiv \int_{-\infty}^{\infty} e^{-\sqrt{k_1^2+\kappa_n^2}|x_2|} e^{ik_1x_1}dk_1 = \frac{2\kappa_n|x_2|}{r}K_1,$$

$$I_{1a} = \int_{-\infty}^{\infty} \frac{e^{(-1)^m\sqrt{k_1^2+\kappa_n^2}x_2}}{\left(k_1^2+\kappa_n^2\right)^{3/2}} e^{ik_1x_1}dk_1 \equiv \int_{-\infty}^{\infty} \frac{e^{-\sqrt{k_1^2+\kappa_n^2}|x_2|}}{\left(k_1^2+\kappa_n^2\right)^{3/2}} e^{ik_1x_1}dk_1.$$

Terms in brackets $\| \ \|$ are operators acting on A_n and others, separately; Π_1 is the value of Π_z for $z = 1$, $r^2 = x_1^2 + x_2^2$ and subscripts $i = 1$ to 3 and $j = 1$ and 2; $K_n[x]$ is the nth-order modified Bessel function usually so denoted and δ_{ij} is the Kronecker delta. We stress that the various integrations (such as in I_{1a} and Π_z) performed in the present study are given for spatial positions satisfying the condition $(-1)^m x_2 < 0$ (*i.e.* $(-1)^{m-1} = \mathrm{sgn}(x_2)$) with $m = 1$ when $x_2 > \xi_n \sin \kappa_n x_3$(half-space 1) and $m = 2$ when $x_2 < \xi_n \sin \kappa_n x_3$ (half-space 2). However, this makes no difference in the elastic fields to first order in ξ_n.

The pair $(\overline{\alpha}_{3b}^{A_n(m)}, \overline{\beta}_{3b}^{A_n(m)})$ is obtained using (13a–d) associated with the displacement. We have (using the similar notations)

$$u_{1b}^{A_n(m)A} = A_n \frac{l_{1b}^{(m)}}{\mu_m} \frac{\partial}{\partial x_1} \left(\kappa_n^2 l_{3b}^{(m)} I_{1a} - \frac{\partial}{\partial |x_2|} \left\{ \frac{\left[(1-\tilde{\Omega}_{1b}) l_{2b}^{(m)} + l_{4b}^{(m)} \right] \Pi_1 - l_{4b}^{(m)} \Pi_{\tilde{\Omega}_{1b}}}{1-\tilde{\Omega}_{1b}} \right\} \right),$$

$$u_{1b}^{A_n(m)B} = A_n \frac{t_{1b}^{(m)}}{\mu_m} \frac{\partial}{\partial x_1} \Big((-1)^m 4(1-\nu_m) \frac{\partial}{\partial |x_2|} \left[-\Pi_1 + \tilde{\Omega}_{1b} \Pi_{\tilde{\Omega}_{1b}} \right]$$
$$-x_2 \left[(\tilde{\Omega}_{1b} - 1) I_0 + \kappa_n^2 \Pi_1 - \kappa_n^2 \tilde{\Omega}_{1b}^2 \Pi_{\tilde{\Omega}_{1b}} \right] \Big);$$

$$u_{2b}^{A_n(m)A} = \frac{A_n l_{1b}^{(m)}}{\mu_m} (-1)^m \left(l_{2b}^{(m)} I_0 + \kappa_n^2 l_{3b}^{(m)} \Pi_1 - \kappa_n^2 l_{4b}^{(m)} \Pi_{\tilde{\Omega}_{1b}} \right),$$

$$u_{2b}^{A_n(m)B} = \frac{A_n t_{1b}^{(m)}}{\mu_m} \left\| 1 + (-1)^{m-1} x_2 \frac{\partial}{\partial |x_2|} \right\| \left((1-\tilde{\Omega}_{1b}) I_0 - \kappa_n^2 \Pi_1 + \kappa_n^2 \tilde{\Omega}_{1b}^2 \Pi_{\tilde{\Omega}_{1b}} \right);$$

$$u_{3b}^{A_n(m)A} = \frac{\partial A_n}{\partial x_3} \frac{l_{1b}^{(m)}}{\mu_m} \left(\kappa_n^2 l_{3b}^{(m)} I_{1a} - \frac{\partial}{\partial |x_2|} \left\{ \frac{\left[(1-\tilde{\Omega}_{1b}) l_{2b}^{(m)} + l_{4b}^{(m)} \right] \Pi_1 - l_{4b}^{(m)} \Pi_{\tilde{\Omega}_{1b}}}{1-\tilde{\Omega}_{1b}} \right\} \right),$$

$$u_{3b}^{A_n(m)B} = \frac{\partial A_n}{\partial x_3} \frac{t_{1b}^{(m)}}{\mu_m} x_2 \left[(1-\tilde{\Omega}_{1b}) I_0 - \kappa_n^2 \Pi_1 + \kappa_n^2 \tilde{\Omega}_{1b}^2 \Pi_{\tilde{\Omega}_{1b}} \right];$$

$$\sigma_{11b}^{A_n(m)A} = -2A_n l_{1b}^{(m)}\left(-\kappa_n^4 l_{3b}^{(m)} I_{1a} + \frac{\partial}{\partial|x_2|}\left\{-l_{2b}^{(m)} I_0 + \kappa_n^2\left[l_{2b}^{(m)} - l_{3b}^{(m)} + \frac{l_{4b}^{(m)}}{1-\tilde{\Omega}_{1b}}\right]\Pi_1 - \frac{\kappa_n^2 l_{4b}^{(m)}\tilde{\Omega}_{1b}}{1-\tilde{\Omega}_{1b}}\Pi_{\tilde{\Omega}_{1b}}\right\}\right),$$

$$\sigma_{11b}^{A_n(m)B} = 2A_n t_{1b}^{(m)}\left((-1)^m 2(2-\nu_m)\frac{\partial}{\partial|x_2|}\left[(1-\tilde{\Omega}_{1b})I_0 - \kappa_n^2\Pi_1 + \kappa_n^2\tilde{\Omega}_{1b}^2\Pi_{\tilde{\Omega}_{1b}}\right] - x_2\left\{(\tilde{\Omega}_{1b}-1)\left\|\kappa_n^2(1+\tilde{\Omega}_{1b}) + \frac{\partial^2}{\partial x_1^2}\right\| I_0 + \kappa_n^4\Pi_1 - \kappa_n^4\tilde{\Omega}_{1b}^3\Pi_{\tilde{\Omega}_{1b}}\right\}\right);$$

$$\sigma_{22b}^{A_n(m)A} = -2A_n l_{1b}^{(m)}\frac{\partial}{\partial|x_2|}\left(l_{2b}^{(m)} I_0 + \kappa_n^2 l_{3b}^{(m)}\Pi_1 - \kappa_n^2 l_{4b}^{(m)}\Pi_{\tilde{\Omega}_{1b}}\right),$$

$$\sigma_{22b}^{A_n(m)B} = 2A_n t_{1b}^{(m)}\left((-1)^{m-1}2(1-\nu_m)\frac{\partial}{\partial|x_2|}\left[(1-\tilde{\Omega}_{1b})I_0 - \kappa_n^2\Pi_1 + \kappa_n^2\tilde{\Omega}_{1b}^2\Pi_{\tilde{\Omega}_{1b}}\right] + x_2(1-\tilde{\Omega}_{1b})\left\{\left\|-\kappa_n^2(1+\tilde{\Omega}_{1b}) + \frac{\partial^2}{\partial|x_2|^2}\right\| I_0 + \kappa_n^4\tilde{\Omega}_{1b}^2\Pi_{\tilde{\Omega}_{1b}}\right\}\right);$$

$$\sigma_{33b}^{A_n(m)A} = 2A_n l_{1b}^{(m)}\kappa_n^2\left(\frac{1}{1-\tilde{\Omega}_{1b}}\frac{\partial}{\partial|x_2|}\left\{\left[l_{2b}^{(m)}(1-\tilde{\Omega}_{1b}) + l_{4b}^{(m)}\right]\Pi_1 - l_{4b}^{(m)}\Pi_{\tilde{\Omega}_{1b}}\right\} - \kappa_n^2 l_{3b}^{(m)} I_{1a}\right),$$

$$\sigma_{33b}^{A_n(m)B} = 2A_n t_{1b}^{(m)}\left\|-\kappa_n^2 x_2 + (-1)^m 2\nu_m\frac{\partial}{\partial|x_2|}\right\|\left((1-\tilde{\Omega}_{1b})I_0 - \kappa_n^2\Pi_1 + \kappa_n^2\tilde{\Omega}_{1b}^2\Pi_{\tilde{\Omega}_{1b}}\right);$$

$$\sigma_{12b}^{A_n(m)A} = 2A_n l_{1b}^{(m)}(-1)^m\frac{\partial}{\partial x_1}\left(l_{2b}^{(m)} I_0 + \kappa_n^2 l_{3b}^{(m)}\Pi_1 - \kappa_n^2 l_{4b}^{(m)}\Pi_{\tilde{\Omega}_{1b}}\right),$$

$$\sigma_{12b}^{A_n(m)B} = 2A_n t_{1b}^{(m)}\frac{\partial}{\partial x_1}\left(\kappa_n^2 2(1-\nu_m)\left[\Pi_1 - \tilde{\Omega}_{1b}\Pi_{\tilde{\Omega}_{1b}}\right] + \left\|3 - 2\nu_m + (-1)^{m-1}x_2\frac{\partial}{\partial|x_2|}\right\|\left\{(1-\tilde{\Omega}_{1b})I_0 - \kappa_n^2\Pi_1 + \kappa_n^2\tilde{\Omega}_{1b}^2\Pi_{\tilde{\Omega}_{1b}}\right\}\right);$$

$$\sigma_{13b}^{A_n(m)A} = 2\frac{\partial A_n}{\partial x_3}l_{1b}^{(m)}\frac{\partial}{\partial x_1}\left(\frac{\partial}{\partial|x_2|}\left\{\frac{\left[l_{2b}^{(m)}(1-\tilde{\Omega}_{1b}) + l_{4b}^{(m)}\right]\Pi_1 - l_{4b}^{(m)}\Pi_{\tilde{\Omega}_{1b}}}{\tilde{\Omega}_{1b}-1}\right\} + \kappa_n^2 l_{3b}^{(m)} I_{1a}\right),$$

$$\sigma_{13b}^{A_n(m)B} = 2\frac{\partial A_n}{\partial x_3}t_{1b}^{(m)}\frac{\partial}{\partial x_1}\left(x_2\left[(1-\tilde{\Omega}_{1b})I_0 - \kappa_n^2\Pi_1 + \kappa_n^2\tilde{\Omega}_{1b}^2\Pi_{\tilde{\Omega}_{1b}}\right] + (-1)^{m-1}2(1-\nu_m)\frac{\partial}{\partial|x_2|}\left\{\Pi_1 - \tilde{\Omega}_{1b}\Pi_{\tilde{\Omega}_{1b}}\right\}\right);$$

$$\sigma_{23b}^{A_n(m)A} = 2\frac{\partial A_n}{\partial x_3}l_{1b}^{(m)}(-1)^m\left(l_{2b}^{(m)} I_0 + \kappa_n^2 l_{3b}^{(m)}\Pi_1 - \kappa_n^2 l_{4b}^{(m)}\Pi_{\tilde{\Omega}_{1b}}\right),$$

$$\sigma_{23b}^{A_n(m)B} = 2\frac{\partial A_n}{\partial x_3}t_{1b}^{(m)}\left\|1 + (-1)^{m-1}x_2\frac{\partial}{\partial|x_2|}\right\|\left\{(1-\tilde{\Omega}_{1b})I_0 - \kappa_n^2\Pi_1 + \kappa_n^2\tilde{\Omega}_{1b}^2\Pi_{\tilde{\Omega}_{1b}}\right\}; \tag{B.2}$$

The couple $(\overline{\alpha}_{1c}^{A_n(m)}, \overline{\beta}_{1c}^{A_n(m)})$ is obtained using (13f, g, l and m) associated with stresses. We obtain

$$u_{1c}^{A_n(m)A} = -A_n \frac{l_{1c}^{(m)}}{\mu_m} \frac{\partial}{\partial x_1}\left(\kappa_n^2 I_{1a} + (1-2\nu_m)\frac{\partial}{\partial |x_2|}\Pi_1\right),$$

$$u_{1c}^{A_n(m)B} = A_n \frac{iQ_b}{2\mu_m} \frac{\partial}{\partial x_1}\left((-1)^m 4(1-\nu_m)\frac{\partial}{\partial |x_2|}\Pi_1 + x_2\left[\kappa_n^2 \Pi_1 - I_0\right]\right);$$

$$u_{2c}^{A_n(m)A} = A_n \frac{l_{1c}^{(m)}}{\mu_m}(-1)^m\left((1-2\nu_m)I_0 - \kappa_n^2 \Pi_1\right),$$

$$u_{2c}^{A_n(m)B} = A_n \frac{iQ_b}{2\mu_m}\left\| 1 + (-1)^{m-1} x_2 \frac{\partial}{\partial |x_2|}\right\|\left(\kappa_n^2 \Pi_1 - I_0\right);$$

$$u_{3c}^{A_n(m)A} = -\frac{\partial A_n}{\partial x_3}\frac{l_{1c}^{(m)}}{\mu_m}\left(\kappa_n^2 I_{1a} + (1-2\nu_m)\frac{\partial}{\partial |x_2|}\Pi_1\right),$$

$$u_{3c}^{A_n(m)B} = \frac{\partial A_n}{\partial x_3}\frac{iQ_b}{2\mu_m} x_2\left(\kappa_n^2 \Pi_1 - I_0\right);$$

$$\sigma_{11c}^{A_n(m)A} = -2A_n l_{1c}^{(m)}\left(\kappa_n^4 I_{1a} + \frac{\partial}{\partial |x_2|}\left\{2(1-\nu_m)\kappa_n^2 \Pi_1 - (1-2\nu_m)I_0\right\}\right),$$

$$\sigma_{11c}^{A_n(m)B} = A_n iQ_b\left((-1)^m 2(2-\nu_m)\frac{\partial}{\partial |x_2|}\left[\kappa_n^2 \Pi_1 - I_0\right]\right.$$
$$\left. - x_2\left\{\frac{\partial^2}{\partial x_1^2} I_0 + \kappa_n^2\left[I_0 - \kappa_n^2 \Pi_1\right]\right\}\right);$$

$$\sigma_{22c}^{A_n(m)A} = 2A_n l_{1c}^{(m)}\frac{\partial}{\partial |x_2|}\left(\kappa_n^2 \Pi_1 - (1-2\nu_m)I_0\right),$$

$$\sigma_{22c}^{A_n(m)B} = A_n iQ_b\left((-1)^m 2(1-\nu_m)\frac{\partial}{\partial |x_2|}\left(I_0 - \kappa_n^2 \Pi_1\right) + x_2\frac{\partial^2}{\partial x_1^2} I_0\right);$$

$$\sigma_{33c}^{A_n(m)A} = 2A_n l_{1c}^{(m)}\kappa_n^2\left(\kappa_n^2 I_{1a} + (1-2\nu_m)\frac{\partial}{\partial |x_2|}\Pi_1\right),$$

$$\sigma_{33c}^{A_n(m)B} = A_n iQ_b\left\|\kappa_n^2 x_2 + (-1)^{m-1} 2\nu_m \frac{\partial}{\partial |x_2|}\right\|\left(I_0 - \kappa_n^2 \Pi_1\right);$$

$$\sigma_{12c}^{A_n(m)A} = 2A_n l_{1c}^{(m)}(-1)^m\frac{\partial}{\partial x_1}\left((1-2\nu_m)I_0 - \kappa_n^2 \Pi_1\right),$$

$$\sigma_{12c}^{A_n(m)B} = -A_n iQ_b\frac{\partial}{\partial x_1}\left(\kappa_n^2 2(1-\nu_m)\Pi_1 + \left\| 3 - 2\nu_m + (-1)^{m-1} x_2\frac{\partial}{\partial |x_2|}\right\|\left\{I_0 - \kappa_n^2 \Pi_1\right\}\right);$$

$$\sigma_{13c}^{A_n(m)A} = -\frac{\partial A_n}{\partial x_3} 2l_{1c}^{(m)} \frac{\partial}{\partial x_1}\left((1-2\nu_m)\frac{\partial \Pi_1}{\partial |x_2|} + \kappa_n^2 I_{1a}\right),$$

$$\sigma_{13c}^{A_n(m)B} = \frac{\partial A_n}{\partial x_3} iQ_b \frac{\partial}{\partial x_1}\left((-1)^m 2(1-\nu_m)\frac{\partial \Pi_1}{\partial |x_2|} - x_2\{I_0 - \kappa_n^2 \Pi_1\}\right);$$

$$\sigma_{23c}^{A_n(m)A} = \frac{\partial A_n}{\partial x_3} 2l_{1c}^{(m)} (-1)^m\big((1-2\nu_m)I_0 - \kappa_n^2 \Pi_1\big),$$

$$\sigma_{23c}^{A_n(m)B} = -\frac{\partial A_n}{\partial x_3} iQ_b \left\| 1 + (-1)^{m-1} x_2 \frac{\partial}{\partial |x_2|} \right\| \big(I_0 - \kappa_n^2 \Pi_1\big); \tag{B.3}$$

The pair $(\overline{\alpha}_{1d}^{A_n(m)}, \overline{\beta}_{1d}^{A_n(m)})$ is calculated from (13h, i and o, p) associated with stresses. We have

$$u_{1d}^{A_n(m)A} = -A_n \frac{l_{1d}^{(m)}}{\mu_m} \frac{\partial}{\partial x_1}\left(l_{2d}^{(m)} \frac{\partial \Pi_1}{\partial |x_2|} + \kappa_n^2 l_{3d}^{(m)} I_{1a}\right),$$

$$u_{1d}^{A_n(m)B} = A_n \frac{t_{1d}^{(m)}}{\mu_m} \frac{\partial}{\partial x_1}\left((-1)^{m-1} 4(1-\nu_m)\frac{\partial \Pi_1}{\partial |x_2|} + x_2\{I_0 - \kappa_n^2 \Pi_1\}\right);$$

$$u_{2d}^{A_n(m)A} = A_n \frac{l_{1d}^{(m)}}{\mu_m} (-1)^m \left(l_{2d}^{(m)} I_0 - \kappa_n^2 l_{3d}^{(m)} \Pi_1\right),$$

$$u_{2d}^{A_n(m)B} = A_n \frac{t_{1d}^{(m)}}{\mu_m} \left\| 1 + (-1)^{m-1} x_2 \frac{\partial}{\partial |x_2|} \right\| \big(I_0 - \kappa_n^2 \Pi_1\big);$$

$$u_{3d}^{A_n(m)A} = -\frac{\partial A_n}{\partial x_3} \frac{l_{1d}^{(m)}}{\mu_m} \left(l_{2d}^{(m)} \frac{\partial \Pi_1}{\partial |x_2|} + \kappa_n^2 l_{3d}^{(m)} I_{1a}\right),$$

$$u_{3d}^{A_n(m)B} = \frac{\partial A_n}{\partial x_3} \frac{t_{1d}^{(m)}}{\mu_m} x_2 \big(I_0 - \kappa_n^2 \Pi_1\big);$$

$$\sigma_{11d}^{A_n(m)A} = A_n 2l_{1d}^{(m)} \left(\frac{\partial}{\partial |x_2|}\left[l_{2d}^{(m)} I_0 - \kappa_n^2 (l_{2d}^{(m)} + l_{3d}^{(m)})\Pi_1\right] - \kappa_n^4 l_{3d}^{(m)} I_{1a}\right),$$

$$\sigma_{11d}^{A_n(m)B} = A_n 2t_{1d}^{(m)} \left((-1)^m 2(2-\nu_m)\frac{\partial}{\partial |x_2|}\big(I_0 - \kappa_n^2 \Pi_1\big) + x_2\left\{\left\|\kappa_n^2 + \frac{\partial^2}{\partial x_1^2}\right\| I_0 - \kappa_n^4 \Pi_1\right\}\right);$$

$$\sigma_{22d}^{A_n(m)A} = -A_n 2l_{1d}^{(m)} \frac{\partial}{\partial |x_2|}\left(l_{2d}^{(m)} I_0 - \kappa_n^2 l_{3d}^{(m)} \Pi_1\right),$$

$$\sigma_{22d}^{A_n(m)B} = A_n 2t_{1d}^{(m)} \left((-1)^{m-1} 2(1-\nu_m)\frac{\partial}{\partial |x_2|}\big(I_0 - \kappa_n^2 \Pi_1\big) - x_2 \frac{\partial^2 I_0}{\partial x_1^2}\right);$$

$$\sigma_{33d}^{A_n(m)A} = A_n 2 l_{1d}^{(m)} \kappa_n^2 \left(l_{2d}^{(m)} \frac{\partial \Pi_1}{\partial |x_2|} + \kappa_n^2 l_{3d}^{(m)} I_{1a} \right),$$

$$\sigma_{33d}^{A_n(m)B} = -A_n 2 t_{1d}^{(m)} \left\| \kappa_n^2 x_2 + (-1)^{m-1} 2\nu_m \frac{\partial}{\partial |x_2|} \right\| \left(I_0 - \kappa_n^2 \Pi_1 \right);$$

$$\sigma_{12d}^{A_n(m)A} = A_n 2 l_{1d}^{(m)} (-1)^m \frac{\partial}{\partial x_1} \left(l_{2d}^{(m)} I_0 - \kappa_n^2 l_{3d}^{(m)} \Pi_1 \right),$$

$$\sigma_{12d}^{A_n(m)B} = A_n 2 t_{1d}^{(m)} \frac{\partial}{\partial x_1} \left(\kappa_n^2 2(1-\nu_m)\Pi_1 + \left\| 3 - 2\nu_m + (-1)^{m-1} x_2 \frac{\partial}{\partial |x_2|} \right\| \left(I_0 - \kappa_n^2 \Pi_1 \right) \right);$$

$$\sigma_{13d}^{A_n(m)A} = -\frac{\partial A_n}{\partial x_3} 2 l_{1d}^{(m)} \frac{\partial}{\partial x_1} \left(l_{2d}^{(m)} \frac{\partial \Pi_1}{\partial |x_2|} + \kappa_n^2 l_{3d}^{(m)} I_{1a} \right),$$

$$\sigma_{13d}^{A_n(m)B} = \frac{\partial A_n}{\partial x_3} 2 t_{1d}^{(m)} \frac{\partial}{\partial x_1} \left((-1)^{m-1} 2(1-\nu_m) \frac{\partial \Pi_1}{\partial |x_2|} + x_2 \left\{ I_0 - \kappa_n^2 \Pi_1 \right\} \right);$$

$$\sigma_{23d}^{A_n(m)A} = \frac{\partial A_n}{\partial x_3} 2 l_{1d}^{(m)} (-1)^m \left(l_{2d}^{(m)} I_0 - \kappa_n^2 l_{3d}^{(m)} \Pi_1 \right),$$

$$\sigma_{23d}^{A_n(m)B} = \frac{\partial A_n}{\partial x_3} 2 t_{1d}^{(m)} \left\| 1 + (-1)^{m-1} x_2 \frac{\partial}{\partial |x_2|} \right\| \left(I_0 - \kappa_n^2 \Pi_1 \right); \qquad \text{(B.4)}$$

The couple $(\overline{\alpha}_{3e}^{A_n(m)}, \overline{\beta}_{3e}^{A_n(m)})$ is obtained from (13j, k, and n, l) associated with stresses. We have (here, we write r_m for r_m^*, $m = 1$ and 2)

$$u_{1e}^{A_n(m)A} = A_n \frac{l_{1e}^{(m)}}{\mu_m} \frac{\partial}{\partial x_1} \left(-\frac{l_{2e}^{(m)} r_2}{1+r_2} \frac{\partial \Pi_{(-r_2)}}{\partial |x_2|} - \frac{l_{3e}^{(m)} r_1}{1+r_1} \frac{\partial \Pi_{(-r_1)}}{\partial |x_2|} + \frac{l_{4e}^{(m)} r_m}{1+r_m} \frac{\partial \Pi_{(-r_m)}}{\partial |x_2|} \right.$$
$$\left. - \left[\frac{l_{2e}^{(m)}}{1+r_2} + \frac{l_{3e}^{(m)}}{1+r_1} - \frac{l_{4e}^{(m)}}{1+r_m} - 2Q_b + 4Q_c \right] \frac{\partial \Pi_1}{\partial |x_2|} + \kappa_n^2 2 Q_b I_{1a} \right),$$

$$u_{1e}^{A_n(m)B} = -\frac{A_n i}{2\mu_m} \frac{\partial}{\partial x_1} \left(t_{1e}^{(m)} \left\| \kappa_n^2 r_m x_2 + 4(1-\nu_m)(-1)^{m-1} \frac{\partial}{\partial |x_2|} \right\| \Pi_{(-r_m)} \right.$$
$$\left. + x_2 (t_{1e}^{(m)} + Q_b) I_0 - t_{2e}^{(m)} \left\| \kappa_n^2 x_2 + 4(1-\nu_m)(-1)^m \frac{\partial}{\partial |x_2|} \right\| \Pi_1 \right);$$

$$u_{2e}^{A_n(m)A} = \frac{A_n l_{1e}^{(m)}}{\mu_m} (-1)^m \left(\kappa_n^2 \left[2Q_b \Pi_1 + l_{2e}^{(m)} r_2 \Pi_{(-r_2)} + l_{3e}^{(m)} r_1 \Pi_{(-r_1)} - l_{4e}^{(m)} r_m \Pi_{(-r_m)} \right] \right.$$
$$\left. + \left[l_{2e}^{(m)} + l_{3e}^{(m)} - l_{4e}^{(m)} - 2Q_b + 4Q_c \right] I_0 \right),$$

$$u_{2e}^{A_n(m)B} = -\frac{A_n i}{2\mu_m} \left\| 1 + (-1)^{m-1} x_2 \frac{\partial}{\partial |x_2|} \right\| \left((t_{1e}^{(m)} + Q_b) I_0 + \kappa_n^2 \left[t_{1e}^{(m)} r_m \Pi_{(-r_m)} - Q_b \Pi_1 \right] \right);$$

$$u_{3e}^{A_n(m)A} = \frac{\partial A_n}{\partial x_3}\frac{l_{1e}^{(m)}}{\mu_m}\left(-\frac{l_{2e}^{(m)}r_2}{1+r_2}\frac{\partial\Pi_{(-r_2)}}{\partial|x_2|} - \frac{l_{3e}^{(m)}r_1}{1+r_1}\frac{\partial\Pi_{(-r_1)}}{\partial|x_2|} + \frac{l_{4e}^{(m)}r_m}{1+r_m}\frac{\partial\Pi_{(-r_m)}}{\partial|x_2|}\right.$$
$$\left.-\left[\frac{l_{2e}^{(m)}}{1+r_2} + \frac{l_{3e}^{(m)}}{1+r_1} - \frac{l_{4e}^{(m)}}{1+r_m} - 2Q_b + 4Q_c\right]\frac{\partial\Pi_1}{\partial|x_2|} + \kappa_n^2 2Q_b I_{1a}\right),$$

$$u_{3e}^{A_n(m)B} = -\frac{\partial A_n}{\partial x_3}\frac{i}{2\mu_m}x_2\left((t_{1e}^{(m)} + Q_b)I_0 + \kappa_n^2\left[t_{1e}^{(m)}r_m\Pi_{(-r_m)} - Q_b\Pi_1\right]\right);$$

$$\sigma_{11e}^{A_n(m)A} = 2A_n l_{1e}^{(m)}\left(\kappa_n^2\left\{\frac{l_{2e}^{(m)}r_2^2}{1+r_2}\frac{\partial\Pi_{(-r_2)}}{\partial|x_2|} + \frac{l_{3e}^{(m)}r_1^2}{1+r_1}\frac{\partial\Pi_{(-r_1)}}{\partial|x_2|} - \frac{l_{4e}^{(m)}r_m^2}{1+r_m}\frac{\partial\Pi_{(-r_m)}}{\partial|x_2|}\right\}\right.$$
$$-\kappa_n^2\left[-\frac{l_{2e}^{(m)}r_2}{1+r_2} - \frac{l_{3e}^{(m)}r_1}{1+r_1} + \frac{l_{4e}^{(m)}r_m}{1+r_m} - 2Q_b\right]\frac{\partial\Pi_1}{\partial|x_2|} + \kappa_n^4 2Q_b I_{1a}$$
$$\left.+\left(l_{2e}^{(m)} + l_{3e}^{(m)} - l_{4e}^{(m)} - 2Q_b + 4Q_c\right)\frac{\partial}{\partial|x_2|}\left\{I_0 - \kappa_n^2\Pi_1\right\}\right),$$

$$\sigma_{11e}^{A_n(m)B} = A_n i\left\|(-1)^{m-1}2(2-\nu_m)\frac{\partial}{\partial|x_2|} - x_2\frac{\partial^2}{\partial x_1^2}\right\|$$
$$\times\left((t_{1e}^{(m)} + Q_b)I_0 + \kappa_n^2\left[t_{1e}^{(m)}r_m\Pi_{(-r_m)} - Q_b\Pi_1\right]\right);$$

$$\sigma_{22e}^{A_n(m)A} = -2A_n l_{1e}^{(m)}\frac{\partial}{\partial|x_2|}\left(\kappa_n^2\left[2Q_b\Pi_1 + l_{2e}^{(m)}r_2\Pi_{(-r_2)} + l_{3e}^{(m)}r_1\Pi_{(-r_1)} - l_{4e}^{(m)}r_m\Pi_{(-r_m)}\right]\right.$$
$$\left.+\left[l_{2e}^{(m)} + l_{3e}^{(m)} - l_{4e}^{(m)} - 2Q_b + 4Q_c\right]I_0\right),$$

$$\sigma_{22e}^{A_n(m)B} = A_n i\left\|(-1)^m 2(1-\nu_m)\frac{\partial}{\partial|x_2|} - x_2\frac{\partial^2}{\partial|x_2|^2}\right\|$$
$$\times\left((t_{1e}^{(m)} + Q_b)I_0 + \kappa_n^2\left[t_{1e}^{(m)}r_m\Pi_{(-r_m)} - Q_b\Pi_1\right]\right);$$

$$\sigma_{33e}^{A_n(m)A} = 2A_n l_{1e}^{(m)}\kappa_n^2\left(\frac{l_{2e}^{(m)}r_2}{1+r_2}\frac{\partial\Pi_{(-r_2)}}{\partial|x_2|} + \frac{l_{3e}^{(m)}r_1}{1+r_1}\frac{\partial\Pi_{(-r_1)}}{\partial|x_2|} - \frac{l_{4e}^{(m)}r_m}{1+r_m}\frac{\partial\Pi_{(-r_m)}}{\partial|x_2|}\right.$$
$$\left.+\left[\frac{l_{2e}^{(m)}}{1+r_2} + \frac{l_{3e}^{(m)}}{1+r_1} - \frac{l_{4e}^{(m)}}{1+r_m} - 2Q_b + 4Q_c\right]\frac{\partial\Pi_1}{\partial|x_2|} - \kappa_n^2 2Q_b I_{1a}\right),$$

$$\sigma_{33e}^{A_n(m)B} = A_n i\left\|(-1)^{m-1}2\nu_m\frac{\partial}{\partial|x_2|} + x_2\kappa_n^2\right\|$$
$$\times\left((t_{1e}^{(m)} + Q_b)I_0 + \kappa_n^2\left[t_{1e}^{(m)}r_m\Pi_{(-r_m)} - Q_b\Pi_1\right]\right);$$

$$\sigma_{12e}^{A_n(m)A} = 2A_n l_{1e}^{(m)}(-1)^m\frac{\partial}{\partial x_1}\left(\kappa_n^2\left[2Q_b\Pi_1 + l_{2e}^{(m)}r_2\Pi_{(-r_2)} + l_{3e}^{(m)}r_1\Pi_{(-r_1)} - l_{4e}^{(m)}r_m\Pi_{(-r_m)}\right]\right.$$

$$+\left[l_{2e}^{(m)}+l_{3e}^{(m)}-l_{4e}^{(m)}-2Q_b+4Q_c\right]I_0\Big),$$

$$\sigma_{12e}^{A_n(m)B}=-A_n i\frac{\partial}{\partial x_1}\left(\kappa_n^2 2(1-\nu_m)\left(t_{1e}^{(m)}\Pi_{(-r_m)}+Q_b\Pi_1\right)+\left\|3-2\nu_m+x_2(-1)^{m-1}\frac{\partial}{\partial|x_2|}\right\|\right.$$
$$\left.\times\left\{(t_{1e}^{(m)}+Q_b)I_0+\kappa_n^2\left[t_{1e}^{(m)}r_m\Pi_{(-r_m)}-Q_b\Pi_1\right]\right\}\right);$$

$$\sigma_{13e}^{A_n(m)A}=-\frac{\partial A_n}{\partial x_3}2l_{1e}^{(m)}\frac{\partial}{\partial x_1}\left(\frac{l_{2e}^{(m)}r_2}{1+r_2}\frac{\partial\Pi_{(-r_2)}}{\partial|x_2|}+\frac{l_{3e}^{(m)}r_1}{1+r_1}\frac{\partial\Pi_{(-r_1)}}{\partial|x_2|}-\frac{l_{4e}^{(m)}r_m}{1+r_m}\frac{\partial\Pi_{(-r_m)}}{\partial|x_2|}\right.$$
$$\left.+\left[\frac{l_{2e}^{(m)}}{1+r_2}+\frac{l_{3e}^{(m)}}{1+r_1}-\frac{l_{4e}^{(m)}}{1+r_m}-2Q_b+4Q_c\right]\frac{\partial\Pi_1}{\partial|x_2|}-\kappa_n^2 2Q_b I_{1a}\right),$$

$$\sigma_{13e}^{A_n(m)B}=\frac{\partial A_n}{\partial x_3}i\frac{\partial}{\partial x_1}\left((-1)^m 2(1-\nu_m)\frac{\partial}{\partial|x_2|}\left(t_{1e}^{(m)}\Pi_{(-r_m)}+Q_b\Pi_1\right)\right.$$
$$\left.-x_2\left\{(t_{1e}^{(m)}+Q_b)I_0+\kappa_n^2\left[t_{1e}^{(m)}r_m\Pi_{(-r_m)}-Q_b\Pi_1\right]\right\}\right);$$

$$\sigma_{23e}^{A_n(m)A}=\frac{\partial A_n}{\partial x_1}2l_{1e}^{(m)}(-1)^m\left(\kappa_n^2\left[2Q_b\Pi_1+l_{2e}^{(m)}r_2\Pi_{(-r_2)}+l_{3e}^{(m)}r_1\Pi_{(-r_1)}-l_{4e}^{(m)}r_m\Pi_{(-r_m)}\right]\right.$$
$$\left.+\left[l_{2e}^{(m)}+l_{3e}^{(m)}-l_{4e}^{(m)}-2Q_b+4Q_c\right]I_0\right),$$

$$\sigma_{23e}^{A_n(m)B}=\frac{\partial A_n}{\partial x_3}i\left\|(-1)^m x_2\frac{\partial}{\partial|x_2|}-1\right\|\left((t_{1e}^{(m)}+Q_b)I_0+\kappa_n^2\left[t_{1e}^{(m)}r_m\Pi_{(-r_m)}-Q_b\Pi_1\right]\right);$$
(B5)

$$\Pi_{(-z)}\equiv\int_{-\infty}^{\infty}\frac{e^{-\sqrt{k_1^2+\kappa_n^2}|x_2|}}{k_1^2-z\kappa_n^2}e^{ik_1x_1}dk_1.$$

References

H. Abdesselam et al., On the crystallographic, stage I like, character of fine granular area formation internal fish-eye fatigue cracks. Int. J. Fatigue **106**, 132–142 (2018)

P.N.B. Anongba, J.-L. Martin, Lomer-Cottrell dislocations after high temperature deformation of [112] copper single crystals, in *Proceedings of the International Symposium on Electron Microscopy in Plasticity and Fracture Research of Materials*, Holzhau near Dresden (1989), pp. 285–290

P.N.B. Anongba, Doctorate thesis No. 826, EPFL, Lausanne (1989)

P.N.B. Anongba, J.-L. Martin, Dislocation clustering and hardening of copper, in *Proceedings of the 9th International Conference of the Strength of Materials (ICSMA—9)*, vol. 1, Haifa, Israel (1991), pp. 203–210

P.N.B. Anongba, J. Bonneville, J. Douin, J.-L. Martin, Lomer dislocations in aluminium and copper after high temperature deformation, in *Proceedings of the 9th International Conference of the Strength of Materials (ICSMA—9)*, vol. 1, Haifa, Israel (1991), pp. 303–310

P.N.B. Anongba, The rate of hardening in Stage IV at high temperatures in [112] copper single crystals. Phys. Stat. Sol. (a) **130**, 357–363 (1992)

P.N.B. Anongba, Subboundary structures in stage IV in [112] copper single crystals deformed at high temperatures, in *Proceedings of the 7th JIM International Symposium (JIMIS—7) on Aspects of High Temperature Deformation and Fracture in Crystalline Materials*, Nagoya, Japan (1993), pp. 11–18

P.N.B. Anongba, J. Bonneville, J.-L. Martin, Hardening stages of [112] copper single crystals at intermediate and high temperature-I. Mechanical behaviour. Acta Metal. Mater. **41**, 2897–2906 (1993a)

P.N.B. Anongba, J. Bonneville, J.-L. Martin, Hardening stages of [112] copper single crystals at intermediate and high temperature-II. Slip systems and microstructures. Acta Metal. Mater. **41**, 2907–2922 (1993b)

P.N.B. Anongba, S.G. Steinemann, Dislocations and plasticity in [113] $CoSi_2$ single crystals between room temperature and 1173 K. Phys. Stat. Sol. (a) **140**, 391–409 (1993)

P.N.B. Anongba, S.G. Steinemann, Slip systems and microstructures in plastically deformed [113] CoSi2 single crystals, in *Proceedings of the 10th International Conference of the Strength of Materials (ICSMA—10)*, Sendai, Japan (1994), pp. 125–128

P.N.B. Anongba, S. Oberli, S.G. Steinemann, Brittle fracture mechanisms of $CoSi_2$ single crystals at room temperature, in *Proceedings of the 10th International Conference of the Strength of Materials (ICSMA—10)*, Sendai, Japan (1994), pp. 431–434

P.N.B. Anongba, S. Oberli, S.G. Steinemann, A study of the brittle fracture characteristics of $CoSi_2$ using laser beam reflections. Acta Metal. Mater. **43**, 2275–2285 (1995)

P. N.B. Anongba, *Mechanics of Steadily Propagating Cracks*,
https://doi.org/10.1007/978-981-96-5703-2

P.N.B. Anongba, The displacement and stress fields due to a sinusoidal edge dislocation, Phys. Stat. Sol. (b), 190, 135–149 (1995)

P.N.B. Anongba, A non-planar crack analysis using continuously distributed sinusoidal edge dislocations and linear elasticity. Phys. Stat. Sol. (b), **194**, 443–452 (1996)

P.N.B. Anongba, V. Vitek, Significance of the deviations of the crack front into the plane perpendicular to the crack propagation direction: I. Crack- front dislocation generation. Int. J. Fract. **124**, 1–15 (2003)

P.N.B. Anongba, Significance of the deviations of the crack front into the plane perpendicular to the crack propagation direction: II. Crack-front vibration. Int. J. Fract. **124**, 17–31 (2003)

P.N.B. Anongba, An analysis of a non-planar crack under mixed mode I+III loading using infinitesimal dislocations with edge and screw average characters, ResearchGate (2008). https://doi.org/10.13140/RG.2.2.14308.60806

P.N.B. Anongba, A study of the mixed mode I+III loading of a non-planar crack using infinitesimal dislocations. Rev. Ivoir. Sci. Technol. **14**, 55–86 (2009)

P.N.B. Anongba, Non-planar crack under general loading: dislocation, crack- tip stress and crack extension force. Rev. Ivoir. Sci. Technol. **16**, 11–50 (2010)

P.N.B. Anongba, J. Bonneville, A. Joulain, Brittle cracks under compression: introducing Poisson's effect. Rev. Ivoir. Sci. Technol. **17**, 37–53 (2011)

P.N.B. Anongba, Electromagnetic fields generated in metals by the quasistatic propagation of elastic-plastic cracks. Rev. Ivoir. Sci. Technol. **20**, 1–23 (2012)

P.N.B. Anongba, Interface crack loaded in tension: Introducing an internal shear stress promoted by Poisson's effect. Rev. Ivoir. Sci. Technol. **23**, 54–71 (2014a)

P.N.B. Anongba, Interface crack under mixed mode I+II loading and internal shear stress due to Poisson effect: theory and confrontation with experiments. Rev. Ivoir. Sci. Technol. **24**, 26–44 (2014b)

P.N.B. Anongba, A study of the elastic fields of interfacial edge dislocations (straight and sinusoidal) using Galerkin vectors with three-dimensional biharmonic functions in Fourier forms. Rev. Ivoir. Sci. Technol. **25**, 34–55 (2015a)

P.N.B. Anongba, A study of the elastic fields of interfacial edge dislocations (straight and sinusoidal) using Galerkin vectors with three-dimensional biharmonic functions in Fourier forms (completed). Rev. Ivoir. Sci. Technol. **26**, 36–75 (2015b)

P.N.B. Anongba, A theory of the fracture of rectangular bars bent by terminal transverse load and couple. Rev. Ivoir. Sci. Technol. **26**, 76–90 (2015c)

P.N.B. Anongba, Non-planar interface crack under general loading I. Glide-type edge and screw sinusoidal dislocations. Rev. Ivoir. Sci. Technol. **28**, 24–59 (2016)

P.N.B. Anongba, Non-planar interface crack under general loading II. Climb-type sinusoidal edge dislocation. Rev. Ivoir. Sci. Technol. **30**, 1–36 (2017a)

P.N.B. Anongba, Non-planar crack under general loading and induced normal stresses due to Poisson effect. Rev. Ivoir. Sci. Technol. **30**, 37–57 (2017b)

P.N.B. Anongba, Non-planar interface crack under general loading III. Dislocation, crack-type stress and crack extension force. Rev. Ivoir. Sci. Technol. **32**, 10–47 (2018)

P.N.B. Anongba, Fracture mechanics in a three-dimensional elastic half-space under the rectilinear contact pressure of a cylinder. Rev. Ivoir. Sci. Technol. **34**, 25–43 (2019)

P.N.B. Anongba, Planar cracks in uniform motion under mode *I* and *II* loadings. Rev. Ivoir. Sci. Technol. **35**, 1–22 (2020a)

P.N.B. Anongba, Non-planar cracks in uniform motion under general loading. Rev. Ivoir. Sci. Technol. **36**, 119–149 (2020b)

P.N.B. Anongba, Non-planar crack with crack-front plastic yielding under general loading. Rev. Ivoir. Sci. Technol. **37**, 15–31 (2021a)

P.N.B. Anongba, Elliptical crack under arbitrarily applied loadings: dislocation, crack-tip stress and crack extension force. Rev. Ivoir. Sci. Technol. **38**, 388–409 (2021b)

P.N.B. Anongba, Rough conoidal crack under general loading: dislocation, crack-tip stress and crack extension force. Rev. Ivoir. Sci. Technol. **40**, 177–206 (2022a)

P.N.B. Anongba, Rough conoidal crack under general loading: crack extension as a function of the average cone angle. ResearchGate (2022b). https://doi.org/10.13140/RG.2.2.13078.01600

P.N.B. Anongba, Circular crack expanding uniformly under general loading. ResearchGate (2023a). https://doi.org/10.13140/RG.2.2.17546.34240

P.N.B. Anongba, Rough conoidal crack growing uniformly under general loading. ResearchGate (2023b). https://doi.org/10.13140/RG.2.2.25066.59841

P.N.B. Anongba, Brittle cracks in a three-dimensional elastic half-space under the rectilinear contact pressure of a cylinder: introducing gravitational forces. Rev. Ivoir. Sci. Technol. **42**, 190–198 (2023c)

P.N.B. Anongba, Conoidal crack with elliptic bases, within cubic crystals, under arbitrarily applied loadings-I. Dislocation, crack-tip stress and crack extension force. Rev. Ivoir. Sci. Technol. **43**, 100–121 (2024a)

P.N.B. Anongba, Conoidal crack with elliptic bases, within cubic crystals, under arbitrarily applied loadings-II. Application to mechanical twinning of [112] copper single crystals. Rev. Ivoir. Sci. Technol. **43**, 122–131 (2024b)

P.N.B. Anongba, Conoidal crack with elliptic bases, within cubic crystals, under arbitrarily applied loadings-III. Application to brittle fracture systems of $CoSi_2$ single crystals. Rev. Ivoir. Sci. Technol. **44**, 75–82 (2024)

P.N.B. Anongba, Conoidal crack with elliptic bases, within cubic crystals, under arbitrarily applied loadings-IV. Application to 1/2<110> {111} cross-slip in FCC materials. Rev. Ivoir. Sci. Technol. **44**, 83–91 (2024)

P.N.B. Anongba, Planar cracks in uniform motion under mode *I* loading in cubic crystals. ResearchGate (2024). https://doi.org/10.13140/RG.2.2.25821.83683

P.N.B. Anongba, Double-cantilever beam specimen bent by pairs of opposite terminal transverse loads. Rev. Ivoir. Sci. Technol. **44**, 92–95 (2024) (ResearchGate). https://doi.org/10.13140/RG.2.2.14626.24001

P.N.B. Anongba, Non-planar cracks in uniform motion, with crack-front plastic yielding, under general loading. ResearchGate (2024). https://doi.org/10.13140/RG.2.2.10602.94400

P.N.B. Anongba, Conoidal crack with elliptic bases, within cubic crystals, Growing uniformly under arbitrarily applied loadings. ResearchGate. https://doi.org/10.13140/RG.2.2.32998.02887

M.F. Ashby, S.D. Hallam, The failure of brittle solids containing small cracks under compressive stress states. Acta Metall. **34**, 497–510 (1986)

J.R. Barber, Elasticity, in *Solid Mechanics and its Applications*, vol. 107. ed. by G.M.L. Gladwell (Kluwer Academic Publishers, Dordrecht, 2002)

G.I. Barenblatt, The mathematical theory of equilibrium cracks in brittle fracture. Adv. Appl. Math. **2**, 55–129 (1962)

B.A. Bilby, A.H. Cottrell, K.H. Swinden, The spread of plastic yield from a notch. Proc. Royal Soc. **A272**, 304–314 (1963)

B.A. Bilby, J.D. Eshelby, Dislocations and the theory of fracture, in *Fracture*, vol. 1. ed. by H. Liebowitz (Academic Press, New York, 1968), pp. 99–182

J. Bonneville, B. Escaig, J.L. Martin, A study of cross-slip activation parameters in pure copper. Acta Metall. **36**, 1989–2002 (1988)

W.F. Brace, E.G. Bombolakis, A note on the brittle crack growth in compression. J. Geophys. Res. **68**, 3709–3713 (1963)

K.B. Broberg, *Cracks and Fracture* (Academic Press, San Diego, 1999)

R.A. Brown, Electron distribution about an edge dislocation in a metal. Phys. Rev. **141**, 568–572 (1966)

J. Buttet, *Cours de physique générale I. Propriétés élastiques des solides et des fluides* (Département de physique (EPFL), Ecublens, 1981)

S.-J. Chang, S.M. Ohr, Dislocation-free zone model of fracture. J. Appl. Phys. **52**, 7174–7181 (1981)

S.-J. Chang, S.M. Ohr, Dislocation-free zone model of fracture, comparison with experiments. J. Appl. Phys. **53**, 5645–5651 (1982)

M. Comninou, A property of interface dislocation. Phil. Mag. **36**, 1281–1283 (1977a)

M. Comninou, The interface crack. J. Appl. Mech. **44**, 631–636 (1977b)
M. Comninou, J. Dundurs, Effect of friction on the interface crack loaded in shear. J. Elasticity **10**, 203–212 (1980)
M. Comninou, An overview of interface cracks. Eng. Fracture Mech. **37**, 197–208 (1990)
M.L. Cooke, D.D. Pollard, Fracture propagation paths under mixed mode loading within rectangular blocks of polymethyl methacrylate. J. Geophys. Res. **101**, 3387–3400 (1996)
B. Cotterell, J.R. Rice, Slightly curved and kinked cracks. Int. J. Fract. **16**, 155–169 (1980)
G.O. Cress, B.T. Brady, G.A. Rowell, Source of electromagnetic radiation from fracture of rock samples in the laboratory. Geophys. Res. Lett. **14**, 331–334 (1987)
M.-C. Desjonqueres, D. Spanjaard, *Concepts in Surface Physics* (Springer, Berlin, 1993)
D.L. Dexter, Conductivity of cold-worked metals. Phys. Rev. **85**, 936–937 (1952)
M.S. Duesbery, P.B. Hirsch, in *Dislocation Dynamics*. ed. by A. Rosenfield (Battelle Colloquium, Seattle-Harrison, 1967) (Mc Graw-Hill Book Co., New York, 1968) p. 57
J. Dundurs, T. Mura, Interaction between an edge dislocation and a circular inclusion. J. Mech. Phys. Solids **12**, 177–189 (1964)
A.H. England, A crack between dissimilar media. J. Appl. Mech. **32**, 400–402 (1965)
F. Erdogan, Stress distribution in a nonhomogeneous elastic plane with cracks. J. Appl. Mech. **30**, 232–236 (1963)
F. Erdogan, G.C. SIH, On the crack extension in plates under plane loading and transverse shear. Trans. Am. Soc. Mech. Engng. **85**, 519–527 (1963)
B. Escaig, L'activation thermique des déviations sous faibles contraintes dans les structures h.c. et c.c. Phys. Stat. Sol. **28**, 463–474 (1968)
J.D. Eshelby, The determination of the elastic field of an ellipsoidal inclusion, and related problems. Proc. Roy. Soc. A **241**, 376–396 (1957)
J.D. Eshelby, The interaction of kinks and elastic waves. Proc. Royal Soc. Lond. **A266**, 222–246 (1962)
R. Estrada, R.P. Kanwal, *Singular integral equations* (Birkhäuser, Boston, 2000)
A.A. Evtushenko, N. Maeno, V.F. Petrenko, I.A. Ryzhkin, J. de Physique C 1, **48**(3), 109–113 (1987)
F.C. Frank, B.R. Lawn, On the theory of Hertzian fracture. Proc. Royal Soc. Lond. A **299**, 291–306 (1967)
L.B. Freund, *Dynamic Fracture Mechanics* (Cambridge University Press, New York, 1990)
V. Frid, A. Rabinovitch, D. Bahat, Fracture induced electromagnetic radiation. J. Phys. D Appl. Phys. **36**(13), 1620–1628 (2003)
J. Friedel, *Dislocations* (Pergamon Press, Paris, 1964)
Y.C. Fung, *Foundations of Solids Mechanics* (Prentice Hall Inc., Englewood Cliffs, N.J., 1965)
A. George, G. Michot, Dislocation loops at crack tips: nucleation and growth—An experimental study in silicon. Mater. Sci. Eng. A **164**, 118–134 (1993)
Y.I. Golovin, T.P. D'yachek, V.I. Orlov, Y.I. Tyalin, Fiz. Tverd. Tela (Leningrad), **27**, 1110–1115 (1985)
I.S. Gradshteyn, I.M. Ryzhik, *Table of Integrals, Series, and Products* (Academic Press, New York, 1980)
A.E. Green, I.N. Sneddon, The distribution of stress in the neighbourhood of a flat elliptical crack in an elastic solid. Math. Proc. Cambridge Philos. Soc. **46**, 159–163 (1950)
A.A. Griffith, The phenomena of rupture and flow in solids. Phil. Trans. Roy. Soc. Lond. **A221**, 163 (1920)
A.A. Griffith, The theory of rupture, in *Proceedings of First International Congress for Applied Mechanics*. ed. by C.B. Biezeno, J.M. Burgers, J. Waltman Jr. (Delft, 1924), p. 55
V. Hadjicontis, C. Mavromatou, D. Ninos, Nat. Hazards Earth Syst. Sci. **4**(5), 633–639 (2004)
J.P. Hirth, J. Lothe, *Theory of Dislocations* (John Wiley, New York, 1982)
E. Hoek, Z.T. Bieniawski, Brittle fracture propagation in rock under compression. Int. J. Fract. Mech. **1**, 137–155 (1965)

H. Horii, S. Nemat-Nasser, Compression induced microcrack growth in brittle solids: Axial splitting and shear failure. J. Geophys. Res. **90**, 3105–3125 (1985)
J.W. Hutchinson, Metal-ceramic interfaces, in *Acta-Scripta Metallurgica Proceedings Series*, vol. 4. ed. by M. Rühle, A.G. Evans, M.F. Ashby, J.P. Hirth (Pergamon Press, 1990) pp. 295–306
G.R. Irwin, Fracture dynamics, in *Fracturing of Metals* (ASM, Cleveland, 1948), pp. 147–166
G.R. Irwin, Analysis of stresses and strains near the end of a crack traversing a plate. J. Appl. Mech. **24**, 361–364 (1957)
V. Jagasivamani, K.J. Iyer, Electromagnetic waves generation in $Ni_{2.14}Mn_{0.81}GaFe_{0.05}$ Heusler alloy at structural phase transition. Mater. Lett **6**, 418–421 (1988)
J.W. Johnson, D.G. Holloway, On the shape and size of the fracture zones on glass fracture surfaces. Phil. Mag. **14**, 731–743 (1966)
F. Kerkhof, H. Richter, in *Second International Conference on Fracture*, Brighton (Chapman and Hall, London, 1969), paper 40
C. Kittel, *Introduction à la physique de l'état solide*, troisième édition (Bordas, Paris, 1958)
C. Kittel, *Introduction to Solid State Physic*, 7th edn. (Wiley, New York, 2018)
L.D. Landau, E.M. Lifshitz, *Theory of Elasticity*, 3rd edn. (Pergamon Press, Oxford, 1986)
R. Landauer, Conductivity of cold-worked metals. Phys. Rev. **82**, 520–521 (1951)
T. Laub, J.D. Eshelby, The velocity of a wave along a dislocation. Phil. Mag. **14**, 1285–1293 (1966)
B. Lawn, *Fracture of Brittle Solids*, 2nd edn. (Cambridge University Press, New York, 1993)
H.M. Ledbetter, E.R. Naimon, Elastic properties of metals and alloys. II. Copper. J. Phys. Chem. Ref. Data **3**(4), 897–935
K.M. Liechti, Y.-S. Chai, Biaxial loading experiments for determining interfacial fracture toughness. J. Appl. Mech. **58**, 680–687 (1991)
K.M. Liechti, Y.-S. Chai, Asymmetric shielding in interfacial fracture under in-plane shear. J. Appl. Mech. **59**, 295–304 (1992)
A.E.H. Love, *A treatise on the mathematical theory of elasticity*, 4th edn. (Dover Publications Inc., New York, 1927)
G.P. Marshall, J.G. Williams, Time-temperature dependent of fracture toughness of PMMA. J. Mater. Sci. **8**, 138–140 (1973)
A. Misra, Electromagnetic effect at metallic fracture. Nature (London) **254**, 133–134 (1975)
A. Misra, A. Kumar, Some basic aspects of electromagnetic radiation during crack propagation in metals. Int. J. Fract. **127**, 387–401 (2004)
A. Misra, R.C. Prasad, V.S. Chauhan, B. Srilakshmi, Int. J. Fract. **145**, 99–121 (2007)
M.I. Molotskii, Soviet. Tech. Phys. Lett. **6**, 22–23 (1980)
Y. Mori, Y. Obata, J. Pavelka, J. Sikula, T. Lokajicek, J. Acoustic Emission **22**, 91–101 (2004)
N.F. Mott, Brittle fracture in mild steel plates. Engineering **165**, 16–18 (1948)
T. Mura, The continuum theory of dislocations, in *Advances in Materials Research*, vol. 3. ed. by H. Herman (Interscience Publications, 1968), pp. 1–108
T. Mura, *Micromechanics of Defects in Solids* (Martinus Nijhoff Publishers, Dordrecht, 1987)
N.I. Muskhelishvili, *Singular Integral Equations* (Dover Publications, New York, 1992)
F.R.N. Nabarro, *Theory of Crystal Dislocations* (Oxford University Press, Oxford, 1967)
S. Nemat-Nasser, M. Hori, *Micromechanics: Overall Properties of Heterogeneous Materials*, 2nd edn. (Elsevier, 1999)
S.M. Ohr, An electron microscope study of crack tip deformation and its impact on the dislocation theory of fracture. Mater. Sci. Eng. **72**, 1–35 (1985)
B. Paul, Macroscopic criteria for plastic flow and brittle fracture, in *Fracture*, vol. 2, ed. by H. Liebowitz (Academic Press, New York, 1968), p.313
M.O. Peach, J.S. Koehler, The force exerted on dislocations and the stress field produced by them. Phys. Rev. **80**, 436–439 (1950)
V.F. Petrenko, On the nature of electrical polarization of materials caused by cracks. Application to ice electromagnetic emission. Philos. Mag. B **67**, 301–315 (1993)
V.F. Petrenko, *Special Report 96-2, US Army Corps of Engineers (CRREL)* (1996), pp. 1–30

D.D. Pollard, P. Segall, P.T. Delaney, Formation and interpretation of dilatant echelon cracks. Geol. Soc. Am. Bull. **93**, 1291–1303 (1982)

C.C. Radon, P.S. Lever, L.E. Culver, Fracture toughness of PMMA under biaxial stress, in Fracture, vol. 3, (UW Press, 1977), pp. 1113–1118

I.S. Raju, J.H. Crews, M.A. Aminpour, Convergence of strain energy release rate components for edge-delaminated composite laminates. Eng. Fracture Mech. **30**, 383–396 (1988)

F.C. Roesler, Brittle fractures near equilibrium. Proc. Phys. Soc. B **69**, 981–1012 (1956)

T. Sakai, Y. Sato, N. Oguma, Characteristic S-N properties of high-carbon-chromium-bearing steel under axial loading in long-life fatigue. Fatigue Fract. Eng. Mater. Struct. **25**, 765–773 (2002)

G. Schoeck, A. Seeger, in *Report of the Bristol Conference on Defects in Crystalline Solids* (Physical Society, London, 1955), p. 340

G.C. Sih, P.C. Paris, F. Erdogan, Crack-tip, stress-intensity factors for plane extension and plate bending problems. J. Appl. Mech. **29**, 306 (1962)

G.C. Sih, H. Liebowitz, Mathematical theories of brittle fracture, in *Fracture*, vol. 2, ed. by H. Liebowitz (Academic Press, New York, 1968), p.67

E. Sommer, Formation of fracture lances in glass. Eng. Fract. Mech. **1**, 539–546 (1969)

R. Wm. Soutas-Little, *Elasticity*, 1st edn. (Prentice-Hall, Inc., Englewood Cliffs, N.J., 1973)

C.T. Sun, C.J. Jih, On strain release rates for interfacial cracks in bi-material media. Eng. Fracture Mech. **28**, 13–20 (1987)

J.C. Stinville et al., Fatigue deformation in a polycrystalline nickel base superalloy at intermediate and high temperature: competing failure modes. Acta Mater. **152**, 16–33 (2018)

H. Wolf, Z. Naturf. (a) **15**, 180 (1960)

Z. Zhao, F. Zhang, C. Dong, X. Yang, B. Chen, Initiation and early-stage growth of internal fatigue cracking under very-high-cycle fatigue regime at high temperature. Metall. Mater. Trans. A **51A**, 1575–1592 (2020)

MIX
Papier aus verantwortungsvollen Quellen
Paper from responsible sources
FSC® C105338

If you have any concerns about our products,
you can contact us on
ProductSafety@springernature.com

In case Publisher is established outside the EU,
the EU authorized representative is:
Springer Nature Customer Service Center GmbH
Europaplatz 3, 69115 Heidelberg, Germany

Printed by Libri Plureos GmbH
in Hamburg, Germany